Wave Physics

R.E.I. Newton

Edward Arnold
A division of Hodder & Stoughton
LONDON NEW YORK MELBOURNE AUCKLAND

First published in Great Britain 1990

Distributed in the USA by Routledge, Chapman and Hall, Inc.,
29 West 35th Street, New York, NY 10001

British Library Cataloguing in Publication Data

Newton, R. E. I.
Wave physics
1. Water. Waves
I. Title
532′.593

ISBN 0-7131-2656-6

Typeset in 10/11 pt Times by J. W. Arrowsmith Ltd, Bristol.
Printed and bound in Great Britain for Edward Arnold, the educational, academic and medical publishing division of Hodder and Stoughton Limited, Mill Road, Dunton Green, Sevenoaks, Kent by J. W. Arrowsmith Ltd, Bristol

Preface

Wave Physics aims to provide a thorough, but also interesting, grounding in the essential features of oscillations, waves, light and sound, together with an introduction to the application of wave ideas to the structure of matter. It covers the full range of material in A-level courses and also a number of advanced topics, and is thus suitable for students in school and for those in the first year of further or higher education.

In writing the book I have tried to show the experimental basis for the concepts introduced, to explain phenomena in terms of their physical mechanism, and to show the mathematical links between ideas where appropriate. Basic ideas are explained with care and throughout the book the particular properties of light, sound and matter waves continue to relate to the general properties of oscillations and waves established in the early chapters.

The approach adopted allows ample scope for the study of applications, so that the student is encouraged to see physics as a study of the real world and not just as a laboratory experience. To this end I have included an unusually large number of photographs, in addition to the many diagrams which assist the development of the student's understanding.

All too often the links between theory and experiment are made to seem so certain it is difficult for the student to appreciate the cut and thrust of scientific debate, or the boldness of the scientist who publishes a new theory. A particular feature of *Wave Physics* is the inclusion of historical material, often from original sources, which demonstrates the way in which key ideas have developed and the interdependence of theoretical progress and experimental work.

Questions are included in the text to ensure that a student understands each point as it is developed. There is also a large and varied selection of questions at the back of the book taken mainly from examination papers; these questions are divided into A, B and C categories to indicate different levels of difficulty. Answers are given to all numerical questions in the text but only to odd-numbered questions at the back of the book; in this way teachers can set exercises where students do not have access to the answers, but students working on their own can check their work as they progress. In selecting suitable questions preference has been given to those which test or develop the student's ability to apply ideas to a wide range of situations.

Throughout the book advanced sections are clearly set apart in smaller print so that they may conveniently be omitted at a first reading. Thus I hope that the book will be suitable for students of a wide range of abilities—taking those to whom ideas are unfamiliar or difficult at a careful pace whilst providing able or more experienced students with additional material which is both challenging and interesting. To provide further stimulus each chapter, or group of chapters, has a detailed bibliography including both books which can consolidate as well as those which can extend.

In conclusion I hope that *Wave Physics* will provide the student who continues with physics with a good grounding in this important subject, on which a more detailed study can then be based. On the other hand the range of applications, and the tracing of the development of wave theory, should enable students who complete their study of physics at this point to take away with them a clear picture both of the importance of waves in the world around them and of their fundamental position in physics as a whole.

Ian Newton
January 1989

Acknowledgements

It would be impossible to thank individually all those by whose help or influence this book has been improved, and it must inevitably include many ideas that are due to others which it is impossible to attribute.

Equally, my ideas have been moulded over the years by reading a number of books, many of which are listed in the bibliographies, and in particular by teaching the Nuffield Advanced Physics course; the treatment of topics given there is implicit in much of my own teaching, and hence in much of the writing of this book.

I am also grateful to many who have given their permission to reproduce or adapt the illustrations listed below, and particularly to those who have given of their time and trouble to help me find the photographs. Officials at the Hydraulics Research Station, the Institute of Oceanographic Sciences, the National Maritime Institute and the Science Museum were particularly helpful in suggesting sources of photographs, and I am grateful to G. E. Foxcroft, OBE and to Derek Power, of Teltron, who took photographs especially to meet my requirements. Permission to reproduce or adapt illustrations was kindly given by the following:

Barber N. N. F. (1969) *Water Waves*, Wykeham: Fig. 3.1
Boulind H. F. (1972) *Light: Waves or particles*, Longman: Fig. 11.15(c)
Bristol General Hospital: Fig. 5.11(e)
British Broadcasting Corporation (*The Listener*, **527**, p. 4): Fig. 4.20
British Museum: Fig. 2.1
Building Research Centre, Watford, Fig. 5.2(d)
Chedd G. (1970) *Sound*, Aldus Books: Fig. 5.11(c,d)
Collieu A. M. and Powney D. J. (1973) *The Mechanical and Thermal Properties of Matter*, Edward Arnold: Fig. 11.4
Davis and Kaye *The Acoustics of Building*, Bell: Fig. 5.2(c)
Educational Development Centre: Figs. 1.6, 3.8(a), 3.16, 4.1, 4.4, 4.9, 4.11, 4.12, 9.7(a,b), 11.17(c), 2.2, 2.3
ESSO (from their film *Ray Optics*): Fig. 6.24
Feynmann R. P. *et al.* (1963) *Lectures on Physics*, vol. 1, Addison-Wesley: Fig. 4.23
Fforde A. W. B.: Fig. 1.3
Gamov G. (1966) *Thirty Years which Shook Physics*, Doubleday/Heinemann: Fig. 11.18
Goss B. C. P.: Fig. 3.9
Griffin and George (1973) *Laser Accessories Kit:* Fig. 7.3(a), 7.21(a–h), 8.9
Guilemin *The Forces of Nature*, MacMillan: Fig. 5.12
Huygens C. (1678) *Treatise on Light*, reprinted Dover/Macmillan: Fig. 9.8
Hydraulics Research Station, Wallingford (*Crown Copyright*): Fig. 3.4(a), 3.19
Illustrated London News, **16**, p. 251, Fig. 1.19
KeyMed: Fig. 6.19
Kodak: Fig. 8.17(c)
Laver F. I. M. (1959), *Waves*, OUP: Figs. 1.25, 3.14, 3.15
Lipson H. S. (1968) *Great Experiments in Physics*, Oliver and Boyd: Figs. 7.29(a), 9.2, 9.3
Llowarch W. (1961) *Ripple Tank Studies*, OUP: Figs. 3.4(c), 3.6, 3.8(b), 3.12
Longhurst R. S. (1973) *Geometrical and Physical Optics*, Longman: Figs. 7.10, 7.11, 7.15(c), 7.16(a–e), 7.19, 7.20, 9.10, 11.16(b)
Mollenstedt, Prof. G., from an article in *Zeitschrift fur Physik*, 145: Fig. 11.16(a)
McMenemy J. D. S.: Fig. 5.4
Nuffield Advanced Science (1971) Teachers Guide, Unit 8, *Electromagnetic Waves*: Fig. 11.8, Unit 10, *Waves, Particles and Atoms*: Longman, Fig. 11.9
Peters Ballistic Institute, Ohio: Fig. 5.1
Revised Nuffield Advanced Science (1984), *Book of Data*, Longman: Fig. 11.1
Rolls-Royce Limited: Fig. 4.18
Rose, Prof. A.: Fig. 10.11
Royal Aircraft Establishment, Farnborough, (*Crown Copyright*): Figs. 3.20, 7.35
Royal Greenwich Observatory, Herstmonceux: Fig. 8.5, 8.11
Science Museum, London: Fig. 8.7
Taylor C. A. (1965) *The Physics of Musical Sounds*, Hodder & Stoughton: Fig. 5.6

Teltron Limited: Fig. 11.15(a,b)
Ultrasonics Limited: Fig. 5.11(a,b)
van Heel A. C. S. and Velsel C. H. F. (1968) *What is Light?* Weidenfeld and Nicolson: Figs. 6.17, 9.4, 9.5

I would also like to thank a number of people who have helped me to prepare the book for the publishers. My colleagues, Dr. P. J. Cheshire, Mr B. Taylor and Mr J. D. S. McMenemy kindly read the typescript and made many valuable comments; encouragement and guidance on difficult points were also given by Mr G. E. Foxcroft, OBE. I am especially grateful to my father, Mr J. Newton, who has not only read both manuscript and proofs, but also checked solutions to many of the problems.

Despite this abundance of help, responsibility for the final product must rest with me, and I would be very pleased to be told of any errors, especially in the solutions to problems.

Permission has been given by the various Examining Boards to reproduced questions from past papers. Each question is individually acknowledged as indicated below:

(O&C)	Oxford and Cambridge Schools Examination Board
(O&C, Nuffield)	Nuffield Advanced Physics Examination
(C)	University of Cambridge Local Examination Syndicate
(JMB)	Joint Matriculation Board
(L)	University of London School Examination Board
(O)	University of Oxford Delegacy of Local Examinations
(S)	Southern Universities Joint Board
(W)	Welsh Joint Education Committee
(Ox. Schol.)	Oxford Colleges Entrance Examination
(Cam. Schol.)	Cambridge Colleges Entrance Examination
(Cam. Step.)	Cambridge STEP Examinations

Note Where answers are given to numerical questions, they are the responsibility of the author and not of the Examining Board concerned.

Finally, none of this would have been possible without the encouragement and patience of my family and my publishers and, not least, my wife's skill in converting my handwriting into typescript.

R. E. I. N.

Contents

Part 1
The nature of waves

1

Oscillations

1.1 Introduction

The basic theme of this book is wave physics, and its applicability to phenomena as diverse as earthquakes, the propagation of light and the structure of the atom. Our understanding of the general properties of waves will enable us to study topics such as these in depth and to appreciate their common nature.

Waves are propagated along a stretched spring by moving the end of it up and down; the cone of a loudspeaker moves alternately in and out when emitting sound waves. A repetitive motion, such as these, is known as an *oscillation* or *vibration*; most waves are caused by oscillations and we shall, therefore, begin by studying them in this chapter.

However, the study of oscillations is itself of importance. The oscillations of bridges or buildings can hazard their safety, and vibrations in a moving car may cause discomfort to the occupants. The oscillation of balance wheels, or quartz crystals, provide methods for the measurement of time, and much of our understanding of the structure of substances is based on an analysis of atomic vibrations.

A motion that is repeated exactly at equal time intervals is said to be *periodic*; the time taken for a complete cycle is called one *period*, *T*. In Fig. 1.1 we see a graph showing how the displacement of an object from its rest position varies with time in a particular oscillation; the maximum displacement is called the *amplitude*.

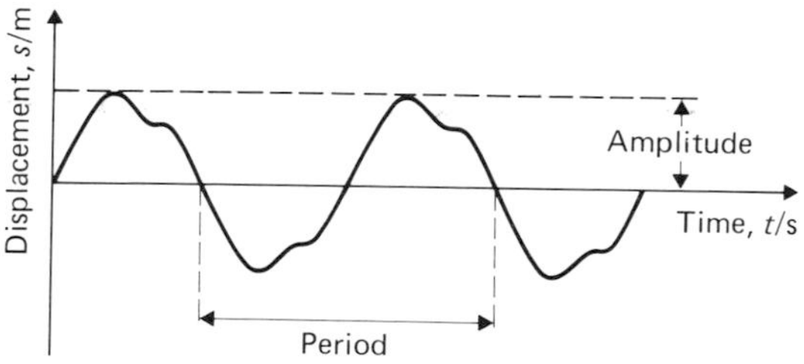

Fig. 1.1

The number of oscillations in one second is called the *frequency*, *f*, where

$$f = 1/T \tag{1.1}$$

Frequency is measured in hertz (Hz), a unit named after the first man to transmit electromagnetic waves, as described in Section 10.1.

Q 1.1

Write down the frequencies of oscillations which have the following periods:

(*a*) 1 s (*b*) 0.1 s
(*c*) 10 s (*d*) 50 s

Write down the periods of oscillations which have the following frequencies:

(*e*) 1Hz (*f*) 0.2 Hz
(*g*) 25 Hz (*h*) 50 Hz

Galileo is recorded as the first person to observe that the period of a pendulum does not depend on the amplitude of its swing, provided that the latter is not too large; he realized this first in 1538 whilst watching a swinging lamp during a service in the Cathedral of Pisa. Any oscillation in which the period does not depend on amplitude is said to be *isochronous*; in order that a clock shall continue to keep good time whilst it is running down, it is important that its balance wheel or pendulum undergoes *isochronous* oscillations.

We must be careful not to assume that all oscillations are isochronous. If a length of curtain rail is bent into a V-ramp, with a sharp curve at the bottom, then a marble rolling backwards and forwards on it will oscillate with shorter and shorter period as the amplitude decreases (Fig. 1.2(*a*)). By contrast, a ball placed on a length curved into a circular arc will oscillate almost isochronously until it comes to rest; the speed at which the ball moves decreases at the same rate as its amplitude (Fig. 1.2(*b*)).

Some oscillations, of course, are completely irregular and a displacement–time graph of one such is shown in Fig. 1.2(*c*); this might represent a blade of grass, blowing in the wind. Unless otherwise stated we shall restrict our discussion to regular isochronous oscillations.

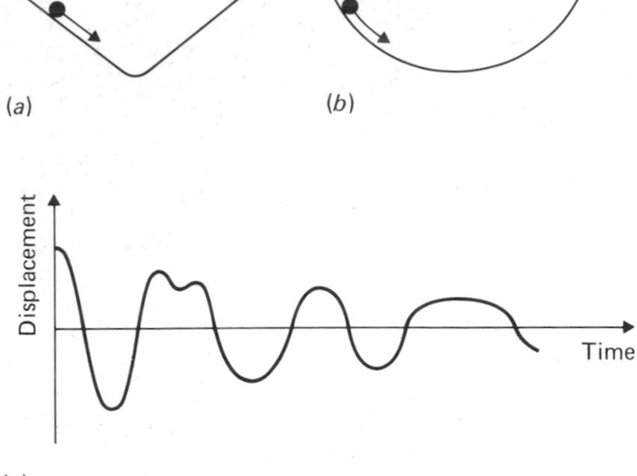

Fig. 1.2 (a) A marble rolling on a V-ramp. (b) A marble rolling on a circular arc. (c) An irregular, non-isochronous motion.

Q 1.2
Draw graphs of distance against time to show:

(*a*) the motion of a ball-bearing rolling from rest down a uniform frictionless slope;
(*b*) the motion of a ball-bearing oscillating under frictionless conditions in a V-ramp;
(*c*) the motion of a ball-bearing oscillating in a V-ramp with reducing amplitude.

Standards of time

Historically the standard of time was provided by the rotation of the earth — once per day. The day was divided into smaller units and a *second*, the unit used in scientific work, was defined as 1/86 400 part of a mean solar day. Oscillating systems, such as the pendulum or watch balance wheel, kept track of the number of seconds or days elapsed.

However, different types of clocks could be shown to disagree with each other. Improvements to the accuracy of clocks were made, especially by Harrison in the eighteenth century, because of their importance in navigation at sea. Sunrise and sunset both occur later as a ship travels to the west and accurate observations of the sun's position, and the time at which it is due south, enable both latitude and longitude to be determined. However, improvements to traditional clocks could never provide a system of measuring time to a greater accuracy than a small fraction of a second.

Recently, the oscillations of crystals have been used as the basis for standard clocks. Vibrations of quartz crystals can be maintained electrically and, if the temperature is kept constant, they provide a time standard with an accuracy of about 1 part in 10^8. The frequency is of the order of 10^5 Hz, providing measurements accurate to 10^{-5} s, a great improvement upon pendulum clocks. Although quartz clocks become less accurate over a prolonged period of time, they were sufficiently stable to provide the first evidence of the effect of tides on the rate of the earth's rotation.

The most modern atomic clocks use as a standard the frequency of a certain radiation from caesium atoms; this frequency is about 9 GHz (9×10^9 Hz). Measurements accurate to 10^{-10} s can be made with such systems, and they are consistent to 1 part in 10^{10} over long periods of time. Since January 1972 *Universal Time*, which is the same as Greenwich Mean Time, has been based on a caesium clock and *the second* defined as 9 192 631 770 periods of a specified radiation from an atom of caesium-133.

Such clocks are now used as the fundamental standard of time, against which chronometers and other clocks can be compared. The Rugby School clock, for instance, shown in Fig. 1.3, is compared regularly with radio signals derived from a master clock at Rugby ratio station. It has been found that even the length of the day varies, becoming as many as 86 430 seconds at the end of the year; this is a consequence of the elliptical shape of the earth's orbit round the sun.

1.2 The motion of an oscillator

In the first section we saw that the simplest type of oscillator is an isochronous one. What type of oscillator is likely to be isochronous?

Any mechanical oscillating system must consist of a mass which is displaced, and a force which will act so as to return the mass to its equilibrium position. In the case of a pendulum this force is provided by gravity and in the case of the balance wheel of a watch by the balance spring.

Let us consider an isochronous oscillation whose amplitude is doubled without change in

Clock mechanism

Radio – used to receive time signals derived from caesium master clock and broadcast continually throughout Europe by Rugby Radio Station. These signals are compared daily with standard clocks elsewhere, and are also used to correct ships' chronometers, and the Rugby school clock.

Adjustment mass – to make small alterations to period of pendulum

Motor for operating hour-chime mechanism

Winding motor

Spring – part of hour-chime mechanism

Driving mass – wound up by electric motor every ten minutes

Pendulum

Fig. 1.3 School House Clock, Rugby School.

period. The mass must now travel twice as far in the same time, and thus its average velocity is twice as large as before. This means that the increase in velocity in a quarter of a cycle from zero to maximum is doubled and, since it occurs in the same time as before, the acceleration of the mass is now twice as great. Hence, for isochronous motions we have shown that:

$$\text{acceleration} \propto -\text{displacement} \qquad (1.2)$$

The minus sign denotes that the acceleration is directed back towards the equilibrium point. Motion in which the acceleration is proportional to the amplitude, and directed back towards the equilibrium position, is always isochronous and is known as *simple harmonic motion* (SHM); a graph of this type of motion is shown in Fig. 1.4(a). It is important because it is common, simple to describe mathematically, and because more complex periodic motions can be expressed as the sum of a number of simple harmonic motions, as discussed in Section 1.7.

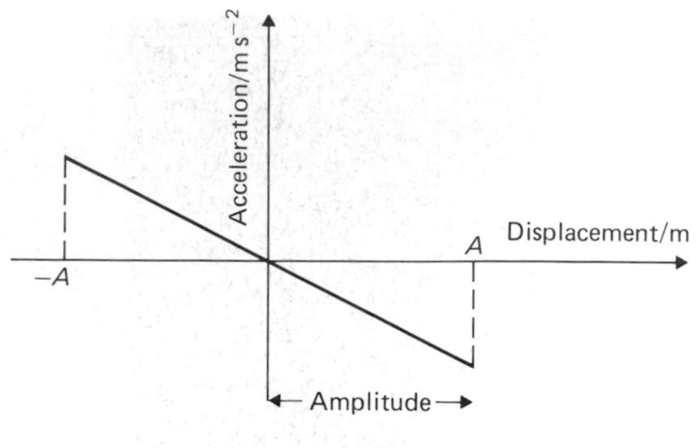

(*a*) Simple harmonic motion

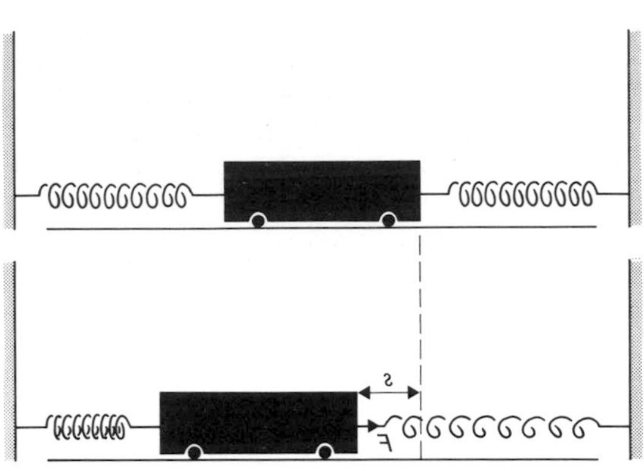

(*b*) A tethered mass

Fig. 1.4

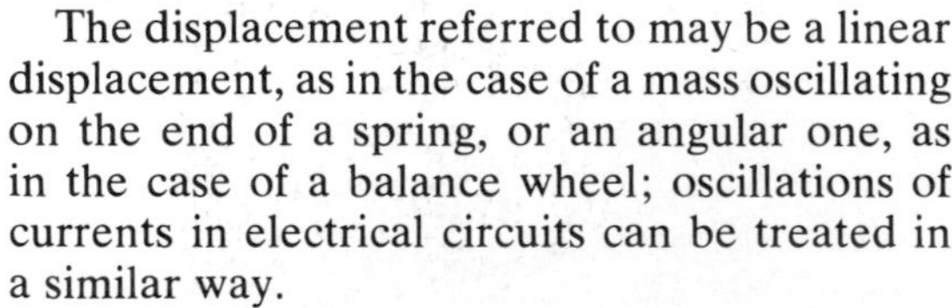

The displacement referred to may be a linear displacement, as in the case of a mass oscillating on the end of a spring, or an angular one, as in the case of a balance wheel; oscillations of currents in electrical circuits can be treated in a similar way.

Q 1.3
Write out an argument, similar to that above, starting from the assumption that

$$\text{acceleration} \propto -\text{displacement}$$

and showing that if this is so, then the period of an oscillation is independent of the amplitude.

The realization of simple harmonic motion

Newton's Second Law states that, for constant mass, the acceleration of an object is proportional to the applied force.

Now for SHM:

$$\text{acceleration} \propto \text{force} \qquad (1.3)$$

$$\text{acceleration} \propto -\text{displacement} \qquad (1.2)$$

Hence $$\text{force} \propto -\text{displacement} \qquad (1.4)$$

The simplest example of this relationship occurs in a stretched spring, since Hooke's Law states that

$$F = -ks \qquad (1.5)$$

where F is the tension in the spring, s the extension and k the force constant, or stiffness; the *stiffness* is the restoring force per unit displacement.

A convenient method of observing SHM is therefore to mount a trolley between two springs as shown in Fig. 1.4(*b*); alternatively a linear air track vehicle may be used, in which case there is less friction.

Let us suppose that the trolley is initially at rest with both springs unstretched. If it is now moved a distance s to the right, the left-hand spring is stretched a distance s and the tension in the spring is proportional to this distance, as shown in Equation (1.5). Thus, the acceleration of the trolley back towards its equilibrium position will be proportional to the displacement, which satisfies our definition of simple harmonic motion. We assume that there is no force due to the compression of the right-hand spring.

However, it is as well to see experimentally whether the acceleration really is proportional to the displacement and to do this we must investigate the way in which the position of the trolley varies with time. A convenient method of observing this is to attach tickertape to the trolley and allow the trolley to move through half a cycle, from one extreme to the other. As the dots on the tickertape are made every 1/50 s, they provide a record of the position, and hence velocity, of the trolley. The velocity can be worked out either from the separation of the dots, or from the slope of the displacement–time graph. The acceleration can then be calculated from the slope of the velocity–time graph.

Note that when the trolley passes through its central, equilibrium position its velocity is a maximum; because the velocity is momentarily constant the acceleration is zero. At each extremity the direction of motion of the trolley is reversed and the velocity is momentarily zero; the velocity is reversing rapidly and the acceleration is a maximum.

It can be seen by comparison of the displacement–time and acceleration–time graphs in Fig. 1.5 that:

$$\text{acceleration} \propto -\text{displacement} \qquad (1.2)$$

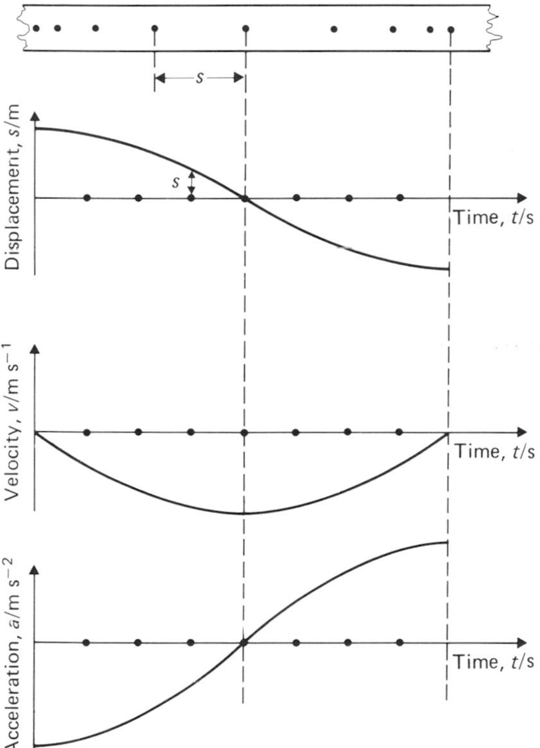

Fig. 1.5 The analysis of a length of ticker-tape, showing the characteristics of simple harmonic motion.

and thus the trolley carries out SHM. Values of *a* taken from the bottom graph can be plotted against the corresponding values of *s* from the top graph, giving a graph such as that shown in Fig. 1.4(*a*).

Similar methods may be used to show that the motion of a pendulum is simple harmonic. Strobe photographs taken at equal time intervals as shown in Fig. 1.6 can be used as a basis for this analysis.

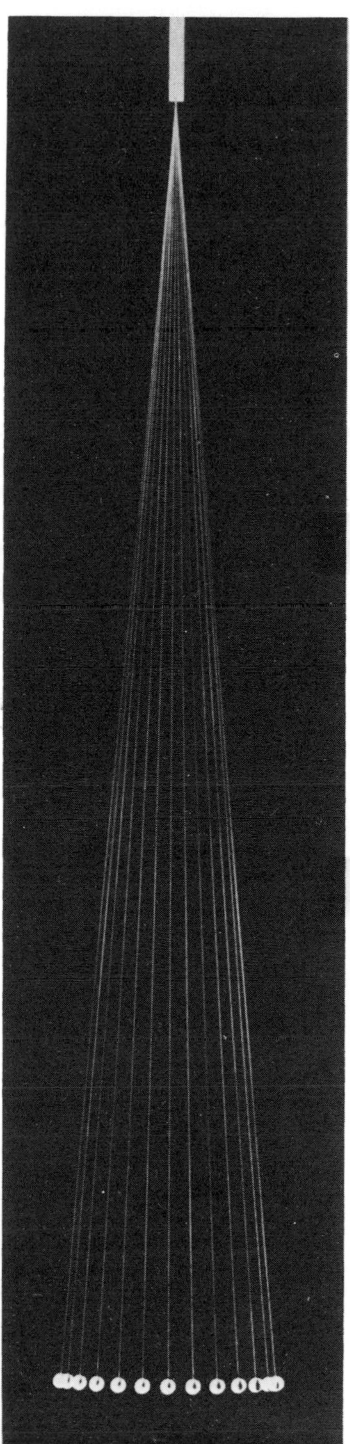

Fig. 1.6 A strobe photograph of a simple pendulum.

Q 1.4
Use a protractor to measure the angular displacements shown in the strobe photograph, and hence show that angular acceleration is proportional to angular displacement by drawing suitable graphs.

The mathematics of simple harmonic motion

In the previous section we defined simple harmonic motion by the proportionality:

$$\text{acceleration} \propto -\text{displacement} \qquad (1.2)$$

The constant of proportionality will depend on factors such as the mass and restoring force. We shall now derive an equation describing the displacement of the object as a function of time.

A suitable equation is suggested by the demonstration shown in Fig. 1.7. The turntable rotates with constant angular velocity and a vertical rod mounted on it is illuminated so that a shadow of its top is cast on a screen; the rod will appear to oscillate from side to side. A pendulum or an oscillating trolley, known to exhibit SHM, is mounted just above the centre of the turntable so that its shadow is also cast upon the screen. The rate of rotation of the turntable is adjusted until the time taken for one complete revolution is exactly equal to the period of oscillation of the pendulum. The two shadows are then seen to correspond exactly in position throughout the motion. The *projection* of the rod thus undergoes simple harmonic motion, just like the oscillating trolley or pendulum.

Now let us consider the geometry of the situation as shown in Fig. 1.7(*b*); R is the rod mounted on the rotating turntable, and P the swinging pendulum. In the experiment the incident light, which is at right angles to AB,

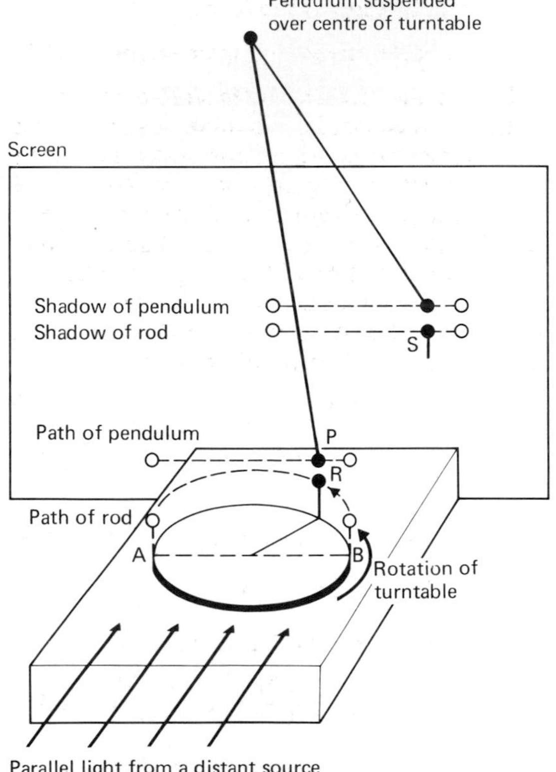

(*a*) The experimental arrangement

(*b*) Plan of arrangement

Fig. 1.7

casts almost coincident shadows of rod and pendulum on the screen at S, and thus the pendulum bob must always be just above the foot of the perpendicular from R to AB as shown; this is called the *projection* of R on to AB. We will assume in this discussion that the pendulum always follows the projection exactly.

The pendulum undergoes SHM and its position is always at the projection of R, R being a point that is moving at a constant speed round a circle; this circle is known as the *reference circle*. We will use our knowledge of the position of R to derive a mathematical expression for the position of P.

R is moving at a constant speed. Suppose that one complete revolution takes a time T. Then, if time t is measured from the instant at which R passes B, the angle θ is given by

$$\frac{\theta}{360}=\frac{t}{T}$$

i.e.

$$\theta=\frac{t}{T}\times 360$$

If θ is measured in radians we have

$$\theta=\frac{t}{T}\times 2\pi$$

since there are 2π radians in 360°. Radian measure is discussed in Appendix 1.

Now we saw in Equation (1.1) that

$$f=1/T$$

where f is the frequency of the rotating turntable measured in hertz.

Hence

$$\theta=2\pi ft \qquad (1.6)$$

Equation (1.6) defines the position of R as a function of time. The position of P is defined from that of R by the equation

$$s=r\cos\theta$$

and hence

$$s=r\cos 2\pi ft \qquad \text{from (1.6)}$$

As the maximum displacement, or amplitude, A, of P occurs when it is at B, a distance r from O, we may rewrite this equation in the form:

$$s=A\cos 2\pi ft \qquad (1.7)$$

It will be seen that this describes the graph of displacement against time in Fig. 1.5, which is a cosine graph.

The acceleration towards the centre of the circle of the point R is given by:

$$\text{acceleration}=-\frac{v^2}{r}$$

where

$$v=(2\pi r)f$$

Therefore:

$$\begin{aligned}\text{acceleration}&=-(2\pi f)^2r\\&=-(2\pi f)^2A \qquad \text{since } A=r\end{aligned}$$

Now the acceleration, a, of the point P is the horizontal component of the acceleration of the point R, and hence:

$$a=-(2\pi f)^2A\cos 2\pi ft \qquad (1.8)$$

Thus, from (1.7):

$$a=-(2\pi f)^2s \qquad (1.9)$$

This is an equation of the form:

$$\text{acceleration}\propto-\text{displacement} \qquad (1.2)$$

and hence the motion of P is simple harmonic.

It is often convenient to think of the analysis of a simple harmonic motion in terms of the projection of an imaginary point moving round a reference circle with constant speed, and the frequency of the oscillation is, of course, equal to that of the motion of the imaginary point.

An alternative derivation of Equation (1.9), using the calculus, enables us to derive an equation for the velocity of point P:

$$s=A\cos 2\pi ft \qquad (1.7)$$

$$v=\frac{ds}{dt}=-(2\pi f)A\sin 2\pi ft \qquad (1.10)$$

$$a=\frac{dv}{dt}=-(2\pi f)^2A\cos 2\pi ft \qquad (1.8)$$

i.e.

$$a=-(2\pi f)^2\,s \qquad (1.9)$$

This equation may also be written:

$$a=-\left(\frac{2\pi}{T}\right)^2 s \qquad (1.11)$$

Q 1.5
Suppose that, in the above analysis, the point P had an amplitude of 10 cm and the rod R made

one complete revolution in 2 s. Derive expressions for:

(*a*) the displacement, s, of the point P as a function of t;
(*b*) the velocity, v, of the point P as a function of t;
(*c*) the acceleration, a, of the point P as a function of t;
(*d*) the acceleration, a, of the point P as a function of s.

A mathematical summary

Thus far, we have derived the following equations, each of which describes simple harmonic motion, but from a different viewpoint:

$$s = A\cos 2\pi ft \qquad (1.7)$$
$$v = -(2\pi f)A\sin 2\pi ft \qquad (1.10)$$
$$a = -(2\pi f)^2 A\cos 2\pi ft \qquad (1.8)$$

i.e.

$$a = -(2\pi f)^2 s \qquad (1.9)$$

The last equation can be written in the form

$$\frac{d^2s}{dt^2} = -(2\pi f)^2 s \qquad (1.12)$$

since

$$a = -d^2s/dt^2$$

It is then known as the differential equation for SHM.

We have introduced experimental evidence above to show that Equations (1.7) and (1.9), in particular, describe isochronous, oscillatory motion. Now it is a standard trigonometrical result, that

$$\sin^2\theta = 1 - \cos^2\theta$$

and thus Equation (1.10) can be rewritten:

$$v = -2\pi fA\sqrt{(1-\cos^2 2\pi ft)}$$
$$= -2\pi fA\sqrt{\left(1-\frac{s^2}{A^2}\right)} \qquad \text{from (1.7)}$$

i.e.

$$v = -2\pi f\sqrt{(A^2 - s^2)}$$

As a square root can be taken to be either positive or negative, this should be written:

$$v = \pm 2\pi f\sqrt{(A^2 - s^2)} \qquad (1.13)$$

We see from this equation that the velocity is zero at each extremity of the oscillation, when $s = \pm A$, and that it is maximum when the object passes through the undisplaced position at $s = 0$. This agrees with the second graph in Fig. 1.5.

Thus, we may summarize the results as shown in Table 1.1, expressing both acceleration and velocity in terms of either the displacement, s, or the time t.

Q 1.6
A particle undergoes SHM of amplitude 20 cm and frequency 0.5 Hz. Calculate the following:

(*a*) the velocity when the displacement is 10 cm;
(*b*) the maximum velocity;
(*c*) the acceleration 0.25 s after leaving one end;
(*d*) the acceleration when the displacement is 5 cm.

Angular frequency, ω

We saw above that the motion of a pendulum, or other object P, undergoing SHM could be described as the projection of a rod R, moving in a circle at constant speed, along the direction

Table 1.1 For a simple harmonic motion with $s = A$ at $t = 0$

	Variation with s	Variation with t	Maximum numerical value	Minimum numerical value
s	—	$s = A\cos 2\pi ft$	$s_{max} = \pm A$ at $t = 0, T/2, T$	$s_{min} = 0$ at $t = T/4, 3T/4$
$v = ds/dt$	$v = \pm 2\pi f\sqrt{(A^2 - s^2)}$	$v = -2\pi fA\sin 2\pi ft$	$v_{max} = \pm 2\pi fA$ at $s = 0$, $t = T/4, 3T/4$	$v_{min} = 0$ at $s = \pm A$, $t = 0, T/2, T$
$a = d^2s/dt^2$	$a = -(2\pi f)^2 s$	$a = -(2\pi f)^2 A\cos 2\pi ft$	$a_{max} = \pm(2\pi f)^2 A$ at $s = \pm A$, $t = 0, T/2, T$	$a_{min} = 0$ at $s = 0$, $t = T/4, 3T/4$

NB: Dependence on the time t can be understood more easily by putting $f = 1/T$. The reader is advised to check that he or she fully understands the maximum and minimum numerical values by working them out.

in which P is moving. In the equations we have derived above, the quantity $(2\pi f)$ appears frequently and we can use the idea of the reference circle to understand its significance.

f is the frequency both of P and of R. Each time the point R moves round the reference circle the line OR must 'sweep out' an angle of 2π, and thus, in one second, it sweeps out an angle of $2\pi f$ which is, therefore, the angular speed of the point R on the reference circle; *it is not the angular speed of the pendulum.* We define the *angular frequency*, or *pulsatance*, ω, of the pendulum as equal to $2\pi f$; it is measured in radians per second.

$$\omega = 2\pi f \tag{1.14}$$

For example, a wheel rotating at 1 Hz has an angular frequency of 2π rad s^{-1}, and a pendulum swinging with a frequency of 1 Hz has an angular frequency of 2π rad s^{-1} also.

Each of the equations given previously can be rewritten using angular frequency instead of frequency:

(1.7)	$s = A \cos \omega t$	(1.15)
(1.10)	$v = -A\omega \sin \omega t$	(1.16)
(1.13)	$v = \pm\omega\sqrt{(A^2 - s^2)}$	(1.17)
(1.8)	$a = -A\omega^2 \cos \omega t$	(1.18)
(1.9)	$a = -\omega^2 s$	(1.19)

We see that the period of the motion, T, is related to the angular frequency, ω, by the equation:

$$\omega = 2\pi f = \frac{2\pi}{T} \tag{1.20}$$

i.e.

$$T = \frac{2\pi}{\omega} \tag{1.21}$$

Q 1.7
Calculate the angular frequency of the following oscillations:

(*a*) frequency 6 Hz;
(*b*) period 2 s.
(*c*) What is the frequency and the period of an oscillation of angular frequency 6π?

Phase

Let us consider again the reference circle that was described above. In Fig. 1.8 the position of a trolley, P, is shown at $t = 0$ and $t = T/8$ for motion described by Equation (1.7).

The graph shows that the motion is begun, as is often the case, by releasing the object, P, at its extreme position, and thus $s = A$ at $t = 0$.

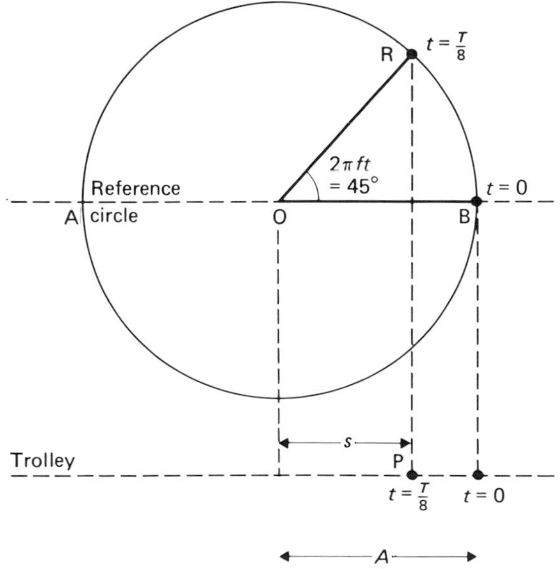

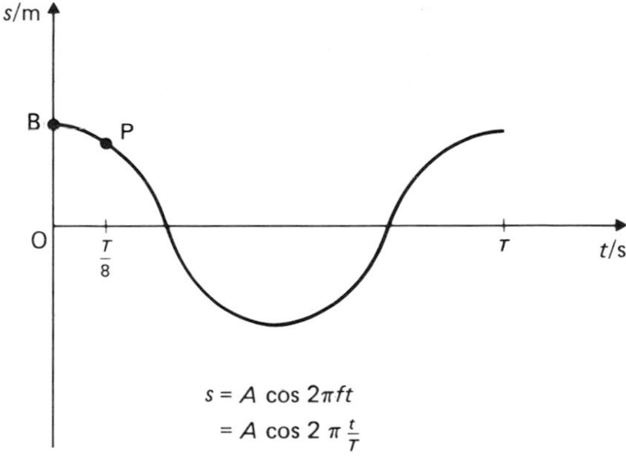

Fig. 1.8

The reference circle can be used to find the position of the trolley after an eighth of a period has elapsed, for the line OR must have rotated through one eighth of a revolution, which is 45° or $\pi/4$ radians. Thus the angle BOR is equal to 45° and hence:

$$2\pi ft = 45°$$
$$= \pi/4 \text{ rad}$$

Since $\cos 45° = 1/\sqrt{2}$

we have $s = A \cos 2\pi ft$
$$= A/\sqrt{2}$$

Now, the quantity $2\pi ft$ represents an angle; this can be seen more easily if we put $f = 1/T$ and obtain $2\pi t/T$. 2π is the number of radians in one complete cycle and t/T represents the fraction of the cycle that has elapsed. The quantity $2\pi ft$ is called the *phase* or *phase angle*; the rotating line OR is called a *phasor*.

Let us consider a similar oscillation of a trolley P′ which starts at a time t' earlier, as shown in Fig. 1.9. This motion will always be 'ahead' of the previous one since an extra time t' has elapsed. Its motion will be described by the equation

$$s = A \cos 2\pi f(t + t') \qquad (1.22)$$

This equation could also be written in the form

$$s = A \cos (2\pi ft + \phi) \qquad (1.23)$$

where $\phi = 2\pi ft'$ (1.24)

In Equation (1.23) the angle ϕ represents the extra angle swept out by the line OR′ because the motion started t' earlier. We say that there is a phase difference of ϕ between the two motions described in Figs 1.8 and 1.9.

Finally, let us consider the motion of a trolley, P″, starting a quarter of a period later than the original oscillation, at $t = T/4$; this is shown in Fig. 1.10. The displacement of P″, at a given time t, is the same as the displacement of P was at a time $T/4$ earlier, so the motion of P″ is described by substituting $(t - T/4)$ in the equation for P; the elapsed time is $T/4$ less. Hence

$$s = A \cos 2\pi f(t - T/4)$$

There are 2π radians in a complete cycle, and hence $\pi/2$ in a quarter cycle, and this equation might, therefore, be written

$$s = A \cos (2\pi ft - \pi/2) \qquad (1.25)$$

The phase difference between P and P″ is $\pi/2$ radians, or 90°. It is a standard result in trigonometry, that

$$\cos (\theta - \pi/2) = \sin \theta$$

and hence Equation (1.25) may be written

$$s = A \sin 2\pi ft \qquad (1.26)$$

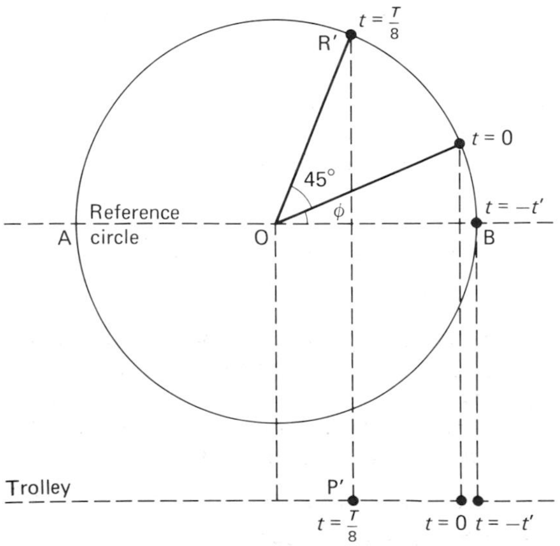

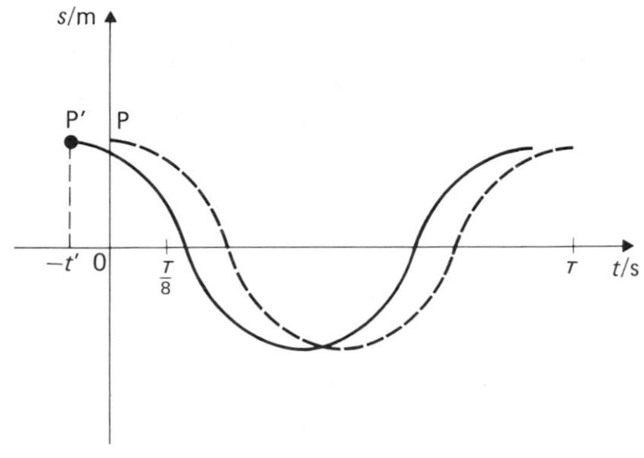

Fig. 1.9

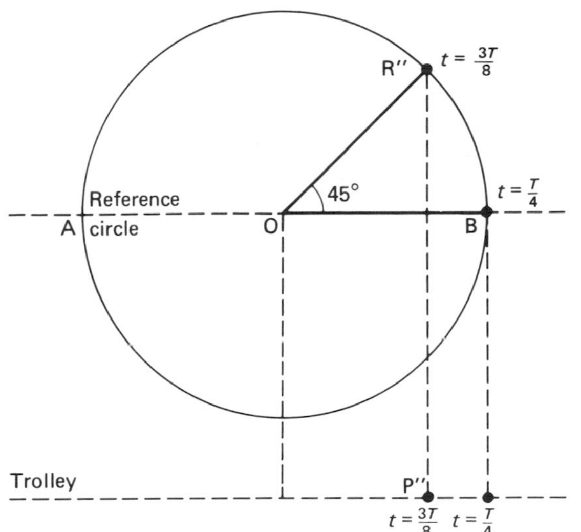

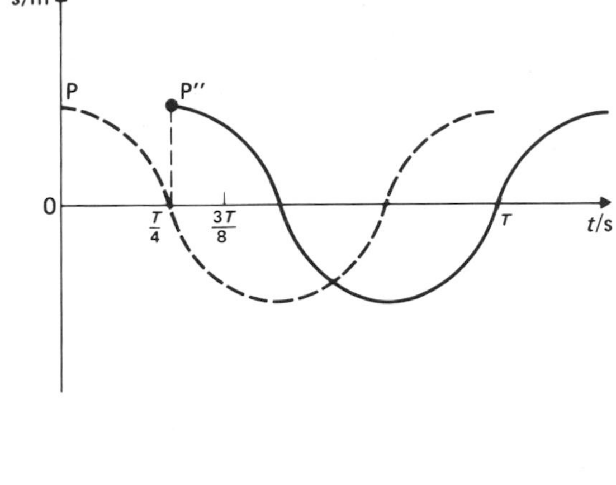

Fig. 1.10

It can be seen that the graph of P'' in Fig. 1.10 is a sine graph with the first quarter cycle missing.

Now, the motions of P'' and P are identical except for the difference in starting time. If P'' had been set moving to the right from the equilibrium position at $t = 0$, it would not have reached the extremity until $t = T/4$ and its future motion would have been predicted by the graph in Fig. 1.10 and by the equation

$$s = A \sin 2\pi ft$$

P, however, was set moving by releasing it at $t = 0$ from an extremity and its motion was described by the graph in Fig. 1.8 and by the equation

$$s = A \cos 2\pi ft$$

The sine equation for P'' can be shown to satisfy the defining equation for SHM as demonstrated below:

$$s = A \sin 2\pi ft$$
$$v = 2\pi fA \cos 2\pi ft$$
$$a = -(2\pi f)^2 A \sin 2\pi ft$$

i.e. $$a = -(2\pi f)^2 s \qquad (1.9)$$

This satisfies the defining equation for SHM, Equation (1.2).

In fact, P and P'' could be the same oscillation. P is viewed with the clock started when the object is at an extremity and the equation is a cosine one; P'' is viewed with the clock started when the object passes to the right through the central point, and a sine equation is derived. Thus, if the starting conditions are not relevant, either form of the equation may be used to describe a given oscillation and we shall use whichever is more convenient.

Q 1.8
Draw distance–time graphs on one set of axes for the following oscillations. Label each graph clearly.

(*a*) $s = 0.1 \cos 4\pi t$;
(*b*) $s = 0.2 \cos (4\pi t + \pi/2)$;
(*c*) $s = 0.3 \sin 4\pi t$;
(*d*) $s = 0.15 \cos 4\pi (t + T/6)$.

1.3 Examples of simple harmonic motion

We have already looked in some detail at the theory of simple harmonic motion and we shall now examine a number of situations in which it can be observed. The reader will recall the conditions describing its occurrence:

$$\text{acceleration} = -(2\pi f)^2 \times \text{displacement} \qquad (1.9)$$

and either $s = A \cos 2\pi ft$ (1.7)

or $s = A \sin 2\pi ft$ (1.26)

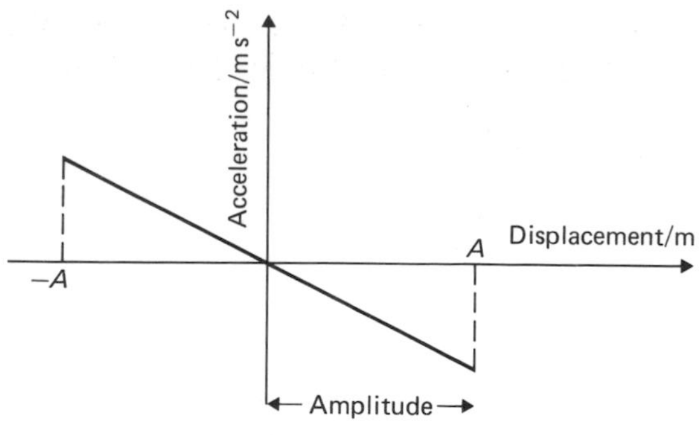

(a)

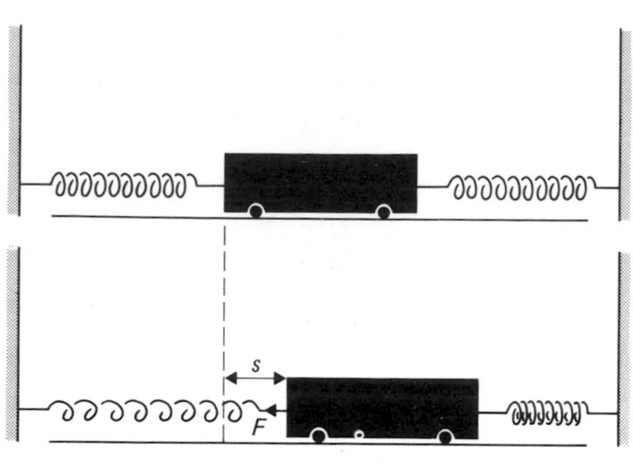

(b)

Fig. 1.11 A tethered mass.

Equation (1.7) is used when the object is at the extremity of its motion at $t=0$, and Equation (1.26) when it is at its equilibrium point at that time. Different equations are used in other conditions.

In this section we shall consider the forces acting on objects undergoing simple harmonic motion and derive relationships for the periods of their oscillations.

Tethered mass

Fig. 1.11 shows again a trolley tethered between two springs. Suppose that when the trolley is at rest as shown in the top diagram there is no tension in either spring. If the trolley is now moved a distance s to the right, the left-hand spring will be extended and the trolley will experience a restoring force, F, whose magnitude is given by Hooke's law:

$$F=-ks$$

where k is the restoring force per unit displacement, or force constant, of the stretched spring. But the restoring force is related to the acceleration of the trolley by Newton's second law:

$$F=ma$$

Hence
$$a=-\frac{k}{m}s \qquad (1.27)$$

where m is the mass of the trolley. We have assumed that the springs are of negligible mass, and that the wheels are of negligible moment of inertia. As this equation is of the general form given in Equation (1.9), simple harmonic motion will occur. The differential equation for simple harmonic motion is

$$\frac{d^2s}{dt^2}=a=-(2\pi f)^2s \qquad (1.9)$$

and therefore, from Equation (1.27), we have

$$(2\pi f)^2=\frac{k}{m}$$

$$f=\frac{1}{2\pi}\sqrt{\frac{k}{m}} \qquad (1.28)$$

and
$$T=2\pi\sqrt{\frac{m}{k}} \qquad (1.29)$$

Therefore, stiffer springs would increase the frequency whilst larger masses would reduce it. If the springs also exert forces when in compression, k becomes the force constant of both springs taken together.

The equation for the motion of the trolley is hence:

$$s=A\cos 2\pi ft$$

i.e.
$$s=A\cos\left(\sqrt{\frac{k}{m}}\right)t \qquad (1.30)$$

Q 1.9
A trolley of mass 3 kg is mounted between springs of total force constant 30 N m^{-1}. Calculate the period, frequency and angular frequency of its oscillation.

The frequency of atomic oscillations

In the above analysis we derived an equation for the frequency of an object oscillating between two springs:

$$f = \frac{1}{2\pi}\sqrt{\frac{k}{m}} \tag{1.28}$$

This result can be applied to the oscillation of ions in a sodium chloride crystal. The ions experience forces of attraction and repulsion due to the electrostatic forces between them, and their mean separation is such that they are in equilibrium. They oscillate about that position in a way analogous to a trolley mounted between two springs.

A study of ionic crystals shows that the force constant of an ionic bond is about $100\ \text{N m}^{-1}$; the average mass of sodium and chlorine atoms is about 5×10^{-26} kg. Therefore

$$f = \frac{1}{2\pi}\sqrt{\frac{100}{5 \times 10^{-26}}}$$

$$= \frac{1}{2\pi}\sqrt{(2 \times 10^{27})}$$

$$\approx 10^{13}\ \text{Hz}$$

The accuracy of this calculation is limited theoretically because we have not considered effects due to the relative motion of different ions.

If light of this frequency were to shine on a sodium chloride crystal, the ions might be stimulated to oscillate and much of the light energy absorbed; this is explained in more detail in Section 1.6 on resonance. The wavelength of the light would be given by:

$$\lambda = \frac{c}{f}$$

$$= \frac{3 \times 10^8}{10^{13}}$$

$$= 3 \times 10^{-5}\ \text{m}$$

This wavelength is in the infrared and absorption does indeed occur at about 6×10^{-5} m; thus our simple analysis yields a result of the right order of magnitude.

Oscillating hydrometer

An example of simple harmonic motion in which gravity is involved is that of a floating object, such as a hydrometer.

Let us consider a tube of uniform cross-section, A, floating with its axis vertical, and with a length h submerged, in a liquid of density ρ, as shown in Fig. 1.12.

$$\text{weight of tube} = mg$$

By Archimedes' Principle:

$$\text{upthrust} = \text{weight of liquid displaced} = Ah\rho g$$

By Newton's second law, since the floating tube has no acceleration:

$$\text{resultant force} = \text{mass} \times \text{resultant acceleration}$$

$$mg - Ah\rho g = 0$$

i.e. $$mg = Ah\rho g$$

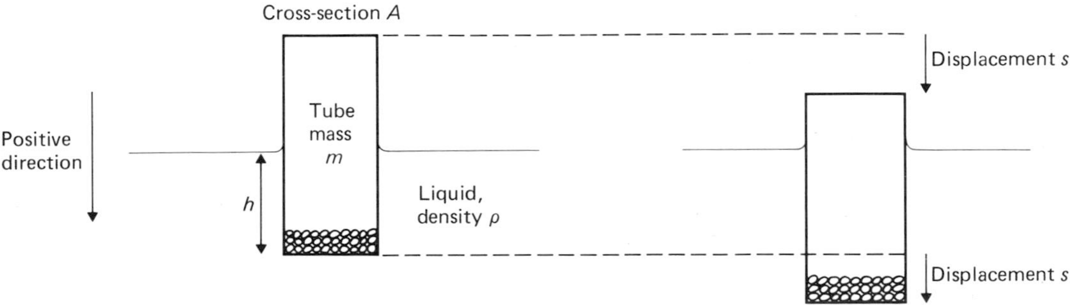

Fig. 1.12 An oscillating hydrometer.

Now suppose that the tube is displaced a distance s downwards, thus displacing an additional volume sA of liquid:

$$\text{weight of tube} = mg$$
$$\text{upthrust} = A(h+s)\rho g$$

By Newton's second law, the acceleration, a, can be calculated from

$$mg - A(h+s)\rho g = ma$$
$$mg - Ah\rho g - As\rho g = ma$$

But $$mg = Ah\rho g$$

and hence $$-As\rho g = ma$$

i.e. $$a = -\left(\frac{A\rho g}{m}\right)s \qquad (1.31)$$

This is an equation of the form

$$\text{acceleration} \propto -\text{displacement}$$

and hence the tube describes simple harmonic motion. The frequency and period are given by the equations

$$f = \frac{1}{2\pi}\sqrt{\frac{A\rho g}{m}} \qquad (1.32)$$

and $$T = 2\pi\sqrt{\frac{m}{A\rho g}} \qquad (1.33)$$

The equation giving the displacement as a function of time is:

$$s = A\cos\left(\sqrt{\frac{A\rho g}{m}}\right)t \qquad (1.34)$$

Now

$$\text{restoring force} = \text{additional upthrust} = (A\rho g)s$$

Therefore $A\rho g$ = restoring force per unit displacement

and hence

$$T = 2\pi\sqrt{\frac{\text{mass}}{\text{restoring force per unit displacement}}}$$

This equation is of the same form as Equation (1.29) derived for the trolley mounted between two springs.

In practice, the motion will not be exactly simple harmonic because of the viscosity of the liquid and the oscillations of the liquid caused by the moving tube.

Q 1.10
A tanker has a draught of 20 m and almost rectangular cross-section. One of its modes of vibration occurs when the whole ship pitches up and down together, and the draught increases and decreases. Use the analysis of an oscillating hydrometer to show that the period of the pitching depends only upon the draught and hence calculate a probable value of the period of this motion. Take $g = 10\ \mathrm{m\,s^{-2}}$.

Simple pendulum

The simple pendulum provides our first example of a system where the displacement is angular rather than linear, and the motion is then only truly simple harmonic for small displacements. A simple pendulum is considered to consist of a point mass m at the end of a light string of length l.

Fig. 1.13 shows the forces acting on a suspended bob that has swung to an angle θ from the vertical. T is the tension in the string and mg the weight of the bob. From diagram (*b*) we see that

$$\theta = s/l$$

where θ is measured in radians, and s is the length of the arc. Now the bob can move only at right angles to the string and if we resolve in this direction we have, by Newton's second law:

$$mg\sin\theta = -m\frac{\mathrm{d}^2s}{\mathrm{d}t^2}$$

where $\mathrm{d}^2s/\mathrm{d}t^2$ is the acceleration along the arc. For small angles $\sin\theta \approx \theta$ provided that θ is measured in radians

and hence $$\sin\theta = s/l$$

Therefore $$mg\frac{s}{l} = -m\frac{\mathrm{d}^2s}{\mathrm{d}t^2} \qquad (1.35)$$

Therefore $$\frac{\mathrm{d}^2s}{\mathrm{d}t^2} = -\frac{g}{l}s \qquad (1.36)$$

which is a differential equation representing simple harmonic motion. It follows that

$$f = \frac{1}{2\pi}\sqrt{\frac{g}{l}} \qquad (1.37)$$

$$T = 2\pi\sqrt{\frac{l}{g}} \qquad (1.38)$$

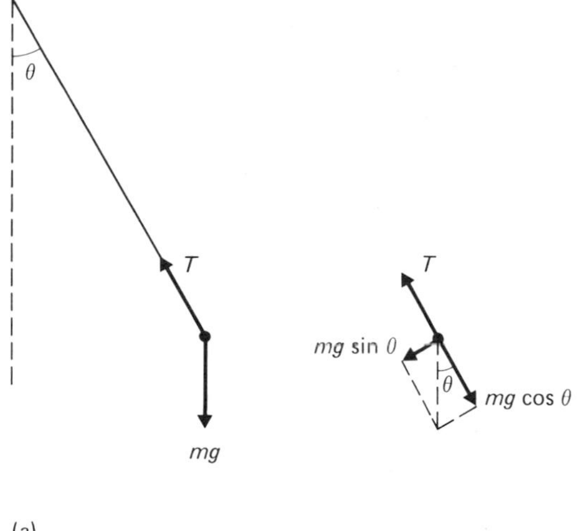

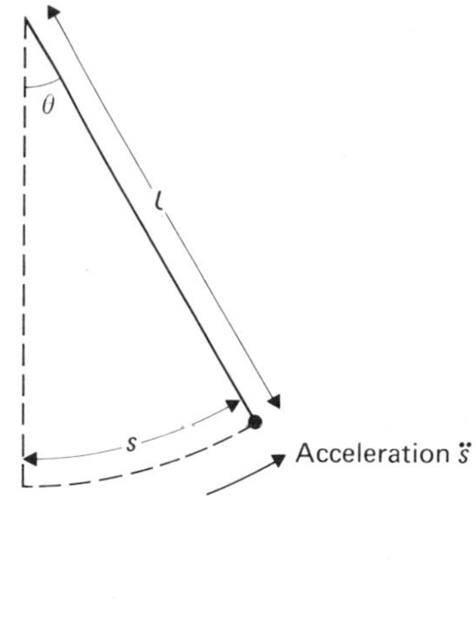

Fig. 1.13 A simple pendulum.

and $$s = A\cos\left(\sqrt{\frac{g}{l}}\right)t \qquad (1.39)$$

Equations (1.36) and (1.39) could be rewritten in terms of the angle θ instead of the displacement, s, along the arc. Note that the period does not depend on the mass of the pendulum.

The equation for the period is

$$T = 2\pi\sqrt{\frac{l}{g}} \qquad (1.38)$$

which may be rewritten in the form

$$T = 2\pi\sqrt{\frac{ml}{mg}}$$
$$= 2\pi\sqrt{\frac{m}{mg/l}}$$

Now $$mg\frac{s}{l} = -m\frac{d^2s}{dt^2} \qquad \text{from (1.35)}$$
$$= \text{restoring force} \qquad \text{by Newton's second law}$$

Hence $$\frac{mg}{l} = \frac{\text{restoring force}}{\text{displacement}}$$

Therefore, we have again

$$T = 2\pi\sqrt{\frac{\text{mass}}{\text{restoring force per unit displacement}}}$$

The equations given above are only good approximations for angles less than 10°, and for larger amplitudes the motion is no longer approximately simple harmonic, and therefore not isochronous.

Measurements of the periods of pendulums may be used as the basis for an accurate determination of g; a compound pendulum, with distributed mass, rather than point mass, is used. Nowadays, however, it is more usual to measure g directly, by accurate free-fall experiments.

Q 1.11
A man's leg with a heavy boot on the end is regarded as a simple pendulum of length 1 m. Calculate the frequency with which it will swing naturally and the speed of walking at this frequency, given that the amplitude is 20°. Will the speed increase or decrease if the man's legs are longer? Explain.

Torsional pendulum

An angular situation which does not involve making trigonometrical approximations is that of the *torsional pendulum*, where an object on the end of a wire rotates first one way and then the other. The wire is twisted as shown in Fig. 1.14. The analysis below will also apply to other situations, such as the balance wheel of a watch which is mounted on a hairspring.

The dynamics of a rotating object are described by an equation similar to Newton's

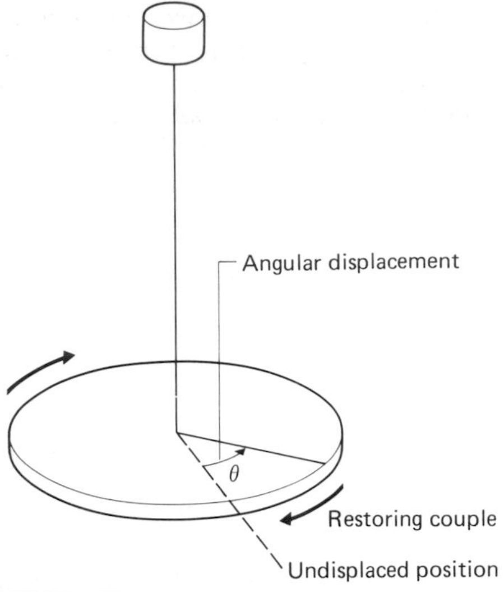

Fig. 1.14 A torsional pendulum.

second law:

$$\text{couple} = \text{moment of inertia} \times \text{angular acceleration} \quad (1.40)$$

instead of

$$\text{force} = \text{mass} \times \text{acceleration}$$

The moment of inertia is a measure of how difficult it is to give an object angular acceleration; it depends not only upon the mass but upon its distribution relative to the axis of rotation. The further away the mass is from the axis, the greater is the moment of inertia; it is measured in kg m^2.

Now, over a considerable range of angles

$$\text{restoring couple} = -c\theta$$

where θ is the angle through which the wire has been twisted and c the torsion constant; that is, c is the restoring couple per unit angular displacement. Hence, if I represents the moment of inertia, we have:

$$-c\theta = I\frac{d^2\theta}{dt^2}$$

where $d^2\theta/dt^2$ is the angular acceleration. This can be written in the form

$$\frac{d^2\theta}{dt^2} = -\frac{c}{I}\theta \quad (1.41)$$

which represents simple harmonic motion. In this case

$$\text{frequency} = \frac{1}{2\pi}\sqrt{\frac{c}{I}} \quad (1.42)$$

and

$$\text{period} = 2\pi\sqrt{\frac{I}{c}} \quad (1.43)$$

These oscillations will be simple harmonic, whatever the amplitude, provided that the restoring torque is proportional to the angular displacement.

We see that Equation (1.43) is also of the form:

$$T = 2\pi\sqrt{\frac{\text{moment of inertia}}{\text{restoring couple per unit angular displacement}}}$$

As the motion is a rotation, mass is replaced by the moment of inertia in this equation.

Torsional suspensions are used to support the coil in many moving-coil galvanometers. Torsional oscillations are also important in understanding the vibrations that can be set up in turbine blades.

General results

Considerable stress has been laid on the general result

$$\text{Period} = 2\pi\sqrt{\frac{\text{inertia factor}}{\text{restoring force or couple per unit displacement}}} \quad (1.44)$$

where the 'inertia factor' may be the mass, moment of inertia or other inertial property, depending upon the type of oscillation. Understanding it enables us to predict detailed equations more readily and also to see a pattern in a variety of results which may otherwise seem disconnected. It may be seen to be inevitable from the following analysis, based on the definition of SHM.

SHM is defined by the following relationship:

$$\text{acceleration} \propto -\text{displacement}$$

$$a \propto -s$$

We have seen that the constant of proportionality is $(2\pi f)^2$ in Equation (1.9) and thus:

$$a = -(2\pi f)^2 s$$

i.e.

$$a = -\left(\frac{2\pi}{T}\right)^2 s$$

However, if the object undergoing SHM is accelerating there must be a restoring force acting on it, whose size is given by Newton's second law:

$$F = ma$$

$$= -m\left(\frac{2\pi}{T}\right)^2 s \qquad \text{from above}$$

Therefore $$T^2 = -(2\pi)^2 \frac{ms}{F}$$

i.e. $$T = 2\pi\sqrt{\frac{m}{(-F/s)}}$$

m is the inertia factor and $(-F/s)$ the restoring force per unit displacement; the minus sign arises because the restoring force and displacement are in *opposite* directions.

1.4 Energy

Let us return to our first example of an oscillation—that of a trolley tethered between two springs. Again we shall assume for simplicity that in the undisplaced position the springs are unstretched and that a force is exerted on the trolley only by the stretched spring, not by the compressed one; we shall also assume that the motion is frictionless so that the amplitude of the resulting oscillations remains constant.

We have already noted, on page 7, that the velocity of the trolley is a maximum as it passes through the central point, and that it is zero when the direction of motion reverses at each extremity. Thus the kinetic energy is zero at each end and maximum in the middle.

When the kinetic energy is less than maximum we would expect, by the Principle of Conservation of Energy, that the balance of the energy would be stored in some other form. As the spring becomes more stretched more energy is stored in it as spring potential energy. We will now examine the situation in more detail to show that the total energy, being the sum of kinetic and potential energies, remains constant throughout the period of the oscillation.

Let us suppose that the trolley is displaced a distance s from its equilibrium position. The force exerted on it by the stretched spring at this displacement is given by:

$$F = -ks \tag{1.5}$$

During the displacement, the force exerted on the trolley by the spring increases from zero to F, and therefore the force exerted on the spring by the trolley must also increase from zero to F. Hence, the work done in stretching the spring is given by:

$$\begin{aligned} \text{work} &= \text{average force} \times \text{displacement} \\ &= \tfrac{1}{2} \times \text{maximum force} \times \text{displacement} \\ &= \tfrac{1}{2} \times F \times s \\ &= \tfrac{1}{2} \times ks \times s \end{aligned}$$

i.e. $$\text{work} = \tfrac{1}{2}ks^2 \tag{1.45}$$

The average force is one half the maximum force because, by Equation (1.5), the force increases linearly from zero to its maximum value as the spring is being extended.

The potential energy stored in the spring, E_p, must equal the work done in stretching the spring, and hence:

$$E_p = \tfrac{1}{2}ks^2 \tag{1.46}$$

Now, the velocity of an object undergoing simple harmonic motion was shown in Equation (1.13) to be given by:

$$v = \pm 2\pi f\sqrt{(A^2 - s^2)} \tag{1.13}$$

where for a tethered trolley:

$$f = \frac{1}{2\pi}\sqrt{\frac{k}{m}} \tag{1.28}$$

Hence $$v = \left(\sqrt{\frac{k}{m}}\right)\sqrt{(A^2 - s^2)}$$

and $$v^2 = \frac{k}{m}(A^2 - s^2)$$

Thus, the kinetic energy E_k is given by:

$$\begin{aligned} E_k &= \tfrac{1}{2}mv^2 \\ &= \tfrac{1}{2}m\frac{k}{m}(A^2 - s^2) \\ &= \tfrac{1}{2}kA^2 - \tfrac{1}{2}ks^2 \end{aligned} \tag{1.47}$$

Therefore, from Equations (1.46) and (1.47) the total energy, E, at a particular moment in the cycle is given by:

$$\begin{aligned} E &= E_p + E_k \\ &= \tfrac{1}{2}ks^2 + \tfrac{1}{2}kA^2 - \tfrac{1}{2}ks^2 \end{aligned}$$

i.e. $$E = \tfrac{1}{2}kA^2 \tag{1.48}$$

The total energy is constant throughout the cycle, and is independent of s. It is important to note that it is proportional to the *square* of the amplitude. These results are shown diagrammatically in Fig. 1.15.

Similar results can be derived for all types of simple harmonic oscillators. In each case, for undamped simple harmonic motion, the total energy can be shown to be constant throughout each cycle, and can be shown to be proportional to the square of the amplitude. In some cases, such as a mass hanging on a spring, we must allow for gravitational potential energy, as well as spring potential energy and kinetic energy.

Q 1.12

A trolley of mass 2 kg is held between springs of force constant 20 N m^{-1}, mounted so that at equilibrium the springs are unstretched; it is set into oscillation with amplitude 20 cm. Calculate the spring potential energy, and kinetic energy, when the displacement is 0 cm, 10 cm and 20 cm and show that the sum of spring potential energy and kinetic energy is constant. Assume that energy is stored in the springs only when they are extended.

The energy of ions in a crystal

We can use our understanding of the energy of oscillations to explain certain properties of ions in a crystal. Fig. 1.16 shows the variation of the potential energy of an ion in a sodium chloride crystal as a function of r, which is the spacing between neighbouring ions. In this case, the graph is *asymmetrical* about the equilibrium position. The zero of potential energy has not been chosen to be at the equilibrium separation, but to be the potential energy when the ions are at an infinite separation. However, there is an ionic separation, r_0, at which the amount of potential energy stored in the ions is a minimum, U_0; in order to increase or decrease the separation more energy must be supplied and thus r_0 is the *equilibrium separation*. If the total energy is U_0, the ion has its minimum possible total energy—consisting of potential energy U_0 and zero kinetic energy—and this corresponds in a simple theory to the absolute zero of temperature. It is equivalent to a trolley stationary in its equilibrium position between two springs.

Energy was supplied to the trolley system by pulling the trolley to one side, and thus storing potential energy in the spring. When the trolley was released this potential energy was converted into kinetic energy, and back again, as the trolley oscillated about the equilibrium position. The energy of the crystal can be increased by heating it; let us suppose that the total energy increases from U_0 to U_1. There is now sufficient energy in the system for the separation to be anything from r_1' to r_1''. At r_1' and r_1'' all the energy must be stored as potential energy but at intermediate separations there can be up to $(U_1 - U_0)$ as kinetic energy. We usually associate the heating of the crystal with this possible increase in kinetic energy.

In the case of the trolley we assumed that it oscillated about a mean position mid-way between the extreme displacements, and this mean position was also the trolley's undisplaced position. However, as noted earlier, the variation of potential energy with separation is not symmetrical about r_0, and hence the new mean position, r_1, is at a greater separation than r_0. It is this fact which accounts for the expansion of solids with an increase in temperature.

The argument above contains the simplifying assumption that the atoms are stationary at

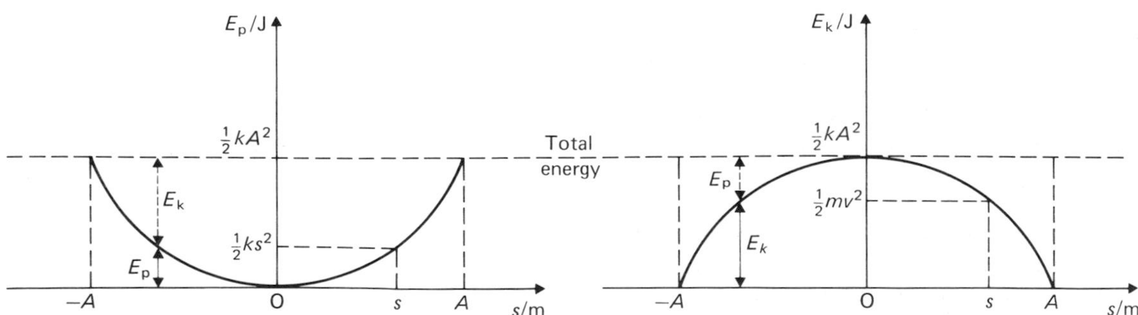

Fig. 1.15 The potential and kinetic energies of an oscillating trolley.

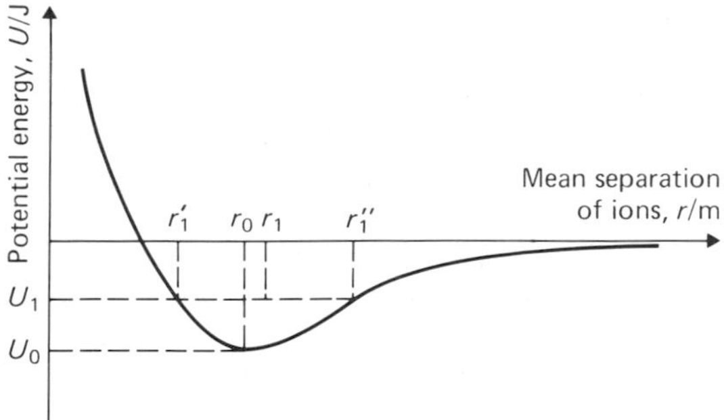

Fig. 1.16 The potential energy of an ion in a sodium chloride crystal.

absolute zero. In fact this is not so; they have a *zero-point energy* whose existence can only be proved by quantum mechanics, as discussed in Section 11.5.

1.5 Damping

At the beginning of the last section we assumed that the energy of the oscillating system remained constant; everyday observation, however, tells us that this is not the case. A simple pendulum will not swing indefinitely; a trolley mounted between two springs will stop oscillating in a fairly short time, although a similar arrangement using a linear air track will keep going for longer. In each case, the amplitude of the oscillation decreases as the energy decreases, because the energy is proportional to $(\text{amplitude})^2$. The energy of the oscillation decreases because work is done against dissipative forces, which cause other forms of energy, such as potential and kinetic, to be transformed into heat.

These dissipative forces may be due to air resistance, or friction; the dissipative forces in the case of a mass hanging on a spring are partly internal, preventing the spring from being perfectly elastic, and warming it gradually as the oscillations decrease in amplitude.

This loss of energy is called *damping* and the examples given above are of *natural damping*, where the damping is inevitable and perhaps a nuisance. Frequently, however, we choose to provide *artificial damping* to reduce an unwanted oscillation. For example, damping is provided by the 'shock absorber' of a car to reduce its oscillation after running over a bump, and the panels of a car are often sprayed with a special compound to reduce vibration, and hence the noise produced, as the car goes along.

The motion of a loudspeaker diaphragm is damped because the loudspeaker radiates sound energy; here the damping is caused by the useful function of the loudspeaker.

Artificial damping sometimes takes the form of electromagnetic damping. This is illustrated in Fig. 1.17(*a*) where an aluminium sheet is attached to a simple pendulum so that it swings between the poles of a magnet. The motion of the sheet through the magnetic field induces eddy currents in the sheet. The resistive heating effect of these currents causes energy to be lost from the system and thus the oscillations are damped. The currents due to the induced e.m.f. in the coil of a moving-coil galvanometer act similarly to reduce the oscillations. Eddy currents are also used to damp the oscillations of

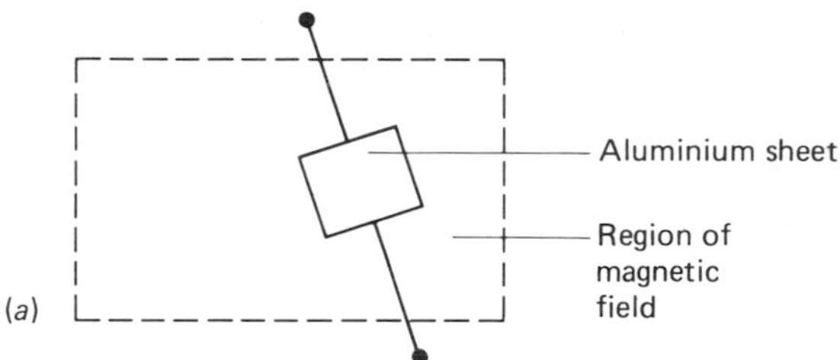

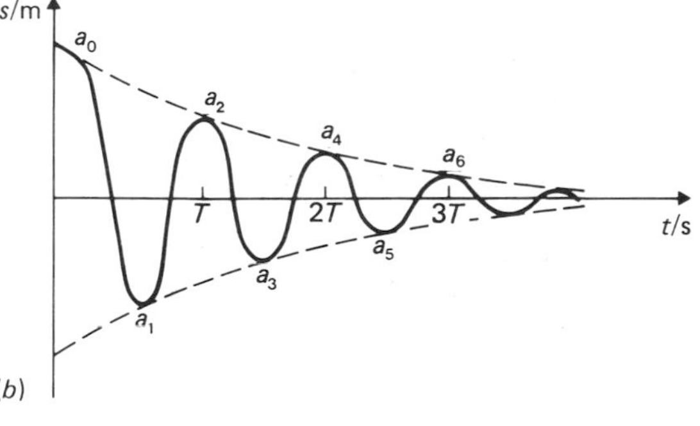

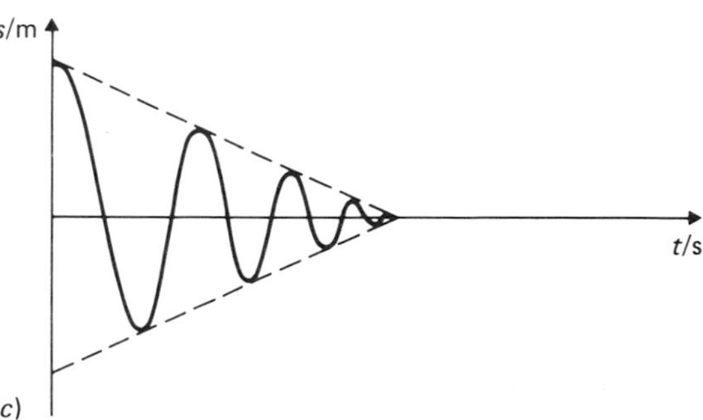

Fig. 1.17

a top-pan balance. The decreasing motion in each case can also be regarded from the point of view of the 'motor-effect' forces exerted on the induced currents by the magnetic field; these forces will act in such a direction as to oppose the motion causing them.

Types of damping

Damping may be investigated by attaching an inked brush to the end of a vibrating lath and pulling a length of paper underneath. Fig. 1.17(b) shows that in this case the amplitude would undergo *exponential decay*; this type of damping would also occur in the case of the oscillating trolley described earlier. If the wheels were taken off the trolley, so that there was merely a wooden block sliding over a rough surface, the amplitude would decrease in a fashion that was more nearly linear, as shown in graph (c).

Detailed examination of the graph (b) reveals two features:

(i) the motion is still closely isochronous;
(ii) the ratio of successive maxima is almost exactly constant.

These two results can be shown, by more advanced mathematical methods, to follow from the fact that the dissipative force is proportional to the velocity of the oscillator. This is generally true where damping is due to air resistance, the viscosity of a liquid or electromagnetic induction.

The constant ratio between successive maxima means that the dotted line in graph (b) must represent exponential decay. We have

$$\delta = \frac{a_0}{a_1} = \frac{a_1}{a_2} = \frac{a_2}{a_3} = \dots \qquad (1.49)$$

where δ is called the *decrement*. The *logarithmic decrement* (log dec), Λ, is defined by the equation:

$$\Lambda = \ln\delta = \ln\frac{a_0}{a_1} = \ln\frac{a_1}{a_2} = \ln\frac{a_2}{a_3} = \dots \qquad (1.50)$$

Now

$$\frac{a_0}{a_n} = \frac{a_0}{a_1} \times \frac{a_1}{a_2} \times \frac{a_2}{a_3} \times \dots \times \frac{a_n - 1}{a_n}$$

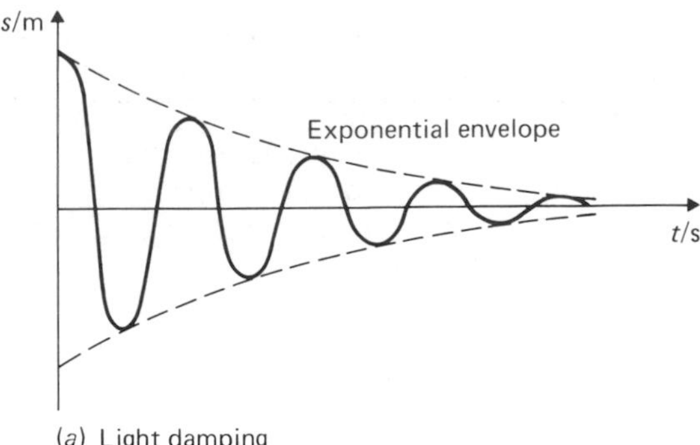

(a) Light damping

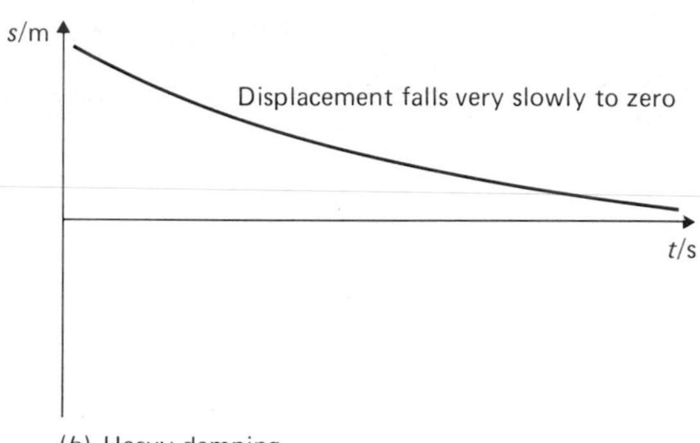

(b) Heavy damping

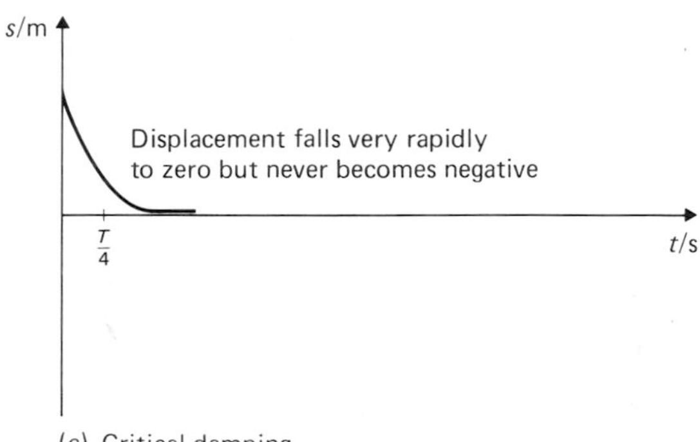

(c) Critical damping

Fig. 1.18

$$\therefore \quad \ln\frac{a_0}{a_n} = \ln\frac{a_0}{a_1} + \ln\frac{a_1}{a_2} + \ln\frac{a_2}{a_3} + \dots$$

$$\dots + \ln\frac{a_n - 1}{a_n}$$

$$= \Lambda n \qquad \text{from (1.50)}$$

$$\therefore \quad \frac{a_0}{a_n} = e^{\Lambda n}$$

i.e. $$a_n = a_0\,e^{-\Lambda n} \qquad (1.51)$$

This is the equation for exponential decay. By analogy with radioactive decay, which is also an example of exponential decay, we might define the *half-life* as being the number of oscillations required for the amplitude to fall to one-half of its original value; the half-life is independent of the initial amplitude.

Q 1.13
A loaded test-tube oscillating in water loses about 16/25 of its energy in each cycle. Draw a graph showing how its *amplitude* changes with time.
Calculate the value of its logarithmic decrement and hence calculate its amplitude after twenty cycles if the original amplitude was 10 cm.

If the damping is very much increased then there may be no oscillations at all, merely a very slow return to the equilibrium position. The condition under which the displacement is reduced effectively to zero in the shortest time is called *critical damping*. This time is usually rather more than a quarter of the undamped period. These various possibilities are shown in Fig. 1.18.

1.6 Forced oscillations and resonance

A child's swing will usually swing naturally with a period of about two or three seconds. We may attempt to set it in motion by taking hold of the ropes or seat and pushing regularly. However, if we do so several times in a second, the resulting amplitude is very small; similarly, if we push only every ten seconds, the amplitude will again be small. As every child knows, in order to set it swinging with a large amplitude, we must push every two or three seconds—in other words, at the swing's own frequency.

These three situations are all examples of forced oscillations, where we are trying to make a system oscillate by applying an external periodic force. The frequency of this applied force is called the *driver frequency*, and the frequency with which the system will oscillate naturally is called the *natural frequency*. The largest amplitude of oscillation is produced when the driver frequency is made equal to the natural frequency; the system is then said to be in *resonance*.

REMAINS OF THE SUSPENSION BRIDGE AT ANGERS, AFTER THE LATE ACCIDENT.

Fig. 1.19 226 soldiers died in 1850, when marching in step across the Angers bridge. (Reproduced, with permission, from the *Illustrated London News.*)

Examples of resonance in everyday life are common. The shaking of car body panels—judged by the resultant noise— is usually worst at one particular car speed, when periodic forces from the engine or road have the same frequency as the natural frequency of the panels. In the United States in 1940 the Tacoma Narrows suspension bridge collapsed when forces set up by wind eddies developed at the same frequency as the natural torsional frequency of the bridge; the short film of this disaster, mentioned in the bibliography, shows the build-up of the oscillations very clearly. An earlier disaster, shown in Fig. 1.19, occurred in 1850 when five hundred French infantry men marched in step across the Angers suspension bridge; when the bridge collapsed the men were plunged into a ravine and two hundred and twenty-six of them were killed. Currents in circuits containing capacitance and inductance will oscillate only at certain frequencies determined by the components in the circuit; they will respond to a narrow band of input signal frequencies and are thus used as tuned circuits in radio receivers.

Barton's pendulum is one simple arrangement for the observation of resonance effects. A series of paper cones are suspended by light threads of various lengths from a horizontal string from which a heavy driving pendulum is also suspended; the length of the driving pendulum is made equal to that of one of the driven

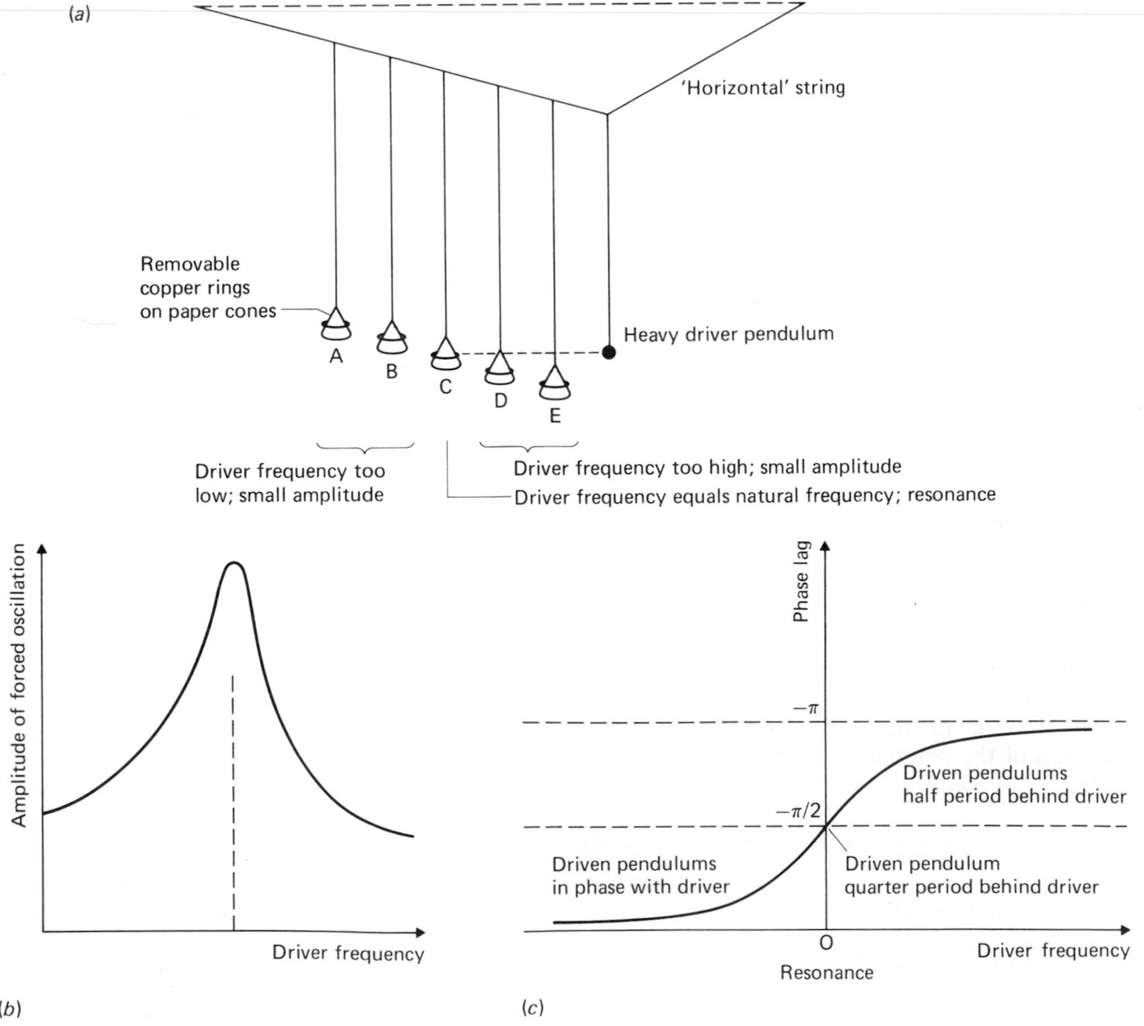

Fig. 1.20 Barton's Pendulum.

pendulums. The arrangement is shown in Fig. 1.20; each of the paper cones is weighted by a copper ring to decrease the relative effect of air resistance by increasing the weight of the cones. We shall discuss the effect on damping of air resistance later.

Pendulum C has the same length as the driver pendulum, and, therefore, the same natural frequency; it develops a large amplitude of oscillation. The other pendulums have different natural frequencies and thus, although they do swing, the amplitudes are small. The surprising observation, however, is that only pendulums A and B are in phase with the driver; pendulum C is a quarter of a period behind the driver and D and E are half a period behind. Each of the pendulums A to E is undergoing forced oscillations; only pendulum C is in resonance. The amplitudes and phase relationships are shown in diagrams (*b*) and (*c*); it is usual to represent them by the variation in amplitude and phase difference as the *driver frequency* is altered.

Damping and Barton's pendulum

The driven pendulums experience forces due to air resistance as well as their weight; in the experiment described previously, air resistance was made relatively less important by increasing the weight of the paper cones with copper rings.

In order to investigate the effect of increased damping these rings are now removed, and the resonance is found to be less pronounced; there is a smaller response from the driven pendulums over a wide range of frequencies instead of a large response over a narrow range. These results are shown graphically in Fig. 1.21; we see that because damping reduces the effective restoring force, the resonant frequency is slightly altered.

The amount of damping, and hence the sharpness of the resonance, is measured by the *quality factor*, *Q*. It is defined by:

$$Q = 2\pi \frac{\text{energy stored in the system at resonance}}{\text{energy lost per oscillation}} \quad (1.52)$$

It can be related to the *log dec* of the oscillation:

$$Q = \frac{\text{resonant frequency}}{2 \times \text{logarithmic decrement}} \quad (1.53)$$

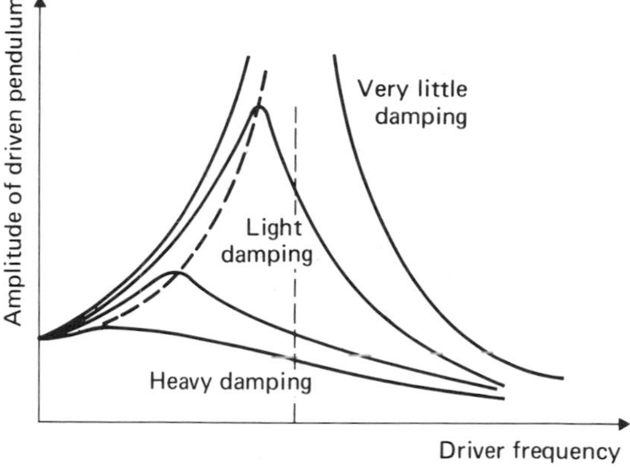

(*a*)

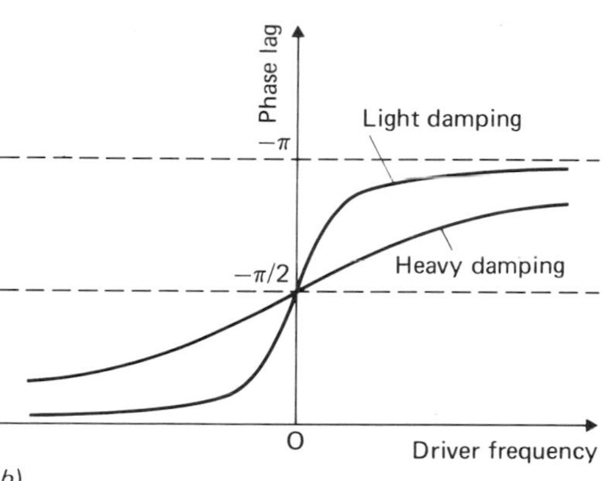

(*b*)

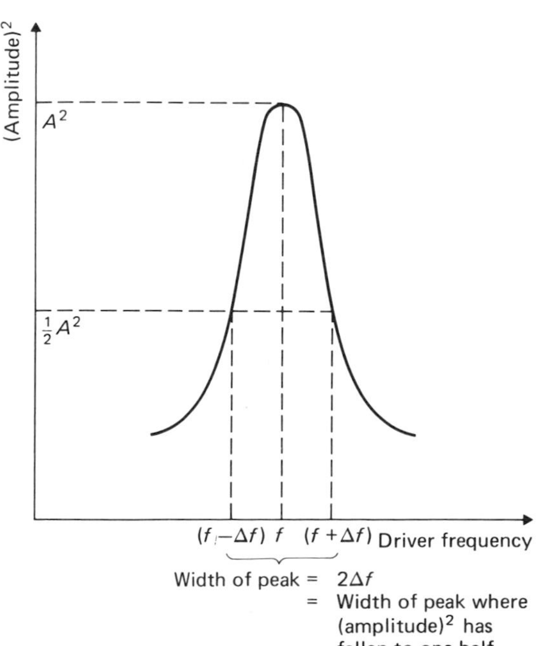

(*c*)

Fig. 1.21 Damping and resonance.

and also to the sharpness of the resonance peak:

$$Q = \frac{\text{resonant frequency}}{\text{width of resonant peak}} \qquad (1.54)$$

The way in which the width of the resonant peak is defined is shown in Fig. 1.21(*c*).

Thus, if there is no damping, and no energy is lost from the system, the quality factor will be infinite and the resonance infinitesimally narrow. A simple pendulum may have a Q factor of about 500, an oscillating hydrometer in water a Q factor as low as 10 and a piano string a Q factor of over 1000. In a radio tuner, circuits are used to select a particular broadcast frequency for amplification and the Q factor here is often about 100; it is determined by the resistance, capacitance and inductance of the circuit.

If the driven system is to oscillate with constant amplitude, the energy input from the driver must equal the loss through damping in each cycle. Thus, the rate at which the energy must be supplied is determined by the damping. We have assumed above that the driver oscillations are maintained by an external supply of energy; if they decrease, complex interactions between the driver and driven pendulums occur.

Molecular vibration spectra

Resonance can occur on an atomic scale, just as it can occur on a macroscopic scale. It can be used to investigate the natural frequencies of atomic and sub-atomic systems; we measure the frequency at which the largest amount of wave energy is absorbed from an incident beam.

Fig. 1.22 shows that when white light is incident on a gas such as hydrogen chloride, light at particular wavelengths will have the same frequencies as those at which the molecules vibrate, and will therefore be strongly absorbed. Although this energy will be subsequently re-emitted when the molecules cease to vibrate only part of it will be re-emitted in the original direction and the intensity in the transmitted beam at this frequency is thus reduced. The detector will show a drop in intensity at the various frequencies at which this process occurs; the lowest frequency at which this occurs is in the infrared. From measurements of the resonant frequency we can determine at what frequency the atoms vibrate, and hence by using

$$f = \frac{1}{2\pi}\sqrt{\frac{k}{m}} \qquad (1.28)$$

we can determine the restoring force per unit displacement of the bond.

Let us assume that the heavier chlorine atom remains stationary whilst the hydrogen atom vibrates, as though it were attached to it by a spring. The lowest frequency that is absorbed by HCl is 8.7×10^{13} Hz, in the infrared. The mass of the hydrogen atom is 1.7×10^{-27} kg. Hence

$$\begin{aligned} f &= \frac{1}{2\pi}\sqrt{\frac{k}{m}} \\ k &= 4\pi^2 f^2 m \\ &= 4\pi^2(8.7^2 \times 10^{26})(1.7 \times 10^{-27}) \\ &= 508 \text{ N m}^{-1} \end{aligned}$$

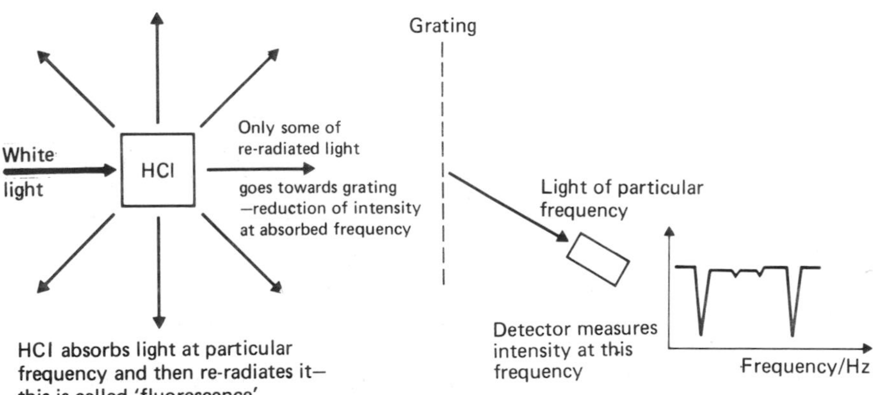

Fig. 1.22 The formation of molecular vibration spectra.

For comparison, a typical 10 N spring-balance has a restoring force per unit displacement of $100\,\mathrm{N\,m^{-1}}$, and a thin steel rod, of cross-sectional area $1\,\mathrm{mm^2}$ and length 1 m, has a restoring force per unit displacement of $2100\,\mathrm{N\,m^{-1}}$ in tension or compression.

1.7 Addition of simple harmonic motions

Lissajous' figures

It is sometimes the case that an object will be subject to two oscillations at once; we will consider here the example where these constituent oscillations are at right angles. In an arrangement known as Blackburn's pendulum, a mass M swings at right angles to the plane of Fig. 1.23(*a*) as a pendulum of length BM, and as a pendulum of length DM in the plane of the figure. The motion of M can be observed if it is a sand-filled funnel leaving a pattern in a large tray; in general, the pattern produced will have a number of loops and the arrangement of these depends upon the ratios of the frequencies of the two oscillations, as shown in Fig. 1.23(*b*).

Fig. 1.23(*c*) shows the patterns obtained with identical constituent frequencies. In general, the shape is an ellipse but if the oscillations are in phase, or π radians (180°) out of phase, a straight line is produced. In the case where the oscillations are $\pi/2$ radians (90°) out of phase, a circle is produced if the amplitudes of the oscillation are equal.

It is possible to explain these results by applying the idea of a reference circle that was used earlier; SHM along a straight line can be regarded as the projection onto a diameter of a point moving round a circle. A particle subject to two simultaneous SHMS must lie on the projection onto one line of points moving round two circles simultaneously. The technique is shown in Fig. 1.24 and the reader is advised to study these diagrams carefully, and to construct others, as in the question below.

Lissajous' figures may be observed more conveniently on a cathode ray oscilloscope. One frequency is connected across the X-plates and one frequency across the Y-plates, and the spot will then move in patterns such as those in Fig. 1.23. This method can be used to compare electrical frequencies; a more detailed discussion is given in Bennet, *Electricity and Magnetism*, p. 232.

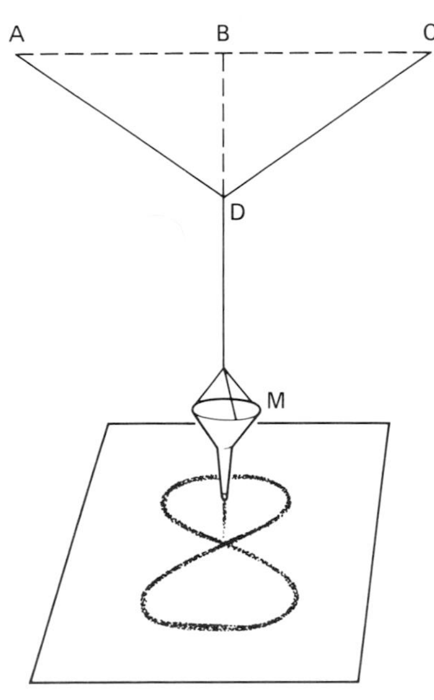

(*a*) Blackburn's pendulum

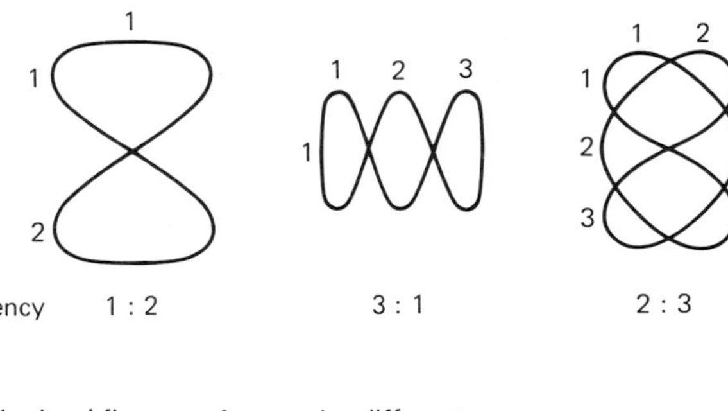

(*b*) Lissajous' figures : frequencies different

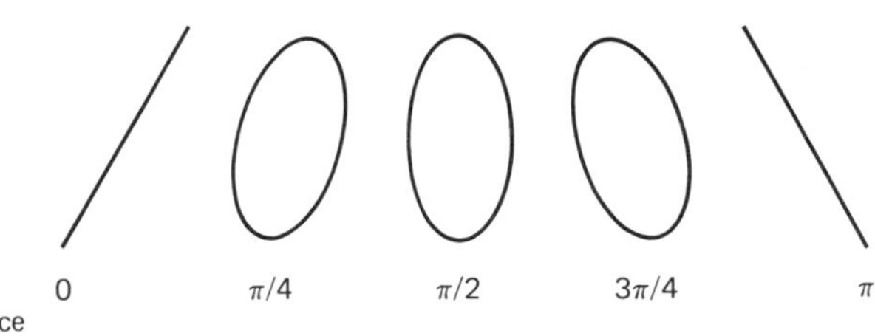

(*c*) Lissajous' figures : frequencies identical, amplitudes different

Fig. 1.23

Q 1.14
Construct a Lissajous' figure for each of the following cases as shown below:

(*a*) frequencies and amplitude identical: 90° out of phase;
(*b*) frequency ratio 2:5, amplitude identical: 90° out of phase.

Fourier synthesis

It is important to remember, as was mentioned in the introduction, that there are many oscillations that are not simple harmonic; a ball rolling on a V-ramp is one example. Fourier, a French mathematician, showed that any such complex

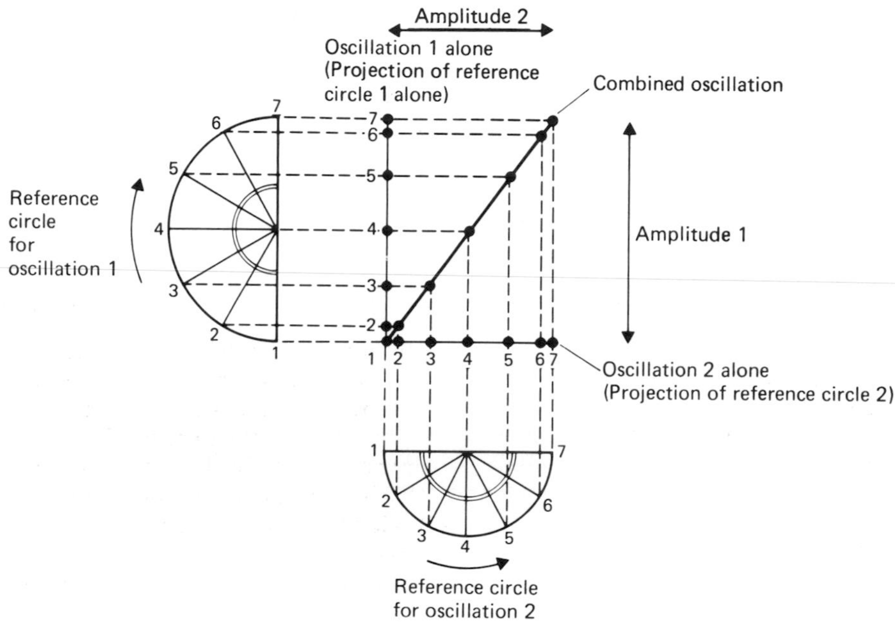

(*a*) Equal frequency, different amplitude, in phase

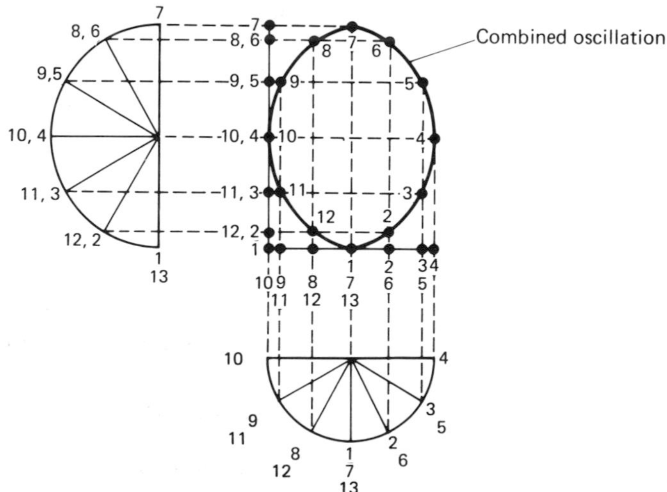

(*b*) Equal frequency, different amplitude, $\pi/2$ out of phase

Fig. 1.24 The construction of Lissajous' Figures.

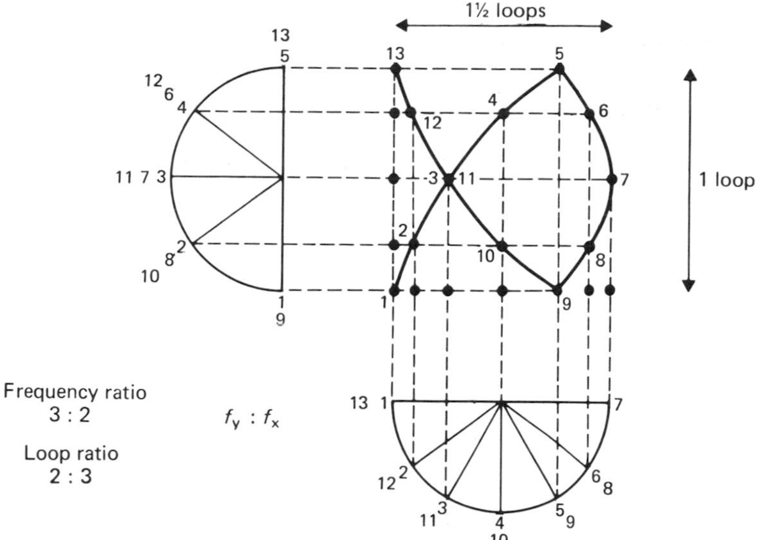

(c) Frequency ratio 3 : 2, equal amplitude

Fig. 1.24 (*continued*)

oscillation could be considered as the sum of a whole series of oscillations, each of which was simple harmonic; in other words, any repetitive function can be represented by the sum of a series of components such as

$$A_0 + A_1 \cos 2\pi ft + A_2 \cos 4\pi ft + A_3 \cos 6\pi ft + \ldots + B_1 \sin 2\pi ft + B_2 \sin 4\pi ft + B_3 \sin 6\pi ft + \ldots$$

The frequencies present are thus a fundamental frequency f, and its harmonics $2f$, $3f$, etc. Fig. 1.25 shows the first three such trigonometric components required for a square wave, and in the third diagram we see the sum of these components and we see that, even with only three such components, it is a reasonable approximation to the required square wave. At any instant the actual displacement is the algebraic sum of the displacement due to each component.

Much of the importance of simple harmonic motion is due to the fact that those oscillations which are not themselves simple harmonic can be expressed as the sum of a number of oscillations which are; thus, simple harmonic motion is fundamental to all oscillation.

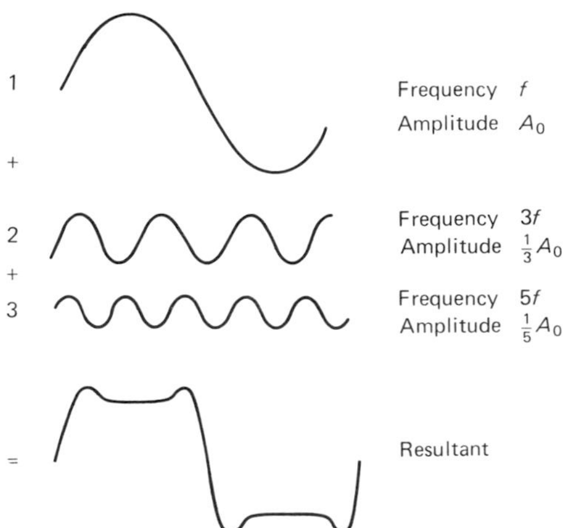

Square wave $A = A_0 \; (\sin 2\pi ft + \frac{1}{3} \sin 6\pi ft + \frac{1}{5} \sin 10\pi ft + \ldots)$

Fig. 1.25 The Fourier synthesis of a square wave.

Q 1.15

Construct the wave of which the following are the components:

(i) a sine wave of frequency f and amplitude A_0;
(ii) a sine wave of frequency $2f$ and amplitude $-A_0/2$ (i.e. out of phase with (i) at $t = 0$);
(iii) a sine wave of frequency $3f$ and amplitude $A_0/3$.

What is this waveform called and in what laboratory instrument is it used?

Bibliography

General accounts

Akrill, T.B. and Millar, C.J. (1974) *Mechanics, Vibrations and Waves.* Murray; Chapter 15 covers oscillation and resonance.

Bolton, W. (1986) *Patterns in Physics* (2nd Edn.). McGraw-Hill; Chapter 6 contains a numerical approach to the analysis of shm as well as some interesting background reading on resonance.

Feynman, R.P., Leighton, R.B., and Sands, M. (1963) *The Feynman Lectures on Physics.* Addison-Wesley; Volume 1, Chapter 9 gives a detailed and clear account of a numerical method for shm; Chapters 21 and 22 give a more advanced study of harmonic oscillation; Chapter 23 discusses resonance.

Revised Nuffield Advanced Physics (1985) *Teachers' and Students' Guide.* Longman; Unit D in Volume 1 covers much of this material, including a wide range of experiments and questions.

Shive, J.N. and Weber, R.L. (1982) *Similarities in Physics.* Hilger; Chapter 6 discusses related phenomena from different parts of physics.

Wenham, E.J., Dorling, G.W., Snell, J.A.N. and Taylor, B. (1984) *Physics: concepts and models* (2nd Edn.). Longman; Chapters 19 and 20 give a thorough and rather more mathematical treatment at a Sixth Form level.

Specific topics

Butler, S.T. and Messel, H. (1965) *Time: selected lectures.* Pergamon.

Dorling, G.W. (1973) *Time.* Longman Physics Topics.

Gould, R.T. (1978) *John Harrison and his Timekeepers* (4th Edn). National Maritime Museum.

Hood, B. (1969) *How Time is Measured.* OUP.

Howse, D. (1980) *Greenwich Time and the Discovery of the Longitude.* OUP.

Time Life (1967) *Time.*

Bishop, R.E.D. (1979) *Vibration* (2nd Edn.). CUP; an introduction to applied vibration based on a series of Christmas lectures to the Royal Institution.

CRAC Science Series (1983) *Vibration.* Hobsons Press; discusses briefly a wide variety of applications in the modern world.

Revised Nuffield Advanced Physics (1986) *Physics in Engineering and Technology.* Longman; contains an article 'Buildings, Bridges and Roads' which discusses resonance issues, in particular.

Walker, J.R. (1978) *The Flying Circus of Physics.* Wiley; contains many interesting questions on resonance.

Wallace, R.H. (1970) *Understanding and Measuring Vibrations.* Wykeham; discusses some engineering applications of vibration, especially in jet-engines.

Revised Nuffield Advanced Physics (1986) *Energy Options.* Longman; contains an article 'Wave Power: the story so far' which discusses the power available from the oscillation of sea waves.

Waller, M.D. (1961) *Chladni Figures.* Bell; a specialized, but readable, account with some fascinating diagrams and photographs.

Bennet, G.A.G. (1974) *Electricity and Modern Physics* (2nd Edn.). Edward Arnold; contains a more detailed discussion of the use of Lissajous' figures to investigate frequencies and phase differences, on p. 232–3.

Reprint (Out of print but available in libraries)

Lyon, H. (1957) Atomic Clocks, *Scientific American* offprint No. 225.

Films

Tacoma Narrows Bridge collapse	Ealing Scientific
Wind induced oscillations	Penguin (no longer in print)
Vibrations of a drum	Ealing Scientific
Soap film oscillations	Ealing Scientific

Software

Longman Micro Software, *Nuffield A-Level Software Pack*, 'SHM'; shows the relationship between shm oscillations and the reference circle, and the significance of phase relationships between two oscillations.

Longman Micro Software, *Dynamic Modelling System*, 'OSCIL'; enables a variety of undamped, damped and forced oscillations to be considered graphically, with a range of parameters input by the student.

2
Wave motion

2.1 Waves

In the first chapter of this book, we considered the physics of oscillations, and we saw that they could act as sources of waves. In the next three chapters, we will examine the phenomenon of wave motion, considering the way in which waves are transmitted and also what happens when they meet.

Waves carry energy and may be capable of great destruction. If an earthquake, or other submarine disturbance, takes place at sea, tidal pulses or *tsunamis* are formed and travel at speeds of up to 800 kilometres an hour, transmitting energy which may cause enormous damage when they reach land. Tsunamis generated in the Aleutian Trench near Alaska on April 1st, 1946, later swept the town of Hilo, in the Hawaiian Islands—costing 159 lives and $25 million damage in property; the waves had travelled 3200 kilometres from their point of generation in about 4 hours. We know, however, that although vast amounts of energy clearly travelled along with these tsunamis, no water at all made the long journey from Alaska to Hilo.

This last point may also be observed by watching a ship at sea riding out a heavy storm. There will be only a slight to and fro movement and the main motion of the ship will be up and down as the waves pass underneath. The ship—and the water—will be in roughly the same place when the storm has finished as they were when it began.

Thus, there is clearly a difference between the motion of objects and the motion of waves. An object and its kinetic energy are transmitted together from one place to another whereas, in the case of a wave, although considerable temporary disturbance of the medium may be caused as the wave or pulse passes, there is no permanent displacement afterwards.

Wave physics is of great importance to us for a number of reasons. In the first place, it enables

Fig. 2.1 The caption in the top left-hand corner of this print by Hokusai, the 19th-century Japanese print-maker, reads: 'The crest of the great wave off Kanagawa'.

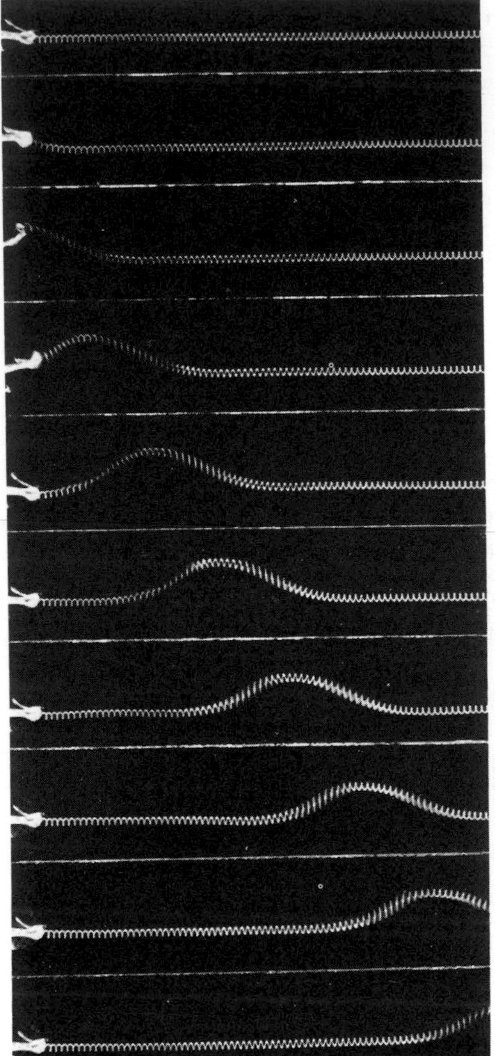

Fig. 2.2 The generation and transmission of a transverse pulse along a spring.

us to understand better various physical phenomena such as the propagation and properties of light and radio waves. Secondly, there are many examples of wave motion that are of practical importance, such as the effect of sea waves on ships and on the coast line. The study of earthquakes and the waves produced by them, both directly in the earth and at sea as tsunamis, can help us to minimize the damage they cause as well as teaching us about the structure of the earth. The wave properties of radar are important in navigation, especially when applied to systems such as the widespread Decca and Loran chains of transmitters.

The theoretical study of waves is of value because all types of wave motion have certain basic properties and share essentially similar mathematical treatments. Thus, having obtained certain results by the analysis of one type of wave motion, we may apply those results to a wide range of other situations.

2.2 Simple wave systems

The continuous oscillation of a source will produce a series of waves; one half oscillation will produce a *pulse*. If we hold the end of a stretched spring and move it up and down once, a pulse travels along the spring as shown in Fig. 2.2.

We see that each point on the spring follows the same kind of motion as its end, but at a slightly later time. When the end of the spring is moved upwards by my hand, this causes the next bit of the spring to be moved upwards as well; this process is repeated at each point along the spring. We say that the pulse moves along the spring; when the pulse has passed a given point, however, the spring is in exactly the same position as it was at the beginning—it has *not* moved from left to right with the pulse. The motion of a particular point on the spring can be seen clearly in Fig. 2.3, because a piece of white material has been attached to it.

As the motion of this point on the spring is at right angles to the direction of motion of the pulse, the pulse is said to be *transverse*. If my hand is moved up and down repeatedly, as in an oscillation, then a succession of identical pulses is sent out which is called a *transverse wave motion*. When transverse waves are propagated through a solid elastic medium, such as the earth's mantle, they are called *shear waves*, since they bend, or *shear*, the mantle. There is a more detailed discussion of earthquake waves in Section 3.7.

A second type of wave motion can be caused by oscillations along the direction of motion of the pulse as illustrated in Fig. 2.8 on page 36. Again we can see how the motion of each trolley and of the source are exactly similar but at different times, and how a pulse is transmitted along the spring. When the motion of the

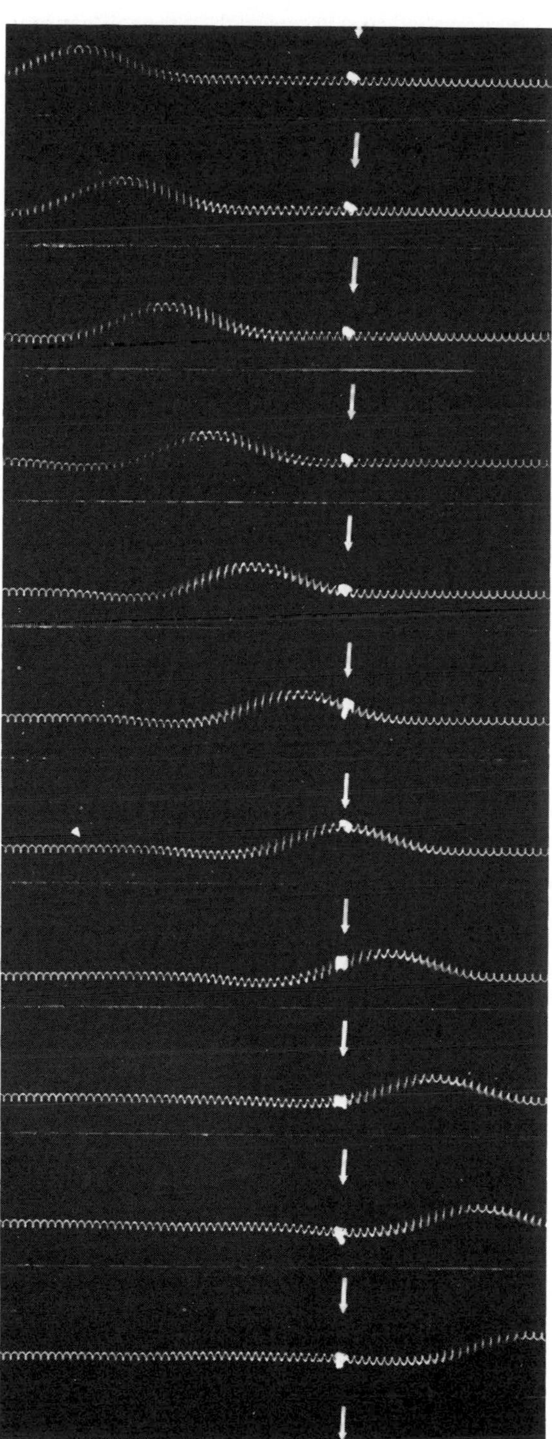

Fig. 2.3 The piece of ribbon shows the motion of a point on the spring; it moves up and down as the pulse passes, but does not travel in the direction of motion of the pulse.

medium is in the same direction as, or opposite to, the motion of the pulse, the pulse is said to be *longitudinal*; again, a succession of identical pulses is called a *longitudinal wave motion*. Sound waves in air are a well known example of longitudinal wave motion; the air molecules have alternately a smaller and larger average spacing than normal.

A third variant of wave motion involves torsional oscillations. Fig. 2.4 represents a series of horizontal bars connected by a taut wire; when the end bar is rotated the torsion in the connecting wire causes the other bars to rotate.

Q 2.1
State which of the following wave motions are transverse and which are longitudinal:

radio waves,
sound waves in a solid,
waves on a violin string,
light,
sound waves in a bassoon tube.

2.3 The description of a wave

The motion of one point in the medium through which a wave passes may be described by a displacement–time graph, as explained for the motion of an oscillator in Section 1.1. Fig. 1.1 shows what is meant by *period*, T, and *amplitude*, A; the same equation as before can be used to relate the period of a wave motion to its frequency f:

$$f = 1/T \tag{1.1}$$

The *frequency* of a wave motion is the number of complete oscillations carried out by a point in the medium in each second which is, of course, also the number of waves passing the point per second. The frequency of the wave motion will be equal to the frequency of the oscillation causing it.

The *velocity* of a wave, c, is the distance travelled by a particular crest or trough per second, measured in metres per second. Because the wave is moving through the medium the displacement at a given time will vary from place to place and this variation is shown by a displacement–distance graph such as that in Fig. 2.5; in this case the source is undergoing simple harmonic motion and the waveform is sinusoidal.

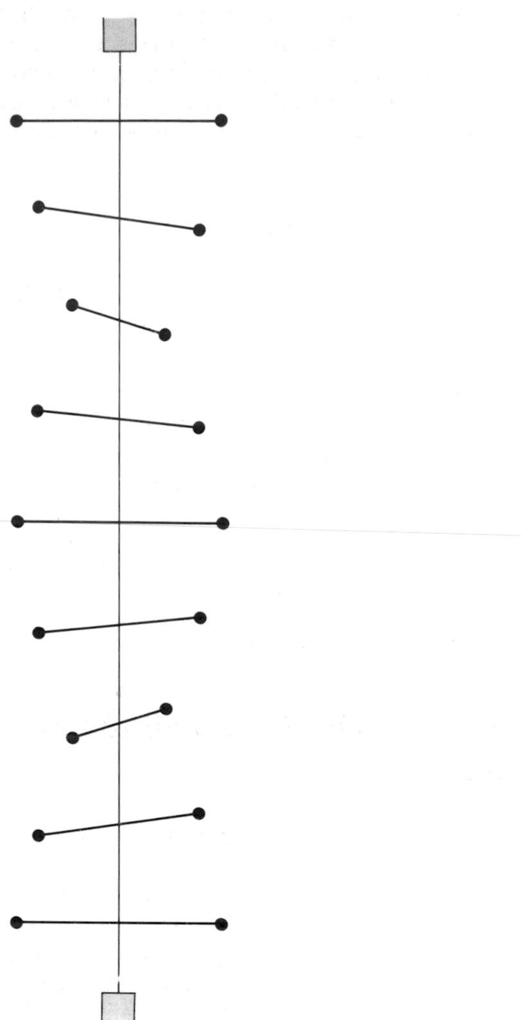

Fig. 2.4 A torsional pulse.

As shown in Fig. 2.5, the *wavelength* of a wave is the distance between any two successive points undergoing simultaneous . identical motion; these may be two crests, two troughs or alternate zeros as shown. It is measured in metres.

If we consider a source beginning to emit waves we can derive an equation connecting their velocity, frequency and wavelength. After a time t, it will have emitted ft waves, since f waves are emitted in each second. The first wave has moved forwards to leave room for each of these waves and, since each wave is of

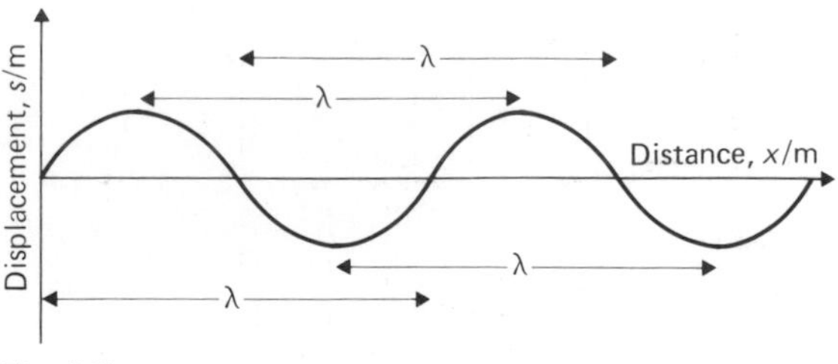

Fig. 2.5

length λ, the first wave has moved a distance $ft\lambda$. Now

$$\text{velocity} = \frac{\text{distance travelled}}{\text{time taken}}$$

$$= \frac{ft\lambda}{t}$$

i.e.

$$v = f\lambda \tag{2.1}$$

This is analogous to saying that the speed at which a man walks is equal to the number of strides per hour, multiplied by the length of each stride.

If the wave is a longitudinal wave, the wavelength is still defined as the distance between two successive points executing identical motion simultaneously, and the amplitude is the maximum displacement each particle experiences. Where the particles of a medium are closer together than usual, there is said to be a *compression*; where they are further apart there is said to be a *rarefaction*; longitudinal waves are, therefore, sometimes referred to as *compression waves.*

A compression wave is shown in Fig. 2.6. It is still possible to plot a graph showing how the displacement of the particles from their undisturbed position varies with distance or time, as shown in the lower half of Fig. 2.6; a positive displacement represents a movement of a particle to the right. It is important to remember that the graph is not a 'picture' of the wave in the same way as it is for transverse waves.

Two waves might have identical amplitudes and frequencies, but be of different shape. The shape of a wave is called its *waveform*; sine, square and sawtooth waveforms are illustrated in Fig. 2.7. A sine wave is emitted by a source whose oscillations are simple harmonic and it is sometimes called a *simple harmonic* or *sinusoidal* wave.

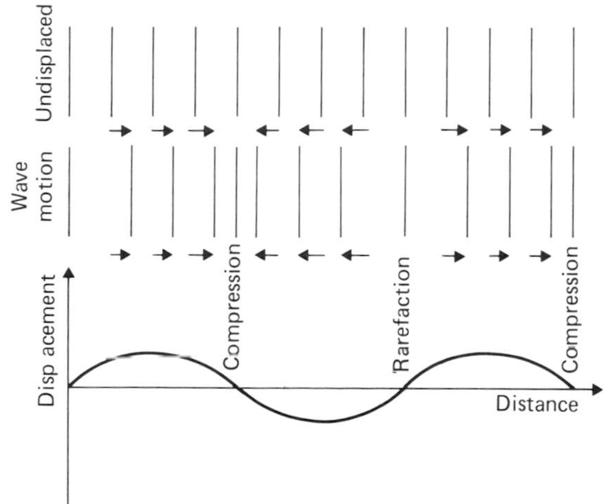

Fig. 2.6 A compression wave.

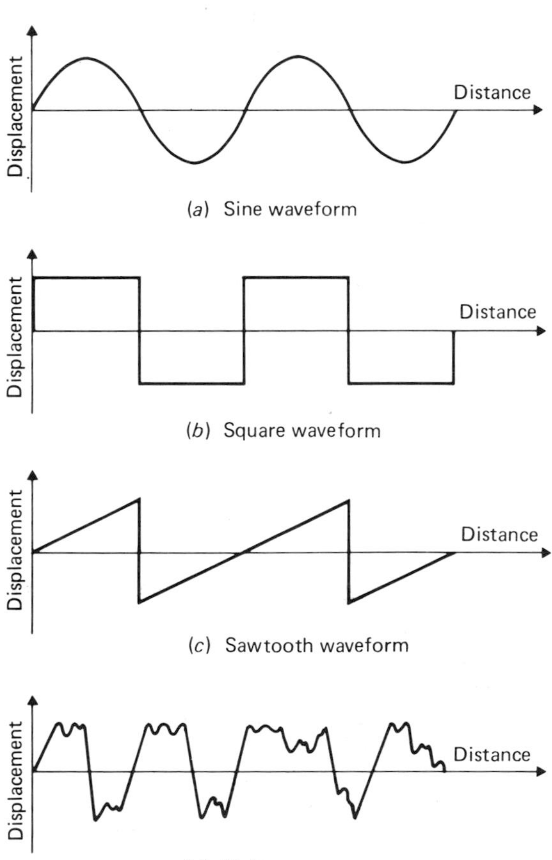

(*a*) Sine waveform

(*b*) Square waveform

(*c*) Sawtooth waveform

(*d*) Noise

Fig. 2.7 Different waveforms.

If the waveform does not repeat itself but follows a random pattern, it is referred to as *noise*; such random fluctuations are always generated in electrical circuits and cause, for example, the background 'hiss' in a radio.

Q 2.2
The speed of electromagnetic waves is about 3.00×10 ms

(*a*) What is the wavelength on which BBC Radio London broadcasts if its frequency is 94.9 MHz?
(*b*) Channel 62 television (u.h.f.) broadcasts at 800 MHz. What is its wavelength?
(*c*) Red light has a wavelength of about 7×10^{-7} m. What is its frequency?
(*d*) X-rays have a wavelength of about 10^{-10} m. What is their frequency?
(*e*) Sound has a velocity in air of 3.3×10^2 m s^{-1}. The lowest note on a bassoon has a frequency of about 65 Hz. What is its wavelength? A bassoon is 2.54 m long; approximately how many wavelengths is this?

2.4 A model for longitudinal waves

It is convenient to begin our detailed discussion by considering longitudinal waves, and the propagation of a wave is seen more easily by considering a system of point masses. We shall examine a system consisting of trolleys linked by springs as this can be demonstrated easily in the laboratory. A system like this is called a '*lumped-mass' system*; in fact, although it is easier to visualize the process of wave propagation in such a system, a full analysis is more complex than for continuous systems. The analysis of the velocity of propagation given later in this section makes a number of simplifying assumptions, and applies only in cases where the wavelength is very much larger than the spacing; it does, however, suggest a method suitable for continuous media, which is discussed in Section 2.5.

We consider, then, a row of trolleys which are linked by springs. When the end trolley is moved backwards and forwards in an oscillation a longitudinal wave is generated, as shown in Fig. 2.8; if the wave is to be of sine form, the oscillation must be simple harmonic, and this

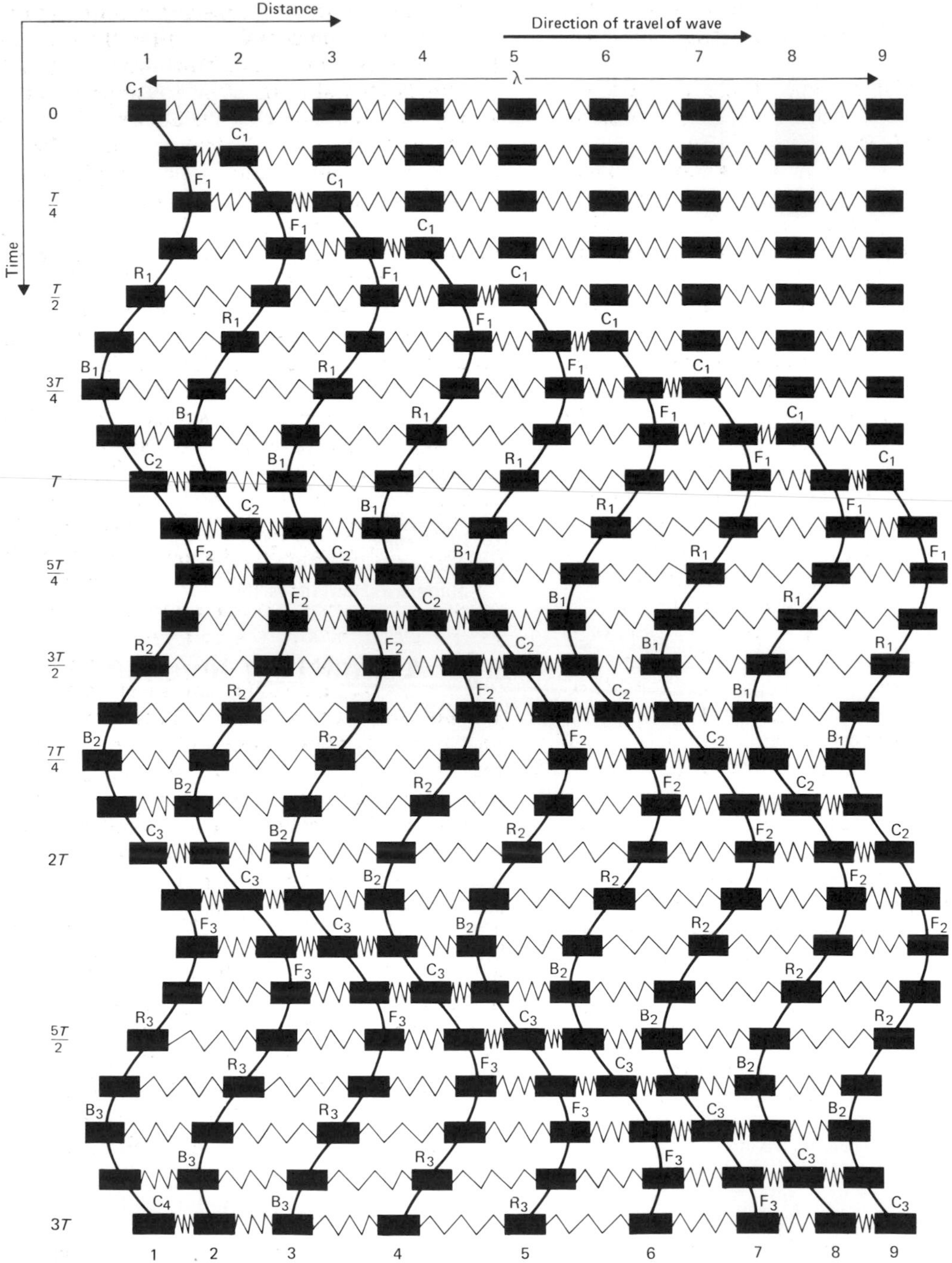

Fig. 2.8 The transmission of a longitudinal pulse down a line of trolleys.

is the case we shall choose to discuss. In the diagram, the trolleys are represented by black rectangles.

The sinusoidal motion of the end trolley, number 1, is seen to repeat itself after successive time intervals T, the period of the oscillation. At times 0, T, $2T$ and $3T$ it has zero displacement but is moving with maximum velocity to the right; at times $T/2$, $3T/2$, $5T/2$ it also has zero displacement and maximum velocity, but this time directed to the left.

When trolley 1 moves to the right, it compresses the spring between trolley 2 and itself and thus trolley 2 also begins to move to the right; this is seen in the second two lines. In the fourth line, trolley 1 has started to move in the opposite direction, slowing down trolley 2 which has by this time begun to push trolley 3 to the right. In this way, each trolley exerts a force varying in a sinusoidal way on the trolley adjacent to it, and each trolley thus oscillates with simple harmonic motion about its equilibrium position. After a time T trolley 9, one wavelength from trolley 1, has just begun to move exactly in step with number 1; the motion of two trolleys one wavelength apart must, of course, by definition be identical. The 'vertical', sinusoidal lines show how each trolley executes the same motion, a little bit out of step with its neighbour.

After a time $T/4$, trolley 1 is at its maximum forward displacement, in the direction of motion of the wave, and this is labelled F_1; the diagram shows how each trolley in turn attains its maximum forward displacement and F_1 moves along the line, reaching a point λ away after a time $5T/4$. Similarly trolley 1 reaches its point of maximum backwards displacement after a time $3T/4$, labelled B_1; B_1 also moves along the line taking a time T to travel distance λ. Later instants of maximum displacement F_2, F_3 and B_2, B_3 are also shown moving down the line.

We may, however, consider a longitudinal wave from a compression, rather than a displacement, point of view. Thus, when trolley 1 is initially moved to the right it compresses the spring between 1 and 2; this has been labelled compression 1, or C_1. As trolley 1 slows down, and trolley 2 begins to move, the spring between 2 and 3 becomes similarly compressed instead; the compression has moved along the row. The diagram shows how C_1 moves along the row, reaching trolley 9 after one period has elapsed; trolley 1 is now creating a second compression, C_2. Thus every T seconds a compression is created which travels a distance λ in the next T seconds.

At a time $T/2$, trolley 1 is moving to the left with maximum velocity and this creates a maximum extension of the spring between 1 and 2; this is labelled rarefaction 1, or R_1. It can be seen that R_1 also travels a distance λ in a time T, after which a further rarefaction is created. Thus we see how a longitudinal wave consists of alternate compressions and rarefactions moving through the medium.

The diagram shows that in a distance λ there will be one compression, one rarefaction and two points of normal spacing. From a displacement point of view, there will be one point of maximum forward displacement, one point of maximum backward displacement and two points of zero displacement. The relationship between these is set out in Table 2.1; the velocity of the trolley is also given.

We have already seen, in Section 1.2, that an oscillator is moving at maximum speed when it is passing through the position of zero displacement, and this must obviously be true also for wave motion. Table 2.1 shows that for a longitudinal wave maximum compressions and rarefactions occur at points of zero displacement; in continuous media the variation in spacing corresponds to changes in density or pressure.

Table 2.1

Point	Spacing	Displacement	Velocity
C	compression	zero	maximum to right
F	normal	maximum to right	zero
R	rarefaction	zero	maximum to left
B	normal	maximum to left	zero

These relationships may also be represented in graphical form; the displacement–distance and spacing–distance graphs are just two ways of looking at the same phenomenon. The symbols used in Fig. 2.9 are the same as those in Table 2.1 and in Fig. 2.8; the graphs correspond to a time of $7T/4$. It is important to note that displacement B_1, which has travelled further, was generated before displacement B_2.

Q 2.3
Draw compression–distance and displacement–distance graphs for the trolleys in Fig. 2.8 at a time $5T/4$. Draw also compression–time and displacement–time graphs for trolley 6. In each case show the actual positions of the trolleys underneath and mark positions C, R, F and B.

As velocity is rate of change of displacement, the reader familiar with the calculus should understand how to derive the velocity column of Table 2.1 mathematically. If

$$v = \frac{dx}{dt}$$

then when x, the displacement, is a positive or negative maximum, the velocity, v, will be zero. Thus the velocity is equal to the slope of a displacement–time graph but it is *not* equal to the slope of a displacement–*distance* graph.

We shall see in Section 2.10 that the motion of a trolley undergoing wave motion can be described by the equation:

$$y = A \sin 2\pi f(t - x/c) \qquad (2.2)$$

The velocity of the trolley is given by:

$$v = \frac{dy}{dt} = 2\pi f A \cos 2\pi f(t - x/c) \qquad (2.3)$$

Thus, the velocity of the trolley will be a maximum $[\cos 2\pi f(t - x/c) = 1]$ when the dis-

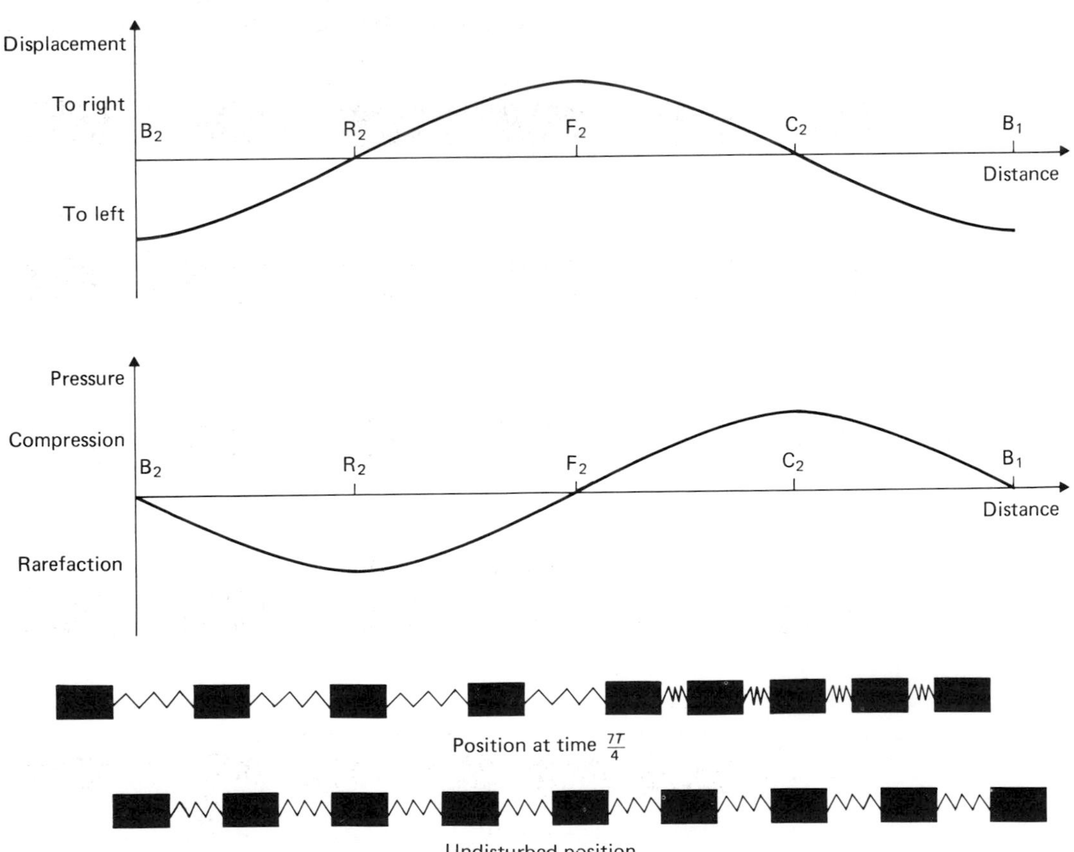

Fig. 2.9 Displacement–distance and pressure–distance graphs for a longitudinal wave.

placement is zero, $[\sin 2\pi f(t-x/c)=0]$, and vice versa.

The velocity of longitudinal waves in a 'lumped-mass' system

A rigorous calculation of the velocity of the waves down a row of linked trolleys, as described above, is complex. Waves of different frequencies will travel at different speeds and there will be an upper limit to the frequency that can be propagated through a given system. However, we shall derive an approximate expression for the velocity because it is easier to visualize the method of calculation than in the continuous-mass case, and because the derived result can be tested in the laboratory by a simple method. We assume that the velocity of a wave, which is a succession of pulses, is the same as that of single pulses; if waves are propagated unchanged this must be the case.

The trolleys are free to move along the line of the springs, as shown in Fig. 2.10, and they are so placed initially that there is no tension in the springs, which are therefore at their natural length. The distance between the centres of the trolleys is x and we shall refer to this as the spacing. It is, of course, equal to the sum of the natural length of the spring and the length of the trolley as shown. The spacing between the trolleys may be altered either by pushing them suddenly closer together, in which case a compression pulse is transmitted down the line, or by pulling them further apart, in which case a rarefaction pulse is transmitted.

The pulse that we shall consider is caused by a constant force being applied to the end trolley, which is made to move; it will cause each trolley in turn to move to the right with constant velocity. As the springs become more compressed they will exert a larger force on the trolleys on either side; the size of this force is

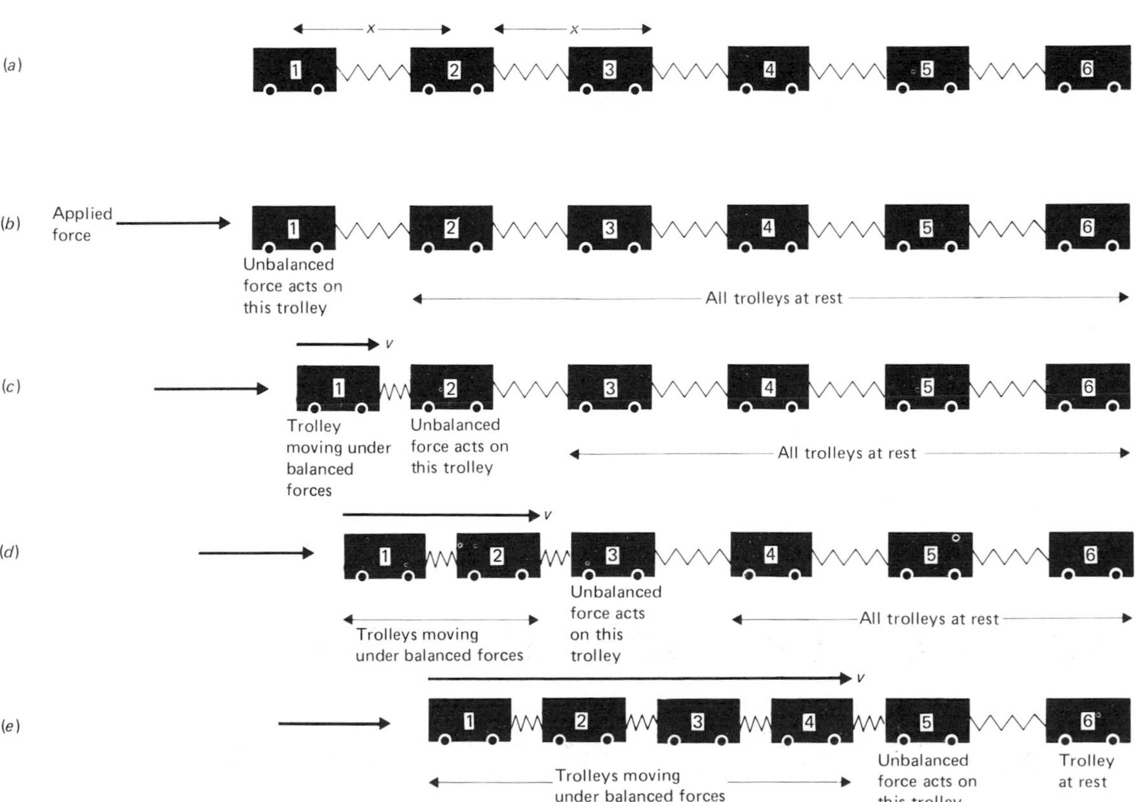

Fig. 2.10 A compression pulse moving along a line of trolleys.

given by Hooke's law which states that:

$$\begin{matrix}\text{force exerted}\\ \text{by spring}\end{matrix} = k\begin{pmatrix}\text{change in length}\\ \text{of spring}\end{pmatrix} \qquad (1.5)$$

where k is the restoring force per unit displacement.

When unequal forces act on a trolley it will accelerate in the direction of the resultant force; this will be the case for trolley 1 in diagram (b), for trolley 2 in diagram (c), and so on. In diagram (b) the unbalanced force is the applied force, and in the later diagrams it is due to the compression of the spring to the left of the trolley being accelerated.

As each trolley is accelerated in turn the spring to the right of it is compressed until the forces acting on the trolley become equal; this is the case for trolley 1 in diagram (c), for trolleys 1 and 2 in diagram (d), and so on. As the resultant force on these trolleys is now zero they move at constant velocity without further acceleration.

Thus each trolley in turn is accelerated to a constant velocity, v, and in diagram (e), for instance, the first four trolleys have been accelerated to this velocity and the fifth is in the process of being accelerated. Thus the pulse has travelled a distance $4x$ in the time between diagrams (b) and (e); to find the velocity of the pulse we divide the distance travelled, $4x$, by the time that has elapsed. The velocity of the pulse, c, will be greater than the velocity of the individual trolleys, v.

In general, let us consider the pulse travelling a distance ct in a time t; in the same time the first trolley will have travelled a distance vt. As the trolleys are a distance x apart to begin with, the number of trolleys which are now moving with a velocity, v, must be given by:

$$n = \frac{\text{distance travelled by pulse}}{\text{trolley spacing}}$$

i.e. $$n = \frac{ct}{x} \qquad (2.4)$$

If each trolley is of mass m the total momentum of the moving trolleys will be given by:

$$\text{total momentum} = n\,(mv)$$
$$= \frac{ct}{x}(mv) \qquad \text{from (2.4)}$$

By Newton's second law of motion we know that:

$$\text{force} = \text{rate of change of momentum}$$
$$= \frac{\text{change in momentum}}{\text{time}}$$
$$F = \frac{ct\,mv/x}{t}$$

i.e. $$F = \frac{cmv}{x} \qquad (2.5)$$

This is the size of the force required to accelerate the trolleys, but it can also be related to the compression of the springs by Hooke's law:

$$\begin{matrix}\text{force exerted}\\ \text{by spring}\end{matrix} = k\begin{pmatrix}\text{change in length}\\ \text{of spring}\end{pmatrix} \qquad (1.5)$$

The change in length of one spring can be calculated from the total change in length of all the springs, which must be equal to the distance moved by the first trolley, vt.

$$\begin{matrix}\text{change in length}\\ \text{of each spring}\end{matrix} = \frac{\text{total change in length}}{\begin{matrix}\text{number of trolleys}\\ \text{accelerated}\end{matrix}}$$
$$= \frac{vt}{ct/x} \qquad \text{from (2.4)}$$
$$= \frac{vtx}{ct}$$
$$= \frac{vx}{c}$$

$$\begin{matrix}\text{force applied}\\ \text{to trolleys}\end{matrix} = \begin{matrix}\text{force exerted}\\ \text{by spring}\end{matrix}$$
$$= \frac{kvx}{c} \qquad \text{from (1.5)}$$

Thus, from equation (2.5) we have:

$$\frac{cmv}{x} = \frac{kvx}{c}$$

which gives:

$$c^2 = \frac{kx^2}{m}$$
$$c = x\sqrt{\frac{k}{m}} \qquad (2.6)$$

It is most important to realize that the velocity that is calculated in Equation (2.6) is that of the pulse along the line of trolleys, and not the velocity of the individual trolleys.

The above argument contains certain simplifying assumptions about the transmission of the pulse. We have assumed that when the pulse has reached trolley 5 all the trolleys to the left of it are moving at velocity v and all the trolleys to the right are stationary. In fact, observation shows that trolley 5 will have begun to move before some trolleys to the left have attained their maximum velocity and that some trolleys to the right will have begun to move before trolley 5 has reached its maximum velocity. However, in terms of the momentum that has been acquired these two effects will tend to cancel out and if the number of trolleys is large it will only involve a small fraction of the total number. When a succession of pulses, making up a wave motion, is transmitted the wavelength must be much larger than the spacing of the trolleys for the equations derived above to apply.

We have also assumed that Hooke's law may be applied. We know that it may be applied to the trolleys and springs provided that the compressions are not too large and will consider in Section 2.6 whether it may be applied to the case of atoms in a solid.

Measurement of the speed of a compression pulse down a line of trolleys

An experiment which can be used to test the result derived above is shown diagrammatically in Fig. 2.11 (*a*). The 1 kHz pulses pass through the dual gate only when the end trolley and the large mass are in contact; the pulses are counted and the digital read-out thus displays the contact time in units of 10^{-3} s.

When the trolley strikes the mass, the other trolleys continue to move towards it, until the compression pulse has reached the far end of the line; their motion then reverses starting with the furthest trolley and eventually the trolley that originally struck the mass is pulled away from it and the contact is broken. During the time for which the trolley and mass were in contact, the pulse travelled twice the distance to the furthest trolley; this is a distance $(2 \times 3x)$.

The velocity with which the pulse moves is calculated from the distance travelled and the time taken. It may be compared with the value predicted by Equation (2.6). The mass and trolley spacing are measured directly; the value of k, the force per unit displacement, is obtained by extending one of the springs with a spring-balance.

The same principle is used in an experiment to measure the velocity of sound in a rod, described on page 43.

Q 2.4

In the experiment described above, four trolleys are used and the distance $3x$ is found to be 110 cm. Pulses are supplied to the digital read-out at 1 kHz and during the time for which the trolley and mass are in contact, the counter reaches an average reading of 1297.

(*a*) Calculate the speed of the pulse down the line of trolleys;

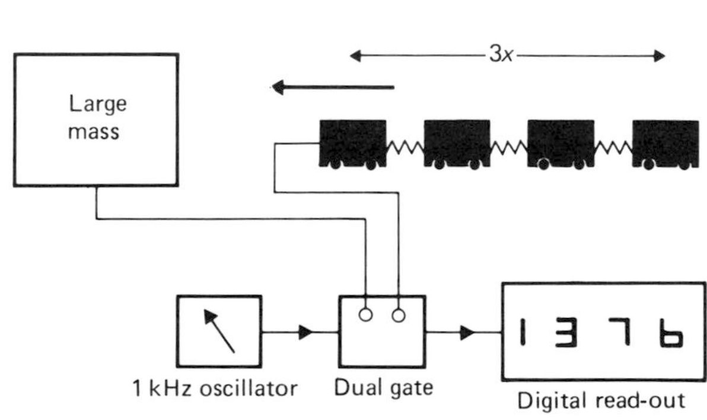

(*a*) Measurement of the speed of a pulse on a line of trolleys

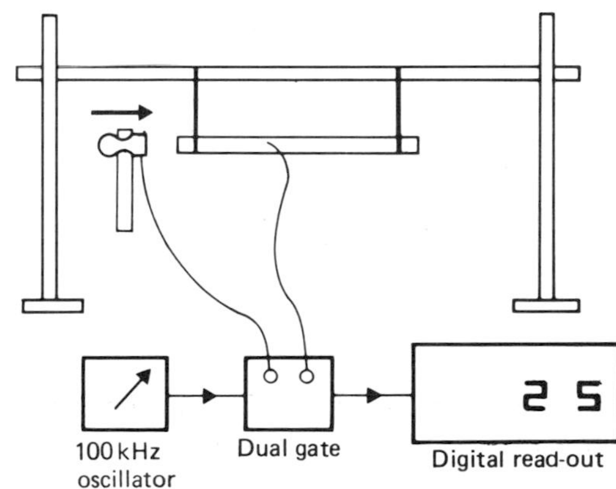

(*b*) Measurement of the speed of a pulse down a rod

Fig. 2.11

(*b*) calculate the theoretical value of the pulse speed from Equation (2.6). The mass of a trolley is 1.85 kg; a force of 4.5 N is required to stretch each spring by 10 cm.

2.5 Longitudinal waves in continuous systems

In the preceding section we discussed how a wave is propagated down a line of trolleys. We saw that the motion of the end trolley affected the motion of the second and so on, with the result that a pulse was propagated down the line. This method was used to derive an expression for the velocity of a pulse down the line; in this section we shall use a similar method to derive an expression for the velocity of a pulse through a solid bar. As before, we shall assume that the velocity of propagation is the same in the case of a sinusoidal wave as it is in the case of the square pulse.

We consider the pulse formed by applying a constant force $\mathcal{F}$ to the left-hand end of a solid bar in such a way that it makes successive sections move at a constant velocity; the bar is considered to be made up of segments as shown in Fig. 2.12. The cross-sectional area of the bar is represented by A.

The end of the bar is set into motion by the applied force and moves to the right with velocity v. This causes the next portion of the bar to move to the right as in the case of the trolleys. Let us suppose that after a time t the furthest point along the bar that is set into motion by the pulse is at Q; this corresponds to trolley number 5 in Fig. 2.10. If c represents the velocity of the pulse then the distance travelled by the pulse must be ct, whereas the distance moved by the end of the bar is vt; from the diagram we see that the distance travelled by the end of the bar is of course less than that travelled by the pulse.

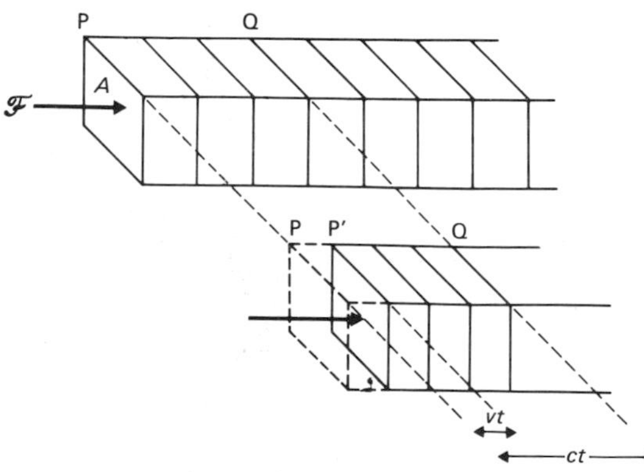

Fig. 2.12

The effect of this pulse in the bar is to compress the length PQ into P′Q; the amount of compression is therefore PP′.

Hence
$$\text{strain} = \frac{\text{change in length}}{\text{original length}} = \frac{\text{PP}'}{\text{PQ}} = \frac{vt}{ct} \quad \text{from the diagram}$$

i.e.
$$\text{strain in bar} = \frac{v}{c}$$

However, the length of the bar PQ has now been given a velocity to the right equal to v. If the density of the bar is represented by ρ we have:

$$\text{mass of PQ} = ctA\rho$$
$$\text{momentum of PQ} = ctA\rho v$$

By Newton's second law:

$$\text{force} = \frac{\text{increase of momentum}}{\text{time}} = \frac{ctA\rho v}{t}$$

i.e.
$$\text{force} = cA\rho v$$

But
$$\text{stress} = \frac{\text{force}}{\text{area}} = \frac{cA\rho v}{A}$$

i.e.
$$\text{stress} = c\rho v$$

By the definition of the Young modulus, E, we have:

$$E = \frac{\text{stress}}{\text{strain}} = \frac{c\rho v}{v/c}$$

i.e.

$$E = c^2\rho$$

$$c^2 = \frac{E}{\rho}$$

$$\therefore \quad c = \sqrt{\frac{E}{\rho}} \qquad (2.7)$$

The measured value of the Young modulus for steel is 2.0×10^{11} N m^{-2} and its density is 7.8×10^3 kg m^{-3}.

$$c = \sqrt{\frac{E}{\rho}}$$

$$= \sqrt{\frac{2 \times 10^{11}}{7.8 \times 10^3}}$$

$$= 5.1 \times 10^3 \text{ m s}^{-1}$$

The velocity of a compression wave in steel quoted from the same data book as the values above is also 5100 m s^{-1}. As in the case of the trolley system we assume a wave motion to be made up of a succession of pulses and to travel with the same speed as one of them.

Q 2.5
The velocity of sound in an aluminium rod is found to be roughly the same as in a steel rod, and its density is 2700 kg m^{-3}. Calculate an approximate value for the Young modulus of aluminium.

Measurement of the speed of sound in a rod

The method for this experiment is almost the same as for the measurement of the speed of a pulse along a line of trolleys. In principle, a rod is swung against a large mass and the time for which they are in contact, which is equal to the time that it takes for a pulse to travel to the far end of the rod and back, is measured. In practice, it is easier to hit the end of the rod with a hammer, as shown in Fig. 2.11(*b*); the two experiments are shown in the same figure to emphasize the similarity of the techniques. As the time interval measured is much smaller than with the line of trolleys, the oscillator must produce pulses at intervals of 10^{-5} s (100 kHz) instead of 10^{-3} s (1 kHz).

Q 2.6
In an experiment to measure the velocity of sound in an aluminium rod of length 50.5 cm, the average number of 100 kHz pulses counted is 24. Calculate the velocity of sound in the rod.

2.6 An atomic model for sound waves in a solid

It is possible to discuss the transmission of sound through a solid in terms of the vibrations of atoms; these atoms experience attractive and repulsive forces whose sizes depend upon the displacement of the atom from its equilibrium spacing. There is, therefore, an analogy with the line of trolleys where the springs exert a force both when extended and when compressed. The velocity of a wave through a block of steel we have seen to be about 5×10^3 m s^{-1} and even for a frequency of 1 MHz the wavelength will be 5 mm, which is very much greater than the interatomic spacing. We thus expect that the result derived above for the line of trolleys:

$$c = x\sqrt{\frac{k}{m}} \qquad (2.6)$$

can be applied to the waves in steel.

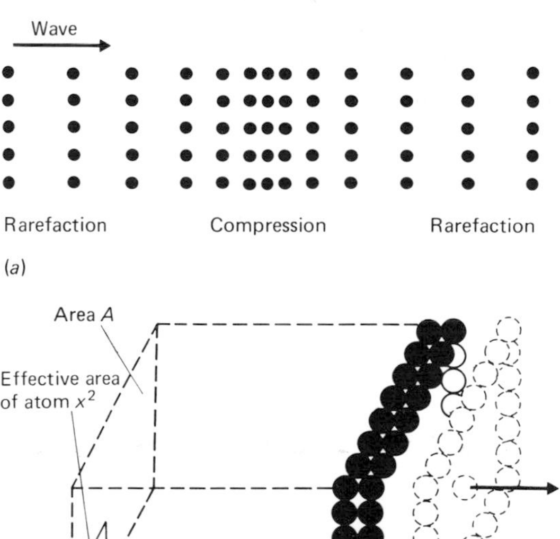

Fig. 2.13 (a) A longitudinal wave moving through an array of atoms. (b) A simplified model of the stretching of a solid.

A block of steel consists of a large number of rows of atoms side by side and we must assume, for our much simplified calculation, that each row behaves like an independent line of trolleys. If the direction of the wave were not parallel to the rows of atoms, the motion would have to be resolved into two components, one parallel to the rows and one parallel to the columns of atoms. Fig. 2.13 (*a*) shows how a wave can pass through an array of atoms with a periodic alteration in the spacing of atoms in the direction of propagation.

To develop an atomic model for the speed of sound waves in a solid we need to relate the applied force to the interatomic forces; we assume that these obey Hooke's law so that we would expect the Young modulus for the sample as a whole to be related to the force per unit displacement, k, for the interatomic forces. In the analysis below we will see what this relationship is, and hence derive an expression for the velocity of sound waves in a solid in terms of *microscopic* parameters.

Let us, therefore, consider steel to consist of atoms spaced a distance x apart; Fig. 2.13 (*b*) shows a force $\mathscr{F}$ applied to a layer of atoms of area A, so that the spacing between the layers increases from x to $x+\Delta x$.

Now if the atoms are a distance x apart the number, n, in the layer of area A will be given by:

$$n=\frac{A}{x^2}$$

since one atom effectively occupies an area of x^2, as shown in the diagram.

Therefore, if $\mathscr{F}$ is the force exerted on the whole area, and F the force exerted on each atom we have:

$$\mathscr{F}=nF$$

i.e. $$\mathscr{F}=\left(\frac{A}{x^2}\right)F \qquad (2.8)$$

We have assumed that interatomic forces are only significant between an atom and its nearest neighbour and have, therefore, ignored forces between other pairs of atoms. The Young modulus, E, is defined by:

$$E=\text{stress/strain}$$
$$=\frac{\mathscr{F}/A}{\Delta x/x}$$

Hence $$E=\frac{AF/x^2A}{\Delta x/x} \qquad \text{from (2.8)}$$

i.e. $$E=\frac{F}{x\,\Delta x} \qquad (2.9)$$

However, if k is the force per unit displacement for interatomic bonds, we may write Hooke's law for a bond in the form:

$$F=k\,\Delta x$$

We therefore have:

$$E=\frac{k\,\Delta x}{x\,\Delta x} \qquad \text{from (2.9)}$$

i.e. $$E=\frac{k}{x} \qquad (2.10)$$

Before the force was applied the atoms were each occupying an effective volume of x^3. In a volume of 1 m^3 there would therefore be $1/x^3$ atoms; if each atom was of mass m the total mass would thus be m/x^3. Hence

$$\text{density, } \rho=\text{mass/volume}$$
$$\rho=\frac{m/x^3}{1}$$

i.e. $$\rho=\frac{m}{x^3} \qquad (2.11)$$

If we now substitute from Equations (2.10) and (2.11) into the equation for the velocity of a compression wave in a solid (2.7), we obtain:

$$c=\sqrt{\frac{E}{\rho}} \qquad (2.7)$$
$$=\sqrt{\frac{k/x}{m/x^3}}$$
$$=\sqrt{\frac{x^2k}{m}}$$

i.e. $$c=x\sqrt{\frac{k}{m}} \qquad (2.6)$$

This is the same equation as that obtained for the velocity of a wave down a line of trolleys.

Thus, the velocity of sound waves in a solid may be calculated either from a knowledge of macroscopic parameters, such as the Young modulus and density, or from microscopic parameters, such as k, x and m. The high velocity of sound in a substance such as steel can be regarded as a consequence of its high Young modulus or of the high force per unit displacement of the interatomic bonds; these two properties are, of course, related by Equation (2.10):

$$E = k/x \qquad (2.10)$$

We showed in Section 2.5 that the velocity of sound in steel is about 5100 m s^{-1} and since the interatomic spacing, x, is 2.5×10^{-10} m and the mass, m, of a steel atom is 9.3×10^{-26} kg we have:

$$c = x\sqrt{\frac{k}{m}} \qquad (2.6)$$

i.e.

$$k = \frac{c^2 m}{x^2}$$

$$= \frac{(5100)^2 (9.3 \times 10^{-26})}{(2.5 \times 10^{-10})^2}$$

$$= 38.7 \text{ N m}^{-1}$$

Note that the value of k for steel is the same order of magnitude as the value of k for the springs used in the trolley experiment, quoted in Q1.9 on page 14.

The argument used above is based on a cubic picture of the structure of steel but, by using more complicated mathematics it can be applied to other structures. A more detailed discussion will be found in books on the structure of materials such as Kittel, *Introduction to Solid State Physics.*

Q 2.7

The velocity of sound in a copper rod is 3813 m s^{-1}; the mass number of copper is 63.6 and the interatomic spacing 2.55×10^{-10} m. Use Avogadro's number, 6×10^{23} mol^{-1}, to calculate the mass of a copper atom and hence find the restoring force per unit displacement for the bond between two copper atoms. If a piece of copper wire is joined to an identical piece of steel wire and a force applied, what will be the ratio of the extension of the copper wire to that of the steel wire? Use the data quoted above.

2.7 A model for transverse waves

The study of longitudinal waves was introduced by considering a line of trolleys linked by springs and we will use the same method for transverse waves. As before, we will derive an expression for the velocity of such waves, but will first consider the mechanism by which the wave motion is propagated. Each trolley can now move perpendicular to the line of springs connecting them, instead of along the line.

Fig. 2.14 shows a transverse wave propagated down the line so that each trolley undergoes identical transverse oscillations at marginally different times. If the wave is to be sinusoidal the source must execute simple harmonic motion, and trolley number 1 is seen to be doing this. As it moves it exerts a force on trolley 2 which causes that trolley to carry out an identical motion an eighth of a period behind number 1.

After one period, T, has elapsed the wave motion has travelled a distance λ and we see

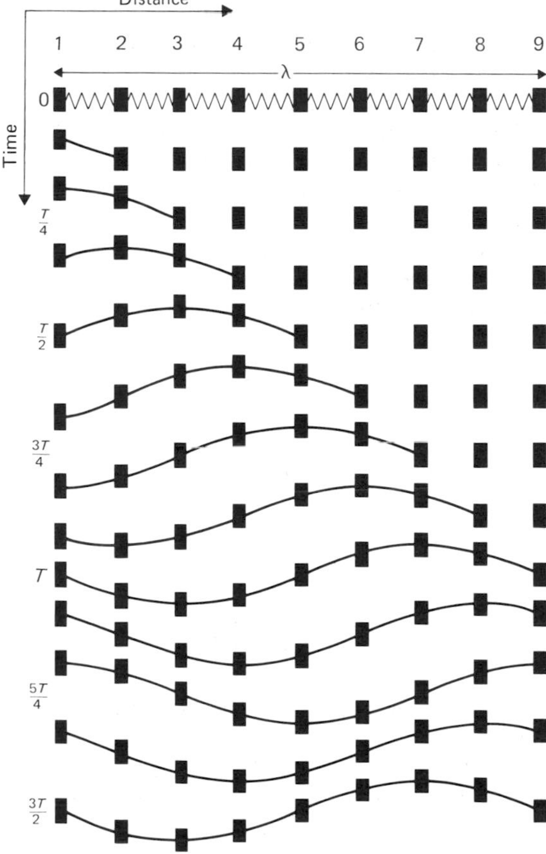

Fig. 2.14 The transmission of a transverse wave down a line of trolleys.

that trolleys a distance λ apart, such as 1 and 9, are executing identical motion at the same instant. Trolley 5, only half a wavelength from 1, has a zero displacement like 1 and 9 but it is moving in the opposite direction. Similarly, at a time $T/2$, trolley 1 has zero displacement, as it does at times 0 and T, but it is moving in the opposite direction.

As with a longitudinal wave it is also worthwhile to examine the motion from a velocity point of view. When trolleys are at their maximum displacement they are momentarily stationary and their direction of motion is changing; as they pass through the point of zero displacement they have maximum velocity. These conclusions are identical to those for oscillations and for longitudinal waves.

These results are all set out in Fig. 2.15; the words 'crest' and 'trough' are only strictly correct for a wave in a vertical plane but are taken to imply 'maximum positive displacement', and 'maximum negative displacement'. Note that, for a transverse wave, the graph of displacement against distance is a 'picture' of the wave motion. The diagram corresponds to time T in Fig. 2.14.

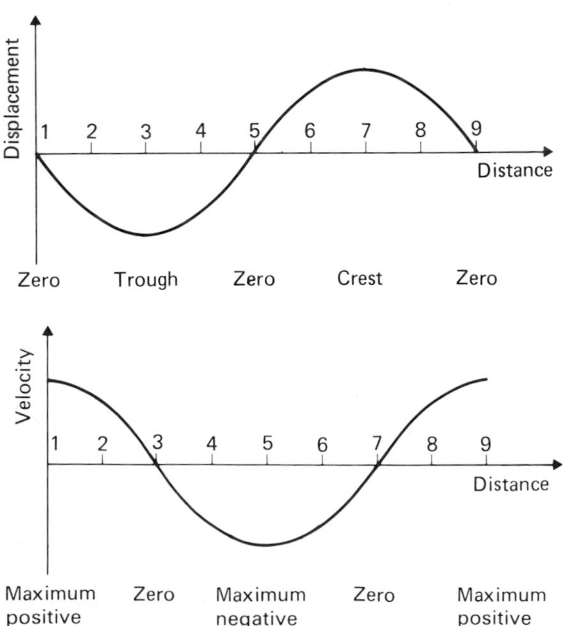

Fig. 2.15 Displacement–distance and velocity–distance graphs for a transverse wave.

Q 2.8
Draw displacement–time and velocity–time graphs for trolleys 1 and 3 in Fig. 2.14.

The velocity of transverse waves in a 'lumped-mass' system

We will now use a method similar to that in Section 2.4 to derive an expression for the velocity of transverse waves along a line of trolleys. We shall again assume that the velocity of a wave is the same as that of a step pulse, and we shall again equate the force generating the pulse to the rate of change of momentum that is caused.

The pulse that we choose to transmit gives each trolley in turn a transverse velocity, v, at right angles to the original line, as shown in Fig. 2.16. The distance between the trolleys increases until the tension in the springs, T, becomes equal to the applied force F. The force F may be resolved into a horizontal and a vertical component, as shown, and by similar triangles we then have:

$$\frac{\text{vertical component of force}}{\text{horizontal component of force}} = \frac{\text{vertical distance}}{\text{horizontal distance}} \qquad (2.12)$$

The diagram shows the situation after a time t; the pulse has just reached trolley 4, which is about to move, and has travelled a distance ct, where c is the velocity of transmission of the transverse pulse. The end trolley is moving with a transverse velocity v, that is different from c, and has moved a distance vt at right angles to the line of trolleys. As before, we assume that the trolleys 1, 2 and 3 are moving at constant velocity under balanced forces, trolley 4 is about to be accelerated by an unbalanced force $T \sin \theta$ and trolleys 5 and 6 are at rest.

The motion of the trolleys at right angles to the line is caused by the vertical component of the applied force.

Now, Equation (2.12) tells us that:

$$\frac{\text{vertical component of force}}{\text{horizontal component of force}} = \frac{\text{vertical distance}}{\text{horizontal distance}}$$

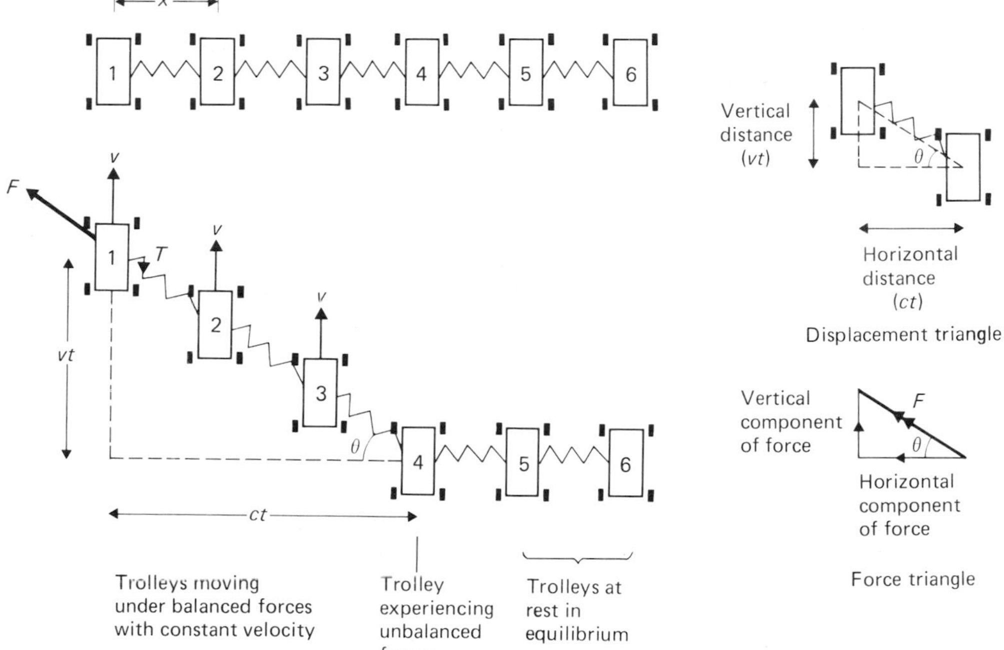

Fig. 2.16 A transverse pulse moving along a line of trolleys.

i.e.
$$\frac{\text{vertical component}}{\text{horizontal component}} = \frac{vt}{ct} = \frac{v}{c}$$

But if the angle θ is small the horizontal component is very nearly equal to T, and therefore we have:

$$\text{vertical component of force} \approx \frac{v}{c}T$$

The distance travelled by the pulse is ct and the spacing of the trolleys is x. Hence

$$\text{number of trolleys moving} = \frac{ct}{x}$$

If the mass of each trolley is m, and its velocity v, its momentum must be mv. Thus, the total momentum after a time t is given by:

$$\text{momentum} = \left(\frac{ct}{x}\right)mv$$

By Newton's second law:

$$\text{force} = \frac{\text{change in momentum}}{\text{time taken}}$$

Therefore
$$\frac{vT}{c} = \frac{mv(ct/x)}{t} = \frac{mvc}{x}$$

$$c^2 = \frac{xT}{m}$$

i.e.
$$c = \sqrt{\frac{xT}{m}} \qquad (2.13)$$

The formula in Equation (2.13) has been derived for a 'lumped-mass' system. However, if we imagine the masses becoming smaller and closer together, the situation becomes closely analogous to that in a continuous medium such as a rope; the trolleys represent atoms and the springs represent interatomic forces. The angle between the rope and the horizontal must be small if the above approximation is to be true, and this restricts the application to small displacements.

The quantity m/x is the mass per unit length and is frequently represented by μ; T becomes the tension in the rope. We then have:

$$c = \sqrt{\frac{T}{\mu}} \qquad (2.14)$$

as the equation for the velocity of transverse waves along a rope, wire or string. We have again assumed that the result in Equation (2.13) applies to waves as well as to pulses and this is justified by considering a wave as a succession of pulses, all travelling with the same velocity.

The formula in Equation (2.14), like that for longitudinal waves, can also be obtained by an analysis which does not require use of the 'lumped-mass' simplification. One method of doing this, using the calculus, is given below.

The velocity of a wave on a rope

Let us consider the motion of a small pulse on a rope as shown in Fig. 2.17; it is assumed that the rope is perfectly flexible, that the hump is an arc of a circle and is sufficiently small not to affect the tension, T, in the rope. Remember that the tension always acts along the rope and thus the forces acting on the hump of rope are as shown in the diagram. Let the mass per unit length be μ and the other quantities be as shown. The displaced hump forms a circular arc of length δl and radius r, and it subtends an angle at its centre given by:

$$2\phi = \frac{\delta l}{r}$$

The resultant of the two tensions acts towards the centre of the circle and is equal to $2T \sin \phi$. Hence:

total force towards centre $= 2T \sin \phi$

$\approx 2T\phi$ if ϕ is small

i.e. $$\text{force} = T\frac{\delta l}{r} \qquad (2.15)$$

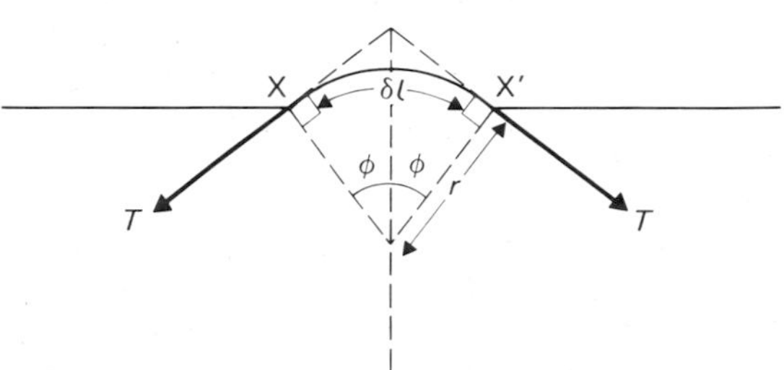

Fig. 2.17

Now let us consider the hump from the point of view of an observer moving along at the same speed; the rope appears to this observer to be moving with a velocity c in the opposite direction. We have altered the frame of reference from one inertial system to another; the laws of physics remain unchanged. We have then at XX' a portion of rope of length δl, and mass $\mu\delta l$, moving with a velocity c in a circle of radius r. Its acceleration towards the centre of the circle must be c^2/r, and by Newton's second law we have:

force = mass × acceleration

Therefore $$\frac{T\delta l}{r} = \mu\delta l\left(\frac{c^2}{r}\right)$$

$$c^2 = \frac{T}{\mu}$$

i.e. $$c = \sqrt{\frac{T}{\mu}} \qquad (2.14)$$

As before this only applies to humps sufficiently low for ϕ to be small.

Q 2.9
A line of trolleys each of mass 1 kg is set up to transmit a transverse wave and the springs are stretched so that the separation, x, is 0.3 m. The tension is then 15 N. At what speed will a wave be transmitted down the line? If the springs are assumed to be massless, what is the average mass per unit length of the system?

A continuous spring, of the same mass per unit length, is also stretched until its tension is 15 N. At what speed will a transverse wave be transmitted along it?

2.8 Velocities of other waves

Let us summarize the results for wave velocities that we have derived so far:

longitudinal wave on a line of trolleys: $c = x\sqrt{\frac{k}{m}}$ $\quad \lambda \gg x \quad$ (2.6)

transverse wave on a line of trolleys: $c = \sqrt{\frac{xT}{m}}$ $\quad \lambda \gg x \quad$ (2.13)

longitudinal wave in a solid bar: $c = \sqrt{\frac{E}{\rho}}$ $\quad$ (2.7)

transverse wave on a perfectly flexible rope: $c = \sqrt{\frac{T}{\mu}}$ $\quad$ (2.14)

In each case the numerator contains a term which gives the size of the force required to distort the medium through which the wave is to pass; the larger this force, the quicker the next section of the medium will respond to a distortion. The denominator is, in each case, related to the inertia of the medium. We will now examine a number of other velocity equations, not derived in this section, and see that in each case the velocity depends on similar parameters.

Water waves

For ripples of very short wavelength the velocity is determined mainly by the surface tension, σ, which determines the force required to distort the surface, the density, ρ, and the wavelength, λ:

short wavelength ripples: $$c = \sqrt{\frac{2\pi\sigma}{\lambda\rho}} \tag{2.16}$$

The speed of long waves in deep water is determined by the wavelength, λ, and the earth's gravitational field, g; the force required to lift the water and form a crest is proportional to the strength of the earth's gravitational field and this is thus a factor in the numerator. This force is also proportional to the density of the water, and as the density would, therefore, appear in both denominator and numerator, it does not appear at all:

long waves in deep water: $$c = \sqrt{\frac{g\lambda}{2\pi}} \tag{2.17}$$

Sound waves

A detailed consideration of the velocity of sound in an ideal gas follows in Chapter 5, but the result proved there is quoted below. The adiabatic bulk modulus of the gas, γP, appears in the numerator where P is the pressure and γ a dimensionless constant. The velocity is given by:

$$c = \sqrt{\frac{\gamma P}{\rho}} \tag{5.5}$$

Pulses in cables

In the analysis of oscillating currents in electrical circuits it is common to draw an analogy between a capacitor and a spring; the capacitance, C, is analogous to the quantity $1/k$ for the spring. The inductance, L, is analogous to the mass on the end of the spring, m. Such an approach is valuable, for instance, in analogue computing where a mechanical situation is represented more conveniently by an electrical circuit so that its oscillations may be observed; it is more convenient to alter the value of capacitors or inductors in a circuit representing the springing or mass of a car than to build an equivalent number of cars.

Thus, a connected series of inductors and capacitors is analogous to the line of trolleys separated by springs. Instead of the equation

$$c = x\sqrt{\frac{k}{m}} = \sqrt{\frac{x^2k}{m}} = \sqrt{\frac{1}{(1/kx)(m/x)}}$$

we have $$c = \sqrt{\frac{1}{\left(\frac{\text{capacitance}}{\text{length}}\right)\left(\frac{\text{inductance}}{\text{length}}\right)}}$$

The reader will recall that in Section 1.3 there is a similar analogy drawn between various expressions for the periods of oscillating systems.

2.9 Dispersion

In the formulae that have been derived so far we have assumed that the velocity of a wave motion through a medium is the same irrespective of the frequency of the wave. It is important to appreciate that this is not always the case; if it is not, the medium is said to be *dispersive*. The frequency of the wave is determined by the source; the velocity of a wave peak is determined by the medium and, in a dispersive medium, by the frequency. The wavelength is then determined from these by using the relation $c = f\lambda$.

We saw earlier that a 'lumped-mass' system such as a line of trolleys does not behave exactly like a continuous medium. In particular, when the wavelength is no longer than the spacing between trolleys, the system is dispersive and waves of different frequency are transmitted at different speeds. The system also exhibits *cut-off*; waves of high frequency are not propagated at all. If the end trolley is oscillated very rapidly, the next one oscillates only a little, the next one less and the oscillations die out very quickly along the line.

Equation (2.17), quoted in Section 2.8, showed that most sea waves are dispersive. Thus, when a tsunami occurs at sea and large waves of a variety of frequencies are generated, the waves of longer period will arrive first. We also know that waves travel along Great Circle paths. It is possible, by analysing when waves of different frequencies arrive, to compute the point from which the waves originated. In one experiment using detectors on the sea bed off the coast of south-west England, surf arriving over a period of three days in 1945 was clearly shown to be derived from a storm that had taken place some time previously off Cape Horn, some 11 000 km away!

Observation of the sea soon shows that most waves are more complicated than simple sine waves. It is possible to show that complex periodic waves may be considered to consist of a large number of sine waves, of different amplitude and frequency, added together; this technique is known as Fourier, or harmonic, synthesis, and was discussed in Section 1.7. If the medium is not dispersive, then all of these component sine waves will be transmitted at equal speeds and the resultant waveform will be unchanged. If, however, the medium is dispersive and all the sine waves that make up its shape are moving at different speeds, then the overall waveform will be constantly changing. This effect may again easily be observed in deep water at sea where a group of waves, travelling in one direction, is constantly changing its shape.

Q 2.10

Fig. 1.25 shows that a square wave, of amplitude 3 m and wavelength 15 m can be considered approximately to consist of three components:

(i) a sine wave of amplitude 3 m and wavelength 15 m;
(ii) a sine wave of amplitude 1 m and wavelength 5 m;
(iii) a sine wave of amplitude 0.6 m and wavelength 3 m.

Use Equation (2.17) to calculate:

(*a*) the velocity of each wave component;
(*b*) the distance each travels in 5 s.
Take $g = 9.8\ \text{m s}^{-1}$.

A square pulse with the above characteristics is emitted by a wave machine. Draw a diagram to show the position of the components of the wave after 5 s, and hence find the shape of the wave at that time.

2.10 The mathematics of sine waves

When discussing the transmission of longitudinal and transverse waves earlier, we showed how an oscillation 'moves through' a medium such as a line of trolleys so that each trolley carries out the same motion a little later than its neighbour. We will now derive equations to describe the transmission of a wave mathematically.

Let us consider an oscillating source, S, for which the equation of the oscillation is found from Equation (1.26) to be:

$$y_S = A \sin 2\pi ft \qquad (2.18)$$

where A is the amplitude and f the frequency. This is the oscillation which is to be transmitted through the medium; it will, therefore, occur at other points in the medium at a later time. In Fig. 2.18 we see a particular cycle that has been emitted by the source; the dotted line shows its position when it has just passed a point P after a time t_P. The oscillation that occurs at P is identical to that at S but all events occur at a time t_P later.

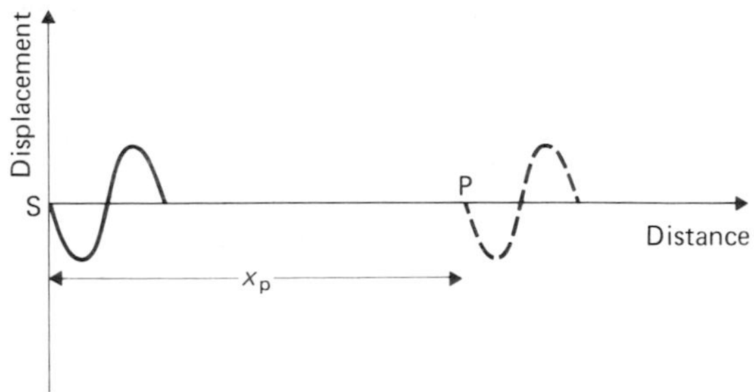

Fig. 2.18

The oscillation at P is identical to that at S, and as the equation of the oscillation at S was given by:

$$y_S = A \sin 2\pi ft \qquad (2.18)$$

the equation of the oscillation at P must be of the form:

$$y_P = A \sin 2\pi ft' \qquad (2.19)$$

Now, if all events at P occur a time, t_P, after events at S, we may write:

$$t' = t - t_P$$

where t_P is the time taken by the wave to move from S to P. Hence

$$t_P = x_P/c$$

because the distance from S to P is x_P. Equation (2.19) can therefore be written:

$$y_P = A \sin 2\pi f(t - t_P) \quad (2.20)$$

$$= A \sin 2\pi f(t - x_P/c) \quad (2.21)$$

Note carefully that the sign in the bracket is a minus sign.

We may rewrite Equation (2.21) for the displacement at any point a distance x from the origin:

$$y = A \sin 2\pi f(t - x/c) \quad (2.22)$$

The displacement, y, is most easily visualized as being at right angles to the motion of the wave, as in a transverse wave. In fact, y could also be the measurement of displacement in a longitudinal wave; therefore, all the equations and results apply just as well to longitudinal waves as to transverse waves.

Q 2.11

The velocity of waves on a rope is 2 m s^{-1} and at time $t = 0$ waves of amplitude 0.1 m and frequency 0.5 Hz ($T = 2$ s, $\lambda = 4$ m) begin to be emitted from one end. Calculate the displacement in each of the following cases:

(*a*) source, $t = 0.5$ s;
(*b*) source, $t = 1.0$ s;
(*c*) source, $t = 1.5$ s;
(*d*) source, $t = 0.8$ s;
(*e*) source, $t = 2.0$ s;
(*f*) $x = 1$ m, $t = 2.0$ s;
(*g*) $x = 2$ m, $t = 2.0$ s;
(*h*) $x = 3$ m, $t = 2.0$ s;
(*i*) $x = 4$ m, $t = 2.0$ s;
(*j*) $x = 2.5$ m, $t = 2.75$ s;
(*k*) $x = 1$ m, $t = 0.5$ s;
(*l*) $x = 4.5$ m, $t = 3$ s.

Use answers (*a*)–(*e*) to sketch a displacement–time graph for the source, and then use answers (*e*)–(*i*) to sketch a displacement–distance graph at $t = 2.0$ s.

There are a variety of different ways of writing Equation (2.22) using two other equations to make substitutions:

$$c = f\lambda \quad (2.1)$$

and $$f = 1/T \quad (1.1)$$

From Equation (2.1)

$$f = c/\lambda$$

and therefore $$y = A \sin 2\pi (ft - fx/c)$$

becomes $$y = A \sin 2\pi \left(\frac{t}{T} - \frac{x}{\lambda}\right) \quad (2.23)$$

The symmetry of the fractions t/T and x/λ emphasizes the fact that it is both the fraction of a period elapsed, and the fraction of a wavelength distance from the origin, that determine the displacement of a wave.

Note that if a wave travelling to the right is represented by:

$$y = A \sin 2\pi f(t - x/c)$$

then a wave travelling in the opposite direction, with negative velocity, is represented by

$$y = A \sin 2\pi f(t + x/c) \quad (2.24)$$

This is because events will occur sooner rather than later than those at the origin; the sign of x must therefore be reversed.

Angular frequency, ω

In exactly the same way as for oscillations, it is sometimes convenient to write equations for waves in terms of the angular frequency, ω, instead of the frequency, f. The relationship between these is:

$$\omega = 2\pi f \quad (1.14)$$

and, for example, Equation (2.22):

$$y = A \sin 2\pi f(t - x/c) \quad (2.22)$$

becomes:

$$y = A \sin \omega (t - x/c) \quad (2.25)$$

This is particularly convenient in more advanced work, but we shall not use this form much here.

Phase and phase relationships

We have shown above that the displacement at a particular point and time for a sinusoidal progressive wave is given by:

$$y = A \sin 2\pi f(t - x/c) \quad (2.22)$$

or by $$y = A \sin 2\pi \left(\frac{t}{T} - \frac{x}{\lambda}\right) \quad (2.23)$$

A sine function can only operate upon an angle and therefore the expressions $2\pi f(t - x/c)$ and $2\pi(t/T - x/\lambda)$ must represent angles; they are called the *phase* or *phase angle*. There are 2π radians in 360 degrees and the factor 2π converts $f(t - x/c)$ and $(t/T - x/\lambda)$ into radian measure. Table 2.2 shows various values of the displacement for different values of the phase.

Table 2.2

	Phase		Displacement, y
$f(t-x/c)$ $=(t/T-x/\lambda)$	$=2\pi f(t-x/c)$ $=2\pi(t/T-x/\lambda)$		$=A \sin 2\pi f(t-x/c)$ $=A \sin 2\pi(t/T-x/\lambda)$
0	0 rad	[0°]	0
1/4	$2\pi \times 1/4 = \pi/2$ rad	[90°]	$+A$
1/2	π rad	[180°]	0
3/4	$3\pi/2$ rad	[270°]	$-A$
1	2π rad	[360°]	0
5/4	$5\pi/2$ rad	[450°]	$+A$

The displacement at a particular instant at the origin is determined by the fraction of a period that has elapsed since $t=0$; if we are at the beginning, middle or end of a period the displacement will be zero. If we are a quarter of the way through, the displacement will be $+A$ and if we are three quarters of the way through the period, the displacement will be $-A$. Thus the displacement at the origin will be determined by the ratio t/T.

The displacement at a point one, two or any other whole number of wavelengths away from the origin will be exactly equal to that at the origin, as the time taken for the wave to travel a distance equal to one wavelength is just one period. However, if the point is half a wavelength from the origin, the displacement at that point will be one half period behind that at the origin. Thus the ratio t/T determines at what point in the cycle the displacement at the origin has reached, and the ratio x/λ determines how far behind the origin is the displacement at another point. Thus, in the equation:

$$y = A \sin 2\pi\left(\frac{t}{T}-\frac{x}{\lambda}\right) \qquad (2.23)$$

the quantity $2\pi x/\lambda$ represents the *phase difference* or *epoch* at a point other than at the origin.

The alternative equation:

$$y = A \sin 2\pi f(t-x/c) \qquad (2.22)$$

provides another way of considering the phase difference at a point relative to that at the origin; we see that this must also be equal to $2\pi fx/c$. The ratio x/c determines the time taken for an oscillation to move from the origin to that point. We might use the expression $(t-x/c)$ to represent the 'local time' at the point in the medium.

We may see from Fig. 2.19 how the phase difference varies through a medium.

In this diagram S represents a source producing waves of wavelength 8 m and P, Q, R and U are points through which the wave passes at distances 2, 4, 6 and 8 m from S. Thus we have a displacement–distance graph showing the displacement at a range of points at one particular time. As shown, P is one quarter of a wavelength distant from S, Q is one half of a wavelength, R three quarters and U one whole

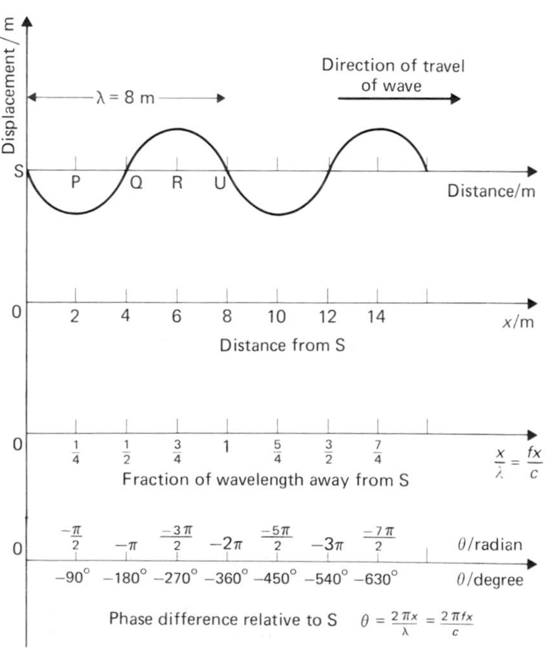

Fig. 2.19

wavelength away. We see that the phases at P, Q, R and U are $\pi/2$, π, $3\pi/2$ and 2π radian respectively behind that at S. If the time is such that the amplitude at S is zero (e.g. $t=0$) we see that the amplitudes are $-A$, 0, $+A$ and 0 corresponding to the values of $A \sin(-\pi/2)$, $A \sin(-\pi)$, $A \sin(-3\pi/2)$ and $A \sin(-2\pi)$.

We frequently refer to two waves arriving at a point 'out of phase' and a number of waves moving through a medium are shown in Fig. 2.20. The first graph is a reference wave and the phase differences of the other graphs with respect to it are detailed below.

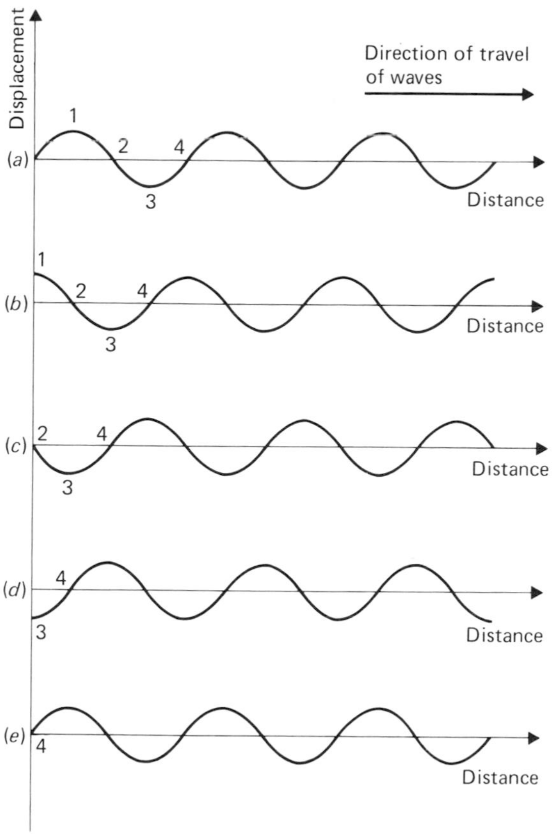

Fig. 2.20 (a) Reference wave. (b) Shifted $\lambda/4$ to left; $\pi/2$ (90°) behind reference wave. (c) Shifted $\lambda/2$ to left; π (180°) behind reference wave (said to be *exactly out of phase* with reference wave). (d) Shifted $3\lambda/4$ to left; $3\pi/2$ (270°) behind reference wave. (e) Shifted λ to left; 2π (360°) behind reference wave (said to be *in phase with* reference wave).

Q 2.12

Waves of period 4 s, wavelength 2 m and amplitude 2 m are emitted from a source. Calculate the phase and hence the displacement, in each of the following cases:

(*a*) source, $t=1$ s;
(*b*) source, $t=2$ s;
(*c*) source, $t=2.5$ s;
(*d*) $x=0.5$ m, $t=4$ s;
(*e*) $x=1.0$ m, $t=4$ s;
(*f*) $x=1.3$ m, $t=4$ s;
(*g*) $x=1.0$ m, $t=6$ s;
(*h*) $x=0.5$ m, $t=2.5$ s.

It was pointed out earlier in the chapter that for longitudinal waves, the compression wave and the displacement wave are 90° out of phase; maximum compression coincides with zero displacement, and so on.

If the displacement, y, in a longitudinal wave is given by:

$$y = A \sin 2\pi f(t - x/c) \qquad (2.22)$$

then the oscillating compression, P, will be given by:

$$P = A_P \sin 2\pi f(t - x/c + 90)$$
$$= A_P \cos 2\pi f(t - x/c) \qquad (2.26)$$

A_P is the amplitude of compression oscillations.

The wave equation

The fact that all waves have similar properties should lead us to suppose that they might have a similar mathematical treatment and the equations above apply in fact to all sinusoidal waves. We may ask whether they may be derived mathematically, as well as conceptually, by a common piece of analysis.

It can be shown that a plane wave propagated in the x-direction will obey the partial differential equation

$$\frac{\partial^2 y}{\partial x^2} = \frac{1}{c^2}\frac{\partial^2 y}{\partial t^2} \qquad (2.27)$$

and all equations purporting to represent a plane wave must be solutions of this differential equation. We may examine one of our equations to see whether it is, in fact, a solution:

Consider $y = A \sin 2\pi f(t - x/c)$

Then $\left(\frac{\partial y}{\partial t}\right)_x = 2\pi f A \cos 2\pi f\left(t - \frac{x}{c}\right)$

$\left(\frac{\partial^2 y}{\partial t^2}\right)_x = -(2\pi f)^2 A \sin 2\pi f\left(t - \frac{x}{c}\right)$

But $\left(\frac{\partial y}{\partial x}\right)_t = -\frac{2\pi f A}{c} \cos 2\pi f\left(t - \frac{x}{c}\right)$

$$\left(\frac{\partial^2 y}{\partial x^2}\right)_t = -\frac{(2\pi f)^2 A}{c^2} \sin 2\pi f\left(t - \frac{x}{x}\right)$$

$$\therefore \quad \frac{1}{c^2}\left(\frac{\partial^2 y}{\partial t^2}\right) = -\left(\frac{1}{c^2}\right)(2\pi f)^2 A \sin 2\pi f\left(t - \frac{x}{c}\right)$$

$$= -\frac{(2\pi f)^2 A}{c^2} \sin 2\pi f\left(t - \frac{x}{c}\right)$$

i.e. $$\frac{1}{c^2}\left(\frac{\partial^2 y}{\partial t^2}\right)_x = \left(\frac{\partial^2 y}{\partial x^2}\right)_t \qquad (2.27)$$

Thus the equations that we derived can be shown to be the solutions of the general one-dimensional wave equation.

Bibliography

The Bibliographies to Chapters 2, 3 and 4 are presented together at the end of Chapter 4.

3

Wave propagation

3.1 Wavefronts

We have seen in the preceding chapter how a transverse wave on a rope consists of a moving series of crests and troughs. As the wave travels in only one direction away from the source it is termed a *wave in one dimension*.

When a drop of water falls into a pond, waves are sent out from the point of impact and they travel all over the pond. They are illustrated in Fig. 3.1 and are called *waves in two dimensions*; a continuous, circular crest, called a *wavefront*, is formed around the splash and this moves outwards, maintaining its shape.

In Fig. 3.2(a), such a crest is seen moving outwards from a source; we see that the wavefront always moves at right angles to itself, in the directions indicated by the arrows. We might instead have focused our attention upon a trough, or on points of zero displacement, and we would have seen that these formed two continuous circular shapes that moved outwards in a similar way. A *wavefront* is thus any continuous region of constant phase. Wavefronts also exist when waves spread out in three dimensions, such as from a radio transmitter in

Fig. 3.1

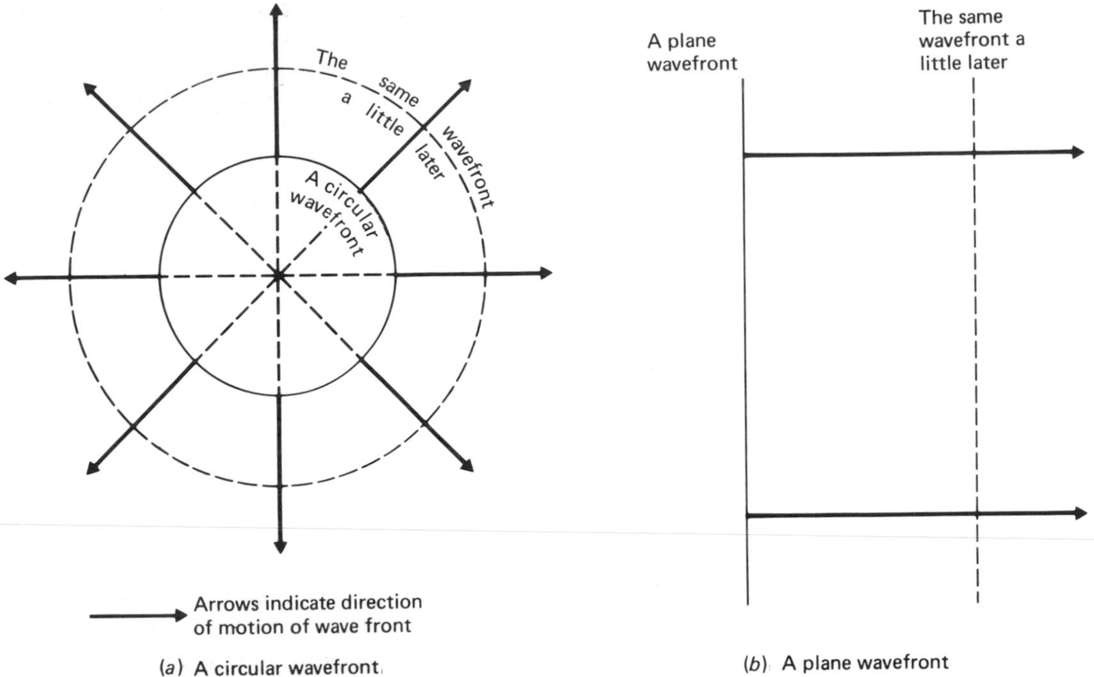

(*a*) A circular wavefront

(*b*) A plane wavefront

Fig. 3.2

space; waves spreading from a point source are called spherical waves and have wavefronts that form the surfaces of concentric spheres.

A plane wavefront may be caused by a linear disturbance, such as a roller moved backwards and forwards in a shallow tank, or by a point disturbance so far away that the wavefronts have effectively become plane; alternatively, plane waves may be produced by 'focusing', as may happen when spherical waves pass through a lens-shaped region of different depth.

Fig. 3.2(*b*) shows how a plane wavefront moves forward in a direction that is at right-angles to the original wavefront. We shall see later that light may be regarded as a wave motion and in this case the arrows on our diagrams will represent the direction of travel of light rays. Thus light rays from a point source are diverging in straight lines and light from a distant source consists of parallel rays. The fact that light rays do travel in straight lines is known as *the rectilinear propagation of light.* Light rays are always at right angles to the wavefronts in the light beam, if the optical properties of the medium are the same in all directions; such a medium is said to be *isotropic*.

3.2 Huygens' principle

Christian Huygens was a Dutch scientist who was a contemporary of Newton; he was also an early advocate of a wave theory of light. As is explained in Chapter 9, his theory was inaccurate in places, and not widely accepted at the time, but by the nineteenth century much of it had been found to be substantially correct. One part of his theory of light attempted to explain why it is that wavefronts move in the way described in the last section; this is known as *Huygens' principle.* It is still valid and can be used to explain and predict a variety of other phenomena. It can be applied to any type of wave motion.

Huygens' principle consists of two basic postulates:

(1) *Each point on a wavefront will act as a source of secondary wavelets.*
(2) *The position of the wavefront at a later time is along a line that just touches each of the secondary wavelets tangentially; this line is known as the envelope of the wavelets.*

Before we apply Huygens' principle to more complicated cases, we will show that in simple situations it gives expected and well known results. In Fig. 3.3(*a*), S is a source of circular waves, and the position of a circular wavefront a time t after leaving the source is shown; the velocity of the waves is assumed to be c. According to Huygens' principle, in order to find the position of the wavefront after a further time Δt, secondary wavelets must be drawn, of radius $c\Delta t$, centred on points on the existing wavefront; these arcs are shown by dotted lines. The new position of the wavefront is then drawn by constructing the envelope to these wavelets. As we would expect, a circular wavefront is obtained, of radius $c(t+\Delta t)$.

Huygens' principle may be applied to the propagation of a plane wave in exactly the same way as was done for a circular wave. As can be seen from Fig. 3.3(*b*), a plane wavefront of unrestricted lateral extent is propagated as a plane wavefront, because the envelope to the secondary wavelets is plane; this is an example of rectilinear propagation.

Huygens' principle is extremely useful but, although it can be shown to produce the right results, it is difficult to justify theoretically. It should not be regarded as a law for it merely provides a geometrical device, which is fortunately right, enabling us to predict the motion of waves. It would, perhaps, be more accurate to call it Huygens' construction, but as the usual term is Huygens' principle, that is what we shall use. It is, in fact, hard to explain why the wavelets only add up along their envelope and also why the wave does not propagate backwards as well as forwards; these difficulties are discussed in more advanced texts.

3.3 Diffraction

Huygens' principle may also be applied to situations where rectilinear propagation does not occur. We are familiar from the study of light with the formation of shadows by obstacles, and this is an example of rectilinear propagation. However, when the wavelength of the waves is longer, the 'shadows' are less marked as seen in Fig. 3.4(*a*).

The photograph shows that the incident waves are approximately plane. We might expect that the breakwater would cause a clean 'shadow' to be formed, as waves travel on the 'open sea' in straight lines. However, circular waves are propagated into the area behind the breakwater from its offshore end; the end appears to act as a source of secondary waves. Fig. 3.4(*b*) shows how this is explained by Huygens' principle; the secondary wavelets formed by the wavefront beyond the end of the breakwater are able to extend into the shadow region because the rest of the wavefront has been blocked by the breakwater.

Fig. 3.4(*c*) shows waves spreading out in a ripple tank after passing through a narrow

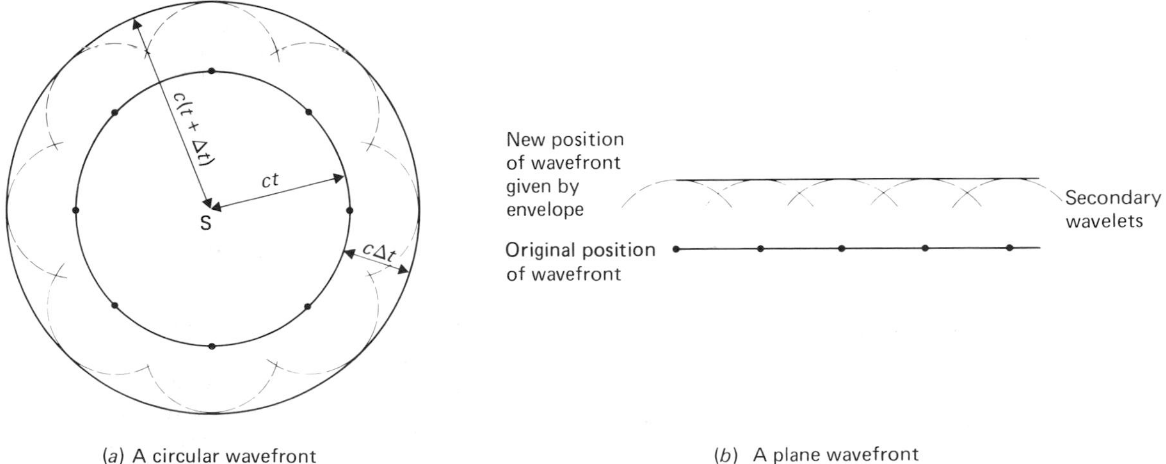

(*a*) A circular wavefront

(*b*) A plane wavefront

Fig. 3.3

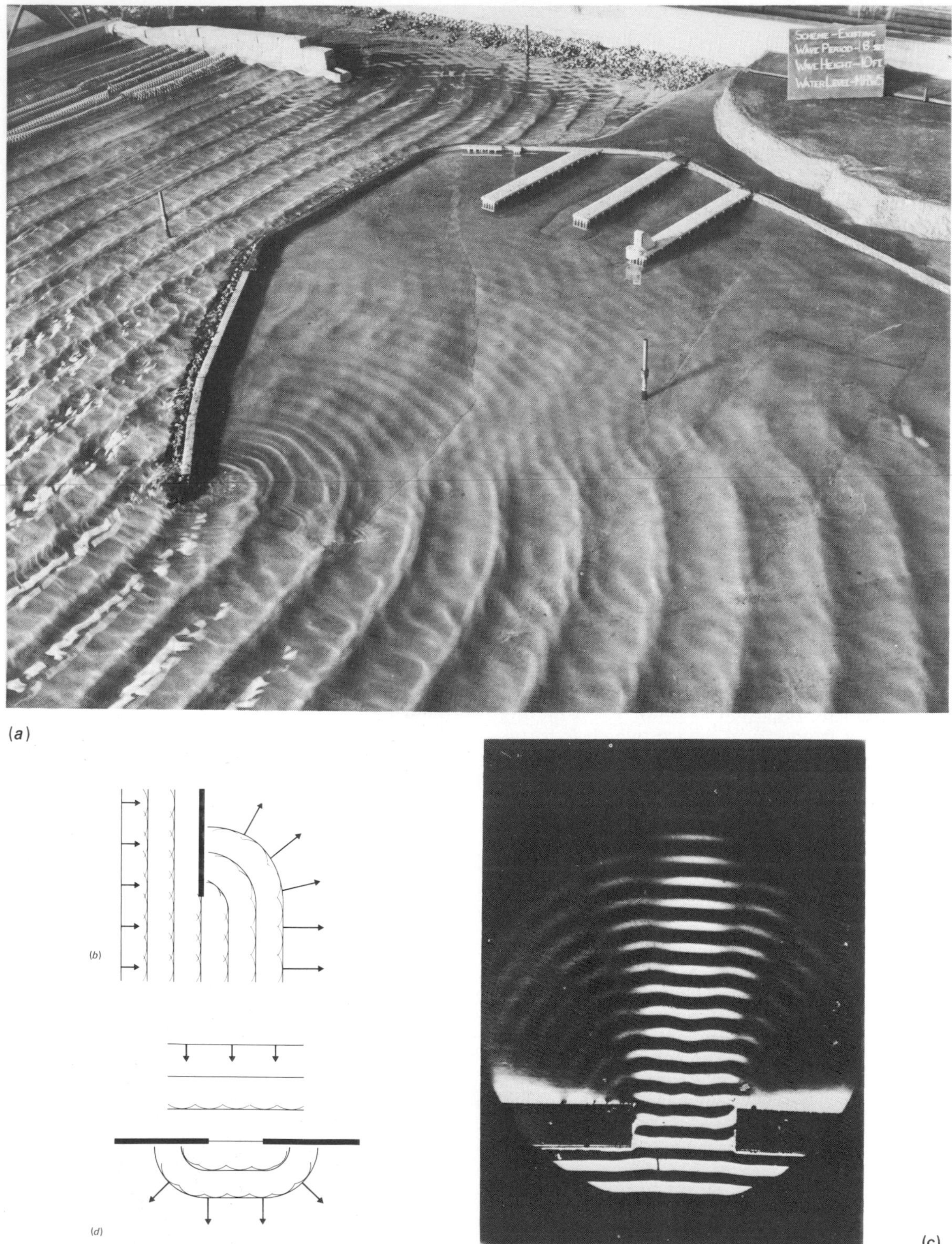

Fig. 3.4 (a) The diffraction of waves at a breakwater in the wave tank at the Hydraulic Research Station, Wallingford. (Crown Copyright). (b) Huygens' principle applied to diffraction at a breakwater. (c) The diffraction of plane waves at openings of different widths. (d) Huygens' principle applied to diffraction at an opening.

opening; this phenomenon is called *diffraction* and we see how it arises in Fig. 3.4(*d*). Note from the photographs that the amount of spreading increases as the hole becomes smaller and that there is a variation of amplitude with direction beyond the hole. This latter point will be discussed in more detail in Chapter 7.

Diffraction effects at a single slit can be seen provided that the size of the slit is no more than one or two orders of magnitude greater than the wavelength of the wave motion. For this reason, it is an easier phenomenon to observe with water waves than with some other kinds of waves. It is, however, a property of all types of wave motion, and in Chapter 7 we shall be discussing the diffraction of light at slits of width around 10^{-6} m.

The diffraction of sound can be observed by listening to a brass band playing round a corner; even if the band cannot be heard by reflection off another wall, diffraction at the corner allows the sound to travel round it. In Chapter 5 we shall see that bass instruments are of lower frequency, and therefore longer wavelength, than treble instruments and we would therefore expect to hear the bass drum and tuba more clearly round a corner than the piccolo or cornet. Try it some time!

Q 3.1
Use Huygens' principle to show what you expect to happen when a wave meets a bridge pier situated in the middle of the wavefront.

3.4 Reflection

The reflection of waves is a phenomenon which may be observed easily, for instance in a ripple tank or on the sea front. Before examining in detail the way in which plane waves are reflected, we will look briefly at the measurement of the angles involved. In optics, it is usual to measure the *angle of incidence*, i, as the angle between a ray and the normal to the surface. When we discuss waves, however, we often measure the angle between the wavefront and the reflecting surface, i', as shown in Fig. 3.5.

In an isotropic medium a wave travels at right angles to its wavefront and this direction, which corresponds to that of a ray in optics, is marked

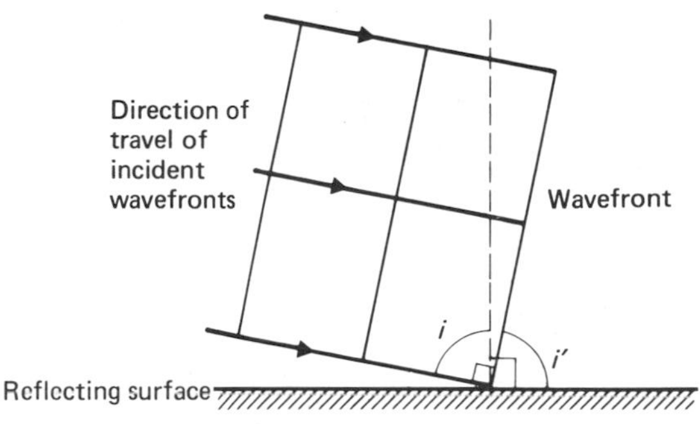

Fig. 3.5 A wave incident at a reflecting surface.

in the diagram. As the angle between the normal and the reflecting surface is 90° and that between the wavefront and its direction of travel is also 90°, we see that:

$$i' = i$$

It is thus immaterial whether by the term *angle of incidence* we mean the angle between the wavefront and the surface, or between the 'ray' (or direction of travel) and the normal to the surface. Clearly, this applies also to measurement of the *angle of reflection*.

Fig. 3.6 shows a plane wave in a ripple tank being reflected from a straight barrier. We see that in this case:

$$i' = r'$$

and this result is, in fact, always true; it corresponds to the *law of reflection* in optics. Notice that the right-hand end of the original wavefront is still moving towards the reflector whilst the

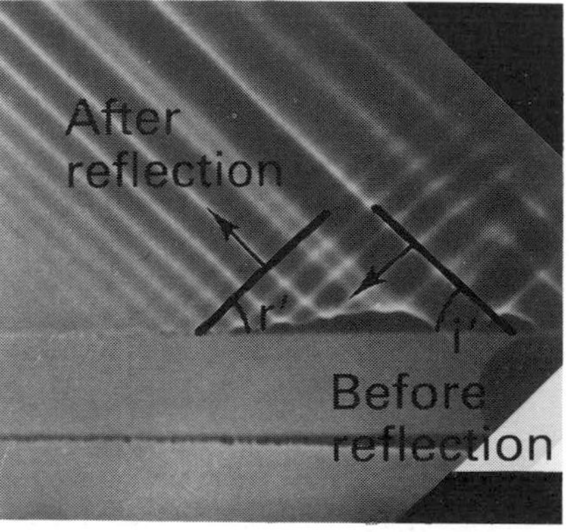

Fig. 3.6 A plane wave reflected in a ripple tank.

left-hand end, after reflection, is moving away from it.

Huygens' principle can be used to relate the law of reflection to the motion of wavefronts, as shown in Fig. 3.7. ABC is a wavefront incident on a surface and A″B″C″ is its position at a time t later, when the last portion of the wavefront has been reflected at C″. The distance CC″ can be measured from the diagram.

To construct the position of the wavefront after reflection, we must draw secondary wavelets and construct their envelope. An arc centred on A and of radius AA″ is drawn; AA″ must be the distance travelled during the time t by that portion of the wavefront that was incident on the surface at A, and as the medium is unchanged, must be equal in length to CC″. The envelope touching the arc and passing through C″ is drawn; A″C″ is the reflected portion of the wave.

As A″C″ is a wavefront, and AA″ is its direction of motion we have:

$$\angle C''A''A = 90°$$

Now in the triangles ACC″ and C″A″A:

$$CC'' = AA'' \quad \text{by construction}$$

$$AC'' \text{ is common}$$

$$\angle ACC'' = \angle C''A''A = 90°$$

and hence the triangles are congruent.

Therefore $\angle C''AC = \angle AC''A''$

i.e. $i = r$

This is the statement of the law of reflection as discussed above.

An intermediate 'ray' BB′B″ may also be constructed, where B′B″ is the distance travelled by that section of the wavefront after reflection, and must therefore be of length (CC″ − BB′). The same argument can be used to show that the triangles B′C′C″ and C″B″B′ are congruent, and therefore as ABC and B′C′

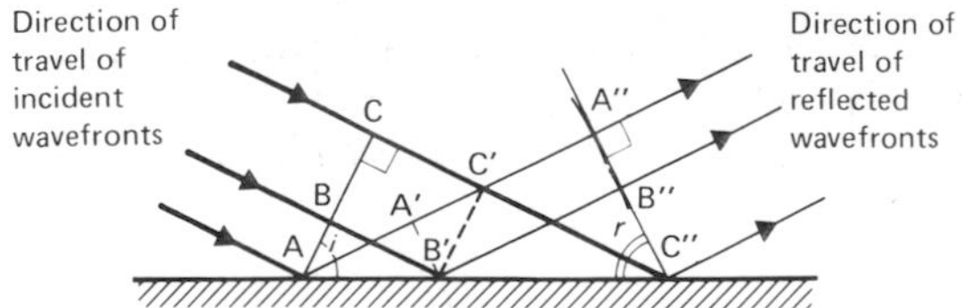

Fig. 3.7 Huygens' principle applied to reflection.

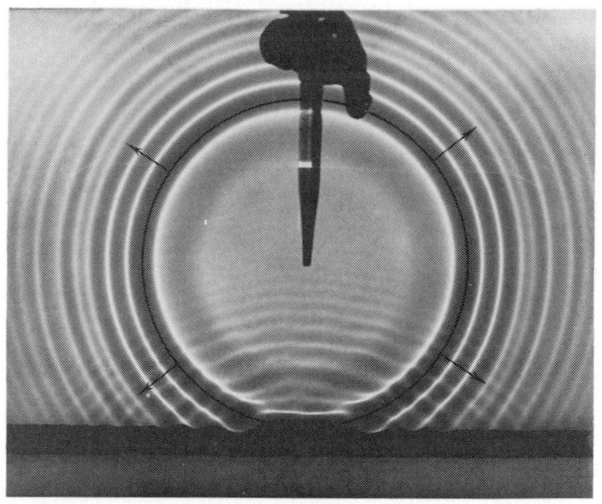

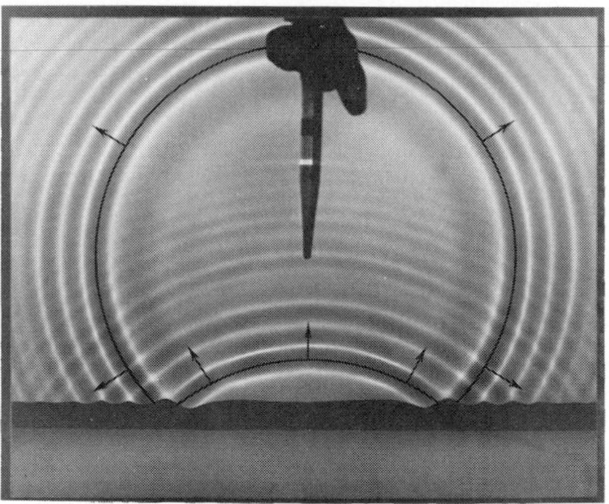

(*a*) The reflection of circular waves from a straight barrier. In the lower photograph part of each wavefront has been reflected.

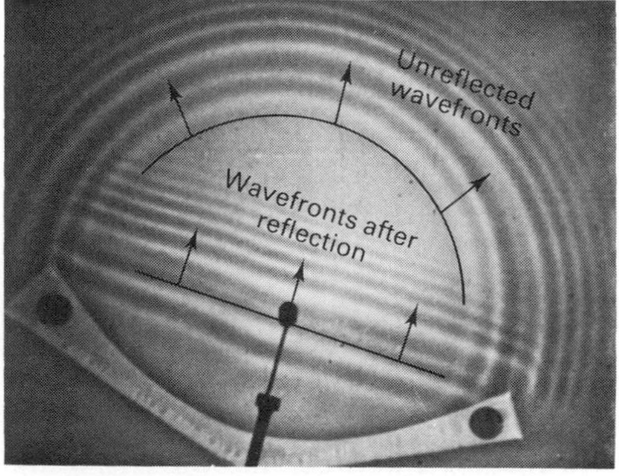

(*b*) The reflection of circular waves from a parabolic reflector; the incident waves are formed at the focus of the parabola and the reflected wavefronts are plane.

Fig. 3.8

are parallel, the triangles C″B″B′ and C″A″A must be similar. It follows that A″B″C″ is a straight line.

Q 3.2
Learn the use of Huygens' principle to construct a reflected wavefront and write it out from memory; it is important!

Two more examples of wave reflection are given in Fig. 3.8. In the first pair of photographs, a drop of water is allowed to fall into a ripple tank and the resulting circular wave is shown before and after reflection at a straight barrier; in the upper photograph a wavefront has just struck the barrier, whilst in the lower one part of it has been reflected. The reflected wavefront is curved and forms an arc of a circle; the point at the centre of this circle is as far behind the barrier as the original disturbance was in front. It is for this reason that the position of an image in a plane mirror is as far behind the mirror as the object is in front.

The final example of the reflection of waves is shown in Fig. 3.8(*b*). Circular waves are produced at one particular point—called the *focus*—near a parabolic concave barrier. After reflection plane waves are produced; the spherical ones beyond them are the unreflected part of the original circular wavefront. A reflector of this type is utilized in a headlight where a powerful lamp is placed at the focus of a parabolic mirror. The technique is used in reverse in a radio telescope where weak, plane-wave radio signals from stars fall on a large parabolic surface and are reflected to a detector placed at the focus; the area of the reflector enables very weak signals to be detected.

Q 3.3
Use your knowledge of the law of reflection as applied to wavefronts to predict what will happen when plane waves are incident on a convex surface.

3.5 Refraction

In water that is much more than one wavelength deep, ocean waves travel at a speed that is independent of the depth; however, when the depth is one wavelength or less, the speed of the wave is proportional to the square root of the depth. Thus, when waves travel at right angles towards the shore they slow down at a rate that depends on the shelving of the sea bed. The reduction in velocity causes a decrease in wavelength and the waves therefore become higher; eventually they become unstable and 'break'.

If the waves approach the shoreline at an angle that is not 90° then the inshore end of a wavefront will reach the shallower water first and therefore slow down before the outer end. As a result, the wavefront will become bent as shown in Fig. 3.9; this change of direction with wave speed is termed *refraction*. The angle between the direction of travel of the waves and the normal to the coast decreases. This is identical to the behaviour of light passing from air into glass; the light ray is bent towards the normal because it travels more slowly in glass.

Again we may use Huygens' principle to formulate a law connecting the angles of incidence, θ_1, and refraction, θ_2. We will assume that the waves cross a distinct boundary between two media in which the wave speeds are v_1 and v_2, respectively. Fig. 3.10 shows a wavefront passing from one such medium to another; the wave speed is greater in the first medium than in the second.

ABC is the position of the wavefront at the moment that its right-hand end meets the interface. Suppose that a time t elapses before the left-hand end of the wavefront meets the interface at C″. In order to find the position of the rest of the wavefront at that time we must, by Huygens' principle, construct suitable secondary wavelets and draw their envelope.

If the velocity of propagation in medium 2 was the same as in medium 1, we would draw an arc centre A and radius CC″. However, the velocity in the second medium, v_2, is less than that in the first, v_1, and the radius of the arc to be drawn is given by:

$$AA'' = \frac{v_2}{v_1} CC''$$

Similarly, if the velocities were identical, we would draw an arc centre B′, length (CC″ − BB′); instead the length of the arc drawn is given by:

$$B'B'' = \frac{v_2}{v_1}(CC'' - BB')$$

Fig. 3.9 The refraction of waves approaching the promenade at Scarborough.

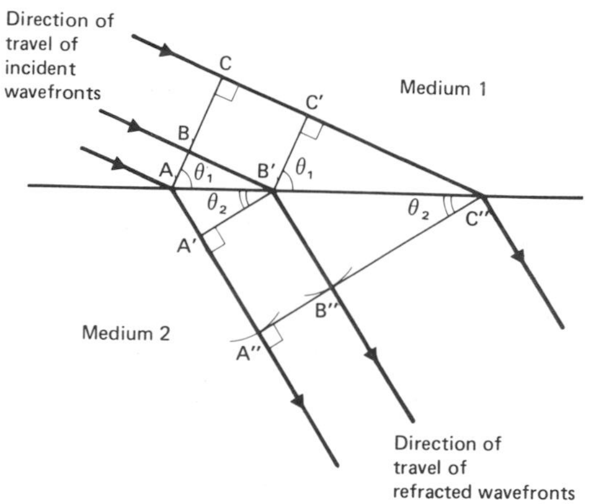

Fig. 3.10 Huygens' principle applied to refraction.

Finally, the new position of the wavefront is found by drawing the envelope to the two arcs; it passes through C″, as shown, since A″B″C″ may be shown to be a straight line by a similar argument to that used in the case of reflection.

Now $\sin\theta_1 = CC''/AC''$

and $\sin\theta_2 = AA''/AC''$

From above $AA'' = \dfrac{v_2}{v_1} CC''$

giving
$$\sin\theta_2 = \frac{v_2}{v_1}\frac{CC''}{AC''} = \frac{v_2}{v_1}\sin\theta_1$$

i.e.
$$\frac{\sin\theta_2}{\sin\theta_1} = \frac{v_2}{v_1} \qquad (3.1)$$

As the ratio of the two velocities is fixed by the nature of the two media we see that sin θ_2 is proportional to sin θ_1; when this law is applied to light it is known as *Snell's law*. It is discussed in more detail in Chapter 6.

Q 3.4
Without looking ahead to the next section, use Huygens' principle to construct the situation where waves pass into a medium in which they move more quickly. What would you expect to happen as the angle of incidence increases?

The refraction of ocean waves in areas of shallow water may have some unexpected effects. Fig. 3.11 shows the effect of shallow water in a bay on both short- and long-wavelength waves. For short waves the depth is much larger than the wavelength and there is therefore little difference in wave speed, and little refraction, in the shallower water. However, for long waves the depth is the same order of size as the wavelength, or less, and a considerable amount of refraction takes place; as a result waves are refracted into the bay.

This sort of process, which can easily cause erosion, has the effect of 'smoothing' out the coastline. It is also possible for a region of shallow water off a headland, or even some distance out at sea, to act as a lens and focus waves on to the headland or some other point on the coast; this can cause concentrated local erosion and freak damage. For example, in 1930 a breakwater at Long Beach, California, was breached after a fairly moderate storm; it was discovered later that the sea was running in such a direction that a small hump on the sea bed 80 metres deep and 10 kilometres out to sea had acted as a 'lens' and caused the wave height to be increased by a factor of four. As the hump varied in height from place to place, the velocity of the waves also varied; this was equivalent to a 'lens' whose 'refracting effect' varies from one part to another. Oceanographers and hydrographers make considerable use of Huygens' principle in solving problems such as these.

Q 3.5
Draw a diagram to show how a small hump on the sea bed can produce the effects described above.

3.6 Total internal reflection

Normally, when a wave of any kind meets a boundary between two media where the wave speed changes, some of the energy is refracted into the second medium whilst some of it is reflected back into the first, obeying the law of reflection. When we are in a room with the curtains open on a dark night, most of the light

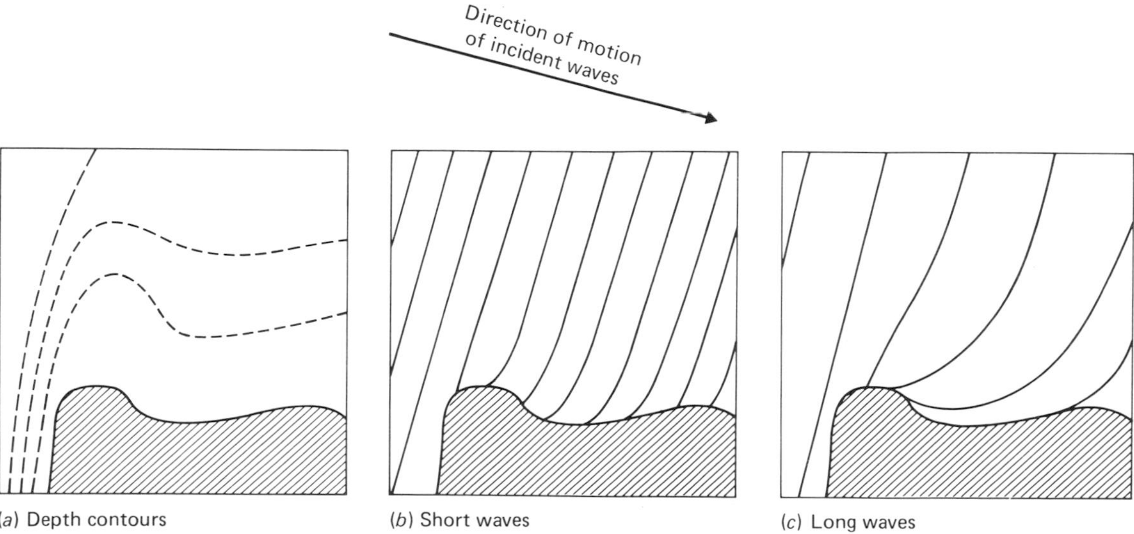

Fig. 3.11 The refraction of waves with changing depth near a headland.

from within passes out through the windows, enabling those outside to see inside, but some of the light is reflected at the window enabling us to see a reflection of ourselves and of the contents of the room.

Fig. 3.12 shows the same effect where water waves in a ripple tank are moving from an area of higher velocity to one of lower velocity. As expected, the direction of travel of the waves changes, this time towards the normal. Careful examination of the photograph, however, shows that a small amount of the wave energy has been reflected. Partial reflection will always occur, whether the wave is moving into a medium of higher or lower wave speed.

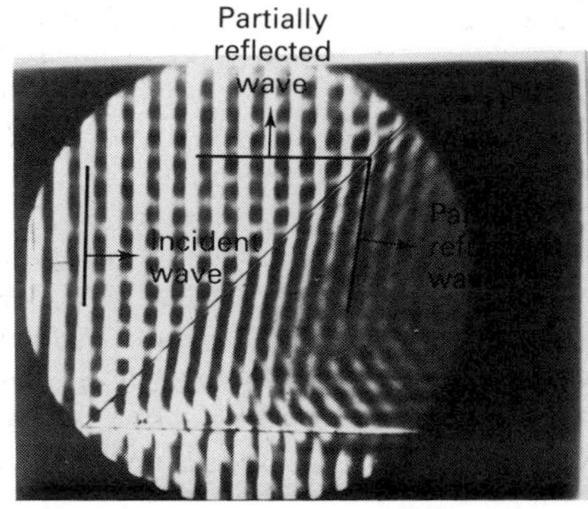

Fig. 3.12 Refraction and reflection in a ripple tank.

However, if the wave is moving, as in Fig. 3.13(*a*), into a region of higher wave speed, and hence the angle between wavefront and boundary in the second medium is larger than that in the first, a further phenomenon may occur. Suppose that the angle in the first medium is increased; then that in the second medium will increase also. As it is the larger angle it will reach 90° before the first angle does; at this stage waves are striking the interface and *on emerging travel parallel to it* as shown in Fig. 3.13(*b*). The angle in the first medium at which this happens is called the *critical angle*, θ_c. At angles larger than this, all the wave energy will be reflected, and *total internal reflection* is said to occur.

In diagram (*a*) the wavefront in the second medium is constructed by drawing secondary wavelets; the envelope A″B″C″ can be drawn in the usual way to give the new position of the wavefront. However, in the second diagram, we see that at the critical angle all the secondary wavelet arcs touch at C″ and therefore the wavefront must be travelling along the interface. At angles larger than the critical angle, it is impossible to draw a wavefront that is an envelope to all the secondary wavelets and therefore there can be no refracted wavefront, as shown in diagram (*c*); refraction into the upper medium is impossible.

In the first two cases, partial reflection will occur in addition to refraction and in the third case, only total internal reflection will occur; for simplicity, no reflected waves at all are shown in these diagrams.

The law of refraction, as derived above, states that:

$$\frac{\sin\theta_2}{\sin\theta_1}=\frac{v_2}{v_1} \qquad v_2>v_1$$

and total internal reflection begins to occur when:

$$\theta_2=90°$$

$$\theta_1=\theta_c$$

Hence $$\frac{1}{\sin\theta_c}=\frac{v_2}{v_1}$$

i.e. $$\sin\theta_c=\frac{v_1}{v_2} \qquad (3.2)$$

Total internal reflection is of great importance in optics as we shall see in Chapter 6. However, it can also be observed with different types of waves, such as electromagnetic waves of non-visible wavelengths.

Q 3.6

An approximate expression for the speed of water waves in a shallow ripple tank is

$$c=\sqrt{gh}$$

where h is the depth of water. A glass plate is placed on the bottom of the tank so that the depth of water is halved and the speed of the waves thus reduced. What is the critical angle at the edge of the plate?

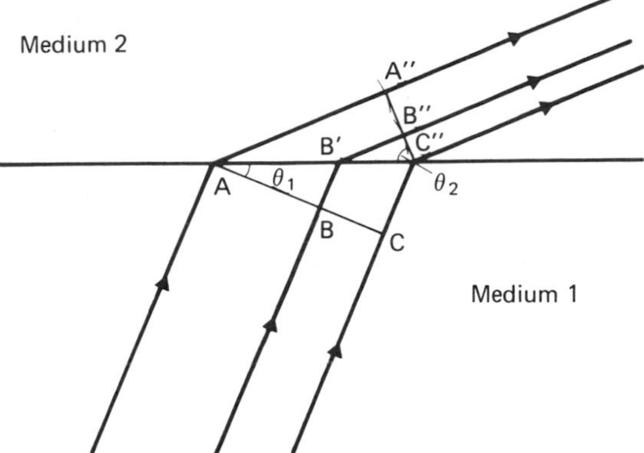

(*a*) Waves refracted into a medium of higher wave speed

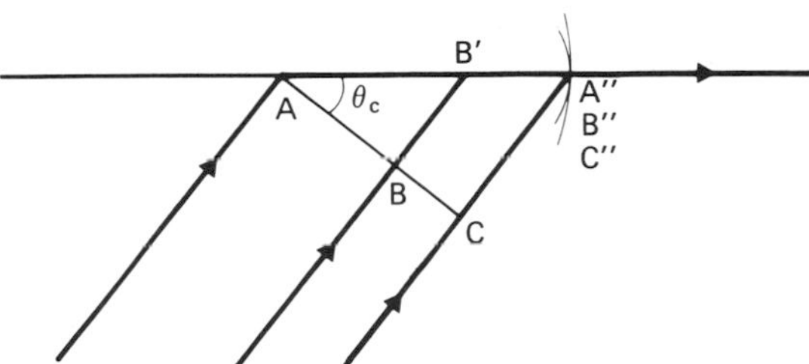

(*b*) Waves refracted along the interface at the critical angle

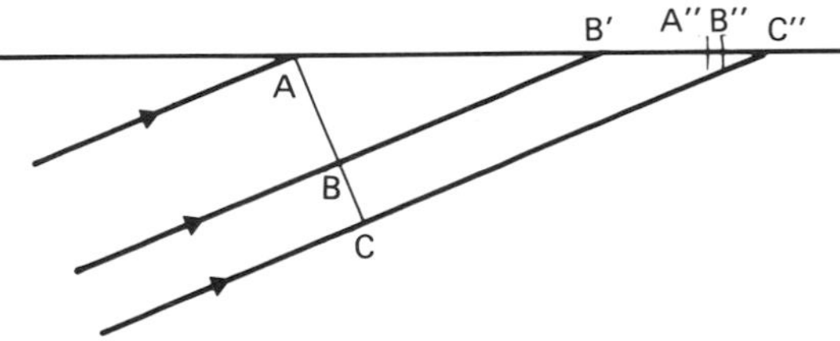

(*c*) The impossibility of refraction at angles larger than the critical angle

Fig. 3.13 Huygens' principle applied to total internal reflection.

3.7 Earthquake waves

We have already mentioned the propagation of sound waves through solids, noting that they are of longitudinal, or compression, type. Transverse waves can also be transmitted through solids and are usually referred to as shear waves; they can be seen by displacing one end of a lump of jelly sideways and observing the resultant motion. It has been discovered that the speed of shear waves through a solid is always less than that of compression waves.

When an underground disturbance causes the ground to be shaken, we call it an *earthquake*; there are in total about ten thousand of these a year, of which about one a week is of any significant size. The smaller ones can only be detected by sensitive instruments called *seismographs* which draw graphical records, termed *seismograms*, of the movements of the earth's surface.

In the simplified seismogram shown in Fig. 3.14, we see that three groups of waves arrive at a recording station from a particular earthquake. The first two groups travel well beneath the surface of the earth and, in order of arrival, are termed *primary* and *secondary*. The primary waves are compression waves and the secondary waves are shear waves. Primary waves travel at about 8 kilometres per second (18 000 mph) and secondary waves travel at just under 6 kilometres per second (13 500 mph). The last, and strongest, group of waves to arrive are termed *long waves* and are also shear waves; they travel through the upper layer of the earth, called the *crust*. They travel at between 1.5 and 5 kilometres per second (3400–11 000 mph) and may have wavelengths from 16 kilometres up to the length of the earth's diameter (13 000 km; about 8000 miles).

Clearly, waves travelling at different speeds arrive at different times and the further they have travelled the greater will be the time interval between arrivals. Thus it is possible, by analysis of the seismogram, to deduce where the earthquake centre was situated. In addition to the obviously useful study of earthquakes and, occasionally, warning of the arrival of shock waves, we can also use seismograms to study the internal structure of the earth. Other applications of seismology include the detection of underground nuclear tests and the use of small-scale explosions or 'seismic shots' to study the structure of rocks that may bear oil.

The density and compressibility of the earth change with depth, and the speed of waves increases as they travel more deeply. Thus they do not travel in straight lines, but undergo continuous refraction in the same way that water waves do when crossing a shelving sea floor; some possible paths are shown in Fig. 3.15.

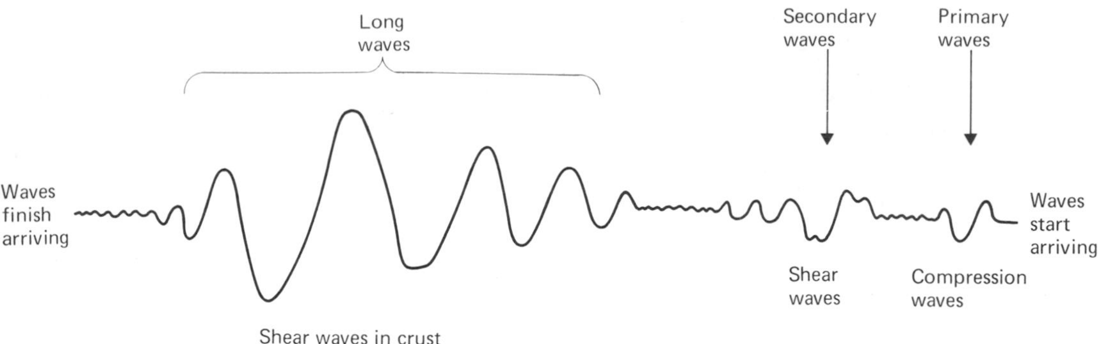

Fig. 3.14 A typical seismogram.

We believe that the upper *crust* of the earth is about 40 kilometres thick and that below this there is a *mantle* of about 3000 kilometres thickness; inside this there is the *core* of radius 3500 kilometres. When the waves are refracted back to the crust, they may be reflected at the interface; there will also be reflections at other discontinuities such as that between mantle and core.

An analysis of the times taken by various types of waves shows that the central core does not transmit shear waves and must therefore be liquid; it must also contain a more dense inner core of about 1300 kilometres radius. However, when a wave strikes an interface at an oblique angle, both shear and compression waves may be generated; thus a shear wave will continue through the core as a compression wave and re-emerge as both a compression wave and a shear wave.

It is a similar analysis of the long waves that has shown, most unexpectedly, that the crust is only about 5.5 kilometres deep below the bottom of the sea; that is, about 11 kilometres below sea level. Under land it is some 40 kilometres below sea level. This discovery has encouraged research workers attempting to drill through to the mantle to make renewed attempts from sea bed drilling platforms instead of on land.

3.8 The Doppler effect

The *Doppler effect* was first predicted in 1842 by an Austrian physicist, Christian Johann Doppler. He argued, on the basis of wave theory, that the frequency of the light in a beam will be increased if either the source or the observer is moving so as to decrease their separation, and will be decreased if their separation is increasing. Scientists were already learning of dark *absorption lines* in the spectra of stars, due to the absorption of certain frequencies of light by gaseous elements in their outer layers; when reference is made to the sun these lines are called *Fraunhofer lines* as discussed on p. 272. Doppler argued that if the star was moving away from the earth, the absorption lines would move towards the red, or low-frequency, end of the spectrum. These predictions were confirmed by experiment in 1868 and are usually referred to as the 'red shift'.

Doppler's work also applies to other types of waves and the first test of his work was, in fact, carried out using sound waves. The test was made in Holland in 1845 and consisted of trumpeters riding on a flat railway wagon past musically trained observers who estimated, by ear, the apparent change in pitch when the wagon moved. We are now familiar with the fact that as a car moves towards us the pitch of its horn is higher than when stationary, and that when it passes us and moves away, the pitch is lower. Thus, there is a distinct drop in pitch as it passes.

The Doppler effect for sound

Although we have referred above to the prediction of the Doppler effect for both sound and

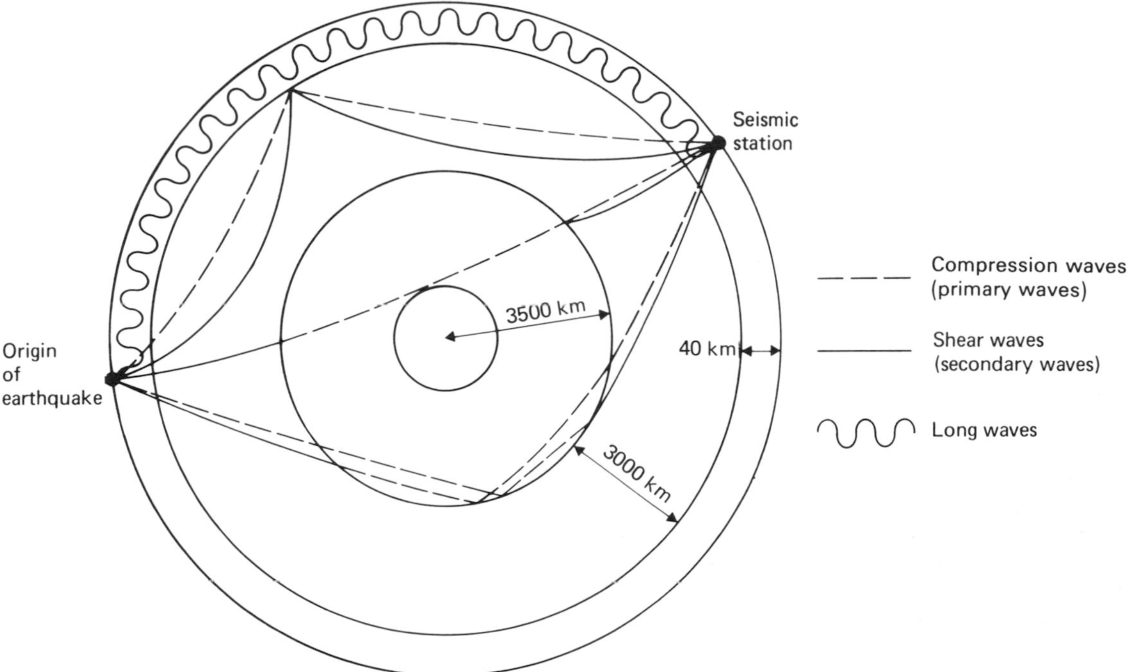

Fig. 3.15 Possible paths for the transmission of earthquake waves through the earth.

light, there are important differences between the two cases; we shall initially confine our attention to cases other than those involving electromagnetic waves. For sound, as for most wave motions, the source determines the frequency of the emitted wave and then this, together with the velocity of sound relative to the source, determines the wavelength in the medium. The velocity relative to the medium is independent of the velocity of the source. The wavelength in the medium, together with the velocity relative to the observer, determines the frequency detected by the observer. The simplest cases occur where either the source or the observer is moving but the argument that follows may easily be extended to the case where both move; Doppler shifts will also be observed even when there is no relative motion between source and observer, but the medium moves relative to both.

The effect of a moving source is shown in the ripple tank photograph in Fig. 3.16. The frequency of the waves emitted is fixed by the vibrator, irrespective of the motion of the source. However, as the source is moving to the right, the waves on that side are spaced more closely together as shown in the photograph. An observer to the right will detect waves of higher frequency but it is important to remember that *the fundamental physical change in the medium is one of wavelength*, and this will occur whether there is an observer present or not. Similarly, on the left side of the source, the wavelength is increased and if an observer is present, he will detect a lower frequency. Waves travelling at right angles to the motion of the source have unchanged wavelength and there is thus no change in frequency; if the direction of motion is at an angle to the line between observer and source, then the component of the velocity along that line must be used. We are assuming throughout that the velocity of motion is less than that of the waves; the case where this is not so is discussed in Section 3.9, which is about bow waves.

We will now compare the wave emitted by a stationary and a moving source mathematically, and these two situations are illustrated in Fig. 3.17. In diagram (a) the source is stationary. Five wavefronts are shown, each moving away from the source with velocity c; wavefront

1 was the first to have left the source, wavefront 2 was the second, and so on. If f_s represents the frequency of the source, and the first wavefront moves a distance c in one second, we see that the wave spacing, or wavelength λ_s, is given by:

$$\lambda_s = \frac{c}{f_s} \tag{3.3}$$

λ_s is the wavelength in the medium and is therefore determined by the frequency of the source and by the velocity of sound relative to the source; because the source is not moving relative to the medium, this velocity is the velocity of sound, c. If the observer is also stationary, he will detect waves of frequency given by:

$$f_o = \frac{c}{\lambda_s} = f_s$$

The observed frequency will, therefore, be the same as that emitted.

Now suppose that the source is moving to the right with velocity, v_s, as shown in diagram (*b*); thus, wave 1 was emitted when the source was at A, wave 2 when it was at B and so on. If f_s represents the frequency of the source, the time interval between the emission of these waves must be $1/f_s$. Therefore, the distances moved by the source between the emission of successive wavefronts must be given by:

$$AB = BC = CD = \ldots = v_s \times \frac{1}{f_s} = \frac{v_s}{f_s}$$

and this must be the amount by which the wavelength has *decreased*. Hence

$$\lambda_s - \lambda_m = \frac{v_s}{f_s} \tag{3.4}$$

where λ_m is the wavelength in the medium when the source is moving.

$$\therefore \lambda_m = \lambda_s - \frac{v_s}{f_s}$$

$$= \frac{c}{f_s} - \frac{v_s}{f_s} \qquad \text{from (3.3)}$$

i.e.

$$\lambda_m = \frac{c - v_s}{f_s} \tag{3.5}$$

But

$$f_s = \frac{c}{\lambda_s} \qquad \text{from (3.3)}$$

$$\lambda_m = \lambda_s \times \frac{c - v_s}{c}$$

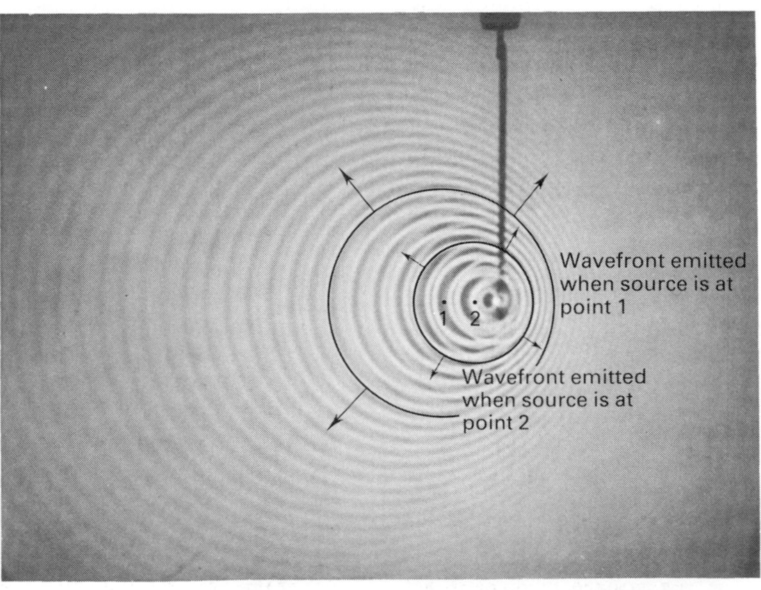

Fig. 3.16 The Doppler effect for a moving source.

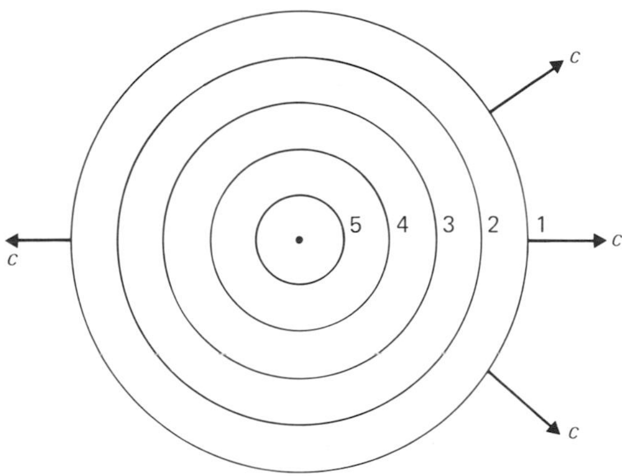

(*a*) Wave pattern from a source at rest

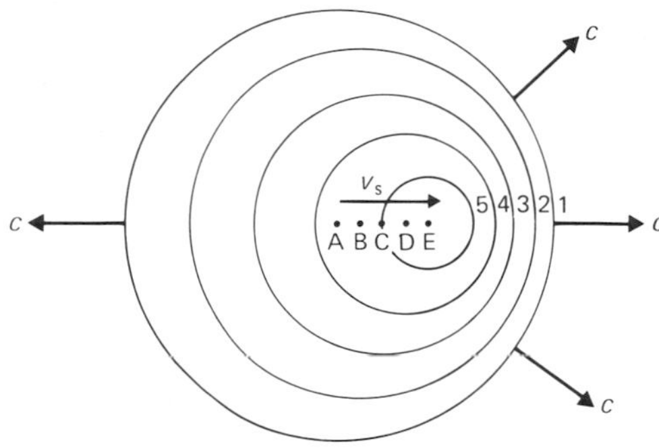

(*b*) Wave pattern from a source moving to the right

Fig. 3.17

i.e. $$\frac{\lambda_m}{\lambda_s} = \frac{c - v_s}{c} \tag{3.6}$$

Thus, for a moving source we may calculate the wavelength of the waves through the medium from the wavelength of the waves that would have been propagated had the source been stationary. This change in wavelength is the physical result of the motion and will occur whether or not there is an observer, as stated earlier.

If an observer is present he will detect the frequency of the sound, not its wavelength, from the wavelength in the medium and the velocity relative to him. If he is stationary this will be the velocity of sound, *c*, since as stated earlier the velocity of the waves relative to the medium is not affected by the velocity of the source; we then have:

$$f_o = \frac{c}{\lambda_m}$$

$$= \frac{c}{(c - v_s)/f_s} \qquad \text{from (3.5)}$$

$$= f_s \times \frac{c}{c - v_s}$$

i.e. $$\frac{f_o}{f_s} = \frac{c}{c - v_s} \tag{3.7}$$

The relative pitch of two notes is determined by the ratio of their frequencies, as discussed in Chapter 5; therefore it is the ratio of the frequencies that is quoted here. It is most important to remember that *the change in the observed frequency is caused by the change of wavelength in the medium.*

If the source is receding, instead of advancing, the sign of the velocity of motion must be reversed and Equation (3.7) then becomes:

$$\frac{f_o}{f_s} = \frac{c}{c + v_s}$$

where values of v_s are substituted as positive numbers. This result may be proved rigorously by a piece of analysis similar to that given above. In fact the formula in Equation (3.7) can be used in both cases provided that the following rule about the sign of v_s is observed:

$$\frac{f_o}{f_s} = \frac{c}{c - v_s} \tag{3.7}$$

If the direction of velocity of the source is in the same direction as that in which wavefronts move from source to observer, a positive sign is used for v_s; if the directions are opposed a negative sign is used.

(Note that the formula used contains a negative sign already in addition to any that may be introduced by the sign rule.)

So far we have restricted our attention to the situation where the observer is at rest with respect to the medium. We will now consider what happens when the observer is moving as shown in Fig. 3.18.

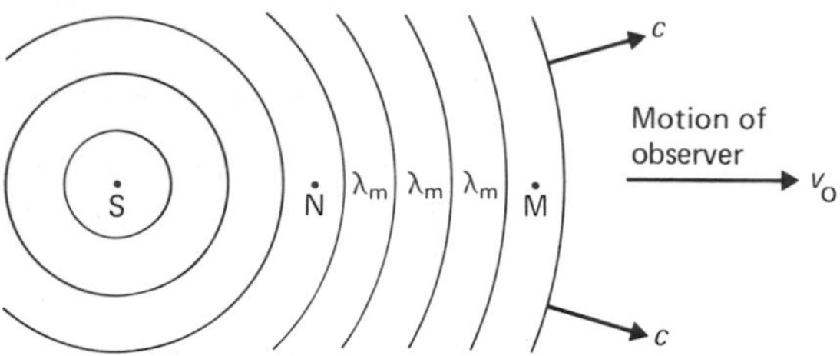

Fig. 3.18 The Doppler effect for a moving observer.

Let us suppose that there are waves, of wavelength λ_m, moving through a medium from a source S towards an observer. It is immaterial at present whether these waves are produced by a stationary source as shown, or by one that is moving. If the observer is stationary, the waves are moving at a speed c towards him and therefore the number of waves, each of length λ_m, that will pass him in t seconds is given by ct/λ_m. Hence, the observed frequency of the waves, f, will be given by:

$$ft = \frac{ct}{\lambda_m}$$

$$f = \frac{c}{\lambda_m}$$

However, if the observer is moving in the same direction as the waves, with a velocity v_o that is less than c, the new observed frequency, f_o, will be given by:

$$f_o t = \frac{(c - v_o)t}{\lambda_m}$$

i.e.

$$f_o = \frac{c - v_o}{\lambda_m} \qquad (3.8)$$

because $(c - v_o)$ is the new relative velocity. Thus *the fundamental change that occurs if the observer is moving is an apparent change in frequency*. If the observer moves in the opposite direction to the wavefronts, the formula can be shown to be:

$$f_o = \frac{c + v_o}{\lambda_m}$$

The first formula can be used in both cases if the same rule as before is used for signs: i.e.

$$f_o = \frac{c - v_o}{\lambda_m} \qquad (3.8)$$

If the motion is in the same direction as the relevant wavefronts, use a positive sign for v_o; if it is in the opposite direction use a minus sign. (Again, this is in addition to the negative sign already in the formula.)

The wavelength in the medium is determined by the frequency of the source as before. If the source is stationary we have:

$$\lambda_m = \frac{c}{f_s}$$

and thus the observed frequency is given by:

moving observer $$\frac{f_o}{f_s} = \frac{c - v_o}{c} \qquad (3.9)$$

This formula should be compared with that given in Equation (3.7) for the case with moving source and stationary observer:

moving source $$\frac{f_o}{f_s} = \frac{c}{c - v_s} \qquad (3.7)$$

The same sign convention is used in both cases.

It is interesting to compare the sizes of the two changes in frequency. Consider a car moving at 30 metres per second (about 65 mph) towards an observer and sounding its horn; the velocity of sound in air is about 330 metres per second (about 730 mph). The formula for moving source and stationary observer is

$$\frac{f_o}{f_s} = \frac{c}{c - v_s} \qquad (3.7)$$

and as the source is moving in the same direction as the waves we take $v_s = +30 \text{ m s}^{-1}$. Therefore

$$\frac{f_o}{f_s} = \frac{330}{330 - 30} = \frac{11}{10}$$

If, on the other hand, the car is stationary but the observer is moving towards it at 30 m s^{-1} we use the other formula:

$$\frac{f_o}{f_s} = \frac{c - v_o}{c} \qquad (3.9)$$

This time the observer is moving in the opposite direction to the waves and we therefore take $v_o = -30\ \mathrm{m\,s^{-1}}$. Therefore

$$\frac{f_o}{f_s} = \frac{330-(-30)}{330}$$

$$= \frac{330+30}{330} = \frac{12}{11}$$

Thus, *although the relative velocity is the same in each case*, the two results are not the same because the medium has a different effect. For comparison, a change in pitch of a minor third involves a frequency ratio of 12/10.

The case where source and observer are both moving may be treated similarly. We start by considering the observer and see that the observed frequency is obtained from the wavelength by the usual equation:

$$f_o - \frac{c - v_o}{\lambda_m} \qquad (3.8)$$

In this case the wavelength in the medium is obtained from the frequency of the source by using Equation (3.5):

$$\lambda_m = \frac{c - v_s}{f_s}$$

Hence

$$\frac{f_o}{f_s} = \frac{c - v_o}{c - v_s} \qquad (3.10)$$

The formula is thus seen to combine the two results obtained above in Equations (3.5) and (3.8). Both v_o and v_s have positive signs if observer and source are moving in the direction of the waves and negative if they are moving in the opposite direction to the waves.

Somctimes when we discuss sound, we have to consider the situation where a wind is blowing. In this case, the values of v_o and v_s used are the velocities relative to the moving air and not the velocities relative to the ground.

Q 3.7
A car is moving along the road at $40\ \mathrm{m\,s^{-1}}$ sounding its horn at a frequency of 300 Hz. Calculate the frequency of the note heard by an observer:

(*a*) towards whom the car is moving;
(*b*) away from whom the car is moving.

The car now stops and sounds its horn at a cow. Calculate the frequency of the note heard by a second car:

(*c*) moving towards the first at $40\ \mathrm{m\,s^{-1}}$;
(*d*) moving away from the first at $40\ \mathrm{m\,s^{-1}}$.

Take the speed of sound in air to be $330\ \mathrm{m\,s^{-1}}$.

The Doppler effect for electromagnetic waves

Unlike sound waves, electromagnetic waves do not require a medium for transmission; attempts to establish the existence of the ether, an all-pervading medium that was thought to be present even in a vacuum, were discredited by the Michelson–Morley experiment, as described in Chapter 9. There is, thus, no physical medium relative to which v_s and v_o can be measured and, in fact, the theory of relativity shows that the velocity of light in a vacuum is the same with respect to all observers, however they are moving.

The laws of relativity can be applied to derive an equation relating the observed and transmitted frequencies:

$$\frac{f_o}{f_s} = \sqrt{\frac{c+v}{c-v}} \qquad (3.11)$$

where v is the velocity of the source towards the observer in his frame of reference. This equation applies only to cases where the relative motion is along the line joining object and source, and it is important to note that other cases cannot be solved merely by using non-relativistic resolution methods. Similarly, if the velocities of the observer and of the source, v_o and v_s, are known relative to a third object, or transmitting medium, it is not sufficient to obtain their relative velocity by subtracting v_o from v_s; relativistic transformations must be used.

When light, or other electromagnetic waves, are detected, the measured parameter is usually the wavelength, and therefore Equation (3.11) is usually written in terms of wavelength:

$$\frac{\lambda_o}{\lambda_s} = \frac{c/f_o}{c/f_s}$$

$$= \frac{f_s}{f_o}$$

i.e.

$$\frac{\lambda_o}{\lambda_s} = \sqrt{\frac{c-v}{c+v}} \qquad (3.12)$$

These changes in wavelength were first observed in the 'red shift' of the light from stars receding at speeds about 1/100 of the velocity of light; they may also be seen in the different wavelengths of light from the east and west sides of the sun due to its spin. A radar speed trap, however, uses the Doppler effect to measure the very much smaller speeds of moving cars. Interferometric methods, in which light from a stationary mirror and light from a slowly moving mirror beat together, can be used to measure speeds as small as 10^{-7} m s^{-1}.

In situations such as these, where the relative velocity of source and observer is very much less than the velocity of light, we have:

$$\frac{f_o}{f_s}=\sqrt{\frac{1+v/c}{1-v/c}}$$

$$\text{where } \frac{v}{c} \ll 1$$

But, by the binomial theorem:

$$(1-v/c)^{-1}=1+v/c,$$

Hence:

$$\frac{f_o}{f_s}=1+\frac{v}{c}$$

$$=\frac{1}{1-v/c}$$

i.e.

$$\frac{f_o}{f_s}=\frac{c}{c-v} \qquad (3.7)$$

This is the same equation as was previously obtained for sound waves in a moving source–stationary observer situation.

Alternatively this equation may be written:

$$\frac{f_o}{f_s}-1=\frac{c}{c-v}-1$$

$$\frac{f_o-f_s}{f_s}=\frac{v}{c-v}$$

We can write Δf as the change in frequency (f_o-f_s) and since $(c-v)\approx c$, we have:

$$\frac{\Delta f}{f_s}\approx\frac{v}{c} \qquad (3.13)$$

The small change in frequency can be used to measure the relative speed, v, by utilizing the principle of beats, discussed in Chapter 4.

In calculations involving speed traps, where a reflector is moving with respect to the source, it is easiest to regard the reflected waves as having been transmitted by the moving image of the source.

Q 3.8
Light of wavelength 6.56×10^{-7} m is emitted by a hydrogen atom in a distant star. The observed wavelength is 6.63×10^{-7} m. Assuming that the motion of the star is directly away from the earth, with what speed is it moving relative to the earth?

Take $c=3\times10^{8}$ m s^{-1}.

Q 3.9
A police speed trap uses radar of wavelength 100 mm; calculate the change in observed frequency when the waves are reflected from:

(*a*) a car approaching at 20 m s^{-1} (45 mph);
(*b*) a car moving away at 40 m s^{-1} (90 mph).

Take $c=3\times10^{8}$ m s^{-1}.

3.9 Bow waves and the sonic boom

In our discussion of the Doppler effect, we made the condition that the velocity of the source must be less than that of the wave motion. Clearly in the case of light transmitted through a vacuum this is always so, but for sound waves or water waves it need not be.

Fig. 3.19 shows a succession of wavefronts emitted by a moving source. In the first diagram the source is moving more slowly than the wave and the effect of the motion is merely a reduction of wavelength in the medium to the right of the source. In the second diagram, however, the source is moving more rapidly than the waves and the series of wavefronts combine along their envelope, or tangent. The resultant is one large wave instead of a succession of small ones. The reader may recognize the formation of a bow wave in diagram (*b*) and this effect is shown in the photograph (Fig. 3.19(*c*)).

Aeroplanes moving through the air also produce waves; these will have a 'steeper' wavefront than for a boat since there is little dispersion of sound waves in air and much dispersion of waves in water. When an aeroplane travels faster than sound, a cone-shaped 'wake' is produced in three dimensions. The wake consists

(*a*) $v < c$ (*b*) $v > c$

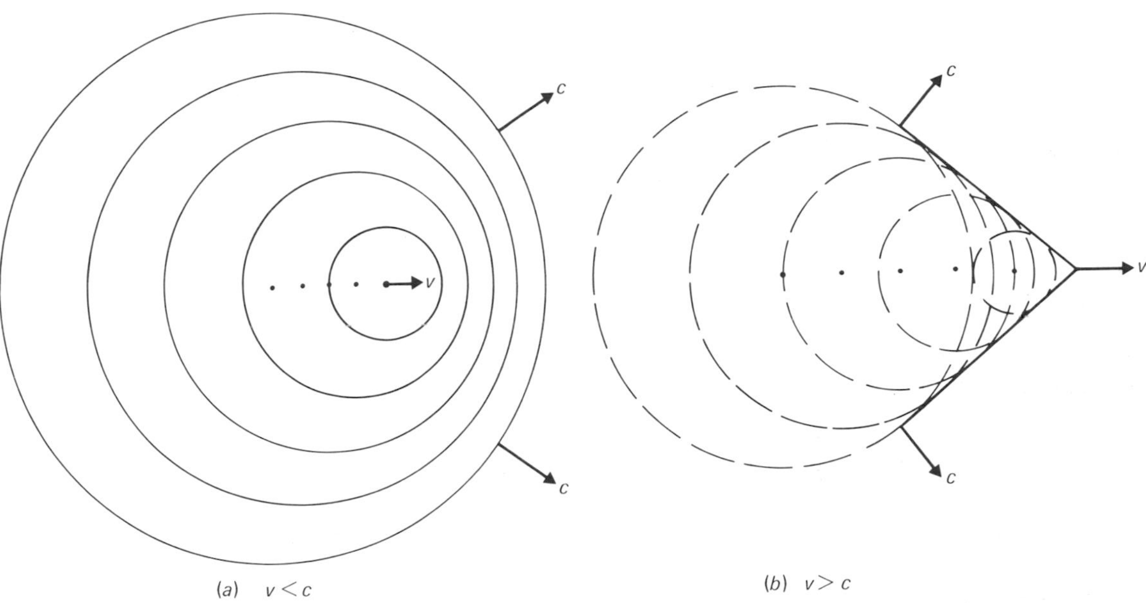

(*c*) A bow wave formed in an experimental tank.

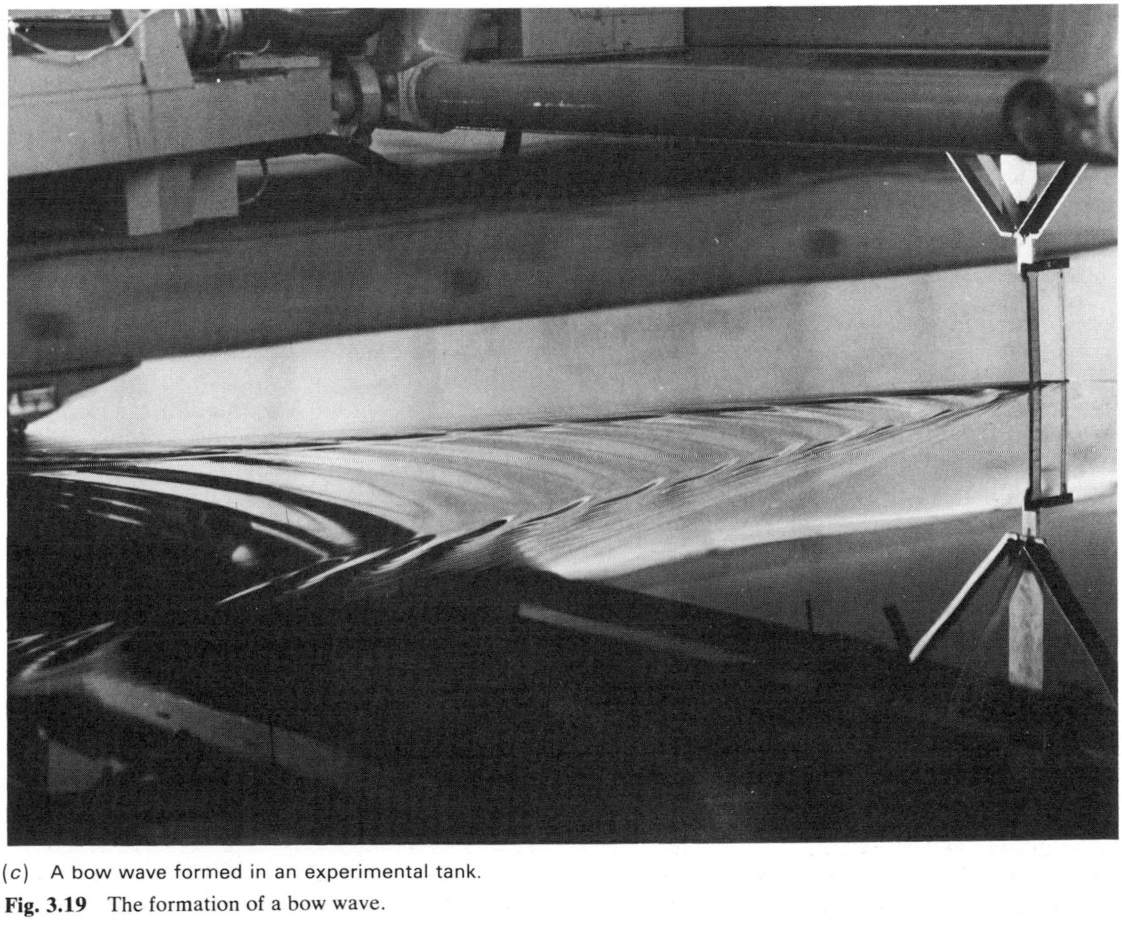

Fig. 3.19 The formation of a bow wave.

Fig. 3.20 The change in air pressure caused by a supersonic aeroplane; the second pressure rise is caused by the movement of the rear of the aeroplane, and results in a 'double-boom' (*Crown copyright*).

of two sharp changes in pressure, one originating at the front of the aeroplane and the other at the rear. It is this double change in pressure that causes a double bang to be heard when a supersonic plane passes overhead. This is illustrated in Fig. 3.20; the 'bow and stern lines' in the photograph mark the changes in pressure caused by the supersonic motion.

Charged particles such as electrons can travel through a transparent medium with a velocity greater than the velocity of light in that medium, and when this occurs an 'optical shock wave' known as *Čerenkov radiation* is formed. This is caused by perturbation of the electric field of the electron and can be used to detect the presence of high-speed electrons in beams of heavier particles. The angle at which the Čerenkov radiation is propagated depends upon the refractive index of the medium, and upon the velocity of the electron; thus the electron velocities can be measured.

Bibliography

The Bibliographies to Chapters 2, 3 and 4 are presented together at the end of Chapter 4.

4

Wave superposition

4.1 The principle of superposition

When two objects collide, the collision considerably alters the subsequent motion of the objects; in this chapter we are going to focus attention on what happens to waves when they 'collide' with each other.

The simplest situation to discuss is that shown in Fig. 4.1. Two pulses are moving towards each other, one from the right and one from the left, along a long spring. In the middle of the spring they meet and a complicated shape is formed. However, the complex shape then vanishes and the two pulses re-appear; they continue to move along the spring as though no collision had taken place at all. The pulse that was moving to the left is still moving to the left with the same shape, and the pulse that was moving to the right, likewise.

We have, therefore, already discovered one fundamental property of waves: that *they move through each other without undergoing any permanent alteration*. This is true for continuous waves as well as for pulses and remains true however many waves meet at the same point in a medium. The situation is thus very different from that when solid objects collide.

At the moment when the waves crossed, a complex shape was formed; the height of this shape was greater than the height of either pulse. Careful measurement shows that *the height of the shape at any particular time and place is equal to the sum of the heights of the two separate pulses, at the same time and place.*

In the third diagram of Fig. 4.2, we can see the effect of this principle. At that particular instant the displacement due to the pulse 1 at point M is y_1 and that due to pulse 2 is y_2; the actual displacement of the spring is therefore $(y_1 + y_2)$. Similarly, at M′ the displacement due to pulse 1 is y'_1, that due to 2 is y'_2, and the actual displacement is $(y'_1 + y'_2)$. In this way, the resultant shape at that instant can be calculated; the shape at any other instant will be different because the two pulses, 1 and 2, will have moved with respect to each other.

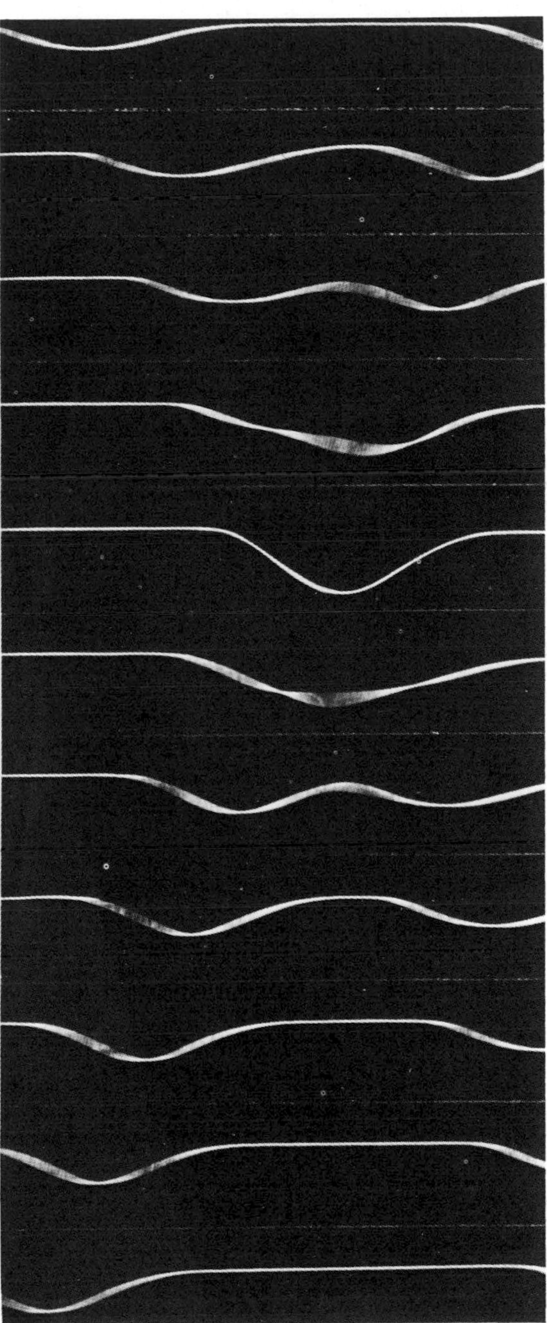

Fig. 4.1 Two similar pulses, travelling in opposite directions, move through each other without alteration. As the pulses have different shapes, they can be identified; when they coincide they 'add up' to produce the total waveform.

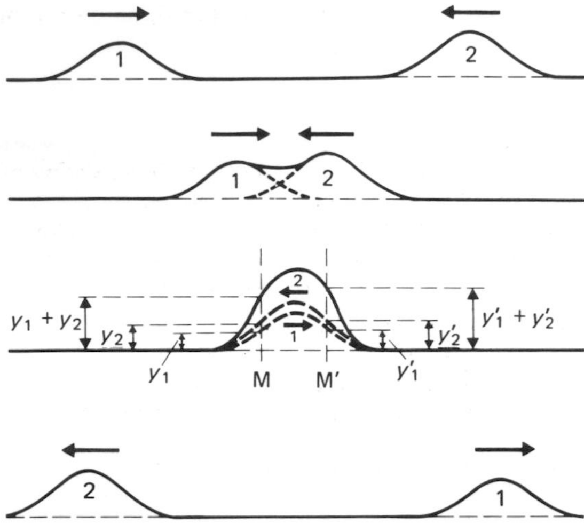

Fig. 4.2 The superposition of two similar pulses.

This principle, like the first, applies to any number of continuous waves meeting at a point; it includes the first, for after pulse 1 has passed through pulse 2 there is nothing to add to either pulse and therefore the pulses move on unchanged. Both ideas may therefore be contained in the statement: *to find the total disturbance at a particular time, we add at each point the displacements due to each pulse or wave passing through the medium*; this is called the principle of superposition.

We may now apply the principle of superposition to a different situation. Suppose that the two pulses starting at each end of the spring are exactly the same size and shape, but one is upside down. We will attempt to predict what happens where they pass by means of the diagrams in Fig. 4.3.

Diagram (a) shows the two pulses travelling on the spring before they meet. In the top half of diagram (b), we have shown the positions that the pulses would occupy just as they begin to superpose. The vertical lines assist in adding up the effects due to the two pulses, and the bottom half of the diagram shows the combined shape at that particular instant. Note that to the left of point X_1 only pulse 1 has displacement and therefore the combined pulse is exactly the same as pulse 1; the same situation occurs to the right of point X_3, where the combined pulse is just the same as pulse 2. At point X_2, however, the effect of pulse 1 is exactly compensated by the opposite effect of pulse 2 and the resultant displacement is zero, as shown.

Diagram (c) shows the situation when the two pulses occupy exactly the same length of spring; the displacements are equal and opposite for all points between X′ and X‴ and therefore the spring is undisplaced, as shown in the lower diagram. Although the displacement is everywhere zero, the situation is not identical to that when there are no pulses at all, for the spring is moving transversely through the position of zero displacement. The energy of the pulses has been stored in the kinetic energy of the spring.

In diagram (d) we see that a few moments later the two pulses have reappeared and are continuing as though they had never met. Each of these stages can be seen in the series of

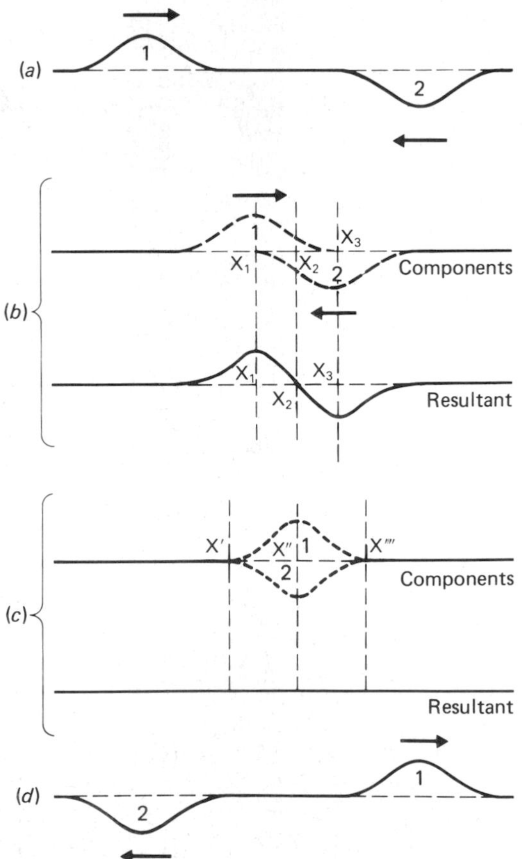

Fig. 4.3 The superposition of two equal, but opposite, pulses.

photographs in Fig. 4.4; in the seventh picture from the top they almost cancel each other out. Of course, if the pulses are not the same size, the cancelling effect is only partial but the resultant pulse produced is smaller than either of the constituent pulses.

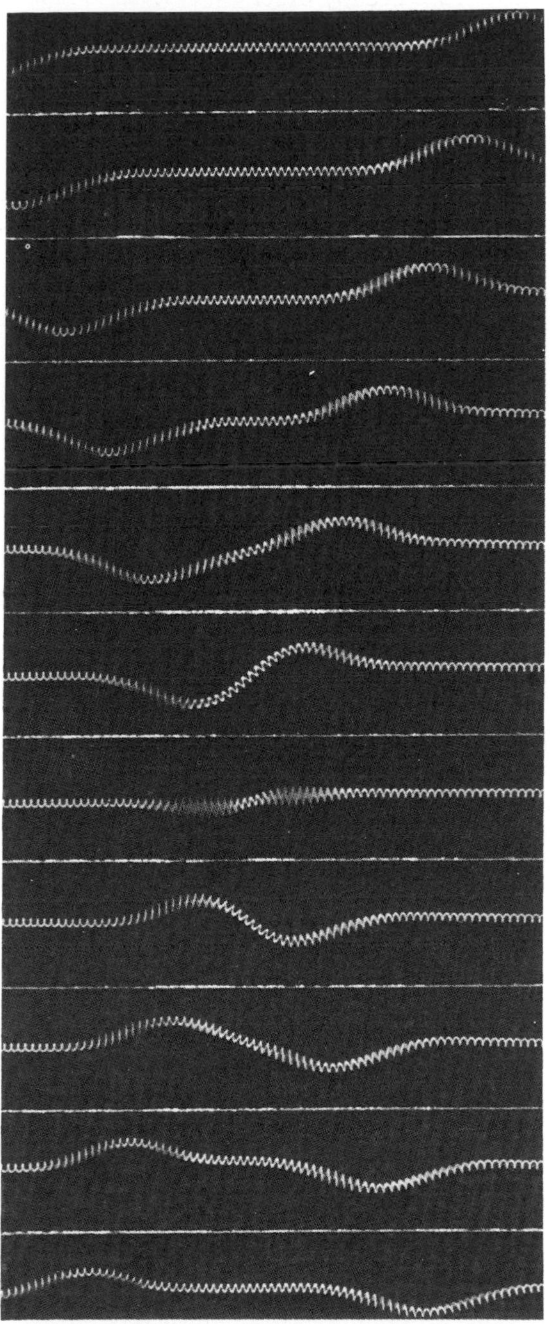

Fig. 4.4 Two equal and opposite pulses, travelling in opposite directions, cancel when they coincide.

Q 4.1
Draw two identical wave pulses, each consisting of one wavelength of a sine wave, moving towards each other along a rope so that two crests meet first. Construct a series of diagrams to show the resultant shape of the rope when:

(*a*) the two crests coincide;
(*b*) the crests coincide with the point of no displacement in the other wave;
(*c*) the crests combine with the trough in the other wave;
(*d*) the two troughs coincide.

Energy

The rate of energy flow per unit area in a wave is called its *intensity*; for a wave of given wavelength and velocity in a given medium we can show that:

$$\text{intensity} \propto (\text{amplitude})^2 \qquad (4.1)$$

We proved, on page 19, that this was also true of simple harmonic motion.

As discussed in Chapter 2, the amounts of energy associated with certain types of wave may be very large. The energy in one metre width of one wavefront of a sea wave 5 metres high, travelling at 7 metres per second, is about 1 000 000 joule; this is sufficient energy to raise a double-decker bus about 12 metres. If the amplitude of the wave is doubled, keeping all other parameters the same, there would be sufficient energy to raise the bus nearly 50 metres. The highest sea wave reliably recorded occurred in the Pacific in 1933 and was almost 40 metres high.

When two waves meet as described above, and produce zero resultant displacement for an instant, the energy is stored in the kinetic energy of the medium which is changing shape; the law of conservation of energy is still obeyed.

4.2 The superposition of wavetrains

In Section 4.1 we discussed how the superposition of pulses causes the transitory creation of either a larger or smaller pulse; regarding a wave as a series of identical and equally spaced pulses, we may apply the same principle to the meeting of two wavetrains as shown in Fig. 4.5. In each group of drawings the top two give the effect of each wave on its own and the bottom gives the total effect; we are considering wavetrains that are identical in every respect except their direction of travel.

Thus we see that at one instant the wavetrains combine to give a resultant that is larger than usual—this is called *constructive superposition*. On the other hand, the moments when the wavetrains are combining to give zero displacement are moments of *destructive superposition*. Sometimes these phenomena are called constructive and destructive *interference* respectively, and the reader should be familiar with these terms, for they are widely used. The term can be confusing because, as we have seen, the waves do *not* interfere but move through each other unaltered; at the place of meeting they each act as though the other wave were not present and the total displacement is obtained by adding together the two separate displacements.

Let us now consider two waves travelling in the same direction along a spring—difficult to achieve but a useful problem to think about. Again, we shall discuss identical waves travelling with the same velocity. It follows that if two crests coincide at one moment they will

(a) Two identical wavetrains approaching

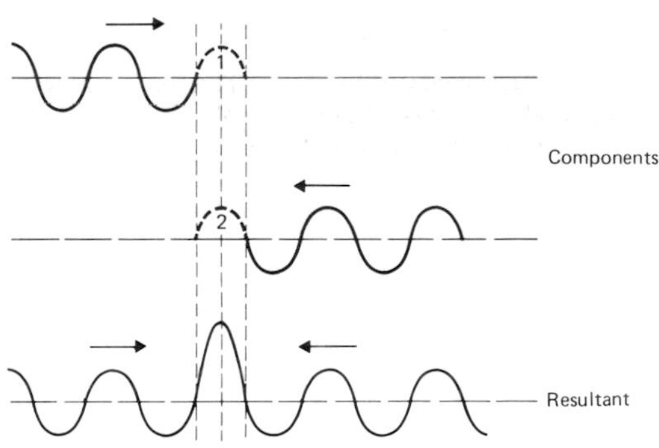

(*b*) Two identical crests coinciding to produce one disturbance of double amplitude.

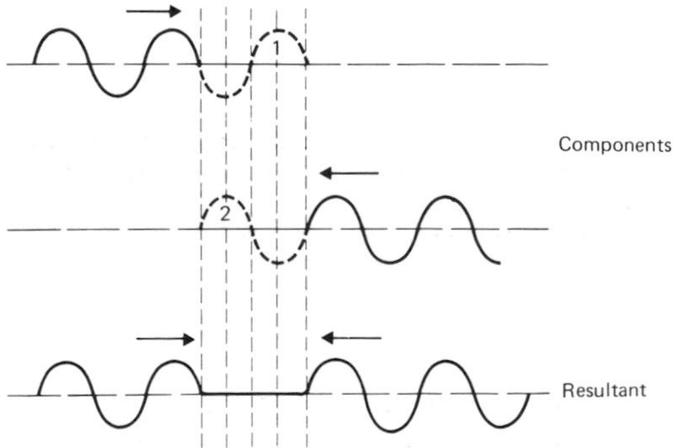

(*c*) The crest in (1) coincides with a trough in (2) and vice versa; in both cases a nil displacement is caused.

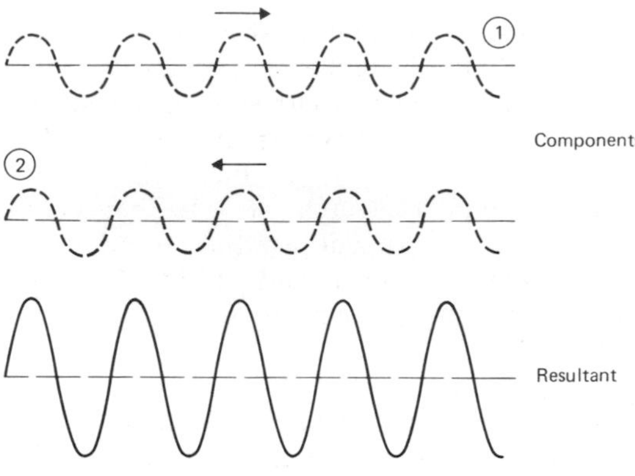

(*d*) All the crests in wavetrain (1) coincide with crests in wavetrain (2), and troughs with troughs; at this instant a wavetrain of double amplitude is formed.

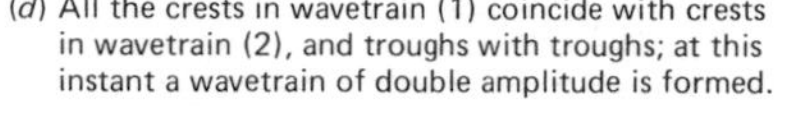

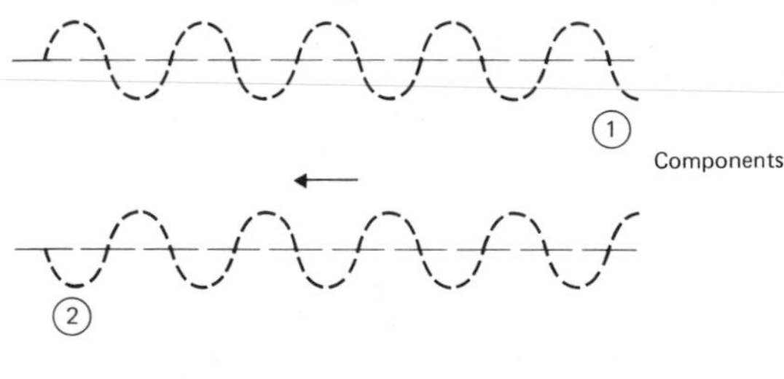

(e) All the crests in wavetrain (1) coincide with troughs in wavetrain (2), and vice versa; they therefore all cancel out and there is no displacement anywhere.

Fig. 4.5 Two coincident wavetrains moving in opposite directions.

always coincide as they move along together, and so will all the other crests and all the troughs. Under these circumstances, constructive superposition occurs permanently, and this is illustrated in Fig. 4.6(*a*). Similarly if one crest coincides with a trough at one particular moment, it will always do so; thus all the crests will coincide with the troughs producing permanent destructive superposition as shown in Fig. 4.6(*b*).

We have considered this problem, not because it can be realized easily in practice, but because it forms a useful introduction to the next section. The problem of applying the conservation of energy to the last example will be discussed at the end of Section 4.3.

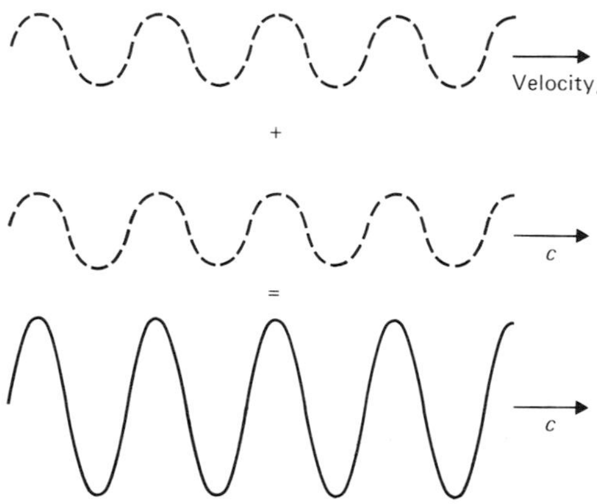

(*a*) Constructive superposition

(*b*) Destructive superposition

Fig. 4.6 Two coincident wavetrains moving in the same direction.

The mathematics of superposition

If $y_1 = A \sin 2\pi f(t - x/c)$ (2.2)

represents a progressive wave of amplitude A and frequency f moving with velocity c, then

$$y_2 = A \sin 2\pi f(t - t_0 - x/c)$$

represents an identical wave going the same way a time t_0 later.

The total effect, y, of these two waves is given by:

$$y = y_1 + y_2$$
$$= A \sin 2\pi f(t - x/c) + A \sin 2\pi f(t - x/c - t_0)$$

Now $\sin P + \sin Q = 2 \sin\left(\frac{P+Q}{2}\right) \cos\left(\frac{P-Q}{2}\right)$

from Appendix 2.

$$\therefore \quad y = 2A \sin 2\pi f(t - x/c - t_0/2) \cos 2\pi f(+t_0/2)$$
$$= 2A \cos(2\pi f t_0/2) \sin 2\pi f(t - x/c - t_0/2)$$
$$= A' \sin 2\pi f(t - x/c - t_0/2) \qquad (4.2)$$

where $A' = 2A \cos(2\pi f t_0/2)$

This represents a progressive wave of the same frequency whose amplitude depends on the value of t_0.

We may consider some typical values of t_0.

If $t_0 = 0$; $\quad 2\pi f\frac{t_0}{2} = 0$; $\quad \cos 0 = 1$; $\quad A' = 2A$

$t_0 = T = \frac{1}{f}$; $\quad 2\pi f\frac{t_0}{2} = \pi$; $\quad \cos \pi = -1$; $A' = -2A$

$t_0 = 2T = \frac{2}{f}$; $\quad 2\pi f\frac{t_0}{2} = 2\pi$; $\cos 2\pi = 1$; $\quad A' = 2A$

Thus, if the two waves are a whole number of periods, or wavelengths, out of phase, they produce constructive superposition.

If $t_0 = \frac{T}{2} = \frac{1}{2f}$; $\quad 2\pi f\frac{t_0}{2} = \frac{\pi}{2}$; $\quad \cos\frac{\pi}{2} = 0$; $A' = 0$

$t_0 = \frac{3T}{2} = \frac{3}{2f}$; $\quad 2\pi f\frac{t_0}{2} = \frac{3\pi}{2}$; $\quad \cos\frac{3\pi}{2} = 0$; $A' = 0$

Thus, if the two waves are an odd number of half periods, or wavelengths, out of phase, they produce destructive superposition.

These results are the same as those produced by graphical methods above.

Q 4.2

Two sinusoidal wavetrains, each of amplitude 0.1 m and frequency 0.5 Hz, travel in the same direction at a speed of 0.5 m s^{-1} through water; the source begins to emit the second wavetrain a time t_0 after the first. Calculate the total displacement a distance 2 m from the source 8 s after it began to emit the first wavetrain for the following values of t_0:

(*a*) $t_0 = 0$ s; (*d*) $t_0 = 1.5$ s;
(*b*) $t_0 = 0.5$ s; (*e*) $t_0 = 2.0$ s;
(*c*) $t_0 = 1.0$ s; (*f*) $t_0 = 1.25$ s.

Check your answers by drawing; it will be useful to calculate the wavelengths first.

4.3 The superposition of plane waves

We must now move on to consider the problem of the superposition of wavefronts. In Fig. 4.7 we have represented two identical wave motions crossing at right angles, wave 1 moving from bottom left to top right and wave 2 moving from bottom right to top left. Each consists of wavefronts such as MNO and XYZ; a wave motion consists of a succession of wavefronts and three are shown for each wave. As different superposition effects occur at different points on the wavefront, the displacements at three particular points on each wavefront, such as M, N and O, have been shown. When trying to visualize the wavefront, it is important to realize that a crest runs all the way along each one and does not exist merely at these three points. As the two sets of wavefronts move through each other, constructive and destructive superposition will occur at different places. In diagram (*a*) we can see that constructive superposition occurs between two crests at points A, C, G and I and produces double-height crests. On the other hand, constructive superposition also occurs at point E where two troughs combine to form a double-depth trough. Destructive superposition is seen at points B, D, F and H where a crest from one wavefront is cancelled

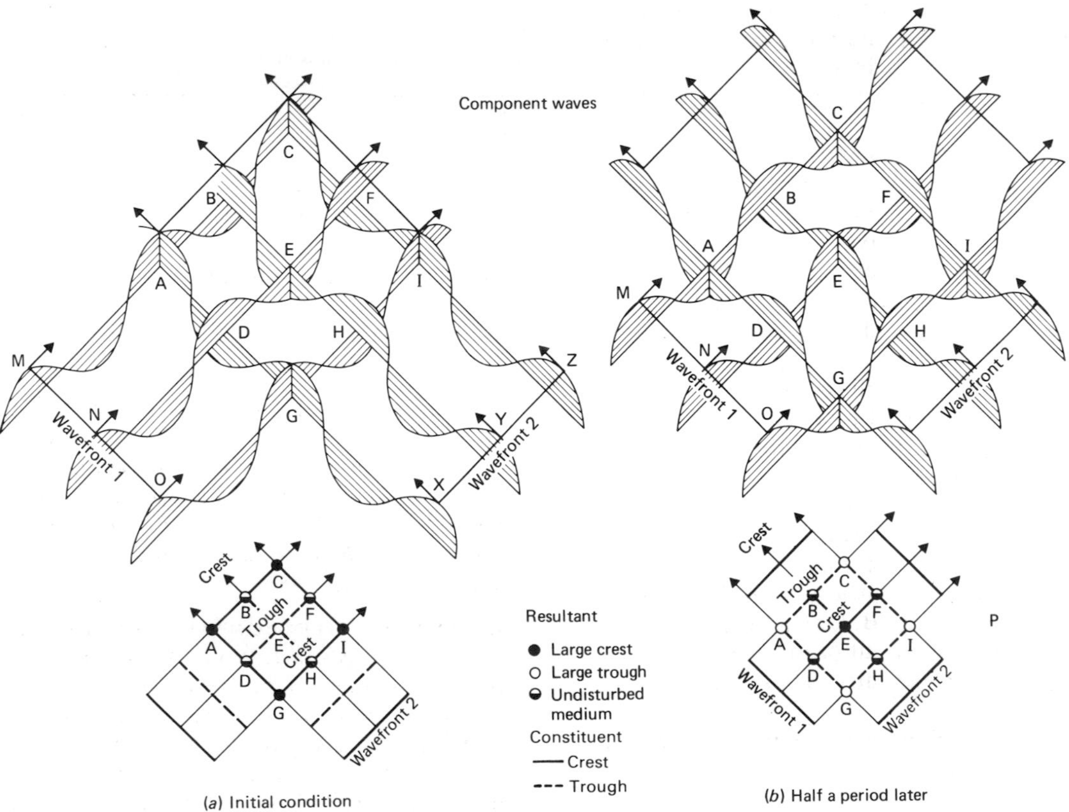

Fig. 4.7 Two identical wave motions crossing at right angles.

out by a trough from another and there is no overall displacement. These effects are marked on the plan diagram as well. At other points there will be neither complete constructive superposition nor complete destructive superposition.

Although all the points of largest crest and deepest trough in the area are marked there will be points, other than those marked, where there is no displacement; the reader may be able to identify such points.

Diagram (*b*) shows what happens when each wave has moved by half a wavelength; in other words, a time equal to half a period has elapsed. The points of coincidence of crests and troughs are now different. For instance, at point A there were previously two crests but now there are two troughs; previously a double crest would have been observed but now there is a double trough. Note, however, that at both times there is constructive superposition; the two waves are always in phase. Similarly at point E where there were previously two troughs forming a double-depth trough, there are now two crests forming a double-height crest. At both times there is constructive superposition; the waves are always in phase.

However, at points B and H, where formerly a trough from wave 1 coincided with a crest from wave 2, we now have a crest from wave 1 and a trough from wave 2; on both occasions destructive superposition occurs. The result is the same at points D and F where initially wave 1 produced a crest and wave 2 a trough, but now wave 1 produces the trough and wave 2 the crest; the result is still no displacement. The waves are always out of phase.

We can see, therefore, that there are points like A, C, E, G and I where waves always arrive in phase. The water (or other medium) at those points will therefore undergo an oscillation of twice the amplitude of that in the two constituent waves; this is illustrated in Figs. 4.8(*a*) and (*b*). At points like B, D, F and H, however, the waves are always out of phase and there is never any displacement; this is illustrated in Fig. 4.8(*c*).

It is now possible to re-define our terms a little more precisely.

Constructive superposition (interference) occurs at a point whenever the two or more waves incident at that point are exactly in phase so that the resultant amplitude of oscillation at the point is a maximum, equal to the sum of the amplitudes of the two or more waves; these points are called antinodes and lines along which all points are antinodes are termed antinodal lines.

Destructive superposition (interference) between two waves occurs at a point whenever the two waves incident at the point are exactly out of phase so that the resultant amplitude of oscillation at that point is a minimum; if the amplitudes of the two waves are equal the resultant amplitude will be zero. These points are called nodes and lines along which all points are nodes are termed nodal lines.

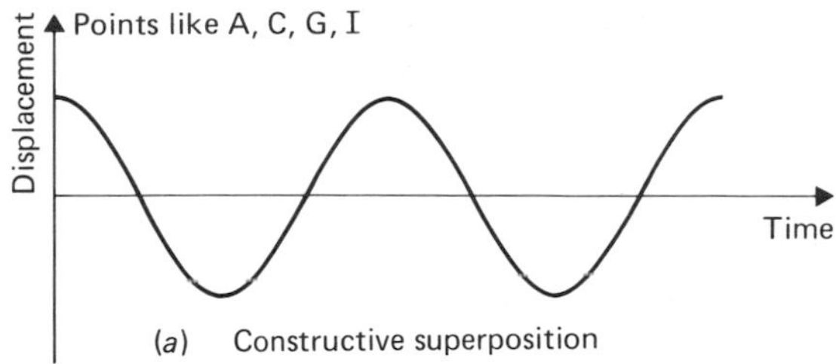

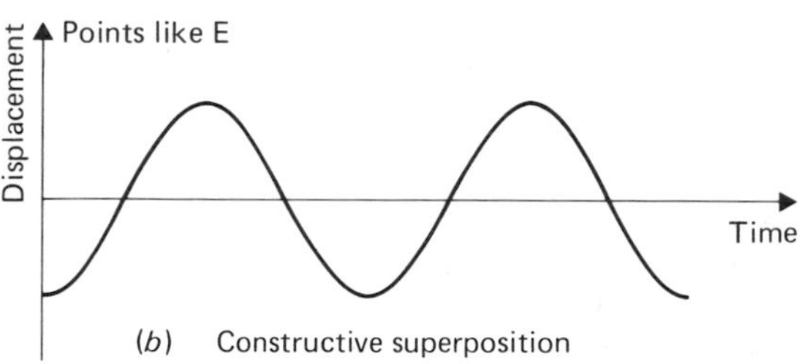

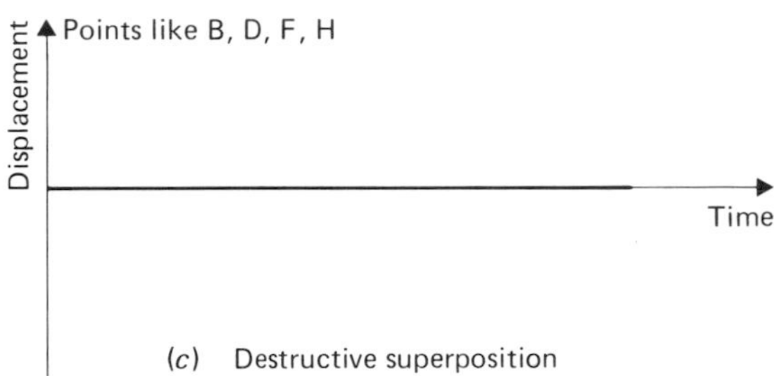

Fig. 4.8 Superposition effects in Fig. 4.7.

These stable states can only occur if the wavelengths and velocities of the waves are the same; cases where this is not so will be discussed later in the chapter. In a region where superposition is occurring, there will, of course, be many places where the type of superposition is intermediate between the two extremes cited above.

The superposition phenomena described earlier in this chapter were essentially transitory in nature, producing at a given point first constructive and then destructive superposition. In the situation just described, there are some points at which constructive superposition *always* occurs and some points at which destructive superposition *always* occurs. It is usual to refer to such *time-independent* applications of superposition as *interference phenomena* and we shall follow this practice from here onwards.

At the points where destructive interference occurs there is no energy stored in the wave; the energy appears at the points where there is constructive interference and the amplitude is thus increased. Accurate calculations show that all the energy is accounted for and the principle of the conservation of energy is upheld.

Fig. 4.9 A two-source interference pattern in a ripple tank.

Q 4.3
Use the 'plan form' of Fig. 4.7 to discover the position of crests and troughs:

(*a*) a quarter of the time between (*a*) and (*b*);
(*b*) half the time between (*a*) and (*b*).

Are the nodes and antinodes still in the same place?

4.4 The interference of circular waves

An example of interference which is easy to demonstrate in a ripple tank, and which is of particular importance in the study of light, is shown in Fig. 4.9. Each source is giving out circular waves of the same frequency; as the waves travel with the same velocity, they also have the same wavelength. The waves are emitted at exactly the same instant and are therefore said to be *in phase.* The large crests that are formed in the resulting pattern appear white in the photograph because they act like lenses and focus light into the camera.

We can see that there is a distinctive pattern of crests, and our work in the last section leads us to suggest that these large crests are observed wherever two of the smaller crests coincide. Similarly, if a trough coincides with a trough, an extra deep trough will be formed and where a crest coincides with a trough, there will be no displacement. The crest pattern may be built up by referring to diagram (*a*) in Fig. 4.10. The same symbols are used here as in Fig. 4.7 to indicate the double crests formed by the superposition of two crests, the double troughs formed by the superposition of two troughs and the regions of zero displacement where a crest and a trough coincide.

A short time later the wave crests will all have advanced a little and the double crests on the dotted line A_0 will have become double troughs and vice versa. However, there will still be constructive superposition at all points, such as X_{A_0}, along this line and it is termed an *antinodal line*. Other antinodal lines, A_1, A_2, and A_3, are marked on either side of A_0; all points on each of these lines also undergo maximum oscillation.

Point X_{N_1} on line N_1 is a point of no displacement because a crest from S_1 arrives simul-

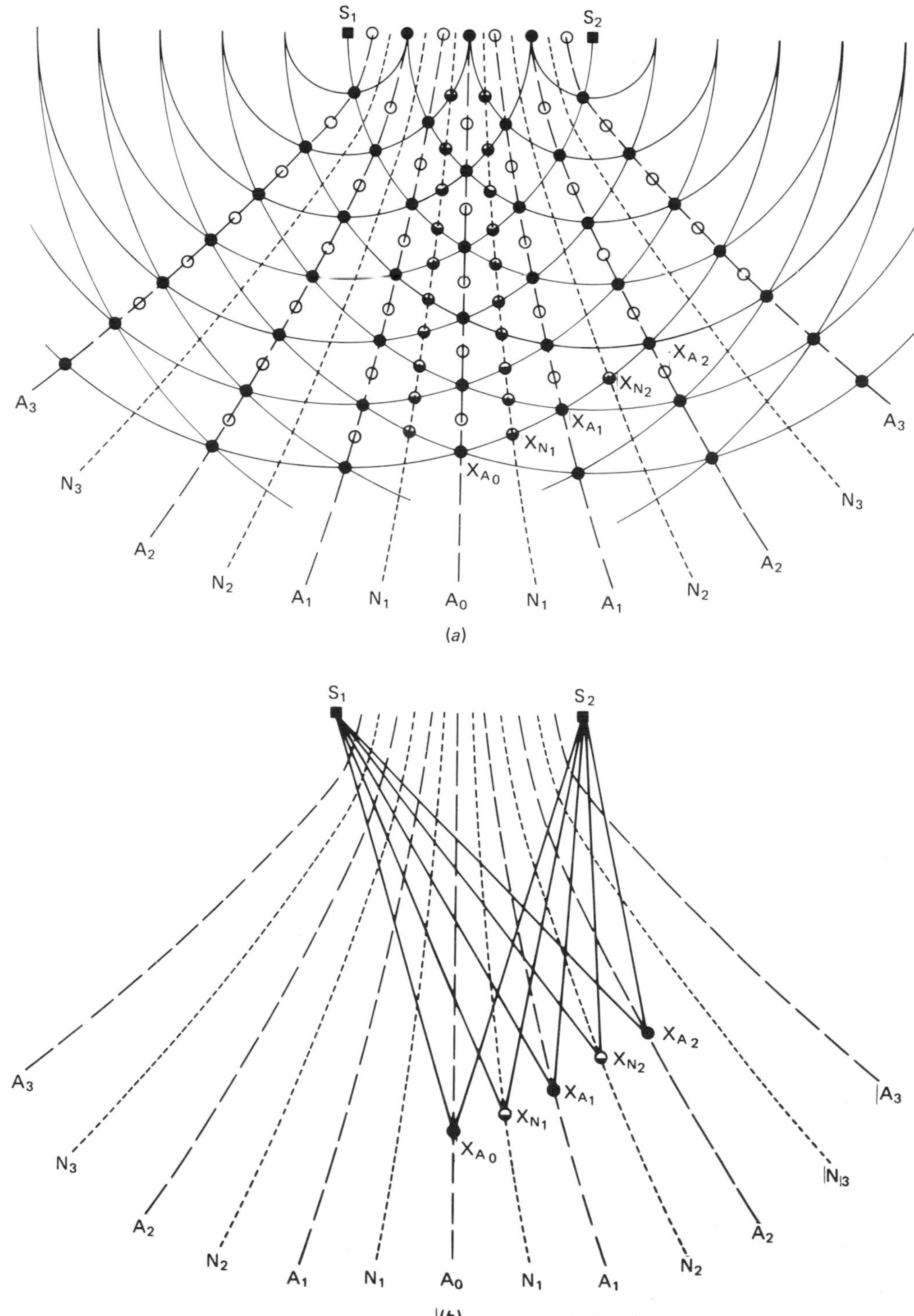

Fig. 4.10 The formation of the interference pattern due to two sources.

taneously with a trough from S_2. A short time later there will be a trough from S_1 but a crest from S_2; again there will be no displacement. This is true for all points on lines N_1, N_2, and N_3 and these are termed *nodal lines*. The reader will appreciate the close similarity between this situation and that described in the last section; again the pattern is *time-independent* and the phenomenon is referred to as *interference*.

Diagram (*b*) shows how these results may be understood in terms of *path differences*. Waves leave S_1 and S_2 at the same instant in phase and as they travel equal distances to points such as X_{A_0} on the antinodal line A_0 they must arrive in phase and therefore constructive interference must occur. However, all points that do not lie on A_0 are different distances from S_1 and S_2 and the waves may not *arrive* in phase, even though they were *emitted* in phase. If the waves arrive at some other point a whole number of periods out of step, constructive interference will occur, but if they are an odd number of half periods out of step, destructive interference will occur instead.

The difference in time taken by the waves to travel from S_1 and S_2 is determined by the difference in distance, or *path difference*. X_{N_1} is a point on nodal line, N_1, and the path difference at points such as X_{N_1} is given by:

$$S_1X_{N_1} - S_2X_{N_1} = \tfrac{1}{2}\lambda$$

The next nodal line is N_2 and at points such as X_{N_2} we have:

$$S_1X_{N_2} - S_2X_{N_2} = \tfrac{3}{2}\lambda$$

The condition for destructive interference is that the path from S_1 should be $\frac{1}{2}$, $1\frac{1}{2}$, $2\frac{1}{2}$, $3\frac{1}{2}, \ldots$ wavelengths longer than that from S_2, and the phase of the waves from S_1 will then be π, 3π, 5π, $7\pi, \ldots$ radians behind the phase of the waves from S_2.

In general the condition for a nodal line is given by:

$$S_1X_N - S_2X_N = (n + \tfrac{1}{2})\lambda \qquad (4.3)$$

where n is any whole number or zero. Nodal lines are situated symmetrically either side of A_0.

The smallest non-zero path difference to give constructive interference must be one wavelength and therefore at points such as X_{A_1} on A_1 we have:

$$S_1X_{A_1} - S_2X_{A_1} = \lambda$$

X_{A_2} also lies on an antinodal line and the path difference must be two wavelengths:

$$S_1X_{A_2} - S_2X_{A_2} = 2\lambda$$

In general, the condition for constructive interference is that the path difference must be a whole number of wavelengths:

$$S_1X_A - S_2X_A = n\lambda \qquad (4.4)$$

where, again, n is a whole number. Antinodal lines are situated symmetrically either side of A_0; the central line A_0 is, of course, also an antinodal line with $n = 0$.

Predicting the resultant amplitude at points between the nodal and antinodal lines is much more complicated. However, the results derived above are sufficient to solve a variety of wave problems; although we have discussed water waves, the results are equally applicable to other forms of wave motion, such as electromagnetic waves or sound. In Chapter 7, we shall discuss them in more detail when applied to optical phenomena.

The conditions for interference are quoted below for future reference:

constructive interference	$S_1X_A - S_2X_A = n\lambda$	(4.4)
destructive interference	$S_1X_N - S_2X_N = (n + \frac{1}{2})\lambda$	(4.3)

where n is any whole number, including zero.

Q 4.4
Redraw Fig. 4.10 at a time $T/4$ and $T/2$ later and show that the positions of nodal and antinodal lines are unchanged.

4.5 The reflection of transverse waves

It is important to consider what happens to a pulse when it reaches a different medium, for example the end of a stretched spring. Fig. 4.11 shows a pulse being reflected from a spring end which is fixed. The pulse is reflected upside down; it is, of course, also travelling in the opposite direction.

It is difficult to show experimentally what happens when a pulse is reflected from an end that is not attached to anything, but we can attach a heavy spring to a very light thread, as in Fig. 4.12. The pulse is still reflected but this time it is not inverted.

The principle of superposition helps us to understand why the pulse reflected from a fixed end is upside down. There are two pulses on the spring—the incident pulse and the reflected pulse—and both must be at the fixed end simultaneously. The fixed end does not move and so the two pulses must cancel out there. Now Fig. 4.3 showed that the motion due to a pulse on a spring is cancelled out by an identical, but inverted, pulse coming from the opposite direction. Thus the reflected pulse at a fixed end must be identical but inverted. The mechanical explanation of the inverted pulse derives from the reaction experienced by the spring when the pulse reaches the fixed end and exerts a force on the support; the support will exert an equal but opposite force on the spring and cause an inverted pulse to be reflected.

If the end of the spring is not fixed then it will, of course, move up and down when the incident pulse arrives; its motion will be in phase with that pulse. Thus it can only act as the source of a pulse that is not inverted and the pulse is reflected unchanged.

A wave is a succession of pulses and, therefore, is reflected in the same way as a pulse. If the end is fixed the displacement of the wave is reversed; if it is not fixed the displacement is unchanged.

In this discussion of transverse waves we refer to free and fixed 'ends' because the particular case we are considering is that of a linear spring. In extended systems, such as those involving water waves or light waves, it is easier to talk about *boundaries*, and this is the more usual scientific term.

The mathematics of the reflection of transverse waves

Let us consider a progressive sinusoidal wave moving with a velocity c towards a fixed end of a spring. Its displacement y_i will be given by:

$$y_i = A \sin 2\pi f(t - x/c) \qquad (2.2)$$

where f is the frequency, and A the amplitude, of the wave.

The discussion above tells us that the reflected wave is inverted, and that it is also travelling in the opposite direc-

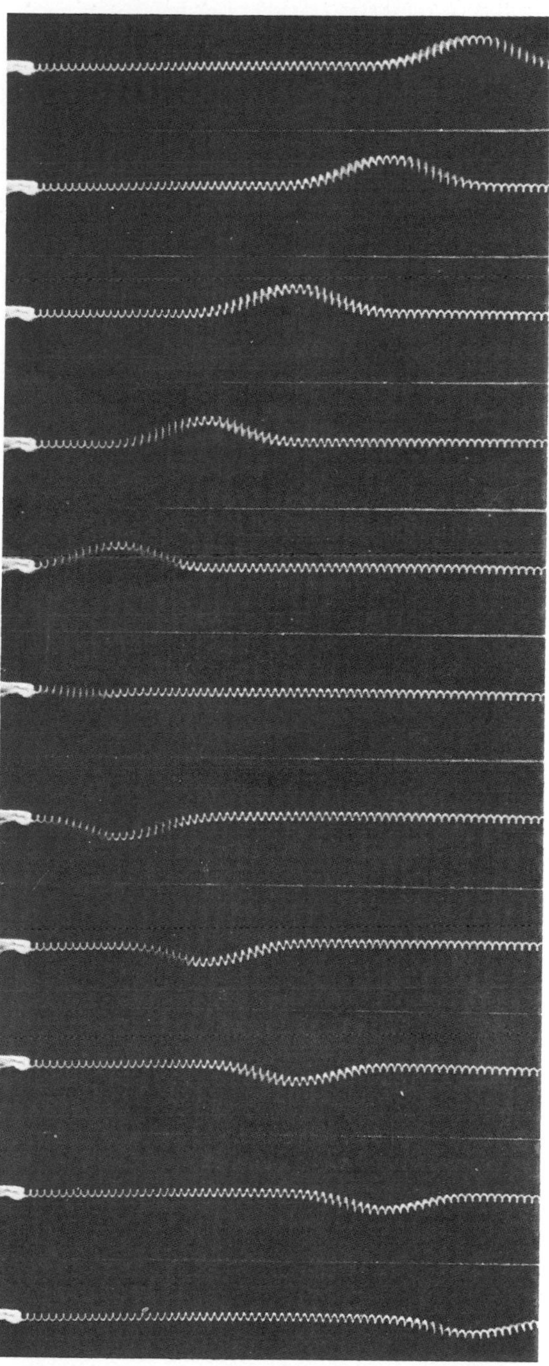

Fig. 4.11 A pulse reflected from the fixed end of a spring is inverted.

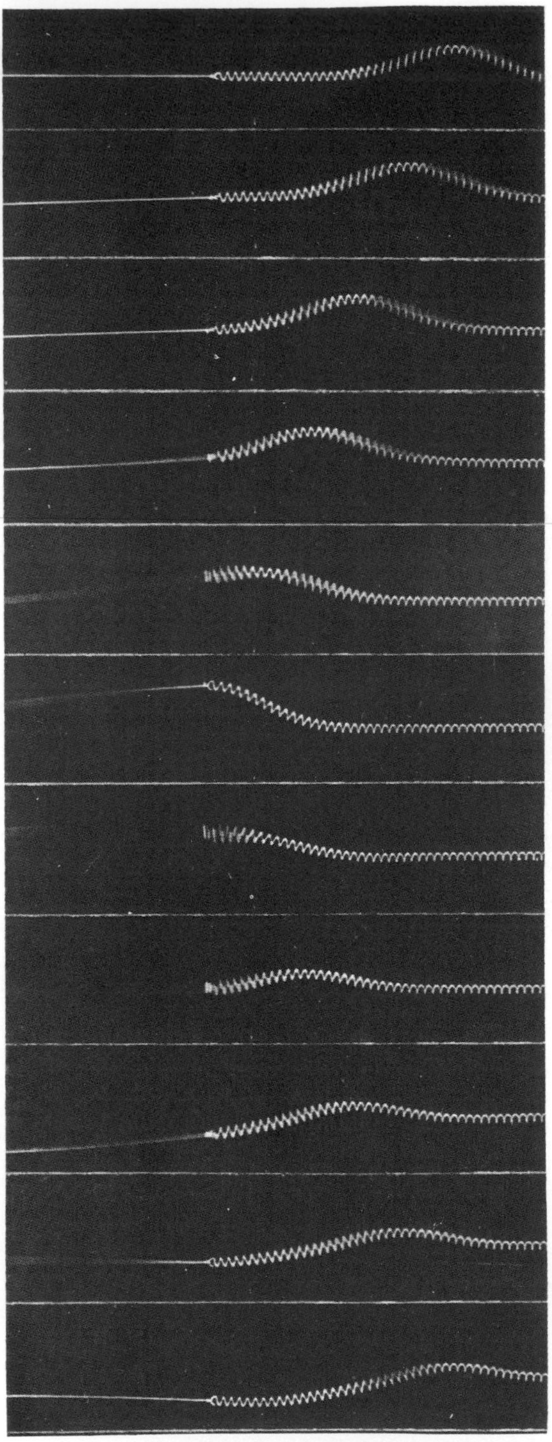

Fig. 4.12 The light thread provides an almost 'free' end to the spring and the pulse is reflected without being inverted.

tion. Hence, from Equation (2.24), the displacement of the reflected wave, y_r, will be:

$$y_r = -A \sin 2\pi f(t + x/c) \qquad (4.5)$$

If, on the other hand, the wave is reflected at a free end the direction of propagation will be reversed, but the displacement will not be inverted; the displacement in this case, y'_r, is:

$$y'_r = A \sin 2\pi f(t + x/c) \qquad (4.6)$$

4.6 Transverse stationary waves

We have seen in Section 4.5 that when waves reach the end of a stretched spring along which they are travelling, they are reflected; their frequency, and often their amplitude, is unchanged. Subsequent reflections at each end alternately give rise to waves travelling in opposite directions along the spring and they must therefore superpose. In general, the resulting motion is complicated and progressive waves can occasionally be identified moving one way or the other; however, at certain frequencies a transverse oscillation takes place without any obvious progressive nature. These waves are termed *stationary* or *standing* waves, as opposed to progressive waves.

The motion of the spring during the period of a typical stationary wave is shown in Fig. 4.13; for comparison a progressive wave of equal wavelength and frequency is also shown. The following comparisons can be made between the two types of waves:

(1) For waves that are simple harmonic both types of waves have clearly definable wavelength and frequency, but a stationary wave has no direction of motion or velocity. Thus, energy cannot be transmitted through the medium, but may be stored in it.

(2) All points on a progressive wave oscillate with the same amplitude if there is no energy loss in the medium. However, in a stationary wave, points such as N, called *nodes*, undergo no oscillation at all, whereas points such as A, called *antinodes*, undergo maximum oscillation.

(3) In a progressive wave all points undergo the same oscillation at a slightly different time from their immediate neighbours; this constantly changing phase means that the wave moves along. In a stationary wave, all the points on the wave in the same half wavelength (X_1X_2) are in phase with each other, but exactly out of phase with the points on the wave in the next half wavelength (X_2X_3); this implies that the wave is stationary since there is no *continuous* change of phase with distance.

It is found by experiment that for a given length of spring, of given tension and mass, there is a set of discrete stationary waves which can be set up; the lowest frequency is called the *fundamental* and the higher frequencies are integral multiples of it. The oscillations can only be maintained if the waves reflected at the source end are exactly in phase with those being sent out by the source itself.

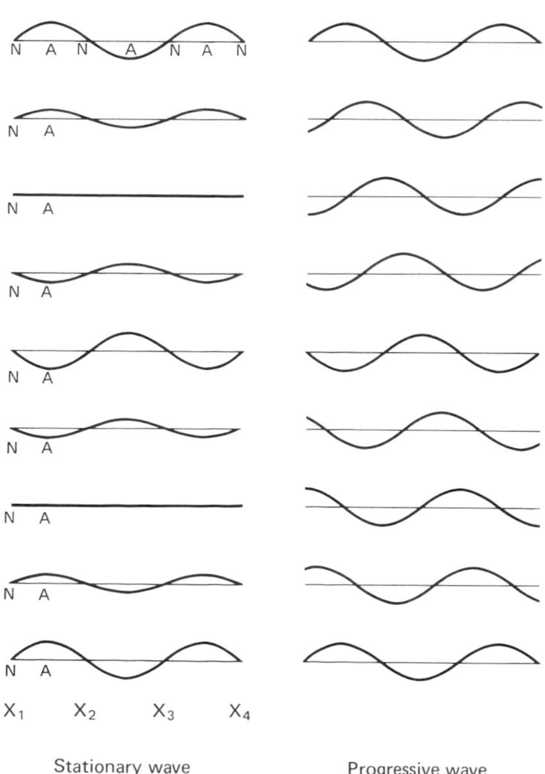

Fig. 4.13 Stationary and progressive waves compared.

Suppose that the spring is of length l, fixed at both ends, and that waves travel along it with speed c. The time taken for the waves to travel to the far end of the spring and back will be $2l/c$, and will be independent of the frequency, f. If the wave that has been reflected at the source is to be in phase with a wave just newly created, the time taken, $2l/c$, must be a whole number of vibration periods, T. We therefore have, where n is any whole number, and λ the wavelength:

$$\frac{2l}{c} = nT$$

$$= n \times \frac{1}{f}$$

$$= n\frac{\lambda}{c}$$

$$\therefore \quad 2l = n\lambda$$

$$\text{i.e.} \quad l = n\frac{\lambda}{2} \qquad (4.7)$$

Thus the length of the spring must be equal to any whole number of half wavelengths if a stationary wave is to be set up as shown in Fig. 4.14.

If the length of the spring is fixed we can calculate the wavelengths of possible stationary waves that can be created, and hence their frequencies:

$$\lambda = \frac{2l}{n} \qquad \text{from (4.7)}$$

$$f = \frac{c}{\lambda}$$

$$\text{i.e.} \quad f = n\frac{c}{2l}$$

$$\text{But} \quad c = \sqrt{\frac{T}{\mu}} \qquad \text{from (2.14)}$$

$$\therefore \quad f = \frac{n}{2l}\sqrt{\frac{T}{\mu}} \qquad (4.8)$$

The fundamental frequency, f_0, is $c/2l$ and, as stated above, the higher possible frequencies ($2c/2l$, $3c/2l, \ldots$) are all integral multiples of the fundamental; a series of frequencies like this is called a *harmonic series*, and we refer to the higher frequencies as the *second harmonic* (twice the frequency of the fundamental), the *third harmonic* (three times the frequency), and so on. The fundamental is the *first harmonic*.

Sometimes these higher notes are referred to as the *first overtone*, *second overtone*, . . . ; this, however, can lead to confusion and on the whole we shall avoid the term *overtone*.

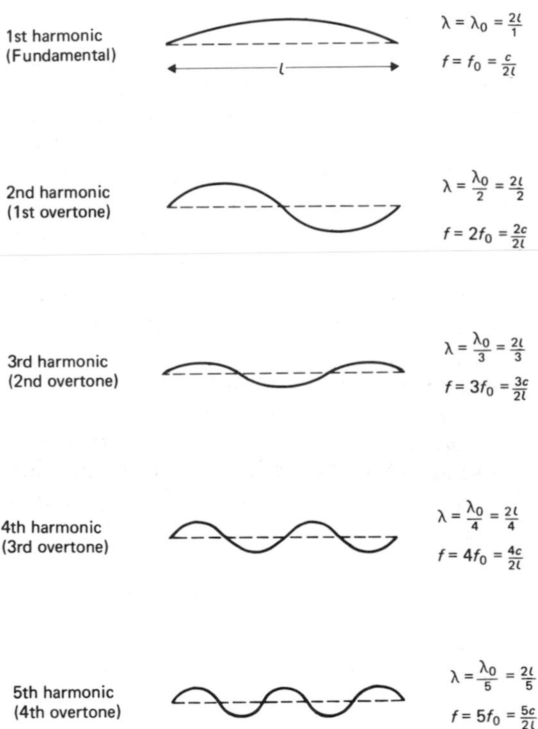

Fig. 4.14 Stationary waves on a stretched spring.

Q 4.5
On a piece of graph paper, draw two progressive waves of the same amplitude and wavelength immediately underneath each other with crest above crest and trough above trough. By adding their amplitudes, construct the resultant shape of the spring on which these waves were moving at that instant. Now draw each progressive wave again, each displaced an eighth of a wavelength in opposite directions; again construct the shape of the spring by adding amplitudes. Repeat a number of times. It should be possible to see that when two identical progressive waves move in opposite directions through a medium, a stationary wave is formed; *this is a most important exercise.*

The mathematics of transverse stationary waves

Suppose the equation of the progressive wave sent out along the spring is:

$$y_i = A \sin 2\pi f(t - x/c) \qquad (2.2)$$

Then since the ends are fixed, the reflected wave, which is travelling in the opposite direction, will have a reversed displacement:

$$y_r = -A \sin 2\pi f(t + x/c) \qquad (4.5)$$

By the principle of superposition, the resultant displacement of the spring is given by the sum of these two:

$$\begin{aligned} y &= y_i + y_r \\ &= A \sin 2\pi f(t - x/c) - A \sin 2\pi f(t + x/c) \\ &= -A\,[\sin 2\pi f(t + x/c) - \sin 2\pi f(t - x/c)] \end{aligned}$$

Now $\quad \sin P - \sin Q = 2 \cos \frac{1}{2}(P + Q) \sin \frac{1}{2}(P - Q)$
from Appendix 2.

$\therefore \quad y = -2A \cos 2\pi ft \sin 2\pi fx/c$

But $c = f\lambda$

and, therefore, rearranging we have:

$$y = -2A \sin \frac{2\pi x}{\lambda} \cos 2\pi ft \qquad (4.9)$$

In Section 2.10 we saw that equations for progressive waves always contain the term $(t - x/c)$, or one derived from it. Thus Equation (4.9) cannot be the equation of a progressive wave as the variables t and x appear in different terms. In fact the equation tells us that a particle at a distance x from the end of the spring executes simple harmonic motion with frequency f and amplitude $2A \sin 2\pi x/\lambda$. Thus, the wave shape does not, as in a progressive wave, move along but remains stationary with each point undergoing simple harmonic motion with an amplitude determined by its position.

As the spring is fixed at each end, there must be a node, or point of zero displacement, there. Thus, from Equation (4.9), at $x = 0$ and $x = l$ we must have:

$$2A \sin \frac{2\pi x}{\lambda} = 0$$

This is clearly satisfied when $x = 0$ but at $x = l$ we must also have:

$$2A \sin \frac{2\pi l}{\lambda} = 0$$

This condition is satisfied if:

$$\frac{2\pi l}{\lambda} = n\pi$$

where n is any whole number.

Hence

$$\begin{aligned} \lambda &= \frac{2\pi l}{n\pi} \\ &= \frac{2l}{n} \end{aligned}$$

i.e.

$$l = n\frac{\lambda}{2} \tag{4.7}$$

as shown before. These possible wavelengths give the frequencies of the fundamental and higher harmonics:

$$f = \frac{n}{2l}\sqrt{\frac{T}{\mu}} \tag{4.8}$$

Thus the mathematical analysis gives the same results as those proved earlier.

4.7 The reflection of longitudinal waves

We will now consider the reflection of longitudinal waves at fixed and free boundaries; as with transverse waves we will discuss the situations first and then analyse them mathematically. In the next section we shall apply these results to examples of longitudinal stationary waves, such as occur in wind instruments.

Fixed boundary

At a fixed boundary, such as the closed end of a pipe, there cannot, by definition, be any displacement. Therefore, the displacement that would have been caused by the incident wave must have been cancelled by the superposition of the reflected wave; the displacement in the reflected wave must be opposite to that in the incident wave, and of equal amplitude. This situation is similar to the reflection of a transverse wave at a fixed boundary; as there is no displacement, there will be a displacement node in the resulting stationary wave.

A longitudinal wave can also be regarded from a compression point of view, as explained in Section 2.4. We stated there that a displacement node is always accompanied by a compression antinode. We might, in any case, have expected that there would be a large change in compression against a fixed boundary, from our experience standing in a crowd pressed against a wall. As there is a compression antinode, the incident wave is not cancelled by the reflected wave at the boundary and thus the compression changes are not reversed on reflection.

Free boundary

At a free boundary the situation is different because there is no hindrance to displacement and thus a displacement antinode occurs. The incident displacement wave is not cancelled at the boundary and thus the wave is reflected without the displacement being reversed.

As there is a displacement antinode there must be a compression node; in order to cancel the incident compression wave at the boundary the sense of the pressure change in the reflected wave must be reversed.

Table 4.1 summarizes the results discussed above; despite the fact that we refer so much to wind instruments, it is important to remember that the reflection of longitudinal waves occurs in solids and liquids as well as in air.

The mathematics of the reflection of longitudinal waves

Let us suppose that the incident displacement wave is represented, as before, by:

$$y_i = A \sin 2\pi f(t - x/c) \tag{2.2}$$

where in this case y_i represents the displacement *in the direction of propagation of the wave*, since it is a longitudinal wave.

When the wave is reflected from a fixed boundary the displacement and direction of propagation are reversed:

$$y_r = -A \sin 2\pi f(t + x/c) \tag{4.5}$$

Table 4.1 The reflection of longitudinal waves

	Displacement wave		*Compression wave*	
	Reflected wave	Behaviour at end	Reflected wave	Behaviour at end
Fixed boundary	displacement reversed	node formed	sense of compression change not reversed	antinode formed
Free boundary	displacement not reversed	antinode formed	sense of compression change reversed	node formed

When the reflection is from a free boundary only the direction of propagation is reversed:

$$y_r' = A \sin 2\pi f(t + x/c) \qquad (4.6)$$

These equations are, of course, identical to those obtained for transverse waves.

The incident compression wave is $\pi/2$ radian (90°) out of phase with the incident displacement wave, as shown in Section 2.4; if A_P is the amplitude of compression changes, and P_i the compression of the incident wave, we may write:

$$P_i = A_P \cos 2\pi f(t - x/c) \qquad (2.26)$$

When the wave is reflected from a fixed boundary there are large compression oscillations, and therefore only the direction of propagation of the wave must have been reversed:

$$P_r = A_P \cos 2\pi f(t + x/c) \qquad (4.10)$$

When the reflection is from a free boundary there are no compression oscillations at the boundary and thus the sense of the compression changes has reversed as well as the direction of propagation:

$$P_r' = -A_P \cos 2\pi f(t + x/c) \qquad (4.11)$$

The superposition of these waves into stationary waves is discussed in Section 4.8.

4.8 Longitudinal stationary waves

Longitudinal stationary waves are formed in exactly the same way as transverse stationary waves; that is, from two waves of the same frequency moving in opposite directions through the same medium. As usual, when discussing longitudinal waves, we must consider two aspects—the compression wave and the displacement wave. Although the examples given relate to waves in air columns, the results are applicable to any situation involving longitudinal stationary waves.

A tube open at both ends

Let us consider the stationary waves that may be set up in an air column open at both ends, for instance an *open organ pipe* with air blown in at the base and an unobstructed top. As explained in Section 4.7, there will be a displacement antinode and pressure node at each end. The lowest frequency that can be set up in this tube occurs when the wavelength is twice the length of the tube, and the fundamental frequency, f_0, is thus $c/2l$, as shown in Fig. 4.15. Additional stationary waves of higher

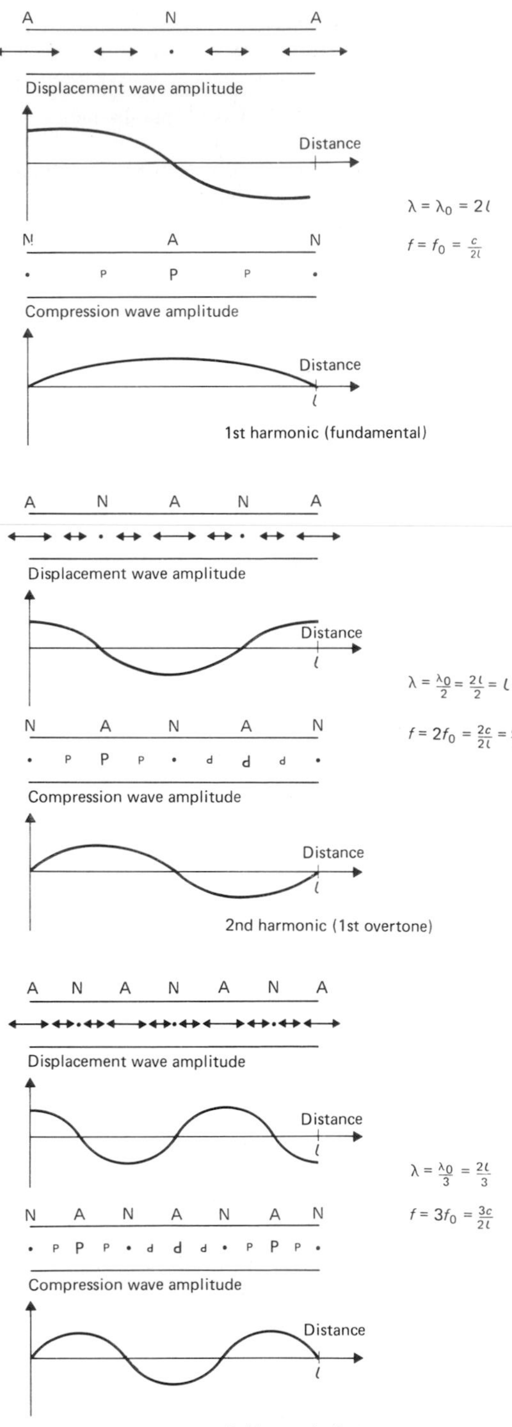

Fig. 4.15 Longitudinal stationary waves in a pipe open at both ends.

frequency, and shorter wavelength, can also be set up and we see that these are the 2nd, 3rd, 4th ... *harmonics*, with frequencies $2f_0$, $3f_0$, $4f_0$ .... In each case, the amplitudes of the variation in displacement and pressure are shown at various points in the tube; the oscillatory motion of the air particles is, of course, along the length of the tube.

A tube closed at one end

Another type of air pipe is closed at one end and open at the other. *Closed organ pipes* have air blown in at the base, as before, but have the top of the pipe filled in. At the open end there will still be a displacement antinode and pressure node, but at the closed end there will be a displacement node and pressure antinode. The various stationary waves possible with this type of tube are shown in Fig. 4.16. The fundamental frequency is half that for a pipe open at both ends, since the tube length is equal to a quarter wavelength, and thus the fundamental wavelength is twice as great. Only odd-numbered harmonics are possible, the 1st overtone being the 3rd harmonic, the 2nd overtone the 5th harmonic, and so on.

It is necessary to point out in closing that, in practice, no situation is as simple as is implied above. Fig. 4.12 shows that some energy is transmitted through the boundary into the light thread and, in the same way, an open end will transmit some energy into the surrounding air; it follows that the displacement antinode there is not quite perfect. Similarly a fixed end does not form a perfect boundary, as sound can be transmitted through it. In musical instruments, these 'imperfections', often determined by the detailed construction, determine the range of harmonics that actually occur, and hence the quality of the sound that is produced. The pitch of the note is determined by the frequency of the fundamental, and the tone by the mixture of harmonics that also occur.

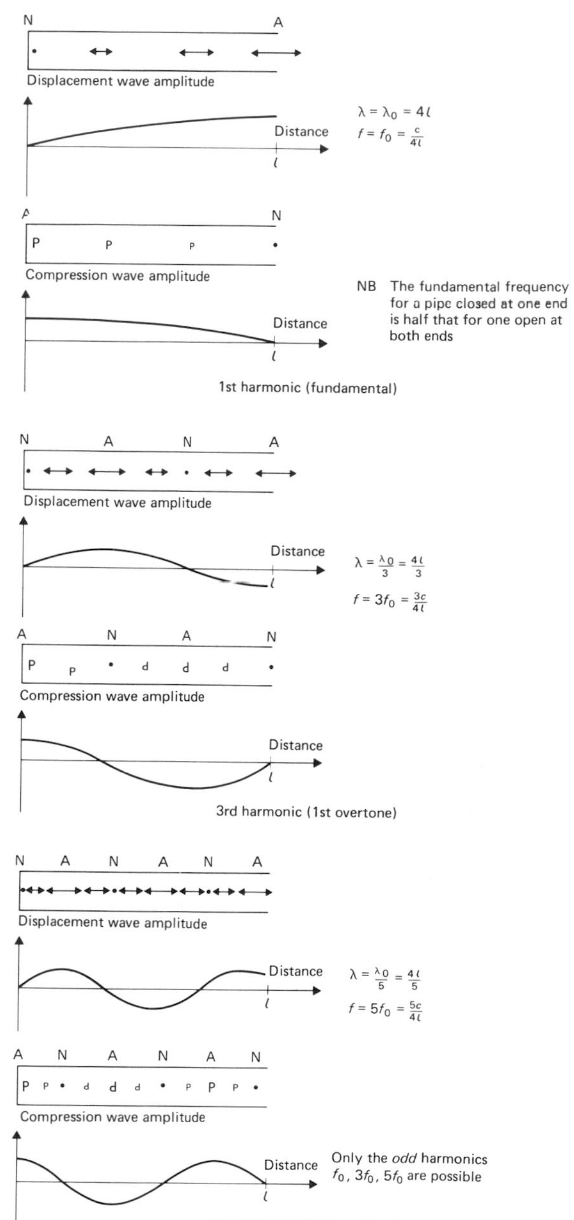

Fig. 4.16 Longitudinal stationary waves in a pipe closed at one end.

Q 4.6
Mathematical equations for the longitudinal stationary waves set up by reflection in pipes can be derived in the same way as the equations for transverse stationary waves in Section 4.6. Use the equations in Section 4.7, together with the appropriate trigonometrical relationships, to find equations for the stationary waves set up in the types of pipe considered. (This question is only suitable for those who have been following the mathematical analyses.)

4.9 Applications of stationary waves

The reader will recall the discussion of resonance in Section 1.6 where single-mass systems, such as the simple pendulum, were found to have one particular frequency at which they could be made to vibrate with a large amplitude. We now see that for systems involving distributed mass—such as a taut wire, or a column of air—there are a whole range of specific frequencies at which 'oscillations' or stationary waves can occur. The lowest frequency is termed the *fundamental* and the higher frequencies *overtones*; in simple cases these are in a harmonic series but in the more complex cases this is not usually so. We term each frequency of vibration, with its own particular stationary wave pattern, a *mode* of vibration; thus Fig. 4.14 shows five modes of vibration for a spring of given length and tension while Fig. 4.15 shows three modes of vibration for a column of air open at both ends. Each mode must 'fit into' the length or other dimensions of the situation; the restrictions caused by this are termed *boundary conditions*. In practice, a number of modes will be excited simultaneously and the resulting stationary wave is the superposition of each of these modes. As with the resonant oscillations of a point mass, if the length of spring is vibrated at a different frequency to that of one of its modes, smaller and less regular motions take place.

Galloping stays

An example of transverse stationary waves is provided by the vibration of telephone wires or mast stays in a wind. The notes, of characteristic frequency, are produced by the setting up of eddies in the moving air stream.

If we move a finger rapidly through a fluid, such as water, small eddies or vortices are set

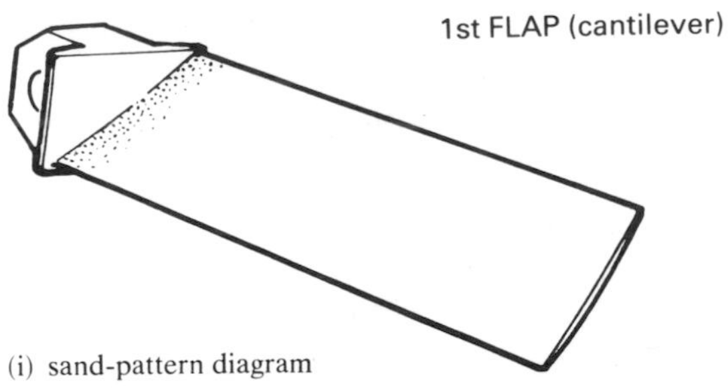

(i) sand-pattern diagram

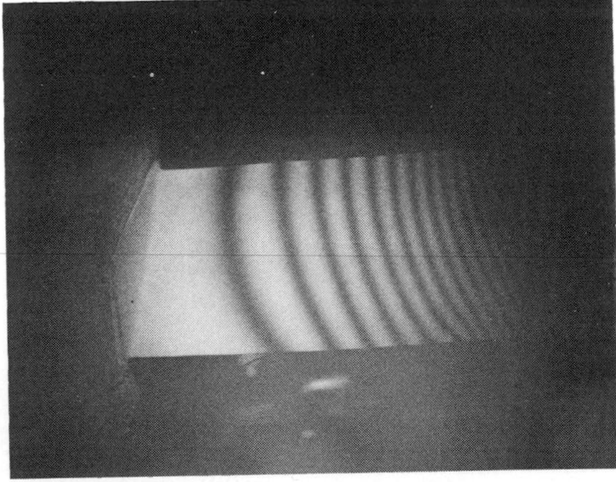
(ii) hologram

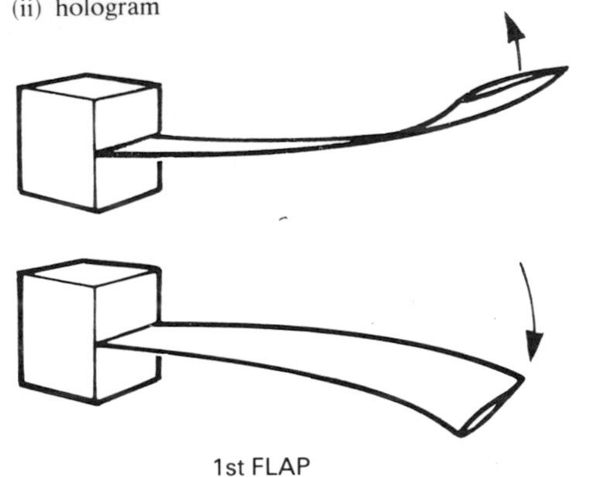

(iii) blade bending mode

(iv) simplified nodal pattern

1st FLAP

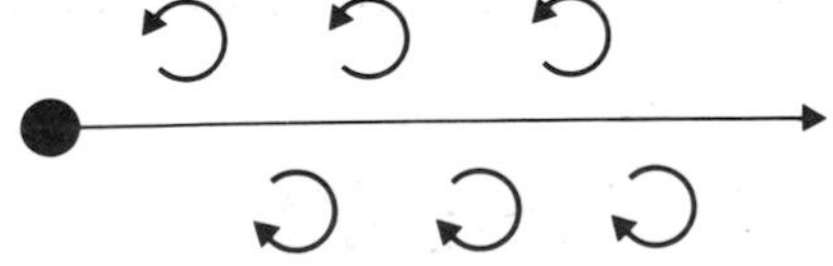
Fig. 4.17 The formation of eddies.

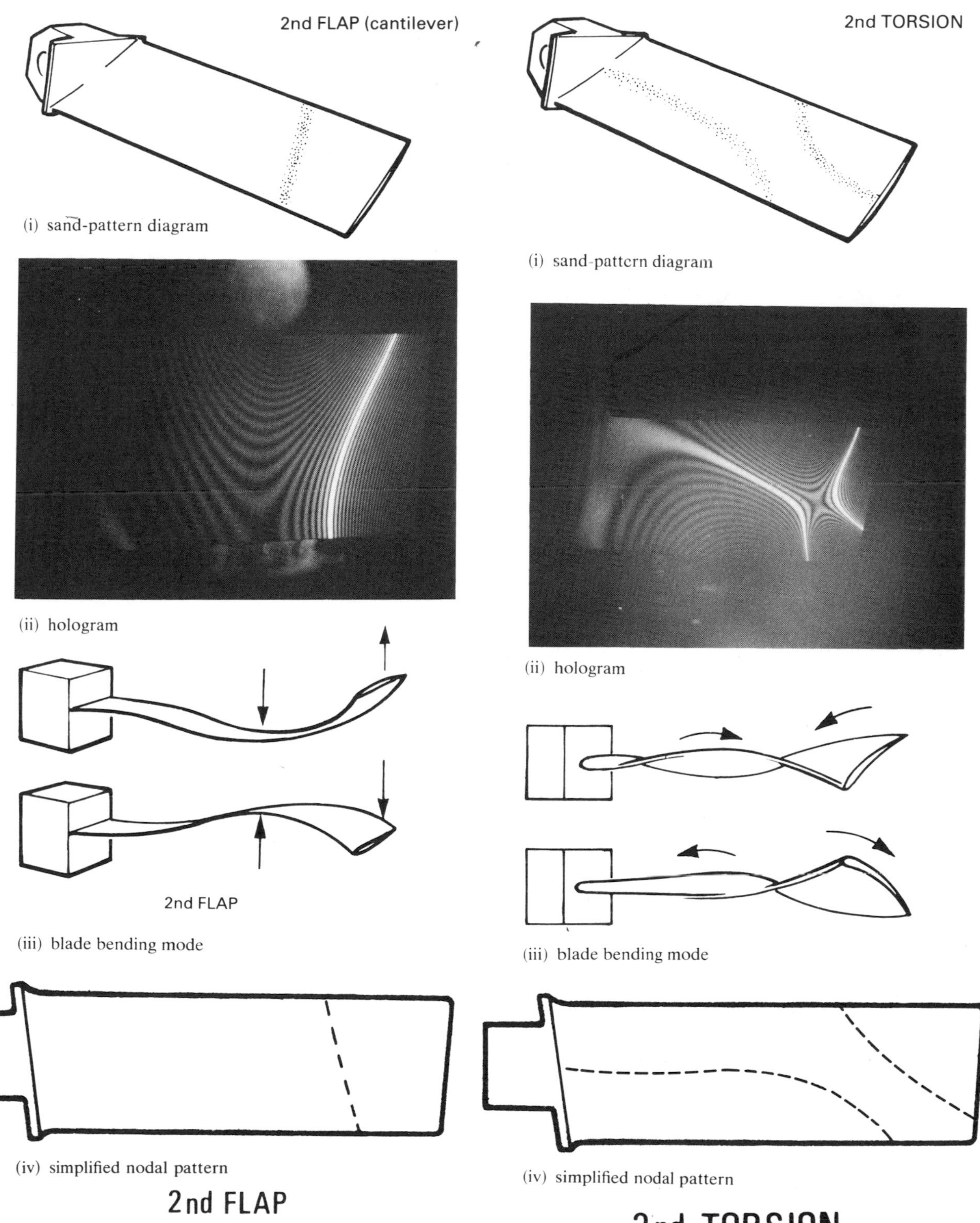

Fig. 4.18 Some possible modes of oscillation for jet turbine blades. (Material by courtesy of *Rolls-Royce Limited.*)

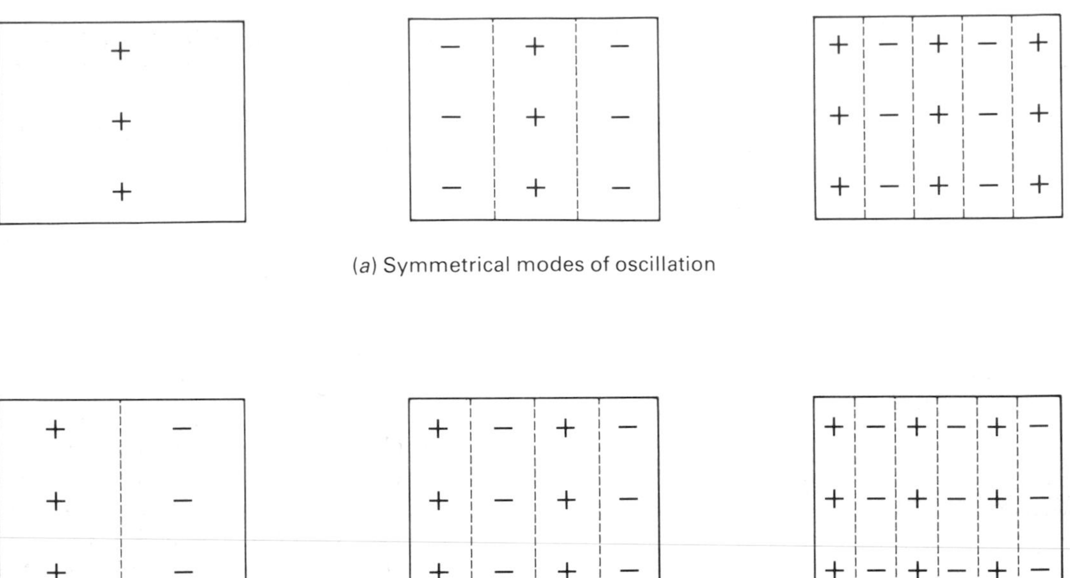

Fig. 4.19

up; these are alternately to the left and right of the path and are in opposite directions of rotation as shown in Fig. 4.17.

Similarly, if a wind blows past a fixed obstacle, eddies will be formed on either side of it and these will rotate in opposite directions. Eventually these eddies will detach themselves from the wire, giving it a thrust and moving off in the air stream. As the eddies depart from alternate sides giving alternate, periodic thrusts, the wire is caused to vibrate. If the frequency of the thrusts is equal to that of one of the modes of the wire then a loud resonant sound is heard; in practice there will be various sets of thrust frequencies set up at once and the tone is caused by the superposition of them.

Jet engine turbine blades

Turbine blades of jet engines are mounted in their supporting drums by joints that only fully tighten when the blades are revolving at full speed; this enables the vibratory stresses to be reduced at low engine speeds but also means that the modes of vibration of the blade may change with engine speed. The effect of the vibration caused in the blade by the engine vibrations and by the air flow is to set up stationary waves in the blade. These are different from those in stays, for one end is fixed and the other free; in addition the blades can twist.

Blade vibrations can be investigated in the laboratory by scattering sand on the stationary, vibrating blades since the sand will collect along the nodal lines; more complex experiments involve making holograms of the moving surface.

The simplest modes are the flexural (or flap) modes as shown in Fig. 4.18; in this situation there is no simple relationship between the frequencies of different modes. The situation is further complicated by the presence of torsional modes; these are caused by air flow at large angles of incidence and can be partially overcome by carefully shaping the end of the blade.

Vibrations of a rubber sheet

A rectangular rubber sheet can be made to vibrate in a variety of modes, exhibiting stationary waves. Fig. 4.19 shows six such modes; the first three are symmetrical and the second three are asymmetrical. The dotted lines represent nodal lines, + sign regions containing a crest, and − sign regions containing a trough; the actual motion will be a superposition of some, or all, of these modes.

Chladni's plate

Chladni's plate is a metal plate which is clamped at its centre so that its edges are free to vibrate. It is set into oscillation by drawing a bow along one edge and the modes of vibration that occur are controlled by touching the edge with a finger, thus causing nodal lines to start from that point. The modes are also affected by the position at which the bow is applied. The modes are observed by scattering sand on the plates; this collects along the nodal lines. Much of the research on Chladni plates was carried out by Mary Waller, Head of Physics at a London Hospital, about 1930 (Fig. 4.20).

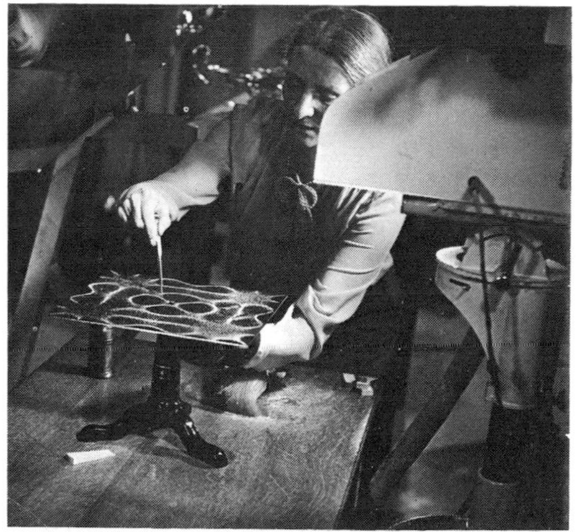

Fig. 4.20 Mary Waller demonstrating one mode of oscillation on a Chladni's plate during a television broadcast in 1937. (BBC copyright photograph: *The Listener*, 1938.)

Earthquake waves

We have already discussed earthquake waves in some detail in Section 3.7 and we saw that they may be either of longitudinal or shear type. The maximum wavelength of earthquake waves observed is about 13 000 kilometres, the length of the earth's diameter, and these waves have a period of about 3400 seconds. These are not really progressive waves but are stationary waves involving the oscillation of the earth as a whole.

A useful analogy is the speaking tube used in ships until quite recently. Normally the wavelength of the sound wave is very much less than the length of the tube and the tube may therefore be used to carry progressive sound waves from place to place. If, however, the pitch of the sound is lowered, and the wavelength lengthened, until the wavelength is of the same order of magnitude as the length of the tube, it begins to act like a musical instrument and stationary waves are set up.

In the same manner the 13 000 kilometre waves are not progressive waves but stationary ones. Since the last century, mathematicians have attempted to predict what the free oscillations of a homogeneous sphere, due to elastic and gravitational forces, would be. They considered three types of vibration: torsional, where the motion of a point on the surface is tangential to the surface, radial and spheroidal. The last two types are illustrated in Fig. 4.21; for each type of oscillation there are of course a whole sequence of more complex modes.

Working on simplified data, computers predicted periods of 42 minutes 30 seconds for the torsional mode, 20 minutes 44 seconds for the radial mode and 56 minutes 44 seconds (3404 seconds) for the spheroidal mode. The measured period (3400 seconds) of the 13 000 km waves agrees closely with that of the spheroidal mode.

Electromagnetic waves

Finally, as a reminder of the universal properties of waves, we will consider stationary waves set up by a form of electromagnetic radiation—radar waves. If a 3 cm radar source is set up opposite a metal plate, the radiation is reflected by the plate and stationary waves are set up between transmitter and plate. This can be demonstrated by moving a detector along a line between the transmitter and plate, when a sinusoidal variation of amplitude will be detected as a $(\text{sine})^2$ variation in intensity. This may

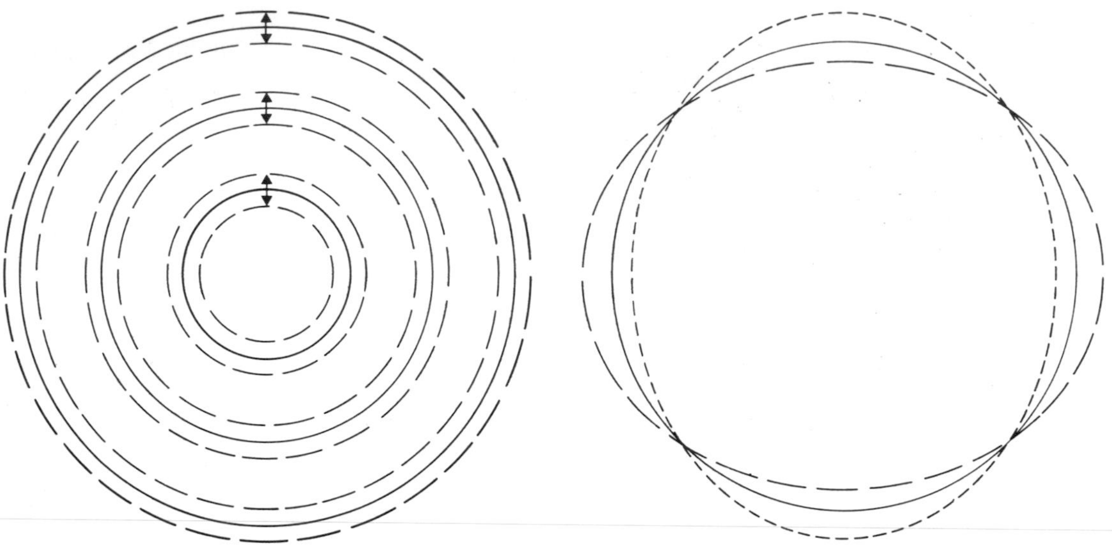

Fig. 4.21 Oscillations of the earth.

be used to measure the wavelength of the radiation, bearing in mind that you pass two nodal points in each wavelength, as seen in Fig. 4.22. This method is used to measure the wavelengths of electromagnetic waves in waveguides.

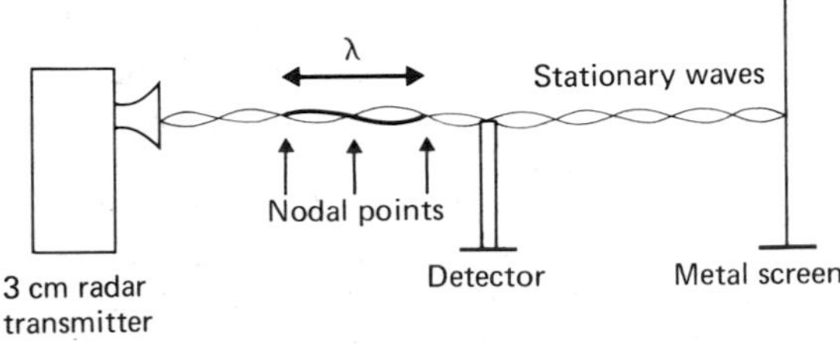

Fig. 4.22 Stationary electromagnetic waves.

Thus we see that all types of wave motion will produce stationary waves in situations where reflection can occur, and where the wavelength is of the same order of magnitude as the basic dimensions of the situation. They can be used to show that a wave motion is present, and sometimes to determine the wavelength of the waves.

4.10 Beats

So far, we have restricted our attention to cases of superposition where the frequencies of the two wave motions have been identical. Clearly superposition must always occur when two waves disturb the same medium, whatever their frequency, but in general the net effect will be complicated. However, one important case with simple results occurs when the two frequencies are different but by a small amount.

If two identical waves are in phase at a particular moment there will be a large resultant amplitude but if they arrive exactly out of phase, there will be zero resultant amplitude. Now if the two waves are of slightly different frequency they will have slightly different periods. Suppose that at one moment they are exactly in step, giving a large resultant. As time progresses, they will get increasingly out of step until eventually the crest of one wave will coincide with the trough of the other and the resultant will then be zero. After a further equal time has elapsed they will be once more in phase and the resultant will be a maximum again. Thus, the resultant amplitude will vary

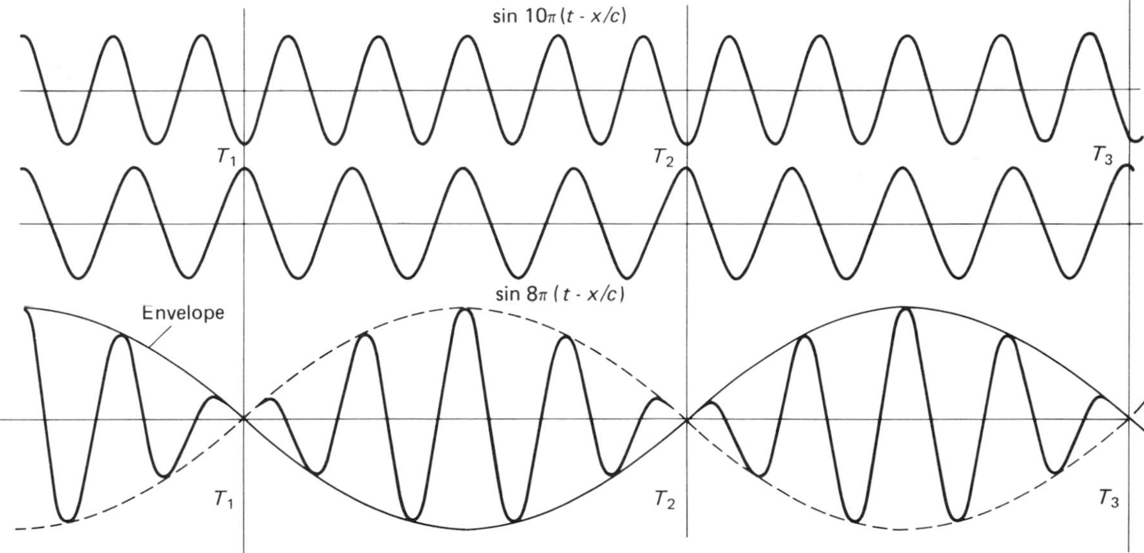

Fig. 4.23 The formation of beats.

from a maximum of twice the amplitude of each wave to zero, and back again. The closer the two frequencies are the slower will be this variation.

This behaviour can be shown graphically, and in Question 4.7 the reader is asked to plot his own graphs to convince himself that the following results are correct; they are shown in Fig. 4.23. The resultant wave is similar to a sine wave except that its amplitude is changing; the *envelope* of the resultant wave has a frequency equal to half the difference between the constituent frequencies. We say that the two waves produce *beats*; if the two amplitudes are not exactly equal the beats will still occur, but less distinctly since the resultant amplitude will never fall to zero.

We have stated above that the frequency with which the envelope varies is equal to half the difference in frequency. Thus, in Fig. 4.23 the top wave is of frequency 5 Hz and the lower wave of frequency 4 Hz, a difference of 1 Hz; the time interval between T_1 and T_2 is 1 second, and contains 5 and 4 complete cycles of the two waves respectively. In this second the envelope varies through only half a cycle and thus the frequency of the sinusoidal variation in amplitude is 0.5 Hz, as shown in the equation underneath. However, if we are concerned with the energy transmitted, as in a sound wave, this is proportional to the square of the amplitude of the resultant wave. The sound will therefore be a maximum half-way between T_1 and T_2, and also half-way between T_2 and T_3; at T_1, T_2 and T_3 there will be silence. Thus the sound fluctuates from zero to maximum and back to zero again once in each second and the 'beat frequency' is 1 Hz, equal to the difference in component frequencies.

Beats may be used to tune the notes produced by two musical instruments or loudspeakers. When the notes are almost in tune, beats will be produced but as the frequencies of the notes are brought closer together the beat frequency will fall, and will become zero when the frequencies of the notes are identical.

Q 4.7

Draw two sine waves, of frequencies 5 Hz and 4 Hz, as accurately as possible, above each other on a sheet of graph paper. Use the principle of superposition to show that the resultant wave will have the form shown in Fig. 4.23.

The mathematics of beats

Let us suppose that the displacements, y_1 and y_2, of the two waves are given by:

$$y_1 = A \sin 2\pi f_1(t - x/c) \qquad (2.2)$$
$$y_2 = A \sin 2\pi f_2(t - x/c)$$

where their frequencies f_1 and f_2 are very nearly equal. Then the resultant amplitude is given by:

$$y = y_1 + y_2$$
$$= A \sin 2\pi f_1(t - x/c) + A \sin 2\pi f_2(t - x/c)$$

and since:

$$\sin P + \sin Q = 2 \sin \tfrac{1}{2}(P+Q) \cos \tfrac{1}{2}(P-Q)$$

from Appendix 2, we have:

$$y = 2A \sin 2\pi\left(\frac{f_1+f_2}{2}\right)\left(t-\frac{x}{c}\right) \cos 2\pi\left(\frac{f_1-f_2}{2}\right)\left(t-\frac{x}{c}\right)$$

But $\dfrac{f_1+f_2}{2} = \bar{f}$, the average frequency,

where $\bar{f} \approx f_1 \approx f_2$ since the frequencies are almost identical.

Hence

$$y = 2A \cos 2\pi\left(\frac{f_1-f_2}{2}\right)\left(t-\frac{x}{c}\right) \sin 2\pi\bar{f}\left(t-\frac{x}{c}\right) \qquad (4.12)$$

This represents a sine wave whose frequency $\bar{f}$, being the average of two almost identical frequencies, is almost unchanged; in audible beats this determines the pitch of the note that is heard. The amplitude of the wave is

$$2A \cos 2\pi\left(\frac{f_1-f_2}{2}\right)(t - x/c)$$

and thus varies between 0 and $\pm 2A$ with a frequency equal to half the original frequency difference. The sound heard will be loud whether the amplitude is $+2A$ or $-2A$ and thus the loudness of a sound produced by beats varies at a frequency equal to the difference in frequencies; mathematically this arises from the fact that the intensity is proportional to the square of the amplitude.

Q 4.8
Use the trigonometric result:

$$\cos^2 P = \frac{1+\cos 2P}{2} \quad \text{from Appendix 2}$$

to prove that, if the amplitude of the superposed waves varies with a frequency $(f_1 - f_2)/2$, then the energy will vary with a frequency $(f_1 - f_2)$.

Amplitude modulation

An interesting application of the idea of beats is used in radio and television broadcasting. A radio station assigned to, say, 800 kHz broadcasts a carrier wave signal at that frequency whenever it is switched on. If someone speaks so that a message is transmitted, the signal from the microphone is used to change, or *modulate*, the amplitude of the carrier wave as shown in Fig. 4.24(*a*).

Suppose that the carrier wave signal, s_c, at the transmitter is represented by:

$$s_c = A_c \sin 2\pi f_c t$$

where f_c is its frequency, and A_c its amplitude. We have omitted the distance parameter because we are considering only the signal at one place.

Now the carrier amplitude is modulated about the value A_c by the modulating signal, which for simplicity we take to be a single frequency; the modulation factor used is of the form:

$$s_m = 1 + A_m \sin 2\pi f_m t$$

where f_m is the frequency, and A_m the amplitude, of the modulating signal s_m.

To obtain the equation for the resultant signal we multiply these two expressions together. Hence:

$$s = s_m \times s_c$$
$$= A_c(1 + A_m \sin 2\pi f_m t) \sin 2\pi f_c t$$
$$= A_c \sin 2\pi f_c t + A_c A_m \sin 2\pi f_c t \sin 2\pi f_m t$$

Now:

$$\sin P \sin Q = -\tfrac{1}{2}[\cos(P+Q) - \cos(P-Q)]$$

from Appendix 2. Thus:

$$s = A_c \sin 2\pi f_c t - \tfrac{1}{2}A_c A_m \cos 2\pi(f_c + f_m)t$$
$$+ \tfrac{1}{2}A_c A_m \cos 2\pi(f_c - f_m)t$$

The modulated carrier wave is, therefore, equivalent to the 'beats' produced by three waves of frequencies f_c, $(f_c + f_m)$ and $(f_c - f_m)$. As the carrier frequency (800 kHz) is much greater than the highest frequency in the audible range (about 15 kHz) we have a signal at 800 kHz together with two closely-spaced side frequencies, as shown in Fig. 4.24(*b*).

In practice, instead of one particular frequency in the modulating signal there will be a whole range up to 10 kHz and possibly beyond. Thus instead of broadcasting at three specific frequencies the transmitter will transmit over two ranges on either side of the carrier, which are called *sidebands*, in addition to the carrier frequency itself. Thus the transmitter and receiver must be able to generate and detect signals in a given frequency band, say between 790 and 810 kHz, and each station that is to broadcast must be allocated exclusively a certain *bandwidth* if there is not to be interference with other transmitters.

4.11 Harmonic synthesis

We have already seen, in Section 1.7, how any complicated oscillation can be considered by Fourier synthesis to consist of an infinite number of sinusoidal oscillations; this principle enables all oscillations to be analysed in terms of one type of oscillation that is particularly

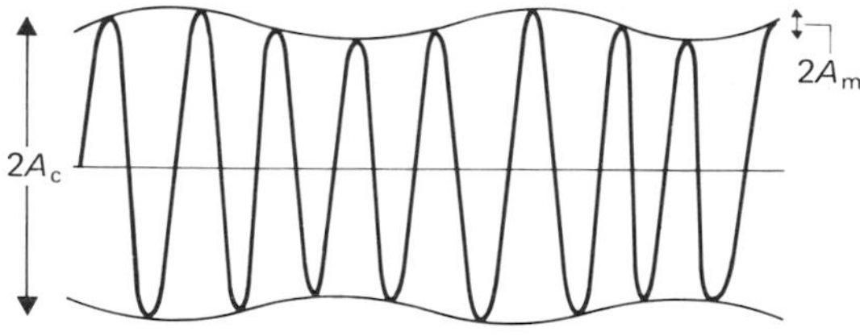

(a) A modulated carrier wave

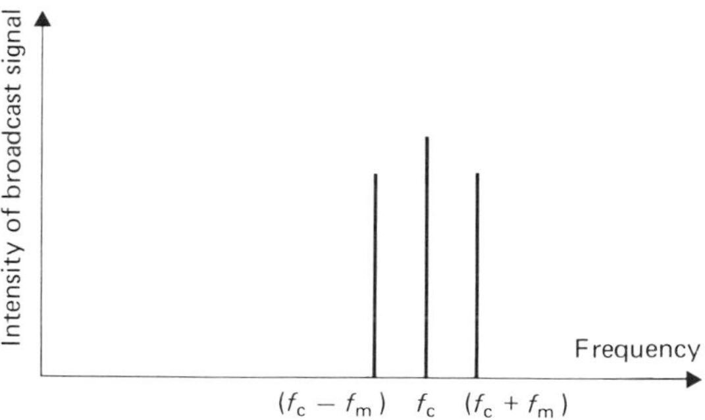

(b) The frequencies produced when a carrier wave of frequency f_c is modulated by a single signal of frequency f_m

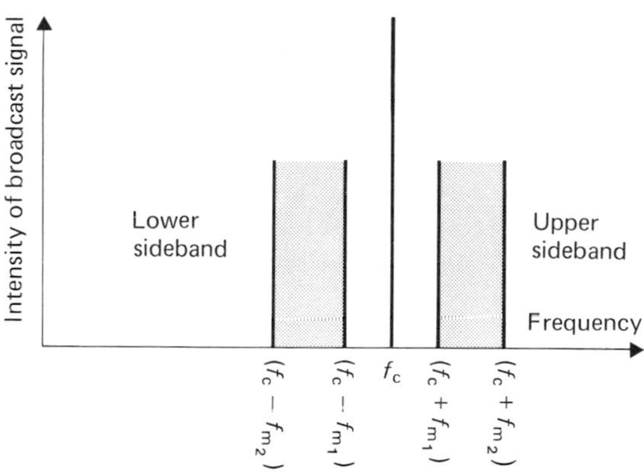

(c) The frequencies produced when a carrier wave of frequency f_c is modulated by signals varying in frequency from f_{m_1} to f_{m_2}

Fig. 4.24 The amplitude modulation of radio waves.

well understood. We shall now briefly apply the same idea to waves.

As has been stated earlier, strings, air columns, metal plates and so on can all form stationary wave patterns in a variety of modes. In more complicated situations the stationary wave patterns that are set up are a combination of a whole series of such modes, all occurring at once. In other words the complex motions of, say, a violin string can be analysed in terms of the superposition of sinusoidal modes, each of which must satisfy the boundary conditions for the particular case.

We cannot, of course, easily 'see' the motion of a violin string. We can, however, use a microphone to detect the sound that is produced and pass its signal to a cathode-ray oscilloscope on which can be displayed a graph of displacement against time like those in Fig. 5.6. A violin string can vibrate in the series of modes illustrated in Fig. 4.14; each mode produces a different note and it is the mixture of modes, and therefore of notes, which gives the violin its distinctive tone. The lowest note present, the fundamental, determines the pitch of the note that is heard. A different instrument, such as an oboe, will have different harmonics present and the relative amplitudes of the range of harmonics will be different. The complex waveform seen on the oscilloscope is formed by adding up all the constituent waveforms.

Bibliography for Chapters 2, 3 and 4

General accounts

Akrill, T.B., Bennett, G.A.G. and Millar, C.J. (1979) *Physics.* Arnold; Chapter 23 deals with superposition and interference effects.

Bolton, W. (1986) *Patterns in Physics* (2nd Edn.). McGraw-Hill; Chapters 7 and 8 cover much of this material.

Chaundy, D.C.F. (1972) *Waves.* Longman Physics Topics; a brief and well-illustrated account of wave motion.

CRAC Science Support Series (1982) *Waves and Sound.* Hobsons Press; discusses briefly a range of modern transducers and their application.

Feynman, R.P., Leighton, R.B. and Sands, M. (1963) *Feynman Lectures on Physics, Volume 1.* Addison-Wesley, Chapters 47–51; these lectures are tough but stimulating, being intended for American undergraduates; they will amply repay selective reading with the omission of the more mathematical sections.

PSSC (1981) *Physics* (5th Edn.). Heath; Chapters 22–24 give a well-illustrated account of wave phenomena.

Shive, J.N. and Weber, R.L. (1982) *Similarities in Physics.* Hilger; Chapters 7, 8 and 9 bring out the similarities between different kinds of wave motion in a wide variety of systems.

Walker, J.R. (1978) *The Flying Circus of Physics.* Wiley; contains thought-provoking examples on interference.

Wenham, E.J., Dorling, G.W., Snell, J.A.N., and Taylor, B. (1984) *Physics: concepts and models* (2nd Edn.). Longman; Chapters 16–18 cover wave motion.

Specific topics

Barber, N.F. (1969) *Water Waves.* Wykeham; a most readable account of the physics of water waves, backed up by a thorough mathematical treatment.

Harvard Project Physics (1970) *Text 3.* Holt, Rinehart and Winston; some good illustrations and background reading, including an article on sonic booms.

Huygens, C. (1678) *Treatise on Light.* Reprinted by Dover; as is often the case, Huygens' own account of his theory is more intelligible than many written since.

Kittel, C. (1971) *Introduction to Solid State Physics.* Wiley; Chapter 4 contains a more rigorous analysis of the speed of waves throughout a continuous solid.

Llowarch, W. (1961) *Ripple Tank Studies,* Oxford University Press; some excellent photographs of waves in a ripple tank.

Open University (1979) Science Foundation Course, Unit 4 *Earthquake Waves and the Earth's Interior.* Open University Press.

Tricker, R.A.R. (1965) *Bores, Breakers, Waves and Wakes.* Mills and Boon; another excellent book on wave phenomena (out of print).

Some books on the musical aspect of wave superposition are listed under Chapter 5.

Reprints (now out of print but available in many libraries)

Bascom, W. (1959) Ocean Waves. *Scientific American*, offprint No. 828.

Bernstein, J. (1954) Tsunamis. *Scientific American*, offprint No. 829.

Bullen, K.E. (1955) The Interior of the Earth. *Scientific American*, offprint No. 804.

Oliver, J. (1959) Long Earthquake Waves. *Scientific American*, offprint No. 827.

Software

Heinemann/Fiveways *Transverse Waves* and *Longitudinal Waves*; enable students to visualize wave motion and observe the effect of changing parameters.

Hutchinson Software *The BBC Microcomputer in Science Teaching*; 'Wave Reflection', 'Wave Superposition' and 'Standing Waves' enable the relevant processes to be observed in slow motion.

5

Sound waves

5.1 The transmission of sound

The wave nature of sound

Suppose that an electronic organ is connected to two loudspeakers, so that the same note is played through each; if you move around the hall in which the organ is situated, you will hear the loudness of the note change from place to place. You are hearing the variation in the intensity of sound due to interference maxima and minima, showing that sound is a wave motion. This experiment may be carried out in the laboratory using a signal generator and two small loudspeakers; the intensity of the sound in the interference pattern is detected by a microphone and displayed on an oscilloscope.

Waves are propagated from oscillating sources; for instance, when the end of a stretched rope is moved up and down, a wave is propagated along it. Similarly the cone of a loudspeaker producing a note can be felt to be oscillating; sound is thus a *wave motion*.

When a wave is sent down a rope, the oscillations of the rope can occur in any direction at right angles to the direction in which the wave is travelling; if the oscillations are restricted to one plane they are said to be polarized. Light waves, which are transverse, can be polarized as explained in Section 7.7. Longitudinal waves cannot be polarized, as the directions of the oscillations are always in the direction of propagation. As no polarization phenomena have ever been discovered for sound waves, we conclude that they consist of *longitudinal waves*. In fact, if very loud sound waves are transmitted through air which contains smoke particles, it is possible to see the resulting longitudinal oscillations of the particles through a microscope.

It is important to note that when a microphone is connected to an oscilloscope, the trace on the screen is a graph of 'longitudinal displacement against time'; the fact that the trace appears to look like a transverse wave does not mean that the sound wave causing it is itself transverse.

If an electric bell is hung in a jar, and air is pumped out, the sound of the bell becomes inaudible; sound cannot be transmitted through a vacuum. Thus, sound is a *longitudinal wave motion, which requires a material medium for transmission*: this medium may be a solid, liquid or gas.

The reader is advised to revise his knowledge of longitudinal waves—their transmission, reflection and superposition—as the topics will not be covered in detail in this chapter. They are covered in Chapters 2, 3 and 4.

The velocity of sound is large, but not as great as the velocity of light, as may be seen by observing a starting pistol fired. The velocity of sound at 20°C is about 330 m s^{-1} in air, 1500 m s^{-1} in water and 5100 m s^{-1} in iron. Methods of measuring the velocity of sound in a solid were described in Section 2.5, and measurements in air are discussed in Section 5.6.

Sound waves, like other waves, can be diffracted, refracted and reflected. The angular width of the diffraction pattern that occurs beyond an opening depends upon the ratio of the wavelength to the slit width; the wavelength of an audible sound wave is of the order of one metre, whereas the wavelength of light is about 10^{-6} m, and therefore the diffraction of sound is noticeable at larger openings than are required for the diffraction of light. Diffraction occurs, for instance, when sound leaves your mouth, spreading the waves throughout a room; a simple conical megaphone increases the size of the aperture at which the diffraction occurs, thus causing less diffraction and producing a more confined beam for use outdoors.

Practical applications of the refraction of sound will be discussed in the section on ultrasonic waves, where they are particularly important. The phenomenon can be observed by filling a lens-shaped balloon with a gas that is heavier than air, such as sulphur dioxide; the velocity of sound in the balloon will be less than in the surrounding air, and thus the balloon will act like a converging lens. If you place your ear close to one side of the balloon, sounds from the other side will appear slightly louder.

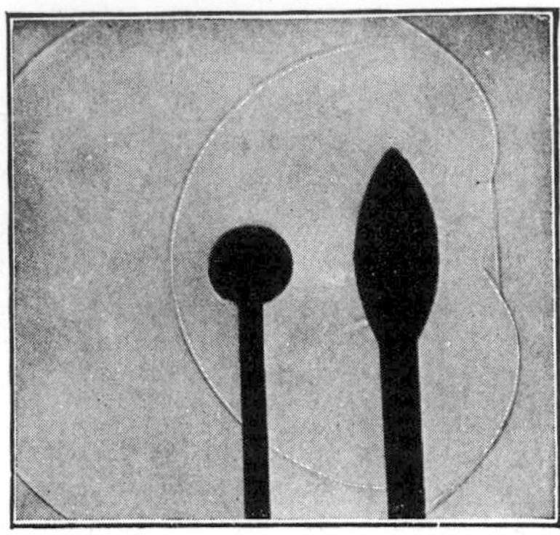

Fig. 5.1 The wavefronts in this photograph are made visible by a special technique known as 'Schlieren' photography; this utilizes changes in the refractive index of air with density to enable the position of wavefronts to be seen.

Fig. 5.1 shows a sound wave, produced behind the black disc, being refracted as it passes through the bag of sulphur dioxide, and producing almost plane waves; the source is at the focus of the lens. Some of the wave that has passed outside the lens has undergone diffraction at its edge and a reflected wave can also be seen. Carbon dioxide is another gas that is heavier than air and a balloon filled with this also forms a converging lens; hydrogen is lighter than air and a balloon filled with it will act as a diverging lens.

The reflection of sound waves, also shown in Fig. 5.1, is of especial importance in the acoustical designs of buildings.

The acoustics of halls

In a hall the reflection of sound from the walls causes echoes and reverberation. An echo may be formed if the sound is reflected from an area of the wall sufficiently continuous in shape to form a strong reflected beam; this used to be the case in the Royal Albert Hall, London, where the front row of the audience would hear a reflected sound nearly a fifth of a second after the direct sound. This problem has been reduced by hanging fibre-glass saucers to absorb sound incident on the roof, as shown in Fig. 5.2(*a*).

Multiple reflections, and the scattering of sound from the walls, cause reverberations as shown in Fig. 5.2(*b*); when the generation of a sound ceases, there is a further period of time during which the sound is dying away because a series of echoes is heard in rapid succession. The time taken for the intensity of the sound to decrease to one millionth of its initial value is called the *reverberation time*.

Q 5.1
Estimate the height of the Albert Hall from the data above. The velocity of sound is about 330 m s^{-1}.

In 1895, Wallace Sabine was commissioned to improve the acoustics in the lecture theatre at the Fogg Art Museum, Harvard, which suffered from a reverberation time of 5.6 s. He carried out a detailed investigation, timing how long the period of reverberation was at the end of an organ note, and adjusting the amount of absorption by importing cushions in large numbers. When he had borrowed 436 cushions, the reverberation time was reduced to 2.0 s; the minimum time he achieved was as little as 1.1 s. He measured the absorbing power in terms of the length of cushions, and he showed that:

$$\text{total absorptive power} \times \text{reverberation time} = \text{constant} \qquad (5.1)$$

In his calculations he had, of course, to allow for absorption by the walls as well as by the cushions. Sabine's original unit of absorption is rather cumbersome and the unit of absorption he replaced it with was the *open window*. An open window is assumed to have a coefficient of absorption of 1.00 because it absorbs all the sound falling on it; the *coefficients of absorption* of other materials can then be measured by comparing their absorption with an open window:

acoustic plaster	0.45
heavy carpet	0.06
unpainted concrete	0.03

As coefficients of absorption vary considerably with frequency, the above data is quoted at 250 Hz, about middle C; the coefficient of absorption of acoustic plaster at 1000 Hz, for instance, is 0.92. The fact that coefficients of absorption vary with frequency means that reverberation times will also be frequency dependent.

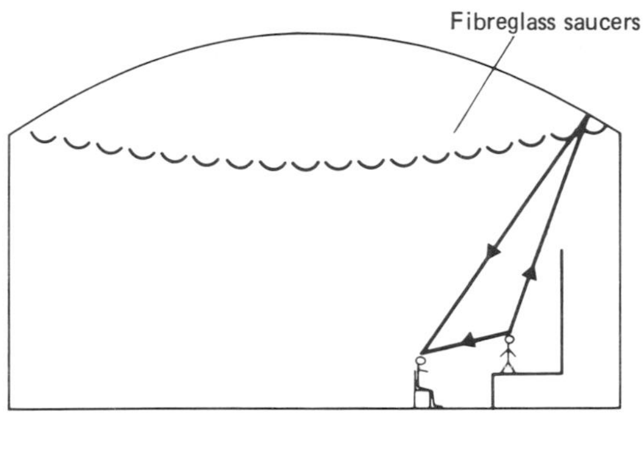

(*a*)

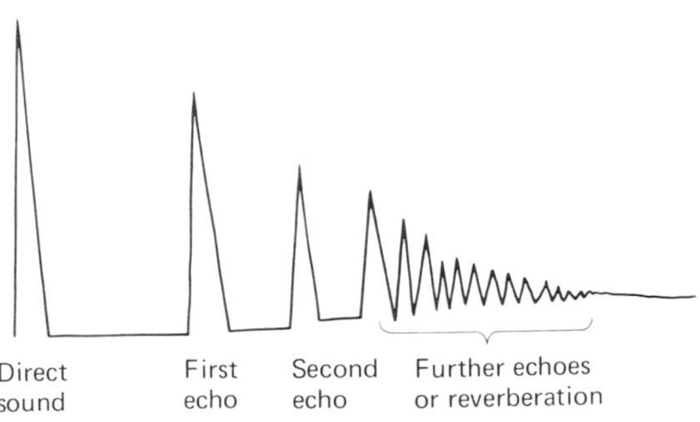

(*b*)

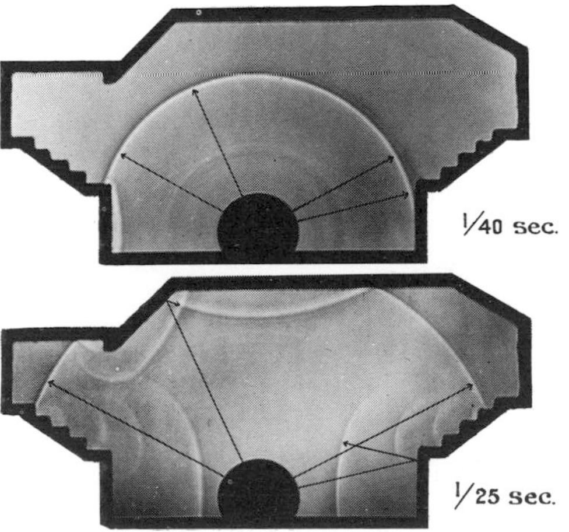

(*c*)

(*d*)

Fig. 5.2 (a) Royal Albert Hall, London. (b) Echoes heard by a listener in a hall. (c) Photographs showing successive stages in the progress of a sound pulse in an auditorium, using a sectional model. (d) Fibre-glass saucers, intended to cut down unwanted reflections in an auditorium, being tested in the Building Research Establishment's anechoic chamber at Garston, Watford.

The ideal reverberation time for a hall depends upon the size and use. In a hall of given size, in Sabine's case about 4000 cubic metres, the following might be suitable:

conference room	0.9 s
chamber music	1.2 s
classical or modern music	1.8 s

The time chosen is, in any case, a matter of personal taste and, if the hall is to serve a variety of purposes, also a matter of compromise.

The presence of an audience will make a substantial difference to the reverberation time; the Mormon Tabernacle, in Salt Lake City, has a very long reverberation time of 4 s at 1000 Hz when empty, an optimum time of 2 s when about 2500 people are present, but becomes rather dull when there are as many as 6000. In modern halls, therefore, it is usual to provide seats whose absorptive properties are the same as a person's, so that conditions in rehearsal and performance are identical.

Waves reflected from the walls might be expected to interfere but this does not usually produce audible variations in intensity; different frequencies, both fundamentals and harmonics, will produce different interference patterns in the hall. It is wise, however, to avoid smooth reflecting surfaces so that reflections are diffuse and the sound spreads throughout the hall as uniformly as possible.

One superposition effect that can cause distortion of the sound is resonance. A hall will have a number of natural frequencies, at which stationary waves can be set up particularly easily; the actual frequencies which are excited will depend upon how close the notes played are to those frequencies, and where in the hall they are played. Resonance may cause some notes to stand out more clearly than intended, or to continue to be heard after others have died away; it may be heard at home when singing in a tiled bathroom.

Sabine's work provided a basis for some theoretical understanding of acoustics, but much work in the subject is still empirical. Building real halls with disastrous acoustics is an expensive way of conducting experiments, and some predictions can be obtained from investigations with models beforehand. Reflections can be observed by making a model plan of the hall in a ripple tank, as shown in Fig. 5.2(*c*). Three-dimensional models can also be constructed, and experiments with microphones and loudspeakers carried out in them; the wavelength of the sound must be scaled down by the same ratio as the dimensions. Such experiments are often carried out in a room which is itself completely free from reverberation, called an *anechoic chamber*; the walls are lined with glass-fibre cones to absorb all the sound incident upon them, as shown in Fig. 5.2(*d*).

5.2 The production of sound by musical instruments

We have, so far, shown that sound is a longitudinal wave motion, with properties such as diffraction, refraction and reflection. In this section, we shall turn our attention to those characteristics which are associated particularly with musical sounds.

Loudness and amplitude

A microphone and oscilloscope can be used to show that louder sounds have larger amplitudes. In Section 1.4 we saw that the energy of an oscillation is proportional to the square of its amplitude and the rate of energy flow per square metre, or *intensity*, of a wave is also proportional to $(\text{amplitude})^2$, at a given frequency; intensity measures power per unit area and has units W m^{-2}.

The loudness of a sound will depend upon the loudness of the source and its distance away; as we move further from the source, the flow of energy is spread over a larger area and the intensity falls. For a point source, the area of the spherical surface over which the power is distributed at a distance r from the source is $4\pi r^2$, and thus if there is no overall loss in power the intensity is proportional to $1/r^2$; it follows that the amplitude decreases as $1/r$.

The amount of power in sound waves is very small. A choir of one hundred thousand—filling Wembley Stadium—singing as loud as possible would only produce as much power as is used by a 100 W light bulb. The loudest sound that can be heard without discomfort has an intensity of 10 W m^{-2}, and the quietest sound that is audible to most people has an intensity of 10^{-12} W m^{-2}; however, the intensity range does not appear to us to be as great as this. The reproduction of the large intensity ranges that are encountered in a symphony concert, or at a pop concert, is impossible to achieve in recordings; fortunately we still get a vivid impression of 'loud' and 'soft', but on a much reduced range.

Pitch and frequency

A microphone and an oscilloscope will also show that notes which we describe as being of *high pitch* have a greater frequency than notes of *low pitch*.

Pythagoras carried out a detailed investigation into notes of different pitch, in the sixth century BC. He used an instrument called a monochord, or sonometer, which consists of a taut string stretched between two supports, as shown in Fig. 5.3. A third support, or *bridge*, may be moved along under the wire so as to divide the string into lengths of any required ratio. He found that the two notes produced were particularly harmonious to the ear, or *consonant*, if the lengths were in a simple ratio. The most consonant arrangement, other than that with equal lengths and pitches, was when the ratio of the lengths was 2 : 1. Pythagoras did not relate the lengths of the string to their frequencies, but we know that as the velocity of sound on the wire must be the same either side of the bridge, the frequencies must be in the ratio 1 : 2.

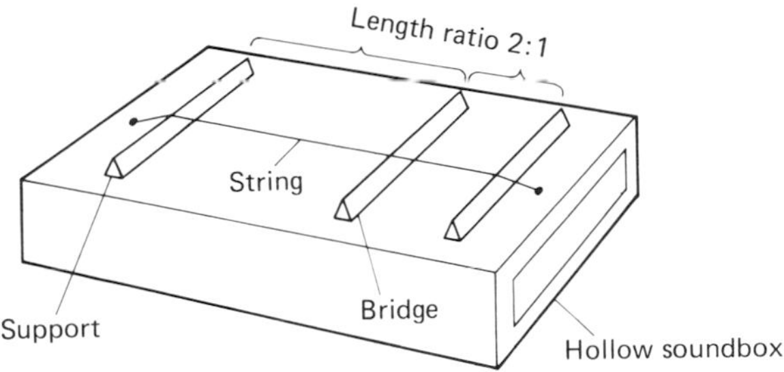

Fig. 5.3

These two notes would today be said to be an octave apart: *any two notes with a frequency ratio of* 1 : 2 *are an octave apart, and sound particularly consonant.*

As all notes which sound an octave apart have a frequency ratio of 1 : 2, we see that *musical intervals depend upon frequency ratios.* In the contemporary western system of naming notes, those an octave apart are given the same letter; for example, the note of frequency 256 Hz is called middle C and the note of frequency 512 Hz, an octave above, is also called C (written C′). These notes are shown in Fig. 5.4.

Pythagoras also found notes closer together that were still fairly consonant. The next most consonant interval occurs when the lengths are in the ratio 3 : 2; the frequencies are in the ratio 2 : 3 and the notes are said to be a fifth apart. For instance, G is five notes above C, counting inclusively, and the interval C–G is termed a *fifth.*

Q 5.2
What is the frequency of the note G above middle C? What is the frequency of the G an octave above that? (G′) Calculate the frequency ratio between C′ and G′.

The question demonstrates that the reason that G′ also sounds as if it is a fifth above C′ is that the *frequency ratio is again* 2 : 3; this will be true for any other pair of notes that are a fifth apart.

Clearly, within the audible range there are an infinite number of frequencies, just as there are an infinite number of points on a line. Those frequencies which are chosen to appear as notes in western music are there because they can be used to make up melodies and chords which, to our ears, sound consonant. The question of why these notes are consonant is a complicated one to which the answer is not really known; other civilisations seem to have chosen different notes to make up their scales and chords. Readers who are interested should consult a good text book on the physics of music; some are listed at the end of the chapter. Any physicist who is also a musician is strongly urged to read one of them!

Mersenne's laws

The work of Pythagoras on stretched strings, probably one of the earliest sets of quantitative experiments ever carried out, was later extended independently by Galileo and by Marin Mersennes, a Minorite friar; Mersennes published the laws given below in 1636:

(1) the fundamental frequency of a stretched string is inversely proportional to its length, l, provided that the tension, T, and mass per unit length, μ, remain constant;
(2) the fundamental frequency is directly proportional to the square root of the tension, provided that the length and mass per unit length remain constant;
(3) the fundamental frequency is inversely proportional to the square root of the mass per unit length, provided that the length and tension remain constant.

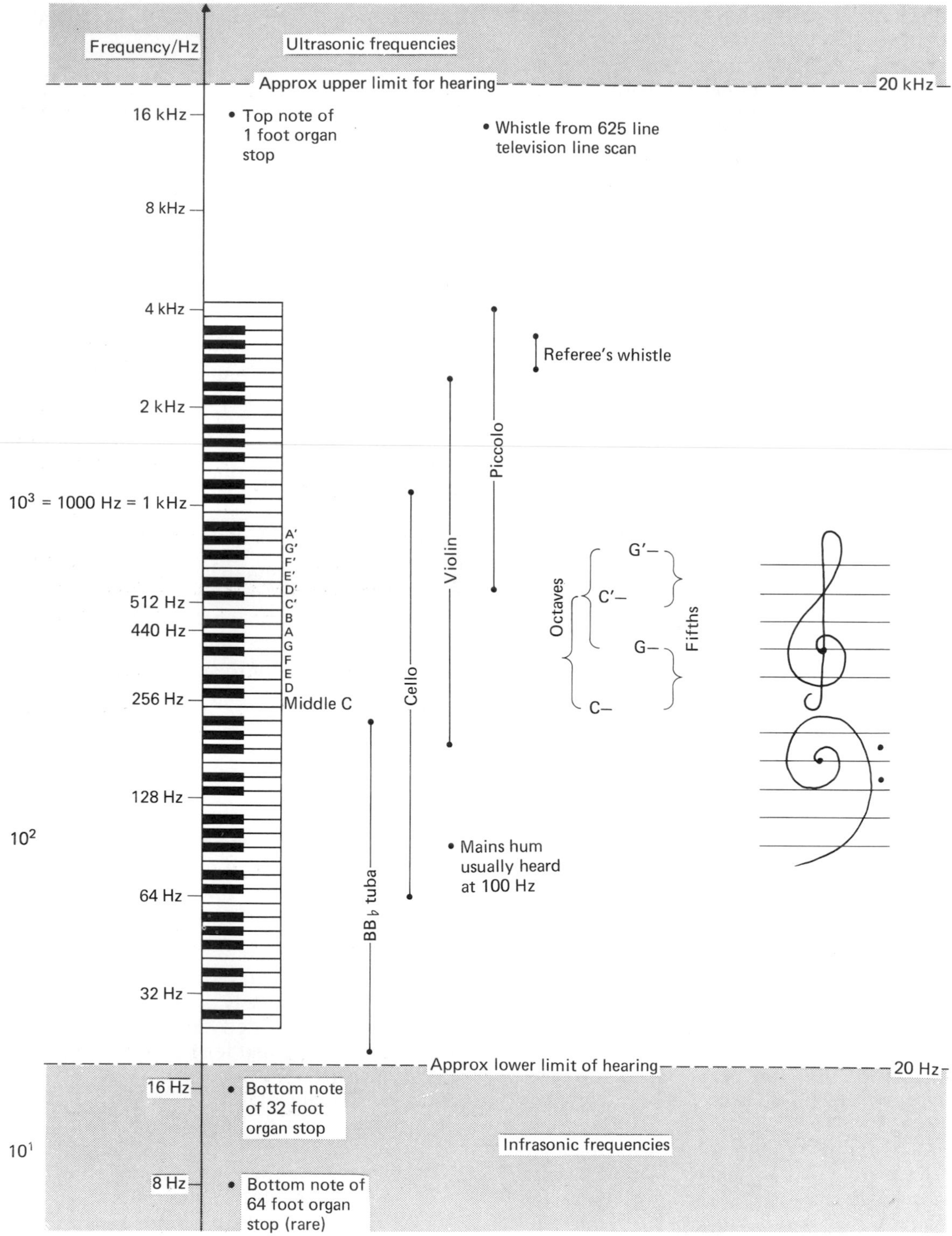

Fig. 5.4 The range of audible frequencies, including approximate ranges for some musical instruments; the keyboard represents the range of a pianoforte. The frequencies on the left are only approximate; they are marked on a logarithmic scale.

These three laws may be combined to give one relationship:

$$f \propto \frac{1}{l}\sqrt{\frac{T}{\mu}} \qquad (5.2)$$

In Section 4.6, we showed that the frequency of the stationary waves that can be set up on a spring, or string, is given by:

$$f = \frac{n}{2l}\sqrt{\frac{T}{\mu}} \qquad (4.8)$$

where n can be any integer depending on the wavelength. The fundamental frequency is the same as that given by Mersenne's law, in Equation (5.2).

Mersenne's laws can be tested by using the sonometer shown in Fig. 5.5. The length can be changed by moving one of the bridges, the tension adjusted by placing known masses in the pan, and the mass per unit length altered by replacing the wire; if the tension in the wire is to be equal to the weight in the pan, it is important that the pulley is well oiled. The frequency is measured by comparison, either with a standard tuning fork, or with the note produced by a signal generator and loudspeaker; the quality of the frequencies can be checked either by ear or by resonance.

Q 5.3
In an experiment using a sonometer, the following results were obtained (the length of the wire was 0.8 m throughout the experiment):

mass in pan/kg	2	4	8	16
fundamental frequency/Hz	210	300	425	600

Plot a graph to show that $f \propto \sqrt{T}$, and calculate the mass per unit length of the wire. Take $g = 10\ \mathrm{N\ kg^{-1}}$.

The audible sound spectrum

The range of sound frequencies that is audible is shown in Fig. 5.4 and the diagram is largely self-explanatory; the upper and lower limits of hearing are only approximate. The lower limit, of about 20 Hz, is difficult to measure because

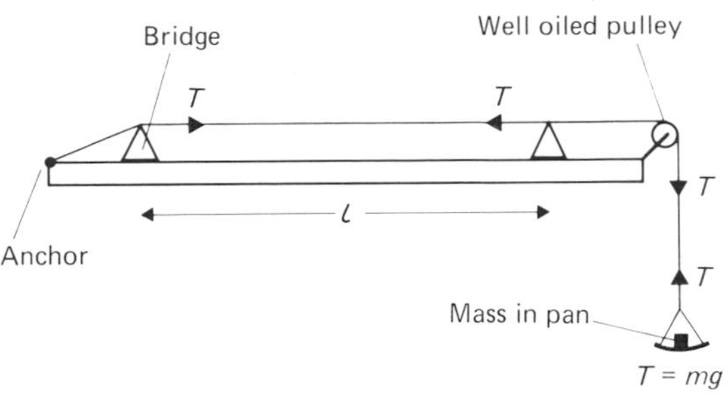

Fig. 5.5

it is extremely difficult to produce pure tones, and if the ear can hear overtones it will 'extrapolate' the fundamental. Both limits vary from person to person, and the upper limit, especially, changes with age; young people can hear up to about 20 kHz, but by middle age, the limit will probably have dropped to about 12 kHz. The ability of the ear to discriminate between notes falls off considerably before the limits of audibility are reached and the extent of the piano keyboard approximately defines the range of notes that are useful musically.

The use of beats to tune an instrument

We have already learnt, in Section 4.10, that when two waves of almost identical frequencies,

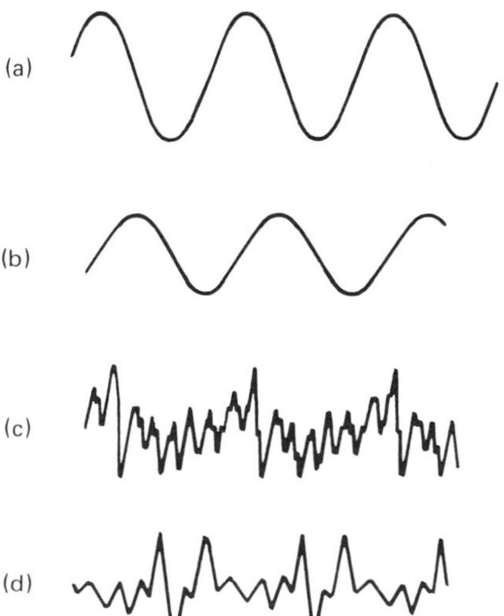

Fig. 5.6 Oscilloscope traces of notes of the same pitch (440 Hz). (a) Pure tone, (b) tuning fork, (c) violin, (d) oboe.

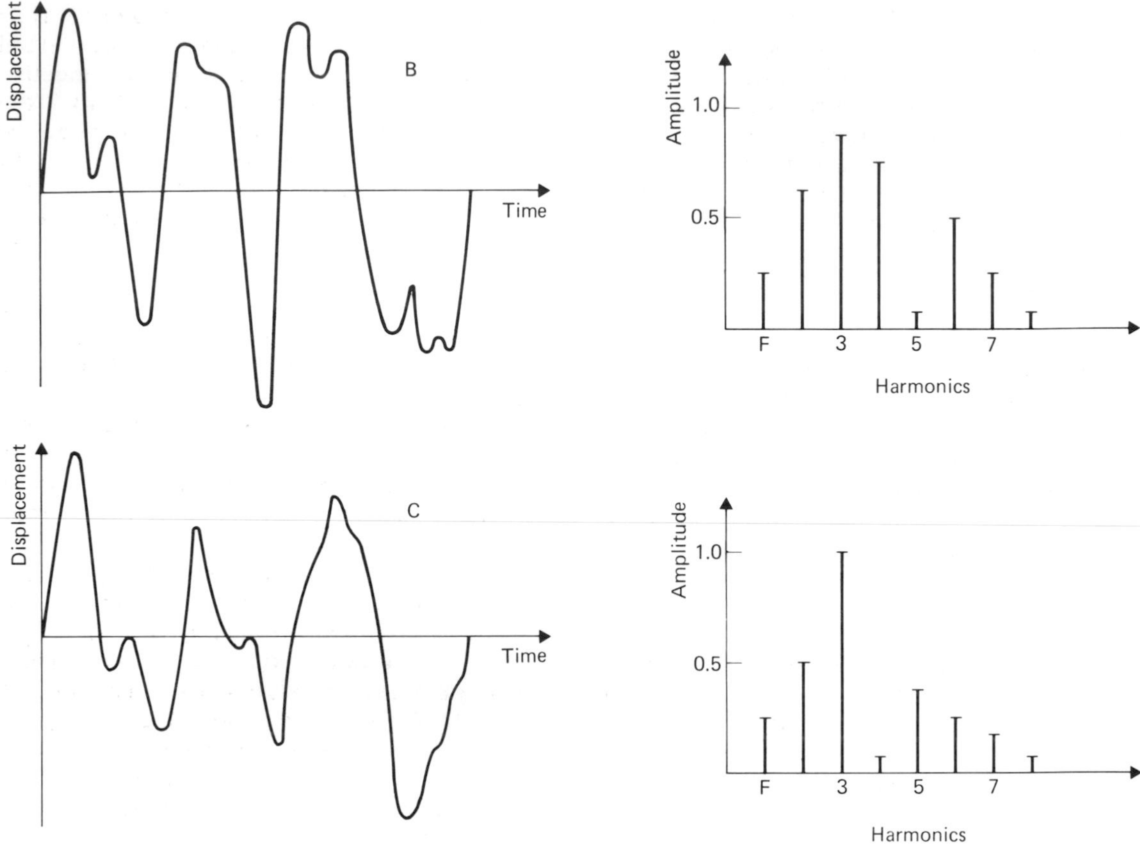

Fig. 5.7 The harmonic spectrum of two adjacent notes played on a bassoon.

f_1 and f_2, move through the same medium, beats are produced; as explained there, the loudness of the sound that is heard varies with frequency $(f_1 - f_2)$.

If the reader is not familiar with this phenomenon, he should take two loudspeakers and connect them to different signal generators; when the frequencies are nearly the same, beats will be heard. As the frequencies are adjusted to be equal the frequency of the beats decreases until no beats are heard. Sometimes two musical instruments or organ pipes are tuned by playing the same note on both instruments and then tuning one of them until the beats disappear.

Tone, or quality

In Chapter 4 we analysed the stationary waves that could be established on strings and in pipes. We saw that there was a wave of lowest frequency, called the fundamental, and an infinite series of waves of higher frequency called harmonics. When a stationary wave is excited some, or all, of these harmonics are produced in addition to the fundamental, and a complex waveform is produced by their superposition; possible shapes can be predicted by harmonic synthesis.

The production of *oscillograms* by a microphone and cathode-ray oscilloscope shows that when different instruments sound a note of identical pitch, they produce different complex waveforms, as shown in Fig. 5.6; the same is true of different players performing on the same instrument, especially if one is good and the other bad! The pitch of the note is determined by the frequency of the fundamental, but the tone, or quality, of it is determined by the shape of the waveform, and hence by the combination of harmonics present. Instrument designers and performers work hard to encourage harmonics that are thought to improve the tone, and to discourage those that make it sound unpleasant.

It is possible to analyse the waveform of a sound to show the relative amplitude of the different harmonics produced; this analysis is called a *harmonic spectrum*. It might be thought, from what has been said already, that different notes on the same instrument would give the same harmonic spectrum, as though that spectrum were characteristic of that instrument. That this is not the case is seen from Fig. 5.7; two oscillograms of adjacent notes (B and C) on a bassoon are shown, together with their harmonic spectra. The harmonic spectrum of a note, in fact, depends on a wide range of factors; they obviously include the particular instrument used and the pitch of the note, but also the loudness, the temperature and the way in which the instrument is played.

Remarkably, the ear can almost always recognize the instrument producing a note, despite these many variations. There is evidence to suggest that the way the sound is built up when the note is first played—the *transient*—is extremely important here. Try playing a tape of a musical instrument backwards to see how different it sounds.

5.3 The recording and reproduction of sound

Transducers

Any system for recording and reproducing sound electronically must contain a microphone and loudspeaker. Both are examples of devices termed *transducers*; a transducer is any device which can transform varying mechanical forces or motions into electric currents or voltages, or vice versa.

One of the most common types is the moving-coil transducer; this can be used either as a microphone or as a loudspeaker. These will usually have different design characteristics, but they operate on the same principle; in simple devices such as baby-alarms, the same transducer may actually perform both functions.

A simple *moving-coil transducer* is shown in Fig. 5.8(*a*). The shape of the permanent magnet is such that there is a cylindrical gap between its pole pieces. Into this gap is placed a small coil wound on a short cardboard tube; the tube is attached to a paper cone, so that when the tube moves the cone moves with it. If this device is used as a loudspeaker, the alternating current from the amplifier is passed through the coil. The magnet then exerts an alternating force on the coil, which is perpendicular to both the magnetic field and the wire in the coil, and as a result, the coil will oscillate in and out of the gap as shown. The resulting oscillations of the cone cause sound waves to be emitted by the loudspeaker.

When the device is used as a microphone, the incident sound waves cause the cone to oscillate, and the motion of the coil backwards and forwards in the field induces an alternating voltage in the coil.

A *capacitor transducer* is shown in Fig. 5.8(*b*); it consists of a flexible metal diaphragm insulated from a metal back plate. When it is used as a microphone, sound waves are incident on the diaphragm making it oscillate so that the distance between it and the back plate changes and as a result, the value of the capacitance varies. If a quantity of charge is initially stored on the plates, and they are then electrically isolated, an alternating voltage between them will be caused by the varying capacitance.

Other types of microphone utilize the small voltage produced when certain types of crystal or ceramics are forced out of shape, as shown in Fig. 5.8(*c*). Such microphones can be inexpensive.

Microphones produce only a small voltage, and hence power; loudspeakers require a much larger power to drive them and produce a reasonable amount of sound. Thus, in any reproduction or transmission system there must be an amplifier to increase the power supplied to the loudspeaker; this extra power is drawn from a power supply.

When a compression is produced in front of a loudspeaker cone, a rarefaction must be produced behind, and if these reach the listener together, they will interfere destructively. This is prevented by mounting the cone in the middle of a *baffle board* or in a cabinet; the design of the baffle board or cabinet are of the utmost importance in designing a loudspeaker system.

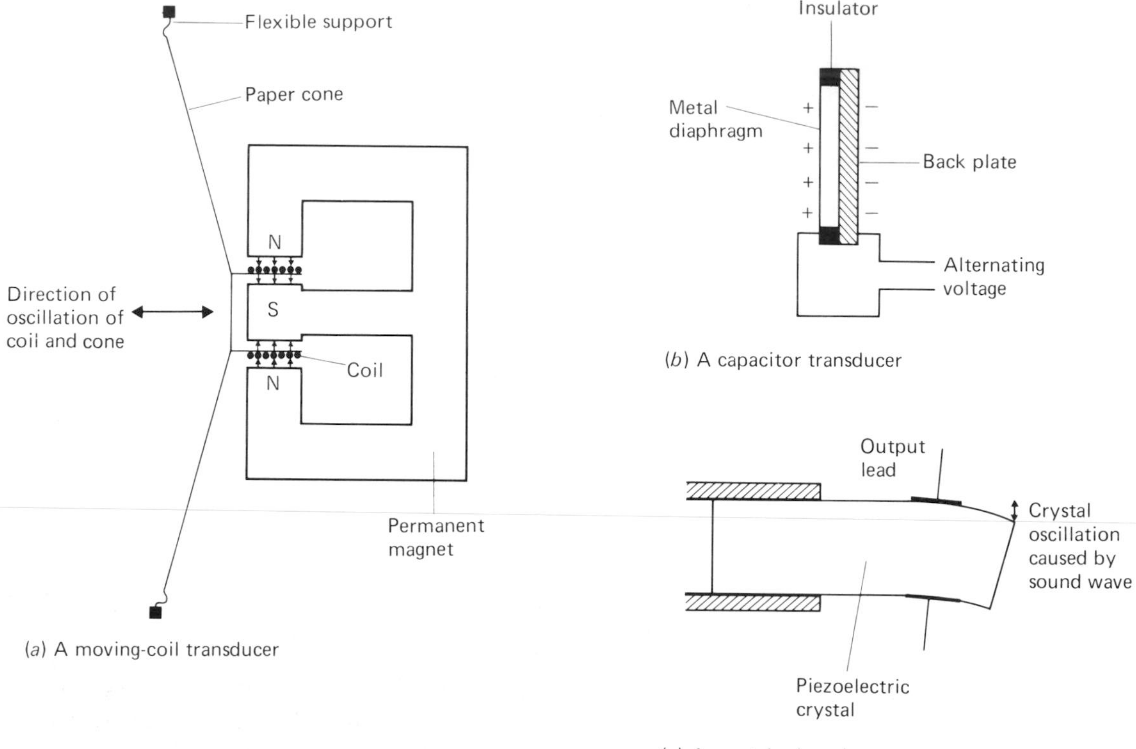

Fig. 5.8

Recording systems

The recording of sound involves the storage of sufficient information to allow the reconstruction of the sound at a later date, and the accuracy of the subsequent reproduction will depend upon the amount of detail which has been stored. Nowadays, hi-fi systems are so good that, if they are played to an audience from behind a curtain, the audience cannot distinguish between the recording and the real thing.

Of the modern systems for recording sound, the gramophone record was developed first. In its simplest form, the sound waves are converted into mechanical oscillations which cause a needle to vibrate sideways in a groove on a rotating wax disc; the oscillations of the needle are recorded as fluctuations in the shape of the groove. If, subsequently, another needle is placed in the groove, and the disc rotated at the same rate, the oscillations of the first needle will be approximately reproduced and these oscillations can be used to drive a loudspeaker cone mechanically. It is important to note that electricity is not required for the basic version of the process, and records had been on the market for thirty years before the electric gramophone became available.

Today, the initial recording of a piece of music is always done on a magnetic tape, since this can be edited and spliced without difficulty. When it is replayed, the information which was stored magnetically on the tape, is converted into mechanical vibrations. These are then used to engrave a groove in an acetate lacquer mounted on a metal disc; this acetate lacquer is the *master disc*. The process of mass-producing copies of the master is technically complicated; the records, of vinyl resin, are pressed out by metal discs made indirectly from the master disc.

In a compact disc system the information recorded is stored in the form of surfaces which

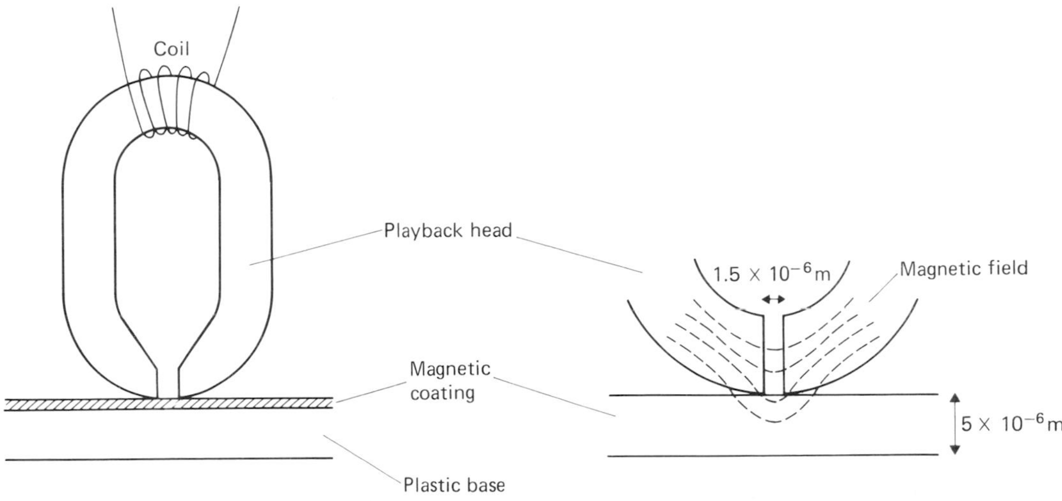

Fig. 5.9 The playback head in a tape recorder.

reflect laser light. The reflections provide a succession of pulses which are either on or off, and which can be used to reconstruct the original sound signal. A *digital* system, using binary code, enables more faithful reproduction than the continuous variations required in *analogue* systems.

Playing back a tape recording

In a tape recording, the sound information is stored on the tape as the induced magnetization of the magnetic coating. The playback head has a gap of about 1.5×10^{-6} m and the tape passes below this, as shown in Fig. 5.9. Now, it is easier for the magnetic flux to pass through the magnetic coating than through the air gap; we say that the air gap has a greater *reluctance* than the coating, because the *magnetic permeability* of air is about 1/1000 that of the coating. The field causes induced permanent magnetization in the coating which fluctuates in a similar way to the original sound. When the tape is passed under the playback head, the changing magnetic field caused by the movement of the tape induces voltages in the coil, which can then be amplified and fed to a loudspeaker.

5.4 Ultrasonics

Many recent important developments in sound technology have been in the field of *ultrasonics*; the frequency range above about 20 kHz in which sounds are no longer audible. As we shall see, ultrasonic waves can be used in applications as diverse as homogenizing emulsions, welding metal and examining unborn babies. Sound waves can, of course, be reflected and refracted in a similar way to light but because of diffraction effects the resolution of detail is much poorer, especially at low frequencies. As the frequency of sound is increased, the wavelength is decreased and resolution improved. Calculations of the resolution produced by optical systems are discussed in Chapter 8, and the principles developed there are applicable also to sound.

Q 5.4
Calculate the wavelength of sound at:

(*a*) 20 Hz
(*b*) 20 kHz
(*c*) 200 kHz.

Take the velocity of sound to be $330\ \mathrm{m\,s^{-1}}$.

Applications of ultrasonic sound can be divided into two types; those that use the beam merely as a source of energy, and are therefore run at high powers, and those that use coherent ultrasonic waves to obtain information. Later, we shall consider some examples of each type of application, but first we shall see how ultrasonic sound is generated.

The generation of ultrasonic sound

Originally ultrasonic sound was produced in specially designed sirens and whistles; the ultrasonic whistle is still used in applications where it is desired to transmit as much sound as possible into a liquid. The whistles are of low power compared with electrical transducers described later, but as the waves are generated directly in the liquid medium the *coupling* between source and medium is very good and little power is wasted.

A typical modern ultrasonic whistle is shown in Fig. 5.10(*a*); the liquid is forced out of the jet past a steel blade and eddy oscillations then occur in the liquid giving rise to sound waves. The note produced depends upon the speed of the jet and the whistle dimensions. The situation is similar to that of galloping stays, discussed in Section 4.9, and to the production of a note by blowing across a pen top. The whistle is simple and is widely used in homogenizing equipment.

In other applications, electrical transducers are more frequently used, and the most common types are the magnetostrictive and piezoelectric oscillators. A *magnetostrictive oscillator* consists of a sample of a suitable ferromagnetic material, such as nickel, placed inside a coil as in Fig. 5.10(*b*); when the current through the coil is changed, the field inside the sample changes and it expands or contracts. Thus, if an alternating current is passed through the coil, the size of the sample will also alternate, with identical frequency, and a sound wave is generated; magnetostrictive oscillators are restricted to frequencies less than 100 kHz.

A *piezoelectric oscillator* utilizes the fact that certain crystals, such as quartz or ammonium dihydrogen phosphate, change in size along one axis when the applied electric field is altered. In certain polycrystalline ceramics, subject to a large but fluctuating permanent electric field, a similar but more efficient effect, called *electrostriction*, occurs and oscillators using these materials are often used in sonar; an appropriately waterproofed example is shown in Fig. 5.10(*c*). The oil helps to couple the oscillations of the ceramic to the surrounding medium, which in sonar applications will be water.

The above account describes these transducers as 'loudspeakers' converting electrical energy into sound energy. These devices will work also as 'microphones' converting sound energy into electrical energy. In distance-ranging equipment, such as echo-sounders and sonar, the same transducer will sometimes act as a loudspeaker to emit pulses of sound but in between these pulses it will act as a microphone.

A particular crystal will have a natural frequency for mechanical vibrations in a number of modes, and at these frequencies resonance occurs between the applied electrical oscillations and the resulting oscillations of the crystal. In a *quartz crystal oscillator*, this effect

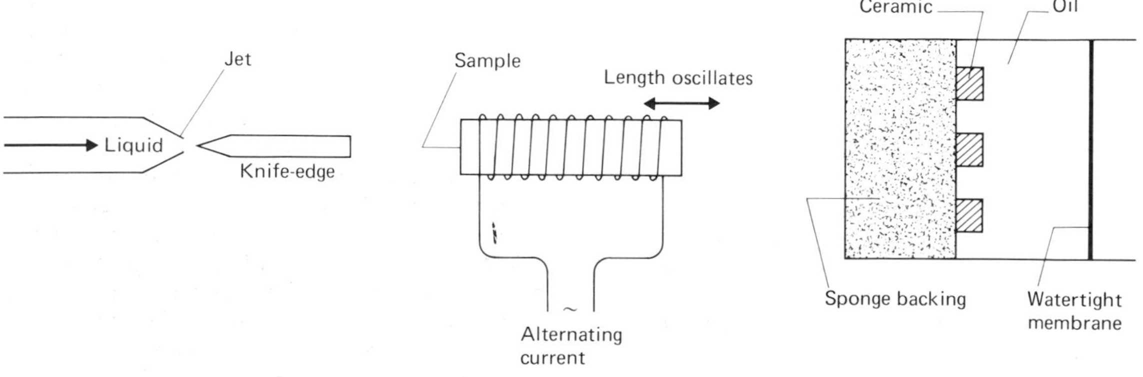

(*a*) An ultrasonic whistle (*b*) Magnetostrictive oscillator (*c*) Electrostrictive oscillator

Fig. 5.10

provides an accurate frequency standard, better than 1 part in 10^8 at a constant temperature; this is the basis of a *quartz clock* and is also used to provide a fixed frequency for radio transmissions.

Ultrasonic sound as a source of energy

When ultrasonic waves are passed through a liquid *cavitation* may occur; a rarefaction in the wave causes a space to form in the liquid which may contain air, liquid vapour or a vacuum. When this bubble collapses, during a compression, the inertia of the liquid moving together to fill the hole will cause a shock wave in the liquid and this may well be heard as a crack.

It is probable that it is these shock waves that cause *ultrasonic cleaning*, one of the most widely used applications of ultrasonics. Dirty articles to be cleaned are lowered into a bath containing a suitable solvent, in which ultrasonic waves are generated by a whistle, as seen in Fig. 5.11(*a*) and (*b*). A suitable frequency is about 20 kHz, as waves of higher frequency do not produce cavitation efficiently.

Cavitation also occurs where machinery such as a ship's propeller moves rapidly through a liquid; the shock waves are a serious cause of erosion of the metal.

An *emulsion* is a fine suspension of one liquid in another, and is formed by the *homogenizing* of two immiscible liquids such as oil and water; traditionally this is done by shaking or stirring the two liquids, but more recently methods have been developed using ultrasonic waves passed through the liquid. The amount of energy required is much reduced if ultrasonic sound is used, and such techniques are now common. Emulsions are of special importance in the paint, food, cosmetics and pharmaceutical industries.

Normal welding methods, involving melting the two surfaces, can often cause unwanted alloys to be formed when different metals are to be joined; the metals may also become brittle. In *ultrasonic welding*, the two metals are pressed together in a beam of low-frequency ultrasonic sound and if they are made to move over each other, welding will occur; less damage is done to the surfaces than by using traditional methods. The sound waves probably clean dirt and oxides from the surface, cause very localized heating and set up vibrations at the interface; plastic flow between the metals is then possible. Spot and seam welds can be carried out in this way, on items ranging in size from sheet aluminium down to miniature electronic circuits. A diagram of a typical welder is shown in Fig. 5.11(*c*).

Ultrasonic sound as a source of information

Ultrasonic waves used for collecting information are of lower intensity than in the applications described so far. To gain maximum resolution frequencies as high as possible should be used, but there is an important restriction in that, for a given amplitude, waves of higher frequency have a decreased range.

If ultrasonic pulses are transmitted into the wall of a pipe, they will reflect from cracks and other imperfections; these reflections will be picked up by the original transducer, or a similar one close by, and this arrangement is termed a *reflectoscope*. The use of pulses, rather than continuous waves, prevents stationary waves being set up in the pipe. A typical application of a reflectoscope is shown in Fig. 5.11(*d*), and ultrasonic pulses can also be used in situations as diverse as testing for cracks in railway lines, measuring the amounts of fat and lean tissue in a cow, and geophysical prospecting.

Finally, the science of ultrasonics has been widely used in medicine in the last ten or fifteen years. As well as giving surgeons another way of heating a tissue, for instance to kill a small number of brain cells, high-frequency sound can be used to take 'X-ray type' pictures of the inside of the body. Such techniques are simpler, and less dangerous, than the use of X-rays, and a typical ultrasonic scan is shown in Fig. 5.11(*e*). Ultrasonic waves reflected from a moving object will undergo a change of frequency which depends upon the object's velocity, as described in Section 3.8; this is known as the *Doppler effect*, and it can be used to investigate moving parts inside the body. The placenta of an unborn baby, for instance, can be located by reflections from the blood corpuscles moving through its capillaries, and the action of the baby's heart can be similarly monitored. The liquid which surrounds the baby acts as an excellent medium for the transmission of the sound.

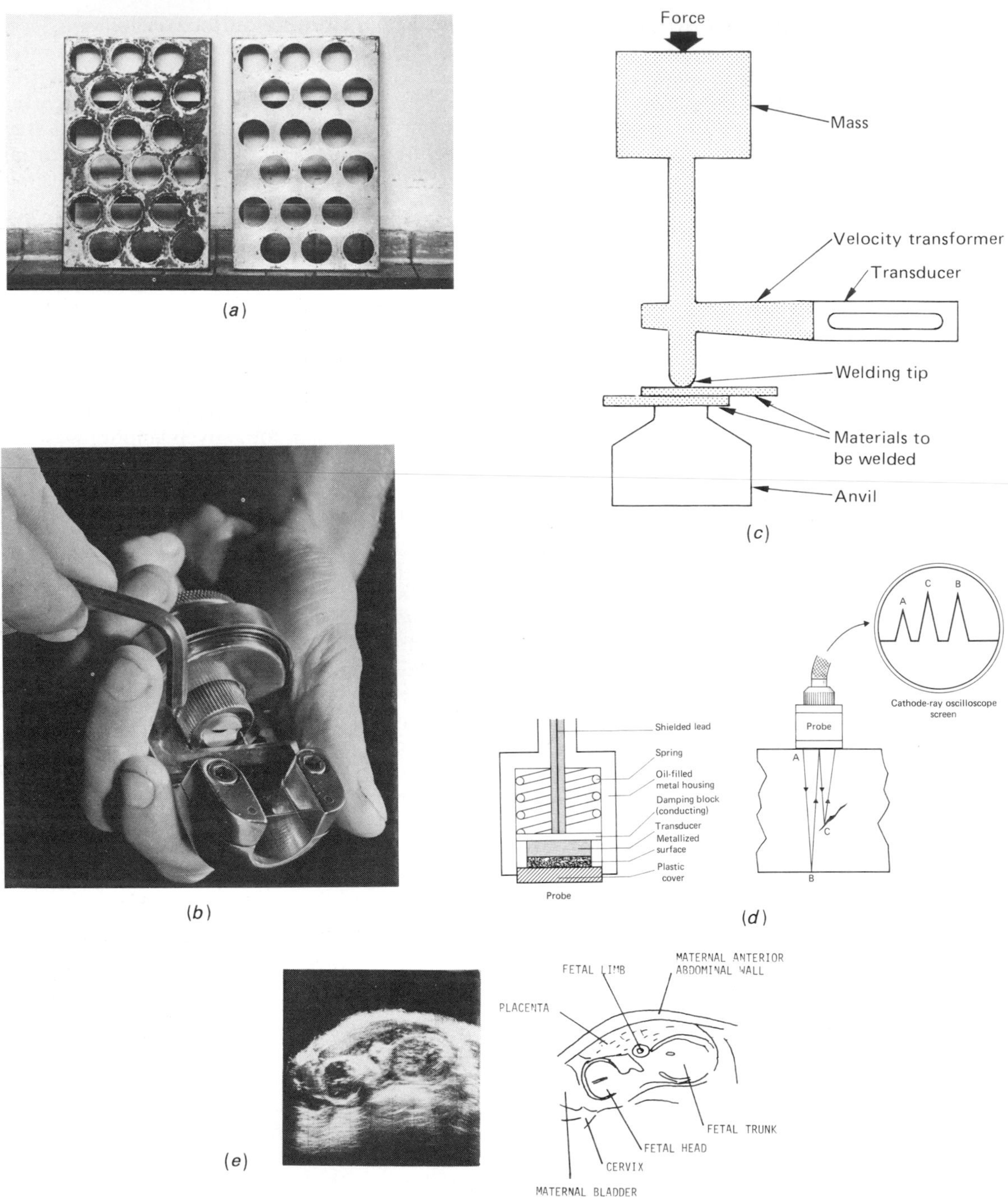

Fig. 5.11 (a) Ultrasonic cleaning in the baking industry. (b) A liquid whistle from a typical ultrasonic homogenizer, manufactured by Ultrasonics Ltd. (c) An ultrasonic welder; the sound is applied via a transducer. (d) An ultrasonic probe used to detect flaws in a material. Pulses (A) are emitted from the transducer and reflected both from the bottom surface (B) and from flaws (C); the reflected pulses are amplified and displayed on a cathode ray oscilloscope, alongside the emitted pulse. (e) A longitudinal cross-section through the abdomen of a pregnant woman lying on her back.

5.5 The velocity of sound

A derivation of the speed of sound in a rod, c, was given in Section 2.5 and we found that:

$$c=\sqrt{\frac{E}{\rho}} \tag{2.7}$$

where E is Young's modulus for the solid, and ρ is the density. In this section, we shall consider briefly those factors which affect the velocity of sound in unbounded gases and liquids.

The velocity of sound in gases

Let us suppose that the velocity of sound in gases depends on similar factors to the velocity of sound in a solid; that is, on a modulus of elasticity and on the density. A longitudinal wave passing through the gas produces expansion and contraction of small regions of gas, associated with changes in pressure; the appropriate modulus is the *bulk modulus*, K, defined by:

$$K=-\frac{\text{change in pressure}}{\text{change in volume/volume}} \tag{5.3}$$

The dimensions of K can be worked out from its definition:

$$[K]=\frac{[\text{force}]/[\text{area}]}{[\text{change in volume}]/[\text{volume}]}$$

$$=\frac{[\mathrm{MLT}^{-2}]/[\mathrm{L}^2]}{1}$$

$$=[\mathrm{ML}^{-1}\mathrm{T}^{-2}]$$

The dimensions of density are:

$$[\rho]=[\mathrm{ML}^{-3}]$$

The velocity might also depend upon the wavelength of the sound, λ, where:

$$[\lambda]=[\mathrm{L}]$$

We shall use the method of dimensions to see how the velocity of sound, c, with dimensions $[\mathrm{LT}^{-1}]$, depends on K, ρ and λ. Suppose

$$c \propto K^x\rho^y\lambda^z$$

$$[\mathrm{LT}^{-1}]=[\mathrm{ML}^{-1}\mathrm{T}^{-2}]^x[\mathrm{ML}^{-3}]^y[\mathrm{L}]^z$$

Hence, by the method of dimensions:

$$\begin{aligned}
&\mathrm{T}: & -1&=-2x & & x=\tfrac{1}{2}\\
&\mathrm{M}: & 0&=x+y & & y=-\tfrac{1}{2}\\
&\mathrm{L}: & 1&=-x-3y+z & &\\
& & &=-\tfrac{1}{2}+\tfrac{3}{2}+z & & \text{from above}\\
&\text{i.e.} & 1&=1+z & & z=0
\end{aligned}$$

$$\therefore c \propto \sqrt{\frac{K}{\rho}}$$

and it can be shown by a more rigorous treatment that

$$c=\sqrt{\frac{K}{\rho}} \tag{5.4}$$

Thus the velocity of sound in a gas is independent of the wavelength; according to this analysis, gases are non-dispersive.

Now, when a compression occurs, the gas is heated and the temperature rises; at a rarefaction the gas is cooled. As little heat flows from a compression to a rarefaction before the wave has moved an appreciable fraction of a wavelength, the temperature changes caused by the passage of the sound wave are hardly affected by heat flow; the conditions approximate to *adiabatic* ones.

For an ideal gas under adiabatic conditions the pressure P and volume V of the gas are related by the equation:

$$PV^{\gamma}=\text{a constant, } k$$

where γ is a constant determined by the atomicity of the gas. It can be shown that:

$$\gamma=\frac{\text{specific heat at constant pressure}}{\text{specific heat at constant volume}}$$

Hence, differentiating with respect to V, we have:

$$P\gamma V^{\gamma-1}+V^{\gamma}\frac{\mathrm{d}P}{\mathrm{d}V}=0$$

$$P\gamma+V\frac{\mathrm{d}P}{\mathrm{d}V}=0$$

i.e.

$$\frac{\mathrm{d}P}{\mathrm{d}V/V}=-\gamma P$$

But

$$K=\frac{-\mathrm{d}P}{\mathrm{d}V/V} \quad \text{from (5.3)}$$

$$=\gamma P$$

and hence, for an ideal gas the velocity of sound, c, is given by:

$$c = \sqrt{\frac{\gamma P}{\rho}} \qquad (5.5)$$

where γP is the adiabatic bulk modulus of the gas.

Now suppose that one mole of a gas has mass M and occupies volume V. The density of the gas, ρ, is given by:

$$\rho = M/V$$

The ideal gas equation states that

$$PV = RT$$

where P is the pressure and T the absolute temperature of the gas, and R is the universal gas constant.

Hence, since: $$c = \sqrt{\frac{\gamma P}{\rho}} \qquad (5.5)$$

we have: $$c = \sqrt{\frac{\gamma PV}{M}}$$

i.e. $$c = \sqrt{\frac{\gamma RT}{M}} \qquad (5.6)$$

We can deduce three results from Equations (5.5) and (5.6), namely:

(1) the velocity of sound in a given gas is independent of pressure, since at a given temperature the density of a gas is proportional to its pressure and thus P/ρ is constant;

(2) the velocity of sound in a given gas is proportional to $\sqrt{T}$;

(3) at a given temperature and pressure, the velocity of sound is inversely proportional to $\sqrt{\rho}$ and thus depends upon the gas. Therefore, in a cavity of given size the frequency of the sound produced is inversely proportional to the square root of the density of the gas, since the wavelength of the sound is determined by the dimensions of the cavity.

It is also of interest to note that the kinetic theory of gases enables us to relate the pressure and density of a gas to the root-mean-square velocity of the molecules, $\sqrt{\overline{v^2}}$. From kinetic theory:

$$P = \tfrac{1}{3}\rho\overline{v^2}$$

Now $$c = \sqrt{\frac{\gamma P}{\rho}}$$

i.e. $$c = \left(\sqrt{\frac{\gamma}{3}}\right)\sqrt{\overline{v^2}} \qquad (5.7)$$

Thus the speed of sound in a gas is proportional to the root-mean-square velocity of the gas molecules.

Q 5.5

The density of air at standard temperature and pressure is 1.29 kg m^{-3}. Air consists predominantly of diatomic molecules for which $\gamma = 1.40$. Take one standard atmosphere to be 10^5 N m^{-2} and calculate:

(*a*) the speed of sound in air under standard atmospheric conditions.

The density of helium at standard temperature and pressure is 0.17 kg m^{-3}. It consists of monatomic molecules for which $\gamma = 1.67$.

Calculate:

(*b*) the speed of sound in helium under standard conditions;

(*c*) the interval through which your voice will change pitch if you breathe in helium instead of air before speaking.

In each case assume that conditions are adiabatic.

The velocity of sound in liquids

The velocity of sound in unbounded liquids can be shown to depend upon the adiabatic bulk modulus, K, and the density ρ, as before:

$$c = \sqrt{\frac{K}{\rho}} \qquad (5.4)$$

As the pressure increases so does the bulk modulus. For most liquids the bulk modulus also decreases as the temperature increases but for water, there is a maximum value at about 50°C.

For sea water, the velocity depends upon the temperature, pressure and salinity, in decreasing order of importance. The velocity of sound

From Guillemin's The Forces of Nature, *by courtesy of Messrs Macmillan and Co. Ltd.*

Fig. 5.12 Colladon and Sturm's experiment.

increases by about 3 m s^{-1} for every 1K temperature rise, and by about 0.2 m s^{-1} for an increase in pressure of 1 atmosphere—due to an increase in depth of 10 m. Changes in the velocity of sound in the sea cause sonar pulses to travel along curved paths, and sometimes to be totally internally reflected at a temperature layer.

Early experiments to measure the velocity of sound through water now seem to us to be archaic, but Colladon and Sturm obtained quite accurate results on Lake Geneva in 1826. Their boats were about 14 km apart and a value of 1600 m s^{-1} was obtained; it is now known that at standard pressure and 20°C, the velocity of sound in water is 1460 m s^{-1}. More recent measurements have been made using depth charges and hydrophones feeding the signal to an automatic timer; the timing of the explosion is transmitted by radio.

5.6 Measurements on sound

The velocity of sound in free air

Modern equipment simplifies the accurate measurement of the velocity of sound in the laboratory; suitable apparatus is shown in Fig. 5.13. Two microphones, placed a known distance apart, are connected to a start/stop gate and digital readout; using a 100 kHz oscillator, this can count to 10^{-5} s. The left-hand microphone is connected to the 'make to count' or 'start' gate, and the right-hand microphone to the 'make to stop' gate. Thus, when a pulse of sound reaches the left-hand microphonc, thc scaler will start counting, and when it reaches the right-hand microphone, it will stop; the elapsed time is shown on the counter. The pulse of sound is a sharp, loud noise—such as a handclap made a few centimetres from the left-hand

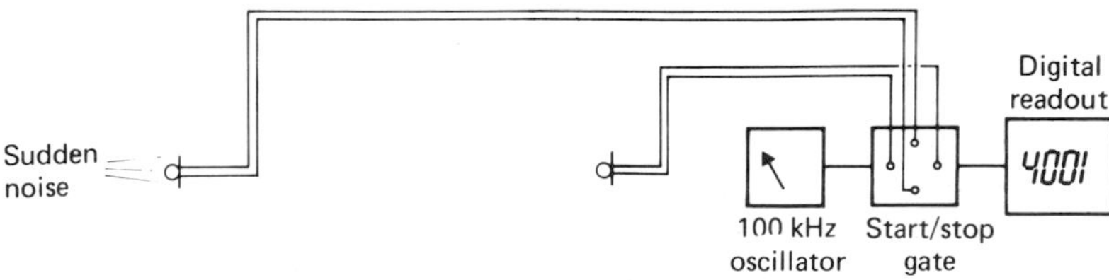

Fig. 5.13

microphone. The distance between the microphones is carefully measured and the velocity calculated from the equation:

$$\text{velocity} = \frac{\text{distance travelled}}{\text{time taken}}$$

To improve accuracy a number of readings at different distances are made and a graph of time against distance is plotted; the velocity is equal to the reciprocal of the slope of the graph. The use of a graph helps to eliminate systematic errors, such as any doubt about the exact position of the microphones in their cases.

Q 5.6
In an experiment to measure the velocity of sound, the following readings are taken: plot a graph and calculate the velocity.

Time/10^{-3} s	3.1	3.9	6.2	7.2	8.9
Distance/m	1.0	1.5	2.0	2.5	3.0

The velocity of sound in a resonance tube

The velocity of sound in a tube could well be different from that in free air; in fact, the accuracy of the experiments described is not sufficient to record any consistent discrepancy. The velocity of sound in a resonance tube is established by calculation from the equation:

$$v = f\lambda \qquad (2.1)$$

The resonance tube shown in Fig. 5.14(*a*) is used to measure the wavelength of the sound produced by a tuning fork of known frequency. In Section 4.8 we saw which waves could be set up in a pipe closed at one end and how the wavelength, λ, was related to the length, l, of a pipe both for the fundamental frequency, and also for a number of higher harmonics.

Fundamental: $\lambda = 4l_1 \qquad l_1 = \lambda/4 \qquad (5.8)$

3rd harmonic: $\lambda = 4l_2/3 \qquad l_2 = 3\lambda/4 \qquad (5.9)$

etc.

The resonance tube may be any tube of adjustable length but it is usual to use a vertical

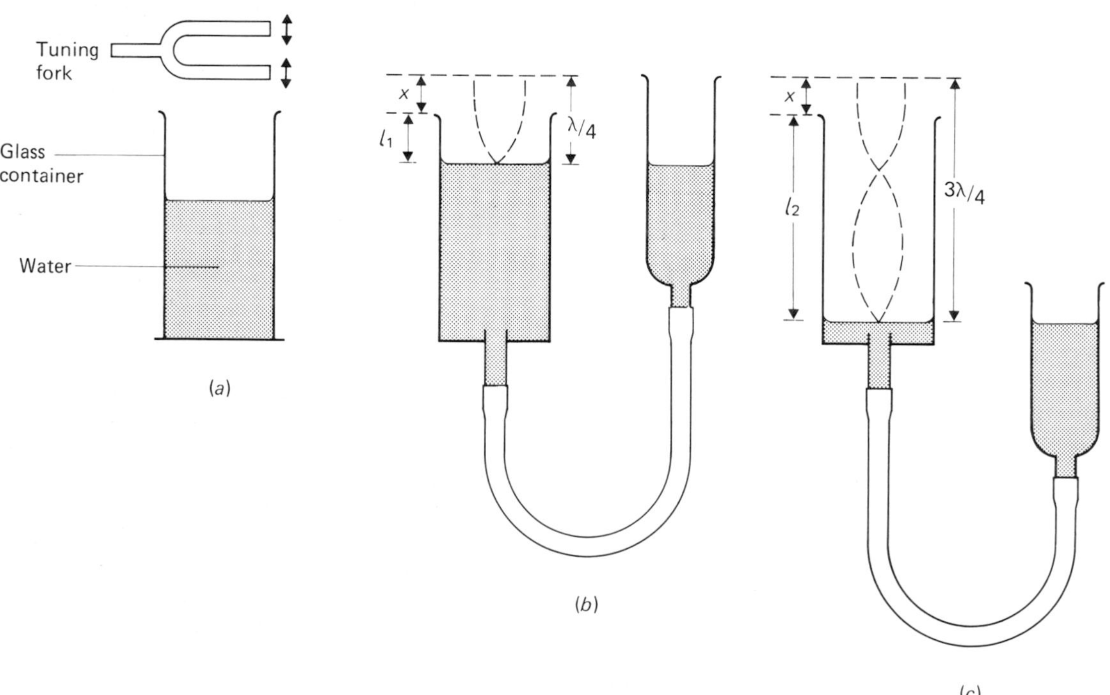

Fig. 5.14

glass tube, closed at the bottom, about 1 m long and 5 cm wide. The length of the air column can be altered by changing the water level in the tube, and this can be done by the arrangement shown in Fig. 5.14(*b*); it can then be changed whilst the note is sounded, by adjusting the level of the right-hand reservoir.

The tuning fork is made to vibrate and held above the tube, as shown. The level is lowered until the first loud resonance is heard and, in theory, the length of the air column will now be one quarter of the wavelength as stated in Equation (5.8). Experiments show, in fact, that the effective length of the tube is greater than the actual length by a constant amount, x, known as the *end correction*. Hence, instead of Equation (5.8), we write:

$$(l_1+x)=\lambda/4 \tag{5.10}$$

The water level is now lowered to the next position at which a resonance can be heard. We now have:

$$(l_2+x)=3\lambda/4 \tag{5.11}$$

as shown in Fig. 5.14(*c*).

The frequency of the note is unchanged and so, therefore, is its wavelength; the length of the tube has been increased to accommodate an extra half wavelength.

As the end correction is constant, we can subtract Equation (5.10) from Equation (5.11):

$$(l_2-l_1)=\lambda/2$$

i.e. $$\lambda=2(l_2-l_1)$$

But $$c=f\lambda$$

Hence $$c=2f(l_2-l_1) \tag{5.12}$$

Q 5.7

A tuning fork of frequency 512 Hz is used with a resonance tube to find the velocity of sound. The first two resonances occur when the lengths of the air columns are 15 cm and 47 cm. Calculate the velocity of sound and the end correction.

The frequency of a sound

The frequency of a sound may be measured, in principle, by using a microphone to display its waveform on a cathode-ray oscilloscope. The timebase control must be in the calibrated position so that the time interval corresponding to 1 cm on the screen is given by the timebase setting, and the period of the wave can be calculated from the number of centimetres it extends across the screen.

The oscilloscope timebase must, of course, be checked against a known frequency. This could be provided, for instance, by a good quality tuning fork or a quartz oscillator.

Q 5.8

A sound of unknown frequency has a period corresponding to 2 cm on the oscilloscope screen. If the timebase is accurately calibrated at 500 μs cm^{-1}, what is the frequency?

5.7 The loudness of sounds

We have seen, on page 81, that the energy incident in any wave motion is proportional to the square of the amplitude; the incident power per square metre is termed the *intensity*. The range of sound intensities with which the human ear can cope is very large indeed, and varies with the frequency of the sound. At any particular frequency, the lowest intensity which can be heard is called the *threshold of audibility* and the intensity at which the sound just begins to become painful is called the *threshold of feeling*. The way in which these two thresholds vary with frequency is shown in Fig. 5.15; a logarithmic scale is used for both intensity and frequency.

The ratio of these two thresholds is greatest at about 3000 Hz, to which frequency the ear is most sensitive; the threshold of feeling is then just over 10^{12} times as intense as the threshold of audibility. However, the brain does not perceive the difference in the loudness of the sounds to be as great as this would suggest, and we must therefore develop a definition of *loudness*, L_N, which approximates more closely to our perception of sound.

The increase in intensity required for us to observe a perceptible increase in loudness is proportional to the original intensity; the increase in loudness observed, ΔL_N, may be written in terms of the increase of intensity, ΔI, as follows:

$$\Delta L_N \propto \frac{\Delta I}{I}$$

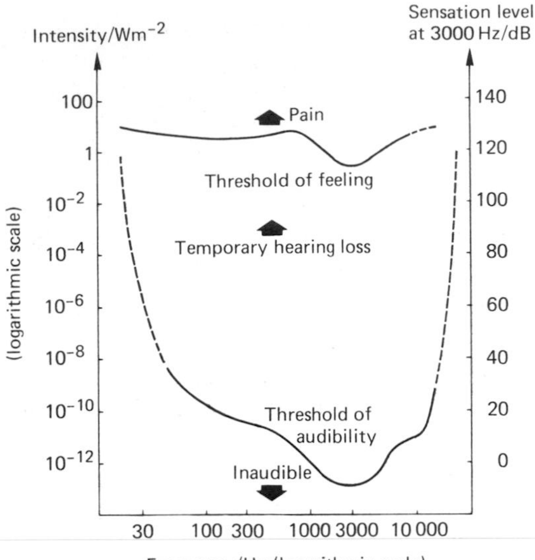

Fig. 5.15

or

$$\Delta L_N = k\left(\frac{\Delta I}{I}\right) \qquad (5.13)$$

This differential equation may be integrated to give:

$$L_N = k \ln\left(\frac{I}{I_0}\right)$$

and written in terms of logarithms to the base 10:

$$L_N = k' \log_{10}\left(\frac{I}{I_0}\right) \qquad (5.14)$$

This equation, known as the *Weber–Fechner* law, tells us that the ear perceives the loudness of sounds on a logarithmic scale; laws of similar form can also be applied to other situations of human physical perception.

Note that the Weber–Fechner law only gives us a means of *comparing* two sounds, of intensity I and I_0. For instance, the loudness of a sound of intensity 10^{-7} W m^{-2} relative to one of intensity 10^{-9} W m^{-2} is given by:

$$L_N = k' \log_{10}\left(\frac{10^{-7}}{10^{-9}}\right) = k' \log_{10}(10^2) = 2k'$$

When we take the value of k' to be one, we say that the loudness is measured in *bels* (which is, of course, a dimensionless unit, being a measure of a ratio); more commonly, we take the value of k' to be ten and the loudness is measured in *decibels* (dB). Thus, the loudness of the first sound was 2 bels, or 20 decibels, above the second.

Q 5.9
An increase in loudness of 3 dB is the smallest which is perceptible by the normal ear. By what proportion does the original intensity have to be increased to cause this?

Clearly it is most useful if the loudness of a sound of a particular frequency is measured relative to the threshold of hearing at that frequency. This method of measuring loudness gives a scale called the *sensation level, SL*, defined by:

$$SL = 10 \log_{10}\left(\frac{\text{intensity of sound}}{\text{intensity of sound of same pitch and waveform at the threshold of hearing.}}\right) \qquad (5.15)$$

Like loudness, sensation level is measured in decibels.

Q 5.10
Use Fig. 5.15 to calculate the sensation level of the threshold of feeling at 100 Hz and at 3000 Hz.

Values of sensation level, calculated in this way, seem to correspond closely with our subjective assessment of the relative loudness of sounds of a particular frequency; the right hand scale in Fig. 5.15 is labelled with values of sensation level at 3000 Hz. Each increase of 10 dB corresponds to an increase in the intensity by a factor of 10. However, when we wish to compare everyday noises, such as those due to machinery, or a busy road, we must allow for the presence of a wide range of frequencies.

The comparison of the loudness of sounds of different frequencies is made difficult partly because, as we have seen, the ear responds differently to the different frequencies. One method uses a frequency of 1000 Hz at an intensity of 4.5×10^{-13} W m^{-2} as a standard. This sound intensity is then increased until it seems to be as loud as the sound, of different frequency, which is being investigated. The increase in loudness of the standard frequency, in decibels, is said to be the *equivalent loudness* of the other sound, which is expressed in *phons*.

Q 5.11
A sound of frequency 300 Hz sounds as loud as a sound of frequency 1000 Hz and intensity 9×10^{-9} W m^{-2}. What is its equivalent loudness, in phons?

The other difficulty that arises with a variety of frequencies is that the degree of irritation caused varies with the pitch of the sound; generally speaking, we find sounds of higher frequencies more unpleasant. As many practical noise measurements are to do with assessing the nuisance caused, for instance, by a motorway or airfield, these subjective factors are clearly of importance. A number of different scales for measuring the loudness of noise consisting of a mixture of frequencies have been developed. Each of these is a logarithmic scale, defined in a similar way to the decibel, but with a weighting in the measuring instrument which accentuates the effect of higher frequencies. In everyday situations, the *A scale*, with units of dBA, is used most often; on the A scale, the weighting network subtracts about 11 dB from the intensity measured at 100 Hz but nothing from frequencies of about 1000 Hz, to which we are more sensitive. Some examples of noise levels are given in the list below; levels above 90 dBA tend to make it difficult to work effectively, and above 95 dBA they may impair hearing.

Live 'pop' music	120 dBA
Boilermaking shop	100 dBA
Busy street	80 dBA
Noisy office	75 dBA
Conversation	60 dBA
Quiet office	40 dBA
Whisper	30 dBA

Modern jet aircraft also produce a lot of noise pollution when they pass close overhead during landing and takeoff, and here measurements are based on the average response of large numbers of people. The loudness is measured as a *perceived noise level*, in units of (PN)dB. Early jets produced a noise of up to 120 (PN)dB on landing and takeoff, but modern high by-pass turbofan engines, such as the Rolls-Royce RB 211, can reduce this to under 100 (PN)dB by making much of the airflow by-pass the combustion chamber. At London's Heathrow Airport, pilots are required to keep below 110 (PN)dB by day and 102 (PN)dB at night, by climbing as rapidly as possible and then reducing the thrust as far as safety allows.

All the scales described above are logarithmic scales, and these are also used by radio and television engineers in power and amplification calculations; the decibel is always related to the ratio of two powers.

$$\text{electrical power level in dB} = 10\log\left(\frac{W_1}{W_2}\right)$$

$$= 10\log\left(\frac{I_1^2}{I_2^2}\right)$$

since $W = I^2R$

$$= 20\log\left(\frac{I_1}{I_2}\right)$$

and similarly

$$\text{electrical power level in dB} = 20\log\left(\frac{V_1}{V_2}\right)$$

since $W = V^2/R$

We can also relate the ratio of sound intensities to the amplitude of the variation in air pressure (the *excess pressure amplitude*, Δp):

$$\text{loudness of sound} = 10\log\left(\frac{I_1}{I_2}\right)$$

$$= 10\log\frac{(\Delta p_1)^2}{(\Delta p_2)^2}$$

$$= 20\log\frac{(\Delta p_1)}{(\Delta p_2)}$$

The advantage of a logarithmic scale is that an increase in the power by a constant factor increases the level in decibels by a constant amount; for example, increasing the intensity by a factor of 10 will increase the level by 10 log 10, which is 10 decibels, irrespective of the initial intensity or level.

Q 5.12
A singer's voice has a power of 4.5×10^{-8} W m^{-2}. What will be the power produced by (i) 10 singers; (ii) 100 singers; (iii) 1000 singers, all singing the same note with the same loudness?

Bibliography

General accounts

Chedd, G. (1970) *Sound.* Aldus; an excellent introduction to many modern developments in sound.

CRAC Science Support Series (1982) *Waves and Sound.* Hobsons Press.

Cracknell, A.P. (1980) *Ultrasonics.* Wykeham; examines a wide range of applications of ultrasonics.

Haines, G. (1974) *Underwater Sound.* David and Charles.

Schools Council Engineering Science Project (1975) *Waves and Vibrations.* Macmillan; includes a discussion of issues relating to sound and loudness, with a variety of practical examples.

SCSST Physics Plus (1985) *Sonar.* Hobsons Press.

Specific topics

Three excellent books which give a thorough introduction to the physics of music and which tackle some of the fascinating but difficult problems relating to consonance and harmony:

Backus, J. (1970) *The Acoustical Foundations of Music.* Murray.

Hutchins, C.M. (1978) *The Physics of Music.* Freeman; a Scientific American book.

Taylor, C.A. (1965) *The Physics of Musical Sound.* English University Press.

Two books which include a discussion of how animals use sound signals, especially at ultrasonic frequencies, for communication.

Griffen, D.F. (1960) *Echoes of Bats and Men.* Heinemann (out of print).

Burkhardt, D., Schleidt, W., and Altner, H. (1967) *Signals in the Animal World.* George Allen and Unwin.

Reprint (now out of print but available in many libraries)

Griffin, D.R. (1958) More about Bat 'Radar', *Scientific American*, offprint No. 56.

Part 2
The properties of light

6
Image formation

6.1 Introduction

In the first part of this book, we have considered the general properties of waves, and related them especially to sound. We now turn to a detailed examination of the properties of light, and in Chapter 7 we shall examine those aspects of the behaviour of light which depend upon its wave nature. The arguments for and against the wave picture of light are presented in the last part of the book.

In this chapter, however, we are concerned with situations in which light can be considered merely to consist of rays, which travel in straight lines; this is sometimes called the science of *geometrical optics*. Much of the behaviour of mirrors, lenses and other simple optical systems can be explained in terms of the *rectilinear propagation of light*. A straight beam of light can easily be seen when passed from a laser through a smoke-filled atmosphere, and Fig. 6.1 illustrates how shadows and eclipses are formed by the obstruction of straight light rays by an opaque object.

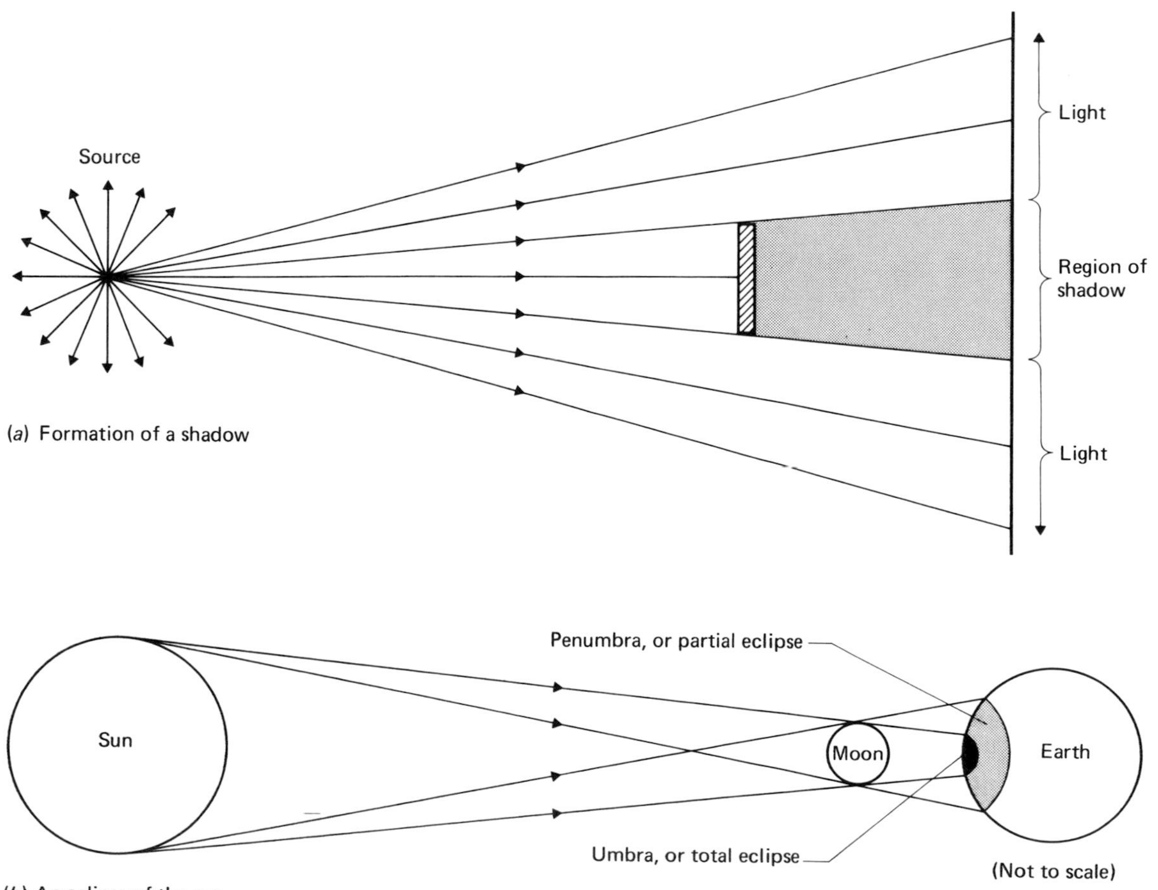

Fig. 6.1

Q 6.1
How accurately does a light beam travel in a straight line? A hair, 1/100 mm in diameter, casts a shadow in a parallel beam of light on a wall 10 m away. Through what angle would the ray passing one side of the hair have to be deflected in order for the shadow just to disappear? Give the answer in radians and in seconds of arc. Ignore diffraction effects.

Approximations

The theory of geometrical optics is frequently simplified by the making of approximations. For instance, a book of trigonometry tables yields the following data:

θ/deg	1	10	30	50
θ/rad	0.017453	0.174533	0.523599	0.872665
tan θ	0.017455	0.176327	0.577350	1.191754
sin θ	0.017452	0.173648	0.500000	0.766044

Thus, for small angles, θ (in *radians*), tan θ and sin θ are nearly identical; even at $\theta = 10°$ (0.17 rad) the errors involved in such assumptions are only about 1%.

These and other approximations frequently made are tabulated in Fig. 6.2; a discussion of radian measure will be found in Appendix 1.

Example	Exact statement	Approximate statement
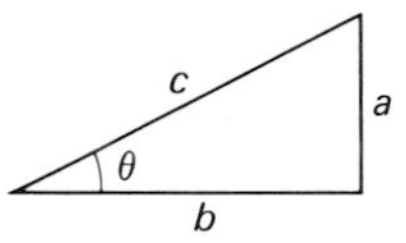 (right triangle with sides a, b, c, angle θ)	$\tan\theta = \frac{a}{b}$ $\sin\theta = \frac{a}{c}$	$\theta = \frac{a}{b}$ $\sin\theta = \tan\theta = \theta$
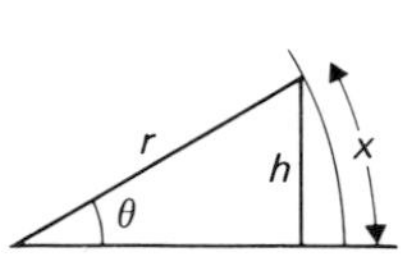 (triangle with radius r, height h, arc x, angle θ)	$\theta = \frac{x}{r}$ $\sin\theta = \frac{h}{r}$	$h = x$ $\theta = \frac{h}{r}$ $\sin\theta = \frac{x}{r}$ $h = r\theta$
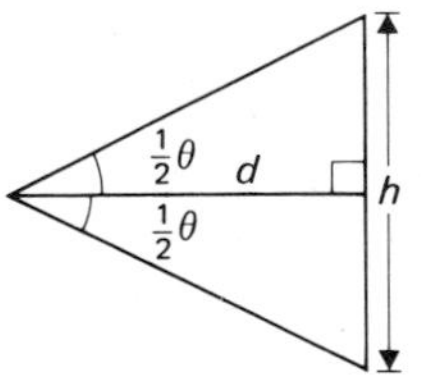 (isosceles triangle with angles $\frac{1}{2}\theta$, $\frac{1}{2}\theta$, distance d, height h)	$\tan\theta/2 = \frac{h/2}{d} = \frac{h}{2d}$	$\tan\theta = \frac{h}{d}$ $\theta = \frac{h}{d}$ $h = d\theta$

NB *In this table θ is expressed in radians.*

Fig. 6.2 Approximations used in optics.

Q 6.2
Calculate the percentage error in making the following assumptions at 1°, 10°, 30°, 50°, to the nearest whole number:

(*a*) $\tan\theta = \theta$;
(*b*) $\sin\theta = \theta$.

In each case θ is measured in radians.

Real and virtual images

When a convex lens forms an image of a distant light bulb on a screen, light rays actually pass through the image—an 'energy-meter' would detect a flux of energy there. The image is said to be a *real image* because light actually passes through it. A film exposed at the image point could be developed to show a clear image of the object. This is illustrated in Fig. 6.3(*a*).

On the other hand, an image formed in a plane mirror is behind the mirror, as shown in Fig. 6.3(*b*). The rays entering the eye appear to diverge from a point that is as far behind the mirror as the object is in front, and that is where the image is located. In this case, no light actually passes through the image, and it is termed a *virtual image*; if a piece of film, or light meter, were placed at the point where the image is formed, it would not, of course, detect any light, since it would be behind the mirror.

Our eyes see a distant object because light rays diverging from it are made to converge by the eye lens and a real image is formed on the retina, as shown in Fig. 6.3(*c*). Similarly, we can see both real and virtual images by making the diverging light converge on to the retina; alternatively, we may take a photograph of the images by forming a real image on the film instead of on the retina.

Lens and mirror systems may produce either real or virtual images depending upon the conditions: it is important to be able to distinguish between them.

A real image is formed when light actually passes through the image.
A virtual image is formed when light only appears to pass through the image.

The method of no parallax

An important technique for determining the position of an image is called *the method of no*

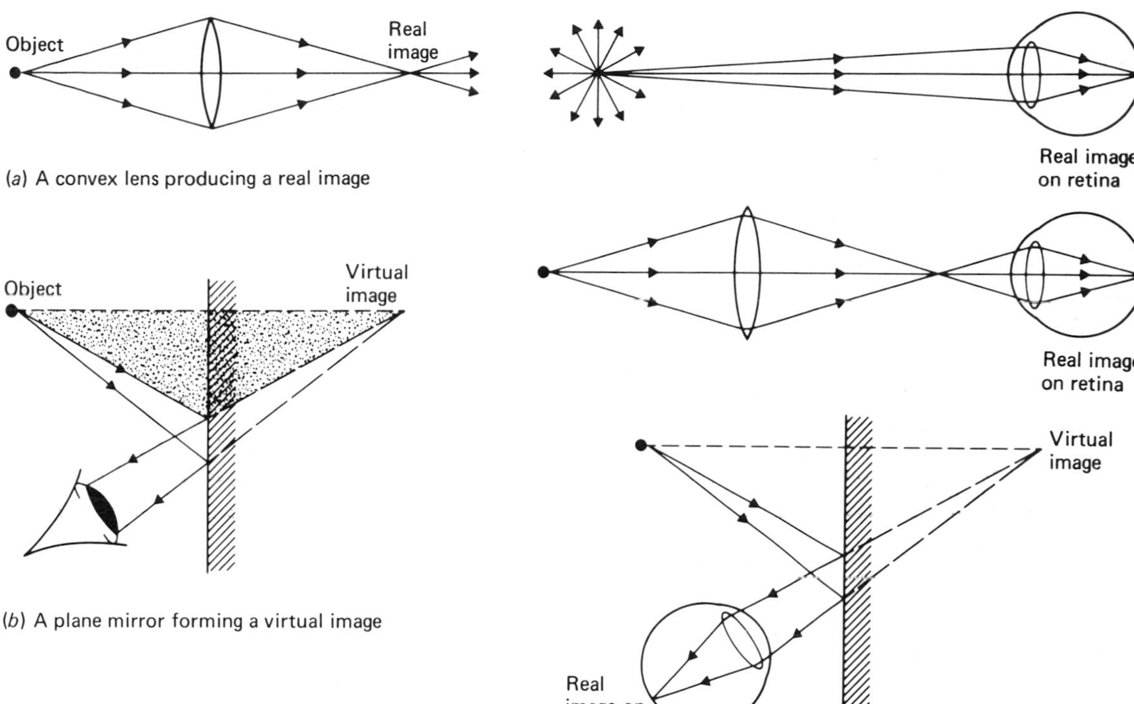

(*a*) A convex lens producing a real image

(*b*) A plane mirror forming a virtual image

(*c*) Real images formed by the eye

Fig. 6.3

parallax. The two objects A and B in Fig. 6.4 appear to be in line when the eye is in the central position in both diagrams (*a*) and (*b*). In diagram (*a*), when the eye is moved to one side, the objects no longer appear to be in line—*parallax* is observed. However, if the two objects were in the same place, as shown in (*b*), they would appear to be 'in line' from wherever they were viewed—we observe *no parallax*.

The method can be used, for instance, to locate the image in a plane mirror. Place an object, such as a candle, in front of the mirror and look at its reflection in the mirror. Now hold a second candle behind the mirror so that both it and the image of the first candle are visible. Move the second candle around until no parallax is observed between the second candle and the image of the first: they must now be coincident, and the second candle's position is the required position of the image. It will be found to be as far behind the mirror as the first candle is in front.

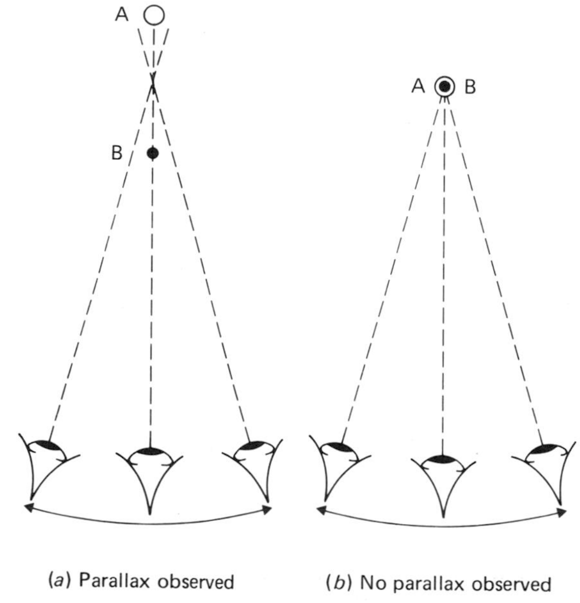

(*a*) Parallax observed

(*b*) No parallax observed

Fig. 6.4

6.2 Curved mirrors

Laws of reflection

Light is reflected from a mirror surface in accordance with simple laws, but before we can formulate these, we must understand how angles are measured in optics. It is usual to measure the angle between a light ray striking a surface and an imaginary line at the point of contact perpendicular to the surface; this line is called the *normal* to the surface and is shown in Fig. 6.5.

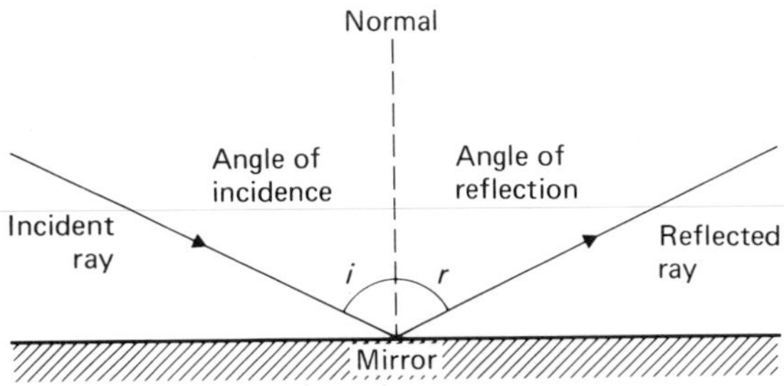

Fig. 6.5

The first law of reflection tells us that the reflected ray must lie in a certain plane:

The incident ray and normal define a plane which contains the reflected ray.

The second law relates the sizes of the angles involved:

The angle of reflection is equal to the angle of incidence.

In fact, the statement now known as 'the second law' was discovered first by the ancient Greeks, and used by Euclid to prove theorems about mirrors; the 'first law' was formulated by the Arab Alhazen in about 1000 AD.

Q 6.3
Use the second law of reflection to show:

(*a*) that the two shaded triangles in Fig. 6.3(*b*) are congruent, and hence that the image in the mirror is along a perpendicular to the surface and as far behind the mirror as the object is in front;
(*b*) that when a ray strikes a mirror the angle through which the ray is deviated is twice that between the ray and the mirror surface.

It is easy to observe that the laws of reflection apply when light is reflected from a mirror or other polished surface, but less easy to see how it applies to the reflection of sunlight from a wall into all parts of a room. The former is termed *specular reflection* and the latter is termed *diffuse reflection*. In the latter case the laws are still obeyed but, because the surface is rough, the normal at each point on the surface is in a different direction, as shown in Fig. 6.6.

The polishing of a mirror involves a succession of processes. Coarse silicon carbide is used for the coarse grinding, and then the fine ground stages are produced by using a very fine abrasive known as 'optical flour'. Polishing is carried out to make the glass surface transparent instead of translucent, and a substance such as red oxide of iron is used for this. Accurate scientific mirrors must not depart from their prescribed shape by much more than one tenth of a wavelength of light.

A highly polished glass surface will reflect about 4% of the light incident normally on it. In 1835, Liebig invented a method for depositing silver on glass chemically and front-silvered mirrors were widely used in scientific work until this century; about 95% of the light is reflected but the silver soon tarnishes and must then be replaced. In many optical instruments the metal

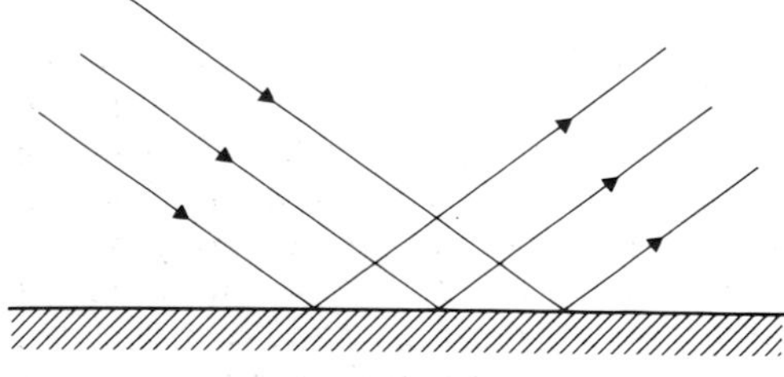

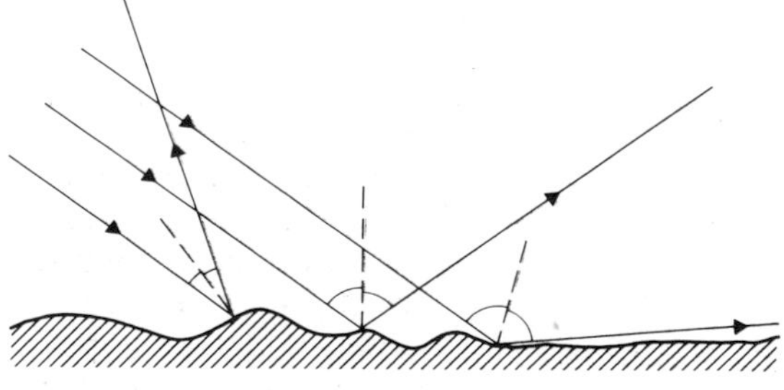

Fig. 6.6

is deposited on the front of the mirror, in order to avoid the confusion which might arise between images formed in both the front and back surfaces of the mirror; in domestic use the reflecting layer is at the back and is covered with a protective coating.

Recently, methods of evaporating aluminium onto a polished glass surface have been developed and this has many advantages; no polishing of the metal surface is required and after a few months it acquires a hard, protective, transparent coating of aluminium oxide and will then last for years. Although the maximum percentage of light reflected is only 85%, the extra durability is a great advantage in many applications.

Concave and convex mirrors

Curved mirrors, both concave and convex, can also form images; these may be either real or virtual depending upon the conditions. Convex spherical mirrors are used, for instance, to produce a large field of view in some types of driving mirror. Concave mirrors produce a magnified image for shaving; they are also used in reflecting telescopes to collect light from a faint, distant object over a wide area and concentrate it in a bright image. On the other hand, in a searchlight or torch, the concave mirror forms the light from a small source into a broad, parallel beam.

Focal points and radii of curvature

In Fig. 6.7 rays are shown striking both concave and convex mirrors; in each case the rays are parallel to the *principal axis*, or axis of symmetry, of the mirror. The *focal point*, or *principal focus*, F, of a concave mirror is defined as that point through which all of these rays of light pass, after reflection; it is the real image of an object at infinity. The focal point of a convex mirror is defined as that point from which all these rays appear to diverge, after reflection; it is the virtual image of an object at infinity. The *focal length*, *f*, of a mirror is the distance from the *pole*, P, to the focal point of the mirror. We shall prove later that all rays parallel to the axis do indeed pass, or appear to pass, through the same axial point wherever they strike the mirror, provided it is not too far from the pole.

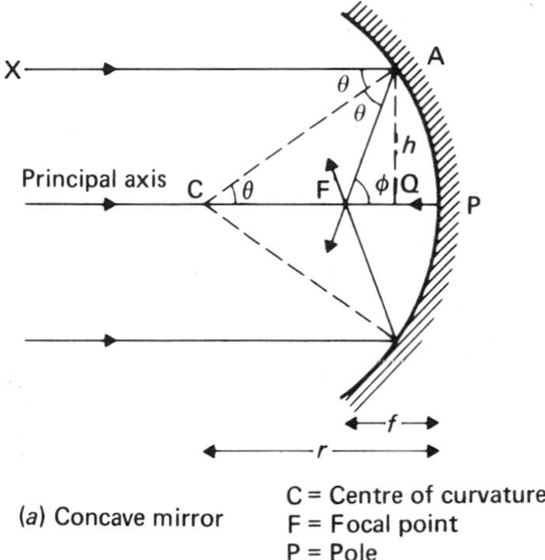

(*a*) Concave mirror

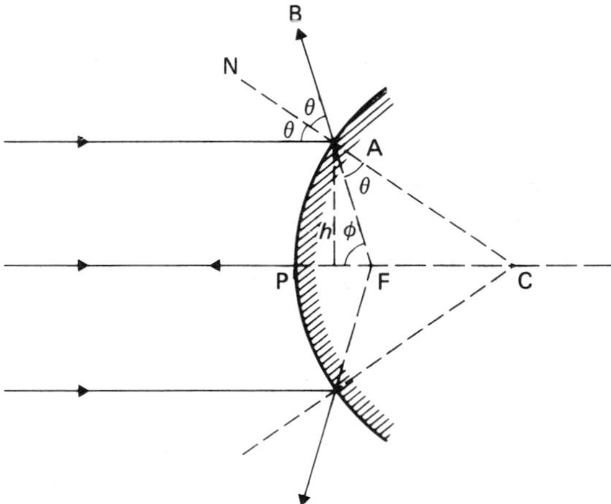

(*b*) Convex mirror

Fig. 6.7

Images of extended objects, formed by curved mirrors, are likely to be distorted unless rays from them strike the mirror at a small angle to the axis, and this is assumed in the approximations made in the proofs which follow; for clarity, however, rays making large angles with the axis are shown in the diagrams. Rays which make small angles with the axis, and which strike a mirror or lens near the middle, are called *paraxial rays.* The types of distortion that may occur when angles are not small are discussed in Section 6.6 on aberrations.

We also assume, in our analysis, that the curved mirrors form small parts of the surfaces of spheres. The *radius of curvature* of the mirror is the radius of the imaginary sphere of which the mirror forms a part; the centre of this sphere is called the *centre of curvature*, C, and is marked on both diagrams. The proof which follows shows that the focal length of curved mirrors is half their radius of curvature.

In the diagram for the concave mirror, we assume that the ray XA is close to the axis, as explained above. The foot of the perpendicular AQ will be close to P and hence we may make the following approximations:

$$FQ \approx FP = f$$
$$CQ \approx CP = r$$

Now, by the angle law of reflection, since CA is a normal to the mirror:

$$\angle FAC = \angle XAC$$
$$= \theta$$

and by alternate angles, since XA and CF are parallel:

$$\angle FCA = \angle XAC$$
$$= \angle FAC$$

i.e. $\angle FCA = \theta$

Hence, if $\angle PFA = \phi$

we have $\phi = 2\theta,$

since the exterior angle of a triangle is equal to the sum of the two opposite interior angles. Now if θ and ϕ are both small:

$$\tan \phi \approx 2 \tan \theta$$

But $\tan \phi = AQ/FQ$

$\approx h/f$ from above

and $\tan \theta = AQ/CQ$

$\approx h/r$ from above

Hence $$\frac{h}{f} = 2\frac{h}{r}$$

i.e. $$r = 2f \qquad (6.1)$$

Thus, we have shown that the focal length of a spherical mirror is half the radius of curvature, and as the result is independent of h, all rays close to the axis pass through F, no matter where they strike the mirror. This can also be shown to be true for a convex mirror.

Q 6.4
Use Fig. 6.7(b) to prove the same result for a convex mirror.

The mirror formula

We shall now derive a relationship between the object distance, u, and the image distance, v, for an image formed by a concave mirror, as shown in Fig. 6.8(a); the proof will also show that all rays from the object that make a small angle with the axis will pass through the image, no matter where they strike the mirror. We restrict ourselves initially to points which lie on the principal axis.

We again assume that the foot of the perpendicular AQ lies close to P, and hence:

$$OQ \approx OP = u \qquad (6.2)$$
$$CQ \approx CP = r \qquad (6.3)$$
$$IQ \approx IP = v. \qquad (6.4)$$

Now, by the angle law of reflection, since CA is normal to the mirror:

$$\angle OAC = \angle IAC = \theta$$

and by the external angles of triangles:

$$\angle PCA = \angle COA + \angle OAC$$

i.e. $\phi = \alpha + \theta$

$$\alpha = \phi - \theta \qquad (6.5)$$

Similarly, for triangle ICA:

$$\angle QIA = \angle ICA + \angle IAC$$

i.e. $$\beta = \phi + \theta \qquad (6.6)$$

Adding Equations (6.5) and (6.6) gives:

$$\alpha + \beta = 2\phi \qquad (6.7)$$

Making our usual approximations for paraxial rays we may write:

$$\tan \alpha + \tan \beta \approx 2 \tan \phi$$

Thus: $$\frac{h}{OQ} + \frac{h}{IQ} = \frac{2h}{CQ}$$

i.e. $$\frac{h}{u} + \frac{h}{v} = \frac{2h}{r}$$ from (6.2), 6.3), and (6.4)

Thus, the equation connecting object and image distances is:

$$\frac{1}{u} + \frac{1}{v} = \frac{2}{r} = \frac{1}{f}$$ from (6.1)

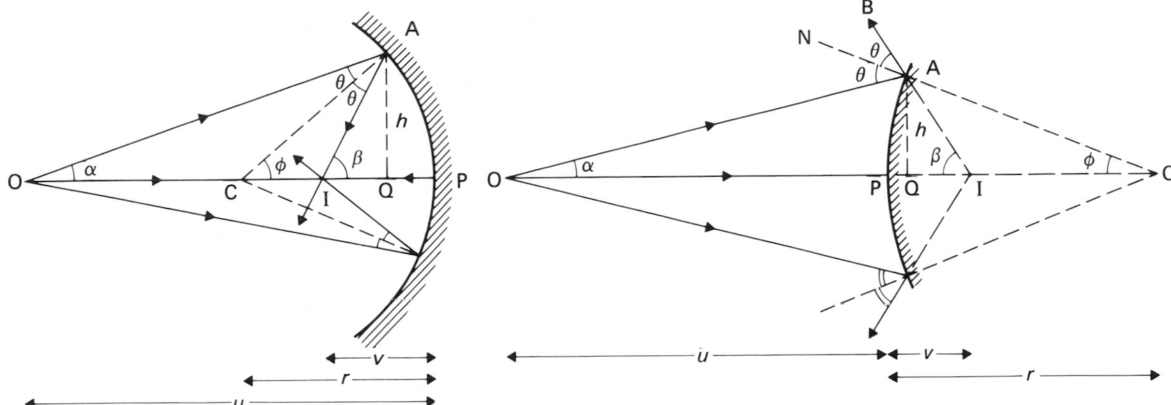

(*a*) An image in a concave mirror

(*b*) An image in a convex mirror

Fig. 6.8

i.e.

$$\frac{1}{u}+\frac{1}{v}=\frac{1}{f} \tag{6.8}$$

Q. 6.5
Use Fig. 6.8(*b*) to derive a similar result for a convex mirror.

The reader's own analysis for a convex mirror should produce the following equation:

$$\frac{1}{u}-\frac{1}{v}=\frac{-1}{f}. \tag{6.9}$$

It would clearly be convenient to use one equation for both types of mirror, and this is made possible by the use of a suitable *sign convention*.

The real-is-positive sign convention

A real object placed in front of a convex mirror produces a virtual image, as seen in Fig. 6.8. The real-is-positive sign convention, which is used throughout this book, distinguishes between real and virtual cases in the following way:

> *When numbers are substituted for object and image distances, positive numbers are substituted for distances to real objects and images, and negative numbers for distances to virtual objects and images.*

We distinguish between concave and convex mirrors as follows:

> *Concave mirrors have a positive focal length and radius of curvature; convex mirrors have a negative focal length and radius of curvature.*

In Equation (6.9), for convex mirrors, if negative numbers are to be substituted for the image distance (since the image is *virtual*) and for the focal length (since it is a *convex* mirror), we must invert the sign of those terms and thus we obtain an identical equation to that for concave mirrors:

Hence
$$\frac{1}{u}-\frac{1}{(-v)}=\frac{-1}{(-f)}$$

i.e.
$$\frac{1}{u}+\frac{1}{v}=\frac{1}{f} \tag{6.8}$$

Worked examples for mirrors

(*a*) An object is placed 0.3 m from a mirror and a virtual image is formed 0.15 m behind it. What is the focal length of the mirror?

$$\frac{1}{f}=\frac{1}{u}+\frac{1}{v}$$
$$=\frac{1}{(+0.3)}+\frac{1}{(-0.15)}$$

since the object is *real* and the image is *virtual*. Thus

$$\frac{1}{f}=-\frac{1}{(0.3)}$$
$$\therefore \quad f=-0.3\text{ m}$$

As the focal length is *negative, the mirror is convex.*

(*b*) An object is placed 0.2 m in front of a mirror and the real image is 0.1 m in front of it. Calculate the focal length, and radius of curvature, of the mirror.

$$\frac{1}{f}=\frac{1}{u}+\frac{1}{v}$$

$$=\frac{1}{(0.2)}+\frac{1}{(0.1)}$$

since the object and image are both *real*. Therefore

$$\frac{1}{f}=\frac{3}{0.2}$$

i.e. $f=+0.067$ m

$=+67$ mm

Since $r=2f$

$r=+134$ mm

As the focal length is a *positive* number, the mirror must be *concave*.

(*c*) An object is placed 0.3 m in front of a concave mirror of focal length 0.1 m. Where is the image?

$$\frac{1}{u}+\frac{1}{v}=\frac{1}{f}$$

$$\therefore \frac{1}{v}=\frac{1}{f}-\frac{1}{u}$$

$$=\frac{1}{(+0.1)}-\frac{1}{(+0.3)}$$

since the object is *real* and the mirror *concave*. Thus

$$\frac{1}{v}=\frac{2}{0.3}$$

i.e. $v=0.15$ m

As the image distance is *positive* the image is *real*, as we would expect.

(*d*) An object is placed 150 mm in front of a convex mirror of radius of curvature 100 mm. Where is the image?

Numerically: $f=r/2$

$=50$ mm

$$\frac{1}{u}+\frac{1}{v}=\frac{1}{f}$$

$$\therefore \frac{1}{v}=\frac{1}{f}-\frac{1}{u}$$

$$=\frac{1}{(-50)}-\frac{1}{(+150)}$$

since the object is *real* and the mirror *convex*. Therefore

$$\frac{1}{v}=\frac{-3-1}{150}$$

$$=\frac{-4}{150}$$

i.e. $v=-37.5$ mm

Since the image distance is *negative* the image must be *virtual*.

It is important to note that other books may use a different convention known as the *New Cartesian*; the reader is advised to refer only to books using the *real-is-positive convention*, to avoid confusion.

Q 6.6

Calculate the image distance in each of the following problems and state whether the image is real or virtual:

(*a*) real object distance 0.15 m; convex mirror focal length 0.10 m;
(*b*) real object distance 250 mm; concave mirror focal length 50 mm.

Calculate the object distance in each of the following problems:

(*c*) real image distance 0.40 m; concave mirror focal length 0.25 m;
(*d*) virtual image distance 70 mm; convex mirror focal length 210 mm.

Calculate the focal length of the mirror in each of the following problems and state whether the mirror is convex or concave:

(*e*) real object distance 100 mm; real image distance 20 mm;
(*f*) real object distance 250 mm; virtual image distance 50 mm.

Extended images in concave mirrors

The position of the image of an extended object in a mirror can be found by drawing, as well as by using the mirror formula. *Three* construction lines are used to check the accuracy of the drawing and these are shown dashed in Fig. 6.9; each line represents a possible ray through the system although, as we see, such a ray might not enter the eye. The rays that do enter the eye are shown separately, as a shaded pencil.

The three construction lines represent the following rays:

(1) a ray parallel to the axis will be reflected through the focal point;
(2) a ray passing through, or appearing to pass through, the focal point will be reflected parallel to the axis;
(3) a ray along a radius of curvature will be reflected back along its own path because it strikes the mirror normally.

As all points in the object plane have an image in the image plane any convenient vertical scale may be used for the construction; it is usual to expand the vertical scale.

The diagrams show that if an object is further from the lens than the focal point, real, inverted images are produced, but if it is closer to the lens than the focal point a virtual, erect image is formed which is larger than the object; this is the arrangement when it is used as a shaving mirror.

A point at infinity on the axis of a mirror produces a pencil of light parallel to the axis, and an off-axis point produces a parallel pencil of light at an angle to the axis. The image of the off-axis point is formed in the *focal plane*, and thus extended objects at infinity have images that are extended in the focal plane, as shown in Fig. 6.9(*d*). Conversely if the object is in the focal plane the image will be at infinity.

Extended images in convex mirrors

Similar rules can be drawn up for constructing diagrams for convex mirrors:

(1) a ray parallel to the axis will be reflected as though it had passed through the focal point;

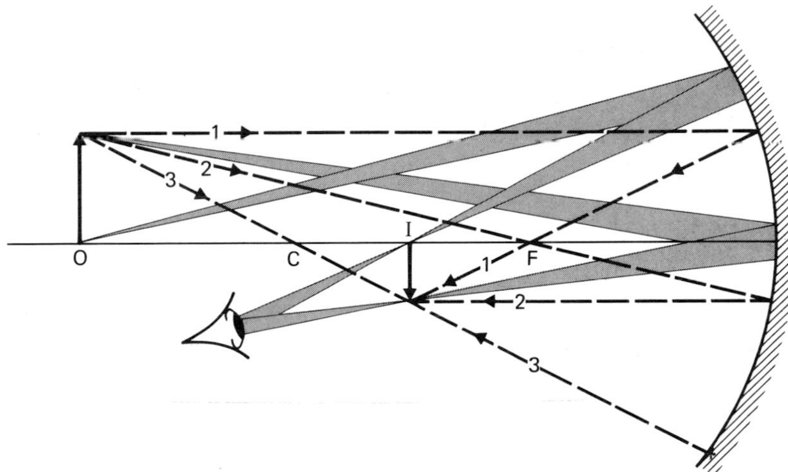

(*a*) A real object ($u > r$) produces a real diminished inverted image

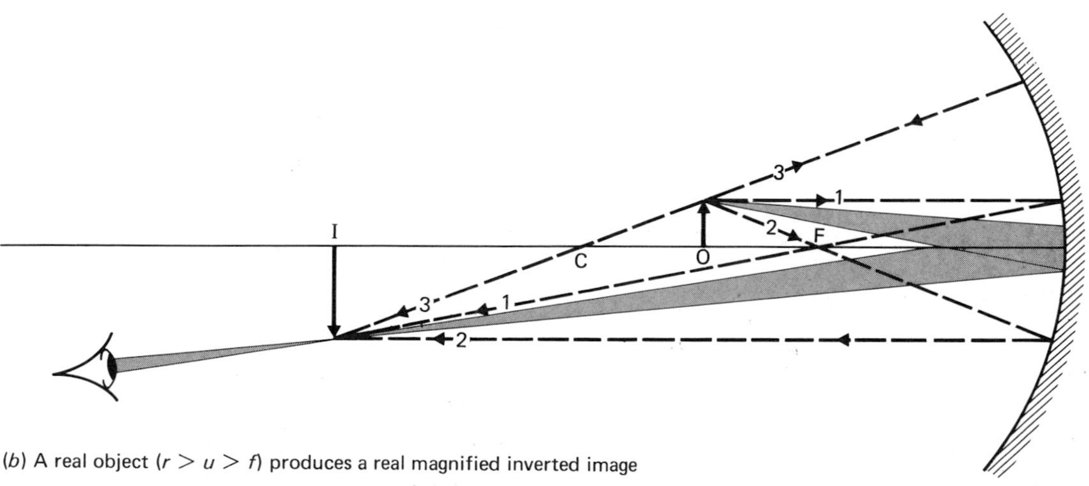

(*b*) A real object ($r > u > f$) produces a real magnified inverted image

Fig. 6.9

Continued overleaf

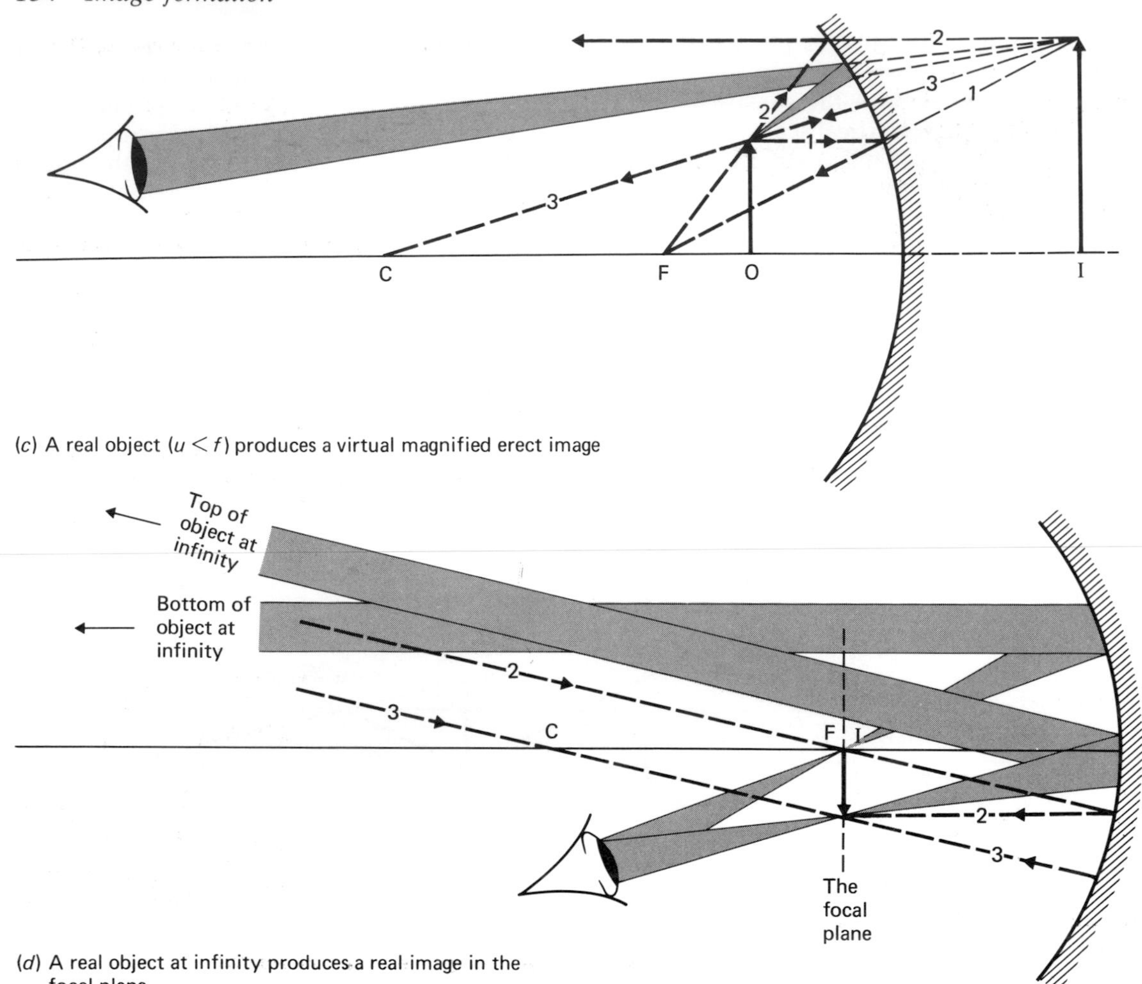

(*c*) A real object ($u < f$) produces a virtual magnified erect image

(*d*) A real object at infinity produces a real image in the focal plane

Fig. 6.9

(2) a ray of light travelling towards the focal point will be reflected parallel to the axis;
(3) a ray along a radius of curvature will be reflected along itself.

An object placed at any position in front of the mirror will form a virtual, diminished and erect image (Fig. 6.10(*a*)). As the image is smaller a large field of view is provided and convex mirrors are often used for driving mirrors for this reason.

Q 6.7
Draw a ray diagram to show how a convex mirror produces an image in the focal plane of an off-axis point at infinity.

Virtual objects

In some types of optical instrument, a real image is formed and then a mirror or lens is interposed in the path of the rays as shown in Fig. 6.10(*b*); the real image is said to become a *virtual object* for the interposed mirror or lens. This is the only circumstance in which a convex mirror can form a real image as shown in Fig. 6.10(*c*).

The object is described as 'virtual' because light does not actually come from it, but merely appears to; in calculations involving virtual objects, the object distance is substituted as a negative number as shown in the example following.

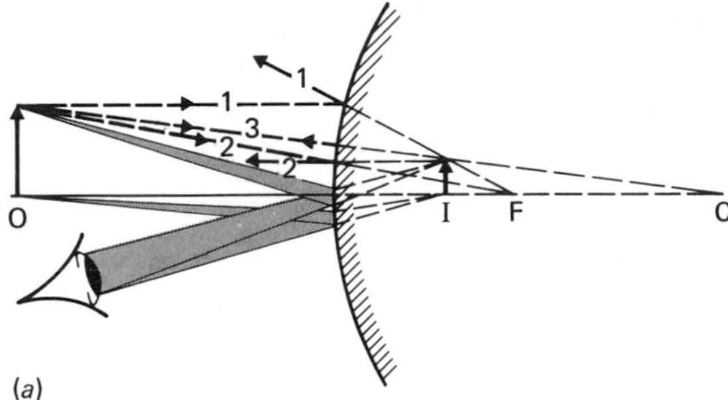

(*a*)
Any real object produces a virtual, diminished, erect image

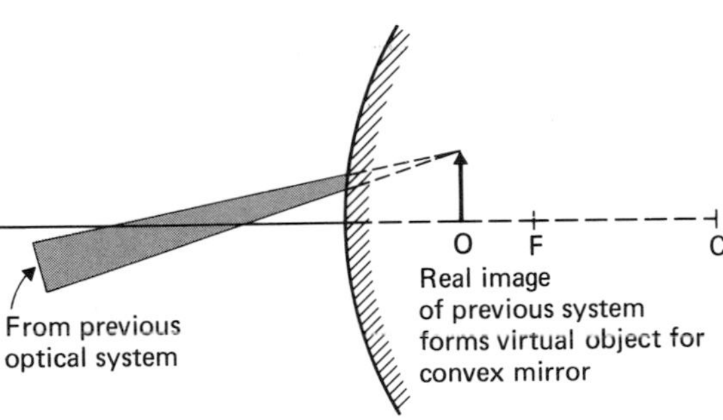

(*b*) Formation of a virtual object

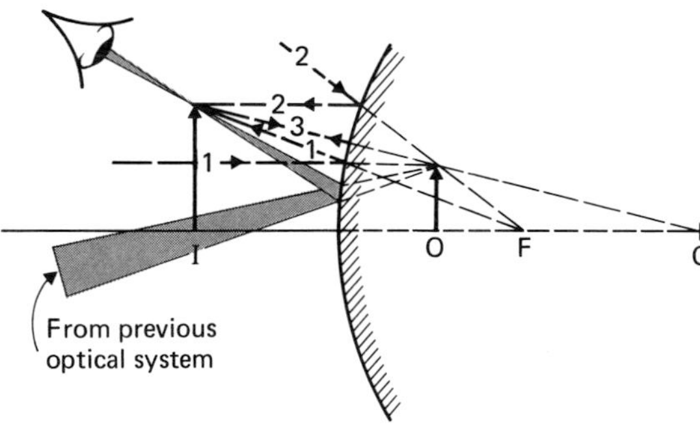

(*c*) A real image of a virtual object formed by a convex mirror

Fig. 6.10

Worked example

A real image is formed by a lens system and then a convex mirror of focal length 100 mm is placed 50 mm in front of the image, so that it becomes a virtual object for the mirror. Where is the final image?

$$\frac{1}{u}+\frac{1}{v}=\frac{1}{f}$$

$$\therefore \quad \frac{1}{v}=\frac{1}{f}-\frac{1}{u}$$

$$=\frac{1}{(-100)}-\frac{1}{(-50)}$$

since the object is *virtual* and the mirror *convex*. Thus

$$\frac{1}{v}=-\frac{1}{100}+\frac{1}{50}$$

$$=+\frac{1}{100}$$

i.e. $\quad v=+100\text{ mm}$

The *positive* sign means that the final image, in this case, is *real*.

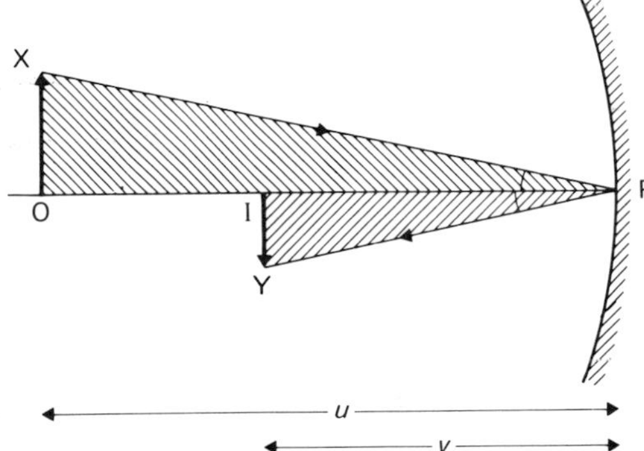

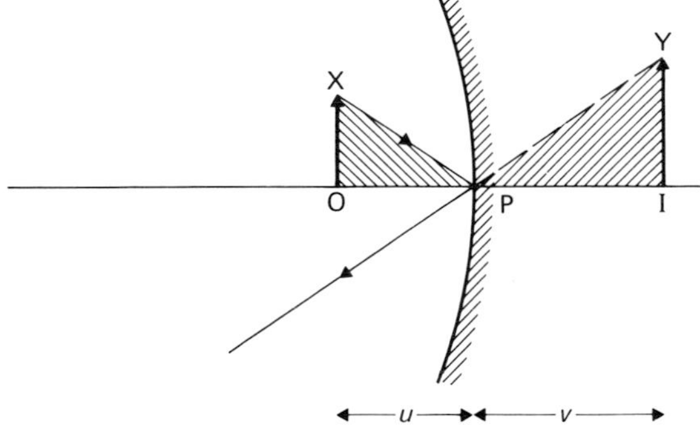

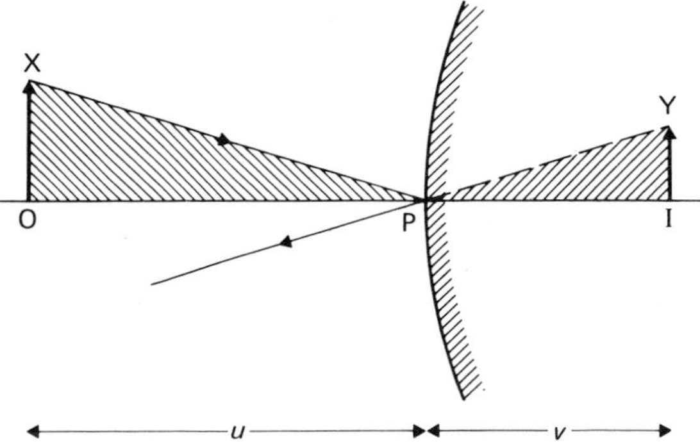

Fig. 6.11 Magnification in curved mirrors.

Magnification

In general, the image produced by a spherical mirror will be a different size from the object. We define:

$$\text{magnification} = \frac{\text{height of image}}{\text{height of object}}$$

In each of the diagrams in Fig. 6.11, the ray reflected at the pole of the mirror is shown, and in each case:

$$\angle XPO = \angle YPI$$

Since the two shaded triangles are similar:

$$\frac{\text{image height}}{\text{object height}} = \frac{\text{image distance}}{\text{object distance}}$$

$$= \frac{v}{u}$$

i.e. $$\text{magnification} = \frac{\text{image distance}}{\text{object distance}}$$

We shall see later that this result holds also for lenses.

Measurements on curved mirrors

Measurement of the focal length of a concave mirror

A quick estimate of the focal length of a concave mirror may be made by casting an image of a distant light or window onto your hand. When the image is in focus, as the object is effectively at infinity, the distance from hand to mirror must equal its focal length.

To obtain a more accurate value, measurements of u and v must be made; a graph of $1/v$ against $1/u$ is plotted as shown in Fig. 6.12(a).

Since $$\frac{1}{u} + \frac{1}{v} = \frac{1}{f}$$

$$\frac{1}{v} = \frac{-1}{u} + \frac{1}{f}$$

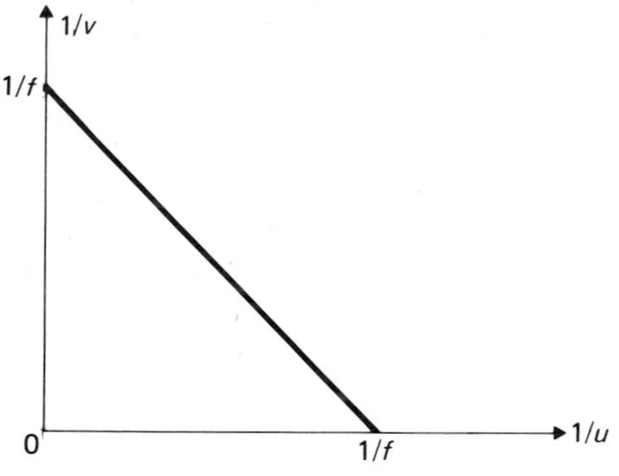

(*a*) Graph plotted to determine *f* for a mirror

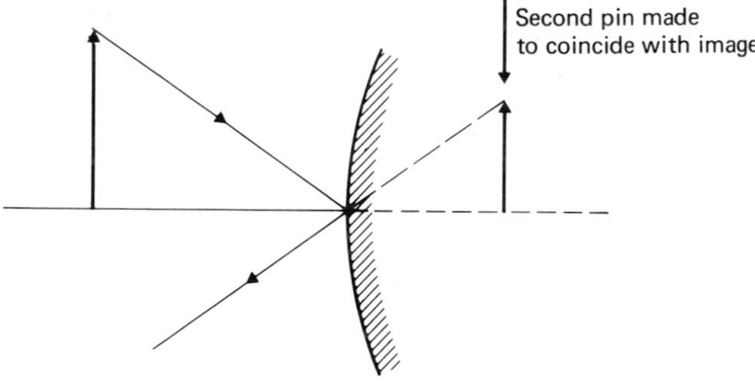

(*b*) Image in convex mirror located by no-parallax

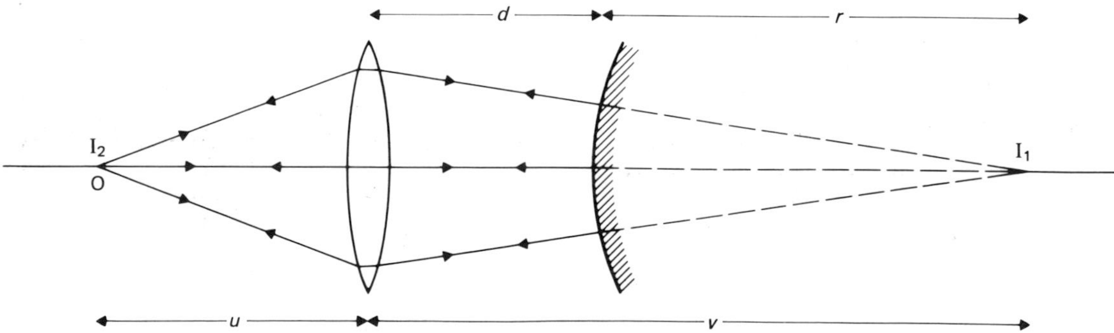

(*c*) Measurement of the radius of curvature of a convex mirror

Fig. 6.12

Thus, the intercept on either axis will be equal to $1/f$. Usually a large pin is used as the object; to locate the image, a second pin is made to coincide with the image of the first by the method of no parallax.

The radius of curvature will be equal to twice the focal length.

Measurement of the focal length of a convex mirror

Measurements on convex mirrors are more difficult because the images are almost always virtual and behind the mirror. It is not possible to cast an image of a distant object on to your hand as with a concave mirror.

The position of the virtual image behind the mirror can be determined as shown in Fig. 6.12(*b*) by the method of no parallax; a second pin is held just above the level of the mirror and moved until it is coincident with the image. A series of readings of u and v are made and a graph plotted as before.

Alternatively, the radius of curvature of the mirror may be measured using an auxiliary convex lens, as shown in Fig. 6.12(*c*). The distance between lens and mirror, d, is adjusted until the final image, I_2, coincides with the object O. The light now retraces its path after reflection from the mirror, and the centre of curvature of the mirror must therefore coincide with the image of O formed by the lens at I_1. The position of the mirror can be marked and, if it is then removed and the position of I_1 observed, the radius of curvature of the mirror can be measured directly. The focal length is half the radius of curvature of the mirror.

Q 6.8
The following readings of object and image distance are made for a convex mirror. Plot a graph and hence determine the focal length.

u/cm	10	20	30	40	50
v/cm	−6.0	−8.5	−10.0	−11.0	−11.5

6.3 Refraction

When light meets an interface between two transparent media, two things occur, as shown in Fig. 6.13; some of the light is reflected, obeying the laws of reflection, whilst the rest of the light is transmitted with a change of direction. The change in direction is called *refraction*; the phenomenon was known to the ancient Greeks, but it was not until the seventeenth century that a quantitative law was deduced.

As in the case of reflection, directions are defined with respect to the normal as shown in Fig. 6.13. The two laws are stated in the form:

The incident ray and the normal define a plane which contains the refracted ray.

For a wave of given frequency (or wavelength), and a given pair of media:

$$\frac{\sin\theta_1}{\sin\theta_2} = \text{a constant, known as the relative refractive index of the two media.}$$

This second law is known as Snell's law. For the direction of travel shown, θ_1 is known as the *angle of incidence* and θ_2 the *angle of refraction*. The relative refractive index for light passing from medium 1 to medium 2 is written ${}_1n_2$

i.e.
$$\frac{\sin\theta_1}{\sin\theta_2} = {}_1n_2 \qquad (6.10)$$

In the diagram the light is bent towards the normal when going from medium 1 to medium 2 and we say that medium 2 is *optically more dense* than medium 1; this is the situation, for instance, when light goes from air into water or glass, or from water into glass.

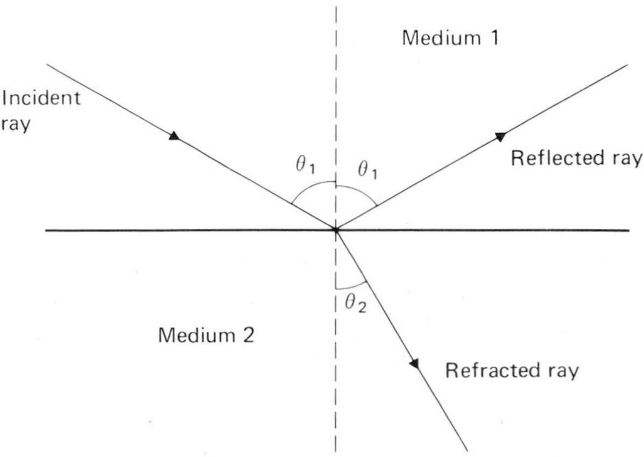

Fig. 6.13

If the ray is passing from medium 2 into medium 1 and the angle of incidence in medium 2 is θ_2, we find that the angle in medium 1 is θ_1; the ray of light is reversible.

Now
$$_2n_1 = \frac{\sin\theta_2}{\sin\theta_1}$$
$$= \frac{1}{_1n_2} \quad \text{from (6.10)}$$

It is convenient to write the relative refractive index, $_1n_2$, as the ratio of two absolute refractive indices, as defined below.

Thus
$$_1n_2 = \frac{n_2}{n_1}$$

where n_1 is the absolute refractive index of medium 1 and n_2 the absolute refractive index of medium 2. This is consistent with the principle of reversibility of light, for we now have:

$$_2n_1 = \frac{n_1}{n_2}$$
$$= \frac{1}{n_2/n_1}$$
$$= \frac{1}{_1n_2} \quad \text{as derived above.}$$

We can now write down a symmetrical form of Snell's law:

$$n_1 \sin\theta_1 = n_2 \sin\theta_2 \qquad (6.11)$$

and this is the version which the reader is advised to use. The *absolute refractive index* of a medium is defined as the value of the constant in Snell's law when light passes from a vacuum into the medium; the absolute refractive index of a vacuum is, by definition, exactly one. Thus the equation:

$$n_v \sin\theta_v = n_2 \sin\theta_2$$

becomes
$$\frac{\sin\theta_v}{\sin\theta_2} = n_2 \quad \text{since } n_v = 1$$

In practice, since the absolute refractive index of air is 1.000 29 we can, for most purposes, measure absolute refractive indices relative to air, instead of relative to a vacuum.

Q 6.9
Use Equation (6.11) to show that when light passes from medium 1 through medium 2 and into a third medium, the direction of the ray in the final medium does not depend upon the refractive index of the intervening medium, provided that the interfaces between the media are parallel.

Q 6.10
(*a*) Light passes into a glass block from air. The angle of incidence is 30°. What is the angle of refraction?
(*b*) Light passes into a water surface from air. The angle of incidence is 30°. What is the angle of refraction?
(*c*) Light passes into a glass block from water. The angle of incidence in the water is 30°. What is the angle of refraction in the glass?
(*d*) Light passes from air, through a parallel-sided glass wall, into a tank of water. The angle of incidence in the air is 30°. What is the angle of refraction in the water?

(Absolute refractive index of this type of glass = 1.50; of water = 1.33.)

Refractive indices and the velocity of light

In Section 3.5 we showed that a wave passing from a medium where its velocity is v_1 to a medium where the velocity is v_2 undergoes refraction; the directions of propagation in the two media are related by the equation:

$$\frac{\sin\theta_2}{\sin\theta_1} = \frac{v_2}{v_1} \qquad (6.12)$$

Snell's law tells us that the direction of propagation of a light ray changes when passing from one medium to another in accordance with the equation:

$$\frac{\sin\theta_2}{\sin\theta_1} = \frac{n_1}{n_2} \qquad (6.11)$$

In Chapter 9 we shall examine the evidence for the wave nature of light. Assuming for the moment that light has been shown to be a wave motion, the two equations above may be combined to give:

$$\frac{n_1}{n_2} = \frac{v_2}{v_1}$$

Now if medium 1 is a vacuum where the velocity of light is c, and medium 2 is a medium of absolute refractive index, n_m, where the velocity of light is c_m, we have:

$$\frac{1}{n_m} = \frac{c_m}{c}$$

Thus $$n_m = \frac{c}{c_m}$$

i.e.

$$\text{absolute refractive index of a medium} = \frac{\text{velocity of light } \textit{in vacuo}}{\text{velocity of light in the medium}}$$

A table of values of n for different media, together with the corresponding values of the speed of light is given below (Table 6.1); it is customary to quote the indices for the light of wavelength 5.893×10^{-7} m, corresponding to one of the sodium D lines. As we shall see later, different wavelengths of light travel with slightly different velocities in transparent media.

Real and apparent depth

Fig. 6.14 shows why the refraction of light at a surface makes a swimming pool, for instance, look less deep than it really is. The light entering the observer's eye comes, of course, from the object, O, at the bottom, but it appears to the observer to come from the image, I. The angles θ_w are equal and so are the angles θ_a, because the normals at X and A are parallel.

Now $$\sin\theta_a = \frac{XA}{AI} \approx \frac{XA}{XI}$$

and $$\sin\theta_w = \frac{XA}{AO} \approx \frac{XA}{XO}$$

since for small angles $\sin\theta \approx \tan\theta$.

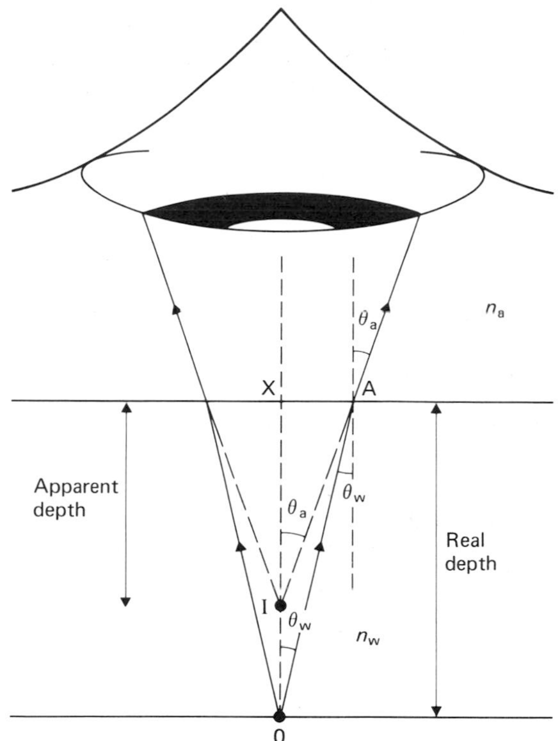

Fig. 6.14 Real and apparent depth.

But $$n_a \sin\theta_a = n_w \sin\theta_w$$

i.e. $$n_a \frac{XA}{XI} = n_w \frac{XA}{XO}$$

Hence $$\frac{XO}{XI} = \frac{n_w}{n_a}$$

Table 6.1 Absolute refractive indices of some materials for light of wavelength 5.893×10^{-7} m

Medium	Absolute refractive index	Speed of light/10^8 m s^{-1}
vacuum	1.000 000	2.997 925
hydrogen, stp	1.000 132	2.993 973
air, stp	1.000 292	2.997 050
borosilicate crown glass†	1.509 70	1.985 78
double extra dense flint glass†	1.927 07	1.555 69
ice	1.31	2.29
Canada balsam‡	1.54	1.95
quartz crystal	1.544 24	1.941 36
diamond	2.419 5	1.239 1

† Chance-Pilkington Optical Works.
‡ *Optica Acta*, 1967, **14,** 401.
Other data from Kaye and Laby (1973) *Tables of Physical and Chemical Constants*, Longman.

In general:

$$\frac{\text{real depth}}{\text{apparent depth}} = \frac{\text{absolute refractive index of lower medium}}{\text{absolute refractive index of upper medium}} \qquad (6.13)$$

or, if the upper medium is air, of refractive index one:

$$\frac{\text{real depth}}{\text{apparent depth}} = \text{absolute refractive index of the lower medium}$$

Q 6.11
An osprey dives into a river to catch a fish 20 cm below the surface. Where did the fish appear to be to the osprey?

The osprey was young and unfamiliar with Snell's law and the fish survived. The fish now leaps to catch an insect which appears to be 6 cm above the water. How far above the water is the insect? (Absolute refractive index of water = 1.33.)

Total internal reflection

When the light passes from an optically more dense medium such as glass, into an optically less dense medium such as water, or air, it is bent away from the normal; in Fig. 6.15, θ_2 is greater than θ_1 because the refractive index of medium 1 is greater than that of medium 2.

In the case of the left-hand light ray, some of the light is reflected in accordance with the laws of reflection and some of the light is refracted. Now, for the refracted ray:

$$n_2 \sin \theta_2 = n_1 \sin \theta_1 \qquad (6.11)$$

or in this case, assuming medium 1 is glass and medium 2 is air:

$$\sin \theta_a = n_g \sin \theta_g \qquad (6.14)$$

Now, as the angle of incidence in the glass is increased, the angle in air also increases; as it is the larger angle, it will reach 90° first. This is the situation of the central ray, where the partially transmitted ray lies actually along the surface, whilst the partially reflected ray obeys the law of reflection as before. In this case, the angle in glass is called the *critical angle*, θ_c, and its size may be calculated from Snell's law as follows:

$$\sin \theta_a = n_g \sin \theta_g \qquad (6.14)$$

At the critical angle:

$$\theta_g = \theta_c$$

and

$$\theta_a = 90°$$

Therefore, since $\sin \theta_a = 1$, we have

$$1 = n_g \sin \theta_c$$

i.e.

$$\sin \theta_c = \frac{1}{n_g} \qquad (6.15)$$

For a glass of refractive index 1.5, the critical angle is about 42°.

For angles in glass larger than the critical angle, there is no refracted ray at all, and all the light is reflected; *total internal reflection* is then said to occur. This is the case for the right-hand ray. Total internal reflection occurs whenever light is incident on an interface with a medium of lower refractive index, at an angle greater than the critical angle.

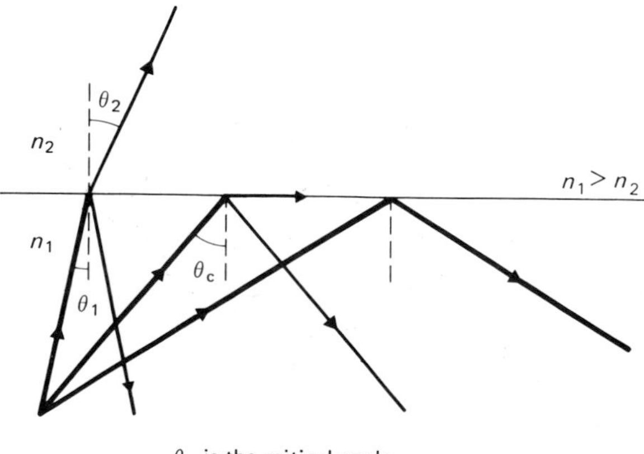

θ_c is the critical angle

Fig. 6.15

Q 6.12
(i) Calculate the critical angles for:

(*a*) a glass–air interface;
(*b*) a water–air interface;
(*c*) a glass–water interface.

(Absolute refractive index of glass = 1.50; of water = 1.33.)

(ii) Draw a diagram to show the field of view of a fish looking upwards out of a clear pond.

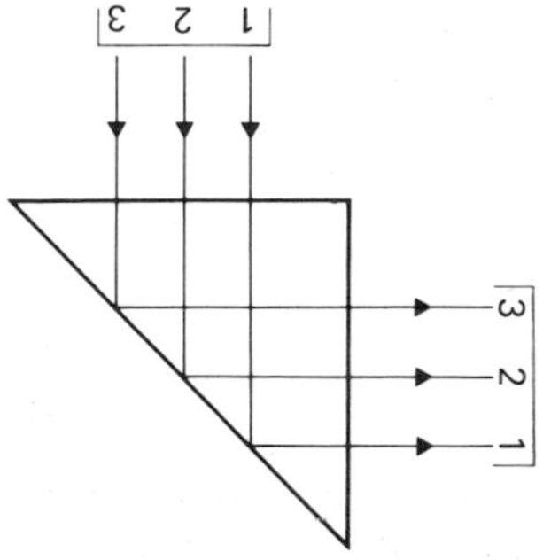

(*a*) Reflection through 90°; beam also laterally inverted (the left and right hand sides of the beam are interchanged)

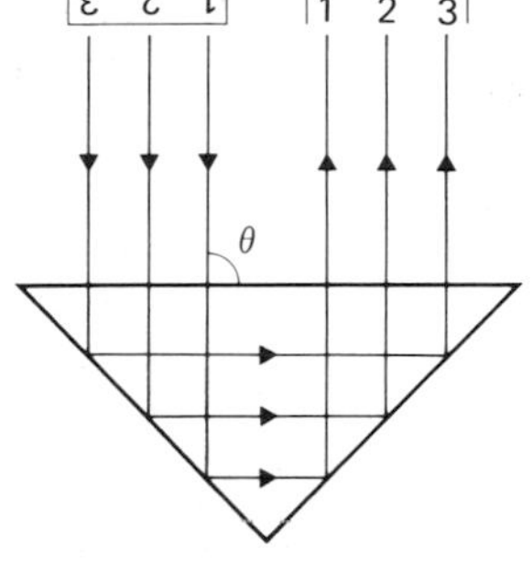

(*b*) Reflection through 180°; no lateral inversion

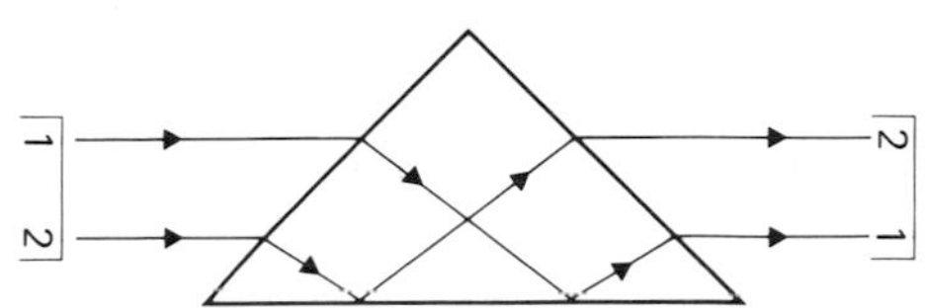

(*c*) Lateral inversion without deviation

Fig. 6.16 Some applications of reflecting prisms.

Reflecting prisms

Total internal reflection in a prism is often preferable to reflection from a silvered or aluminized mirror. There are no multiple images, such as occur in a back-reflecting mirror, and no silver or aluminium surface to be scratched or tarnished like those in front-reflecting mirrors. Although 100% of the light reaching the reflecting surface will be reflected, a total of about 8% of the total light intensity will be lost as the light passes into and out of the other two faces of the prism. It is therefore a more efficient reflector than a front-surface aluminized mirror but less efficient than a front-surface silvered mirror. It will, however, retain its reflecting power indefinitely providing the reflecting surface is kept clean from dirt or grease, which would alter the critical angle. When the application warrants it, the loss of intensity at the first and last faces of the prism can be reduced to a negligible amount by the use of anti-reflection coatings on these surfaces.

Reflecting prisms can be used to turn beams of light through 90° and 180°, and also to invert a beam without deviating it. There are also more complex applications but these three basic ones are shown in Fig. 6.16.

Fig. 6.17 illustrates the use of reflecting prisms in prism binoculars. Convex lenses are used in binoculars as they give a greater field of view than lens systems incorporating concave lenses, but they produce a final image that is upside down and inverted left for right; in prism binoculars this is corrected by using two reflecting prisms to invert the image both vertically and laterally.

Q 6.13
Considering Fig. 6.16(*b*), show that the emergent beam is parallel to the incident beam even if the prism is rotated in the plane of the paper, so that the angle θ is no longer 90°. What is the importance of this in the design of a dinghy's radar reflector?

Light pipes

The light pipe is another application of total internal reflection and in recent years *fibre optics* has become of increasing importance,

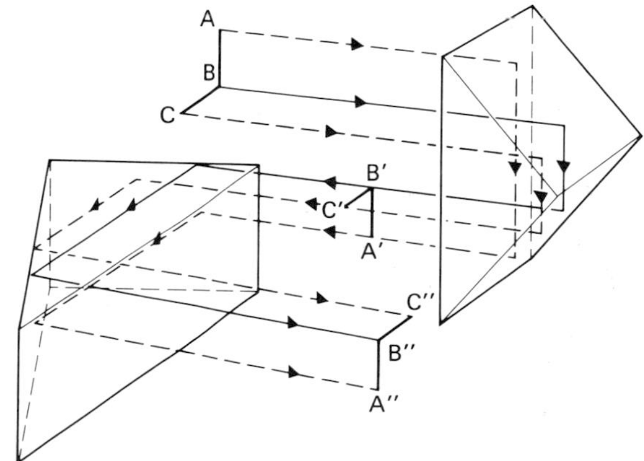

Fig. 6.17 A pair of reflecting prisms, as used in prism binoculars to invert the final image.

especially in medicine and telecommunications. The principle of a light pipe can be understood by considering a long, thin solid glass rod with polished ends. Light is passed in through one end and the rod is bent; the light will now strike the walls, as shown in Fig. 6.18, at angles greater than the critical angle and total internal reflection will occur at a number of places. Clearly the thinner the rod, and the more it is bent, the larger the number of reflections is likely to be. In a glass fibre about 1/20 of a millimetre thick, there will be about 4000 separate reflections in a length of 25 centimetres; nevertheless about 97% of the original light energy will emerge at the far end if the outside of each fibre is clean.

A bundle of glass fibres can be used to transmit light into inaccessible places. Moreover, it is also possible to use a carefully constructed bundle to transport an image, for example of the inside of a person's body. Each fibre will transmit the light intensity at its end and therefore, to build up a detailed image, a large number of very thin fibres is required. Where two identical glass surfaces touch, there will be leakage of light from one to the other and to prevent this each fibre is coated with a glass of significantly lower refractive index. This can be done by drawing down glass rods coated with the different glass, and fibres of about 1/100 of a millimetre diameter have been produced by this method. As long as the relative orientation of the fibres at the end is the same as the relative orientation at the beginning it does not matter how the fibres are tangled up in between, and a bundle of very thin fibres is very flexible indeed. Some light is lost, of course, by slight leakage or absorption in the glass, or where there are broken fibres, but a system two metres long will transmit up to 25% of the incident light. A photograph taken by using fibre optics is shown in Fig. 6.19.

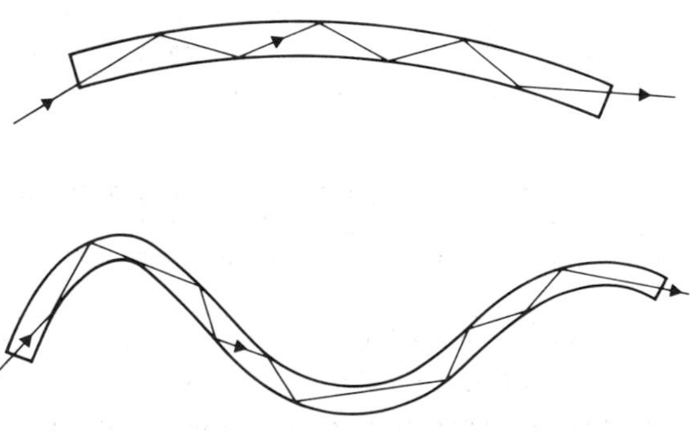

Fig. 6.18

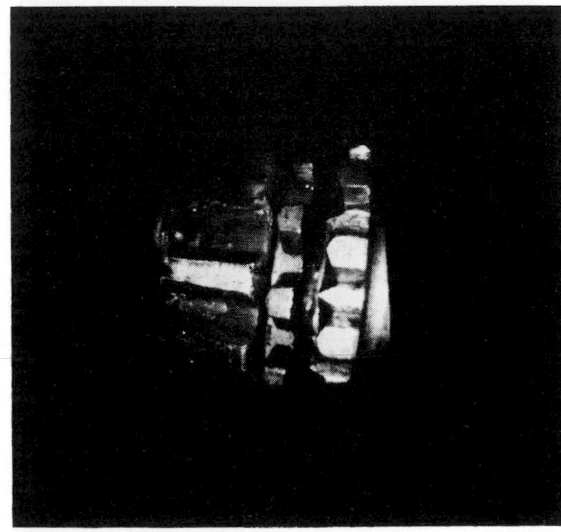

Fig. 6.19 A photograph of the inside of a gearbox taken through an industrial fibrescope, manufactured by Keymed.

Measurement of refractive index

There are many methods of measuring refractive index, and the most accurate require the use of precision instruments; we shall consider here some simple experiments which will yield results with an accuracy of about 1% to 10%.

It is possible, of course, to measure the refractive index of any transparent solid by tracing rays through it and measuring angles of incidence and refraction. Moreover, the thin parallel-sided walls of a container produce no additional deviation, as seen in Question 6.9, and similar measurements can therefore be made upon a liquid if it is placed in a perspex container; the rays are plotted with pins as shown in Fig. 6.20(*a*), and the very small sideways displacement produced by the walls is ignored. A graph of $\sin\theta_a$ against $\sin\theta_l$ has a slope equal to the absolute refractive index of the liquid, if the absolute refractive index of air is taken to be one.

Alternatively, measurements of real and apparent depth can be used to determine refractive indices, and for this purpose a travelling

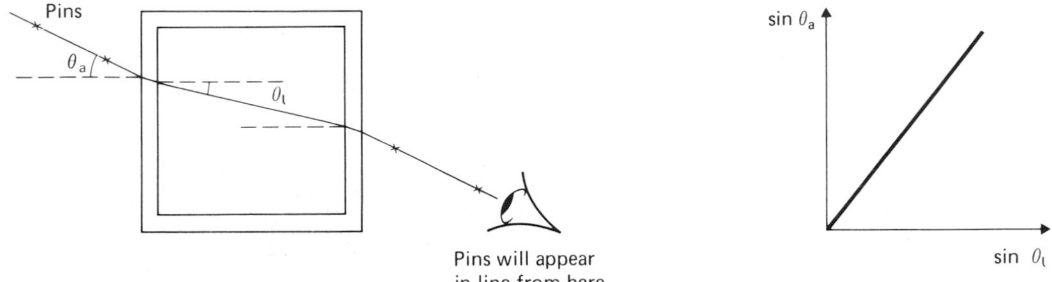

(*a*) Measurement of the refractive index of a liquid (or solid)

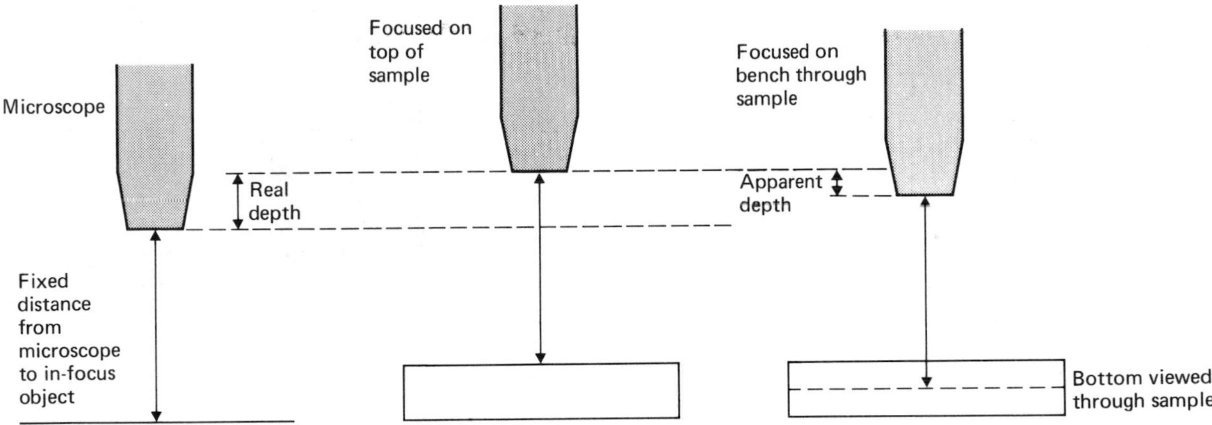

(*b*) Measurement of refractive index by travelling microscope

Fig. 6.20 Methods of measuring refractive index.

microscope is used, as shown in Fig. 6.20(*b*); microscopes have a very small depth of focus and therefore a very small movement of the microscope towards or away from an object will put the image out of focus. The microscope is focused first on the bench, and then on some specks of dust which are scattered on top of the sample; the distance it must be moved to re-focus the image is equal to the real thickness of the sample. The microscope is again focused on the top surface of the sample and then on the bench, viewed through the sample; the distance moved by the microscope this time is equal to the apparent thickness of the sample. Hence, the refractive index of the sample may be calculated from the expression:

$$\text{absolute refractive index of lower medium} = \frac{\text{real depth}}{\text{apparent depth}} \qquad (6.13)$$

Again, we assume that the absolute refractive index of air is equal to one. This method may be used for liquids, which are placed in a small container. As travelling microscopes are accurate instruments, this technique may with care give good results. The theoretical basis for a more accurate method using a spectrometer is given in the section on the minimum deviation of a ray in a prism, on page 145.

Q 6.14

A travelling microscope is used to determine the refractive index of a sample. The following readings of the position of the microscope are taken:

top surface of sample in focus	25.62 cm
surface of bench without sample in focus	30.21 cm
surface of bench viewed through sample in focus	27.49 cm

Calculate the refractive index of the sample and estimate the percentage error in the result if each reading is correct to ±0.01 cm. Note that some glasses have refractive indices very different from 1.5.

6.4 Prisms

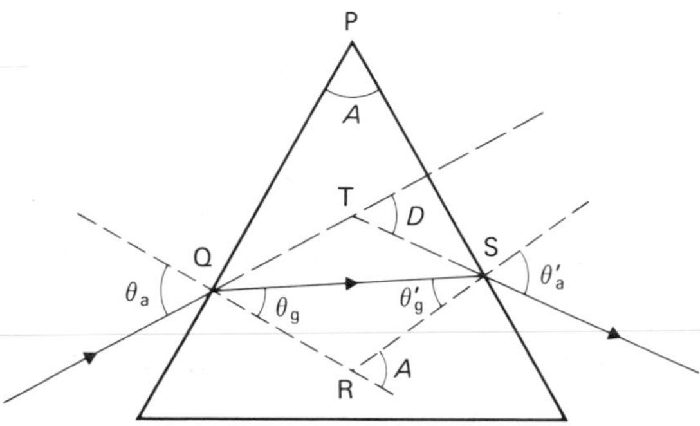

Fig. 6.21

We have already considered how total internal reflection of a light ray may occur in a prism and we will now examine the deviation of a ray which is transmitted through it.

Fig. 6.21 shows a light ray passing through a prism, in which it is deviated by an angle D. The *refracting angle* of the prism is the angle A—in this case 60°; most prisms for spectroscopic use have angles of nominally 60°. The prism angle determines the relative orientation of the two refracting faces.

Now in the diagram, PQRS is a cyclic quadrilateral, as QR is normal to QP and SR is normal to SP. Thus, the angle marked at R is equal to A, the angle of the prism. Moreover, this angle is also the external angle of the triangle RQS.

Hence $$A = \theta_g + \theta'_g \tag{6.16}$$

We can also derive an expression for the deviation, D, in terms of θ_a and θ_g:

$$D = \angle TQS + \angle TSQ$$

where $\angle TQS = \theta_a - \theta_g$

and $\angle TSQ = \theta'_a - \theta'_g$

Therefore $$D = (\theta_a - \theta_g) + (\theta'_a - \theta'_g) \tag{6.17}$$

Minimum deviation

The variation of the angle, D, can be plotted against the angle of incidence, θ_a, as shown in Fig. 6.22; it is found that the deviation is a minimum for one particular value of θ_a. Now, if a ray incident at an angle θ_a leaves the prism at a different angle θ'_a then, as light is reversible, a ray incident at θ'_a would leave at θ_a; the angle of deviation would then be the same as before. However, the *minimum* value of D can only occur at *one* value of the angle of incidence, by definition of *minimum*, and hence under these conditions $\theta'_a = \theta_a$. The condition for the deviation to be minimum is therefore that for which the ray passes through the prism symmetrically, as shown in Fig. 6.22.

Readers are advised to take a prism and to shine a ray of light across the top of a drawing

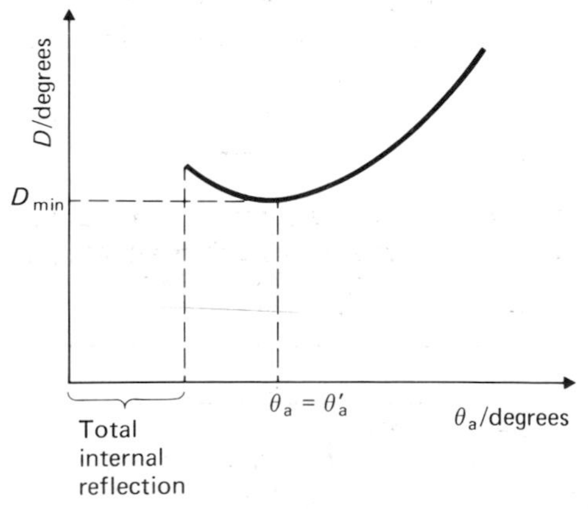

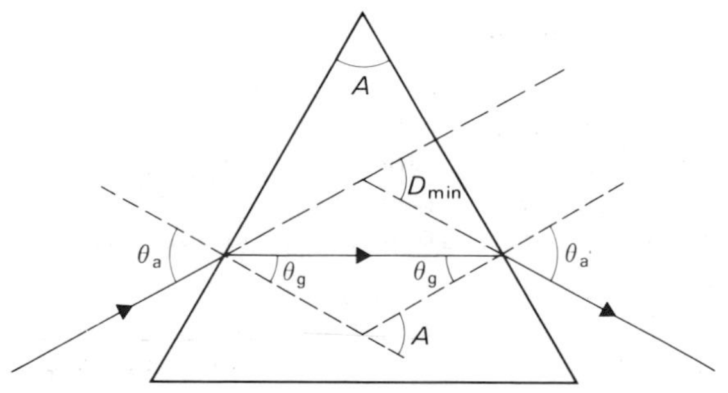

Fig. 6.22 Minimum deviation in a prism.

board; place the prism in the beam and rotate it, observing that there really is a minimum deviation and that it occurs when the ray passes symmetrically through the prism.

Now, since for a symmetrical arrangement, we have:

$$\theta_a = \theta'_a$$

it follows that:

$$\theta_g = \theta'_g$$

Thus, for minimum deviation, Equations (6.16) and (6.17) become:

$$A = 2\theta_g \quad (6.18)$$

and $$D_{min} = 2(\theta_a - \theta_g) \quad (6.19)$$

Hence $$D_{min} = 2\theta_a - A$$

i.e. $$\theta_a = \frac{A + D_{min}}{2}$$

But $$\theta_g = \frac{A}{2} \quad \text{from (6.18)}$$

and from Snell's law we therefore have

$$n_g = \frac{\sin\theta_a}{\sin\theta_g} = \frac{\sin[(A + D_{min})/2]}{\sin(A/2)} \quad (6.20)$$

This equation is used as the basis of an accurate method of measuring refractive index using a spectrometer.

Q 6.15

In a spectrometer experiment, the angle of a prism is found to be 59° and the minimum deviation through the prism is 35°. Calculate the refractive index of the prism glass.

Small-angle prisms

In Section 6.5 we will show that lenses can be regarded as being made up of a number of prisms of small angle, and we shall derive here a result which will be useful in that analysis. By a small-angle prism we mean one with an angle less than 0.1 rad (about 6°) and, for the analysis below, we will also assume that the angle of incidence is less than this.

We start with two equations we derived for all prisms:

$$A = \theta_g + \theta'_g \quad (6.16)$$

and $$D = (\theta_a - \theta_g) + (\theta'_a - \theta'_g) \quad (6.17)$$

From Snell's law we have:

$$\sin\theta_a = n_g \sin\theta_g \quad (6.14)$$

and $$\sin\theta'_a = n_g \sin\theta'_g \quad (6.21)$$

But, for small-angle prisms, if θ_a is small, θ_g, θ'_g and θ'_a will also be small because the total deviation will be small; in the limit for a parallel-sided glass block the total deviation is always zero.

Hence, since all angles are small, we can write approximately:

$$\theta_a \approx n_g\theta_g \quad \text{from (6.14)}$$

$$\theta'_a \approx n_g\theta'_g \quad \text{from (6.21)}$$

Equations (6.16) and (6.17) then become:

$$A = \theta_g + \theta'_g$$

$$D = (n_g\theta_g - \theta_g) + (n_g\theta'_g - \theta'_g)$$

$$= (n_g - 1)(\theta_g + \theta'_g)$$

Hence $$D = (n_g - 1)A \quad (6.22)$$

This shows that *for a small-angle prism at small angles of incidence the deviation is independent of the angle of incidence*. The extent to which this approximation is true may be judged from Table 6.2.

We see from Table 6.2 that a rapid increase in the deviation occurs once the angle of incidence becomes larger than the angle of the prism.

Dispersion in prisms

In our discussion so far, we have assumed that there is a unique refractive index for white light

Table 6.2

Prism angle 5°, $n = 1.5$										
θ_a/(deg)	0	1	2	3	4	5	6	7	8	10
θ_g/(deg)	0.000	0.667	1.333	1.999	2.665	3.331	3.996	4.660	5.324	6.648
θ'_g/(deg)	5.000	4.333	3.667	3.001	2.335	1.669	1.004	0.340	0.324	1.648
θ'_a/(deg)	7.512	6.508	5.505	4.503	3.503	2.504	1.506	0.510	0.486	2.472
D/(deg)	2.512	2.508	2.505	2.503	2.503	2.504	2.506	2.510	2.514	2.529

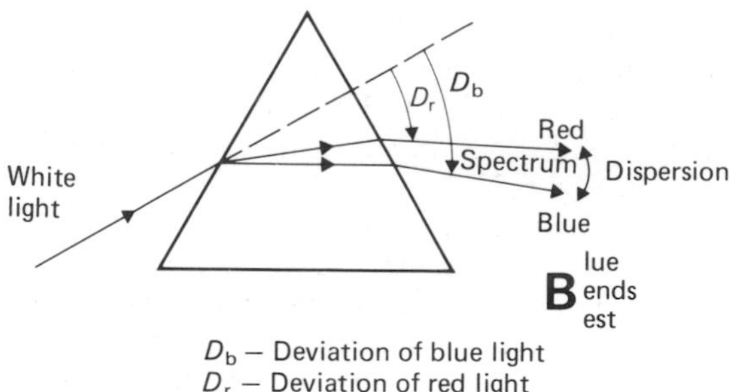

Fig. 6.23

passing through a given type of glass. In fact, this is not so; each colour is deviated by a slightly different amount causing a spectrum to be formed. There is a discussion of the significance of Newton's discovery of the spectrum in Chapter 9 and we will here restrict ourselves to calculations involving deviations.

Fig. 6.23 shows that the deviation of blue light, D_b, is greater than that of red, D_r; the greater deviation of blue light means that the refractive index for blue light, n_b, must be greater than that for red, n_r. It therefore follows from Equation (6.12) that blue light must travel more slowly in glass than does red light. The angular spread of the spectrum produced is called the *dispersion*, and depends upon the *dispersive power* of the glass; the spectrum extends between the colours shown, and beyond blue to violet.

6.5 Thin lenses

Fig. 6.24 shows a converging lens producing a real image of an object. The reader can see that all the rays which enter the lens pass very nearly

Fig. 6.24

through the same point, and a real image is formed there. Lenses have been used since the fourteenth century, first for spectacles and then, since the sixteenth century, for microscopes and telescopes as well. In this section, we will discuss how a lens forms a distinct image and derive equations to enable us to predict where that image will be.

Lens shapes

The simplest lenses have either two spherical surfaces or one spherical surface and one plane surface. Fig. 6.25(*a*) shows three different types of converging, or convex, lenses. In Fig. 6.25(*b*) a real image is formed by a convex lens, similar to that in the photograph. Finally, we see that three prisms could also produce a converging beam of light; in our analysis of lenses we shall consider a thin lens to be made up of a whole sequence of small-angle prisms as shown on the right of Fig. 6.25(*c*).

Fig. 6.26 consists of a similar set of diagrams showing how a virtual image is formed by a diverging lens, in this case of biconcave shape.

We will now prove, as is assumed in these diagrams, that a correctly angled sequence of prisms will cause all the rays from the object to cross the axis at the same point, and in doing so we will derive an equation enabling us to predict where that image will be formed; this is called the lens equation.

The lens equation

When rays of light pass through the *optical centre* of a thin lens they emerge undeviated, as illustrated in Fig. 6.27(*a*); the sides of the lens are parallel at its middle.

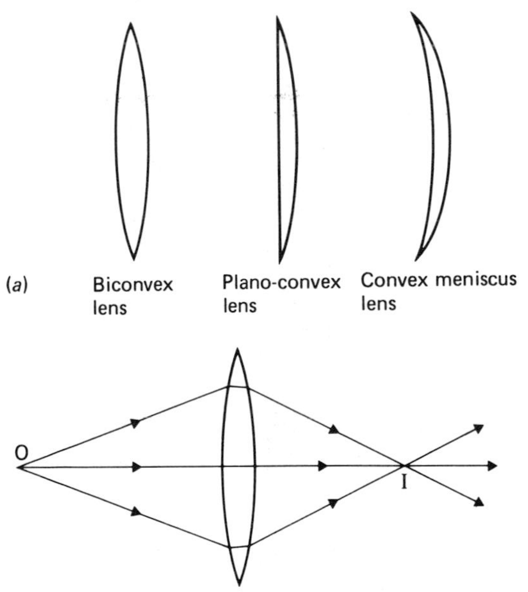

(*b*) A convex lens forming an image

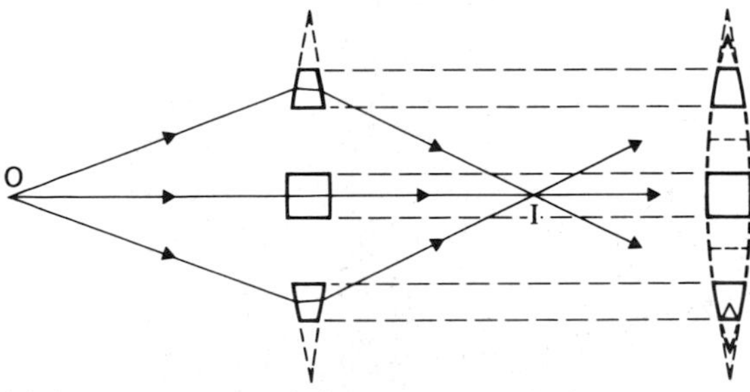

(*c*) A convex lens regarded as a sequence of prisms

Fig. 6.25

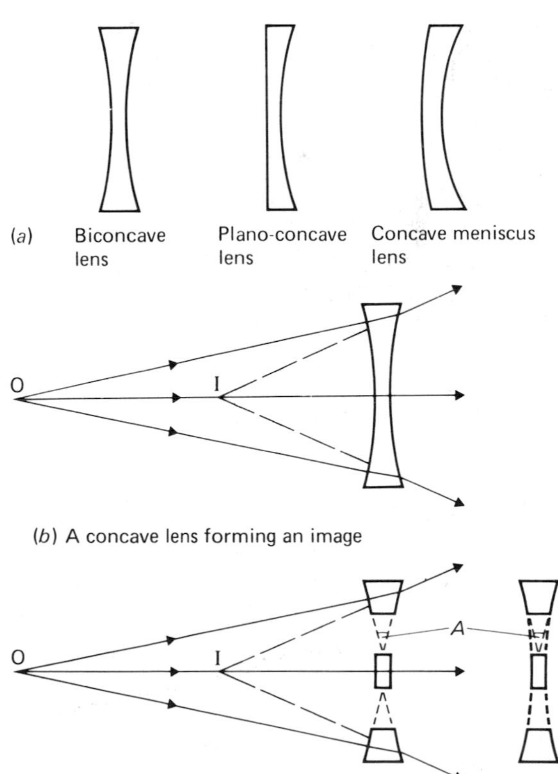

(*c*) A concave lens regarded as a sequence of prisms

Fig. 6.26

When rays of light pass through one of the *focal points* of a converging lens, they emerge parallel to the axis; if they strike the lens parallel to the axis they pass through a focal point after refraction by the lens. Thus there are two focal points, as shown in Fig. 6.27(*b*), and for a thin lens these are equidistant from the optical centre of the lens. The focal points of a lens are sometimes referred to as the *principal foci*; the distance from the optical centre to each principal focus is called the *focal length* of the lens.

The focal points of a diverging lens are similarly illustrated in Fig. 6.27(*d*).

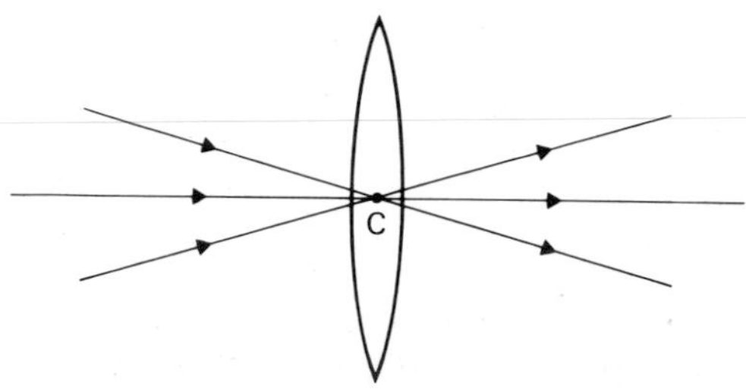

(*a*) The optical centre, C, of a thin lens

A thin converging lens may be considered as a sequence of small-angle prisms and the 'shape of the prism' at a distance h from the axis is marked in the diagrams of Fig. 6.27. In the proof which follows, we shall assume that the lens is very thin, and hence A is small; in addition, the heights at which the light enters and leaves the lens are assumed to be identical and equal to h.

Now the deviation, D, of a ray entering or leaving the lens parallel to the axis is shown in Fig. 6.27(*b*) and by opposite angles:

$$\angle YXF = D$$

But, because XY and CF are parallel, by alternate angles:

$$\angle CFX = \angle YXF = D$$

as marked in Fig. 6.27(*b*).

Hence $\tan D = h/f$

Now, as the angle of the 'prism' is small, the deviation of a ray passing through it is independent of the angle of incidence. In Fig. 6.27(*c*)

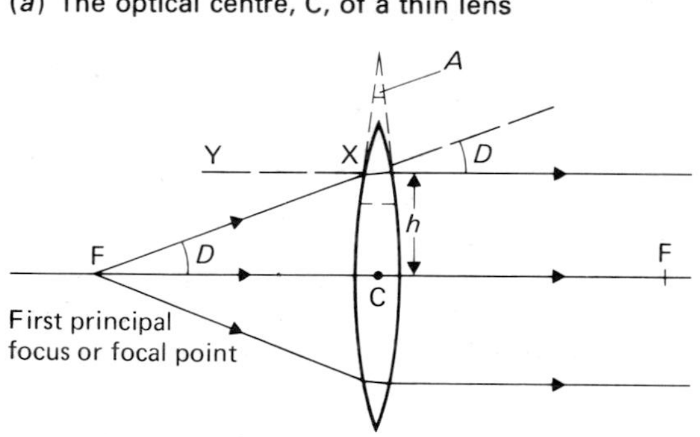

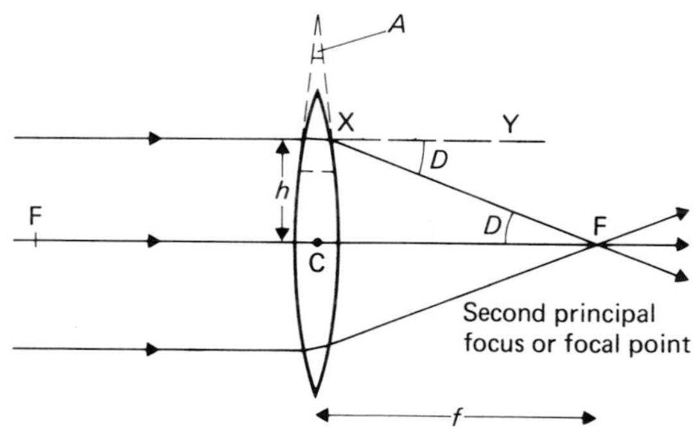

(*b*) The focal length of a converging lens

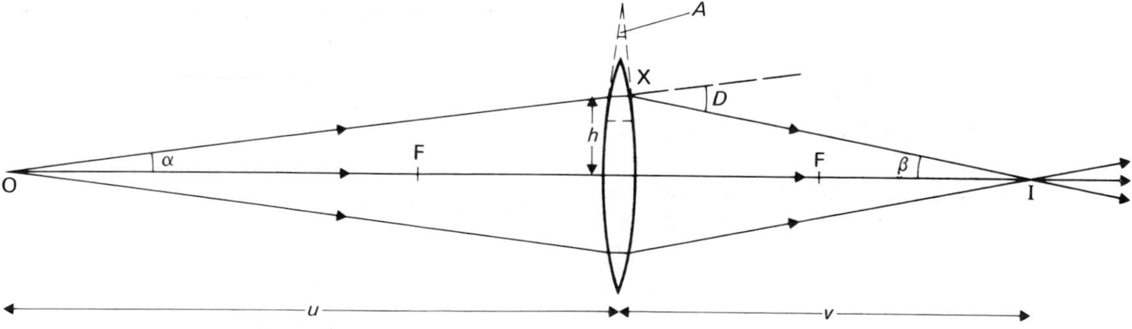

(*c*) Image formation in a converging lens

Fig. 6.27

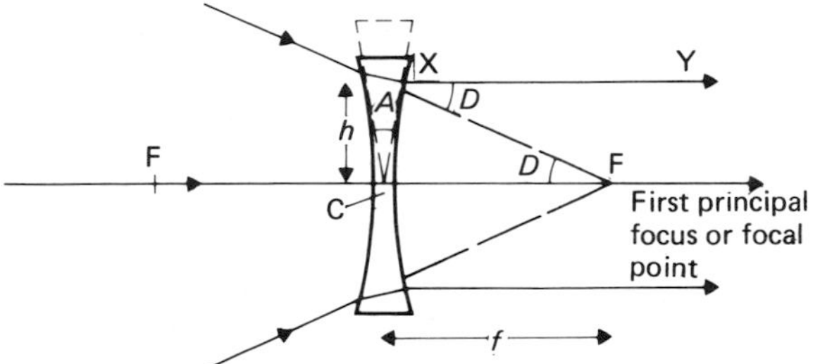

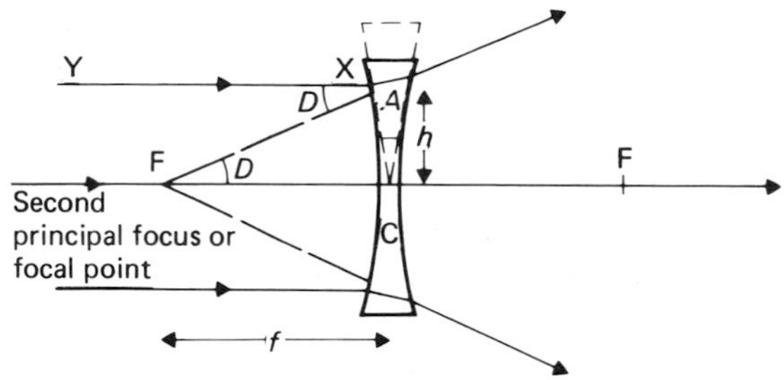

(*d*) The focal length of a diverging lens

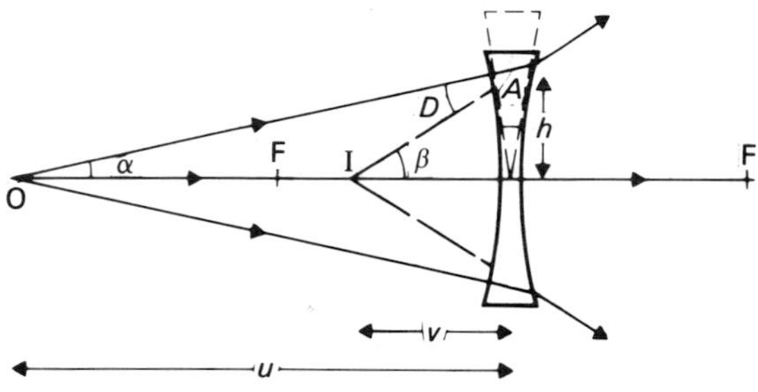

(*e*) Image formation in a diverging lens

Fig. 6.27

the incident ray makes an angle α with the axis, but its deviation is still D. Therefore, since D is the exterior angle of the triangle OXI.

$$D = \alpha + \beta$$

where $\tan \alpha = \dfrac{h}{u}$

and $\tan \beta = \dfrac{h}{v}$

Now if $D = \alpha + \beta$

and all angles are small:

$$\tan D \approx \tan \alpha + \tan \beta$$

since $D \approx \tan D$, etc.

Hence $\dfrac{h}{f} = \dfrac{h}{u} + \dfrac{h}{v}$

i.e. $$\frac{1}{f} = \frac{1}{u} + \frac{1}{v} \qquad (6.23)$$

This is an equation identical to that derived for spherical mirrors.

If the same analysis is carried out for the diverging lens, using Figs. 6.27(*d*) and (*e*) we obtain the equation:

$$\frac{-1}{f} = \frac{1}{u} - \frac{1}{v} \qquad (6.24)$$

Q 6.16

Derive the above result for a diverging lens.

To use Equation (6.23) for both types of lens, a real-is-positive sign convention is used almost identical to that used for mirror calculations.

When numbers are substituted for object and image distances, positive numbers are substituted for distances to real objects and images, and negative numbers for distances to virtual objects and images.

Converging lenses have a positive focal length; diverging lenses have a negative focal length.

If this convention is used the equation:

$$\frac{1}{f} = \frac{1}{u} + \frac{1}{v} \qquad (6.23)$$

applies to both converging and diverging lenses as well as to spherical mirrors.

The relationship between the focal length of a lens, and the radii of curvature of its faces, is more complicated than the equivalent relationship for a spherical mirror, and will be derived later, on pages 153–5.

Worked examples for lenses

(*a*) An object is placed 20 cm in front of a lens and a real image is formed 10 cm behind the lens. What is the focal length of the lens?

$$\frac{1}{f} = \frac{1}{u} + \frac{1}{v}$$

$$= \frac{1}{(+20)} + \frac{1}{(+10)}$$

since the object and image are *real*. Thus

$$\frac{1}{f} = \frac{3}{20}$$

i.e. $$f = +\frac{20}{3}\text{ cm}$$

As the focal length is *positive*, the lens must be *converging*.

(*b*) An object is placed 25 cm from a lens and a virtual image is formed 5 cm from it. What is the focal length of the lens, and what type of lens is it?

$$\frac{1}{f}=\frac{1}{u}+\frac{1}{v}$$
$$=\frac{1}{(+25)}+\frac{1}{(-5)}$$

since the object is *real* and the image *virtual*. Thus

$$\frac{1}{f}=\frac{1}{25}-\frac{1}{5}$$
$$=\frac{1-5}{25}$$
$$=\frac{-4}{25}$$

i.e. $$f=\frac{-25}{4}\text{ cm}$$

As the focal length is *negative*, the lens must be *diverging*.

(*c*) An object is placed 30 cm from a converging lens of focal length 5 cm. Where is the image; is it real?

$$\frac{1}{u}+\frac{1}{v}=\frac{1}{f}$$
$$\frac{1}{v}=\frac{1}{f}-\frac{1}{u}$$
$$=\frac{1}{(+5)}-\frac{1}{(+30)}.$$

since the object is *real* and the lens *converging*. Therefore

$$\frac{1}{v}=\frac{6-1}{30}$$
$$=\frac{5}{30}$$

i.e. $$v=+\frac{30}{5}\text{ cm}$$
$$=6\text{ cm}$$

Since the image distance is a *positive* number, the image is *real*.

(*d*) An object is placed 20 cm from a diverging lens of focal length 10 cm. Where is the image; is it real?

$$\frac{1}{u}+\frac{1}{v}=\frac{1}{f}$$
$$\frac{1}{v}=\frac{1}{f}-\frac{1}{u}$$
$$=\frac{1}{(-10)}-\frac{1}{(+20)}$$

since the object is *real* and the lens *diverging*. Thus

$$\frac{1}{v}=-\frac{1}{10}-\frac{1}{20}$$
$$=\frac{-2-1}{20}$$
$$=-\frac{3}{20}$$

i.e. $$v=-\frac{20}{3}\text{ cm}$$

As the image distance is a *negative* number, the image is *virtual*.

Q 6.17

Calculate the image distance in each of the following problems and state whether the image is real or virtual:

(*a*) real object distance 0.2 m; converging lens focal length 0.1 m;
(*b*) real object distance 0.15 m; diverging lens focal length 0.05 m.

Calculate the object distance in each of the following problems:

(*c*) virtual image distance 50 mm; diverging lens focal length 100 mm;
(*d*) real image distance 50 mm; converging lens focal length 100 mm.

Calculate the focal length of the lens in each of the following problems and state whether the lens is converging or diverging:

(*e*) real object distance 140 mm; real image distance 70 mm;
(*f*) real object distance 180 mm; virtual image distance 60 mm.

Extended images produced by lenses

It is possible, as in the case of mirrors, to find the position of the image of an extended object produced by a lens using ray-construction methods. The three rays used to construct images produced by lenses are as follows.

(1) A ray parallel to the axis. This will pass through the focal point after refraction in a converging lens, or appear to come from it in the case of a diverging lens.
(2) A ray passing through the focal point, or which has a path which can be extrapolated through the focal point. It is refracted so as to be parallel to the axis.
(3) A ray through the optical centre of a thin lens. It is undeviated because the centre portion of the lens behaves like a thin, parallel-sided, block of glass.

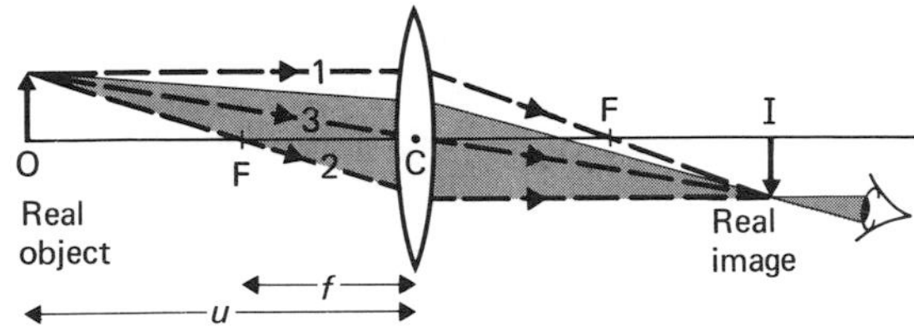

(*a*) A real image is formed from a real object if $u > f$

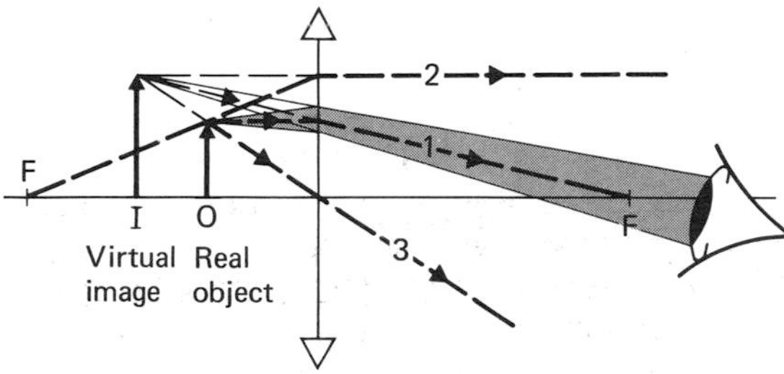

(*b*) A virtual image is formed from a real object if $u < f$

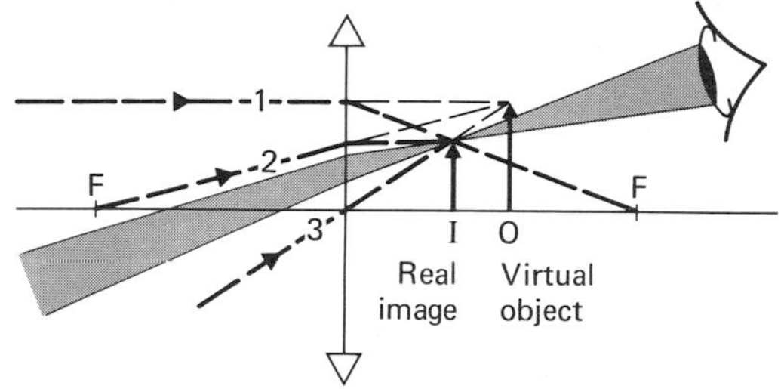

(*c*) A real image can be formed from a virtual object

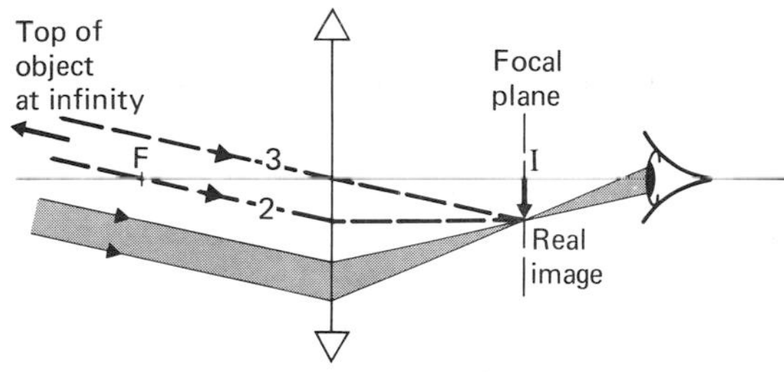

(*d*) A real image in the focal plane is formed of an object at infinity

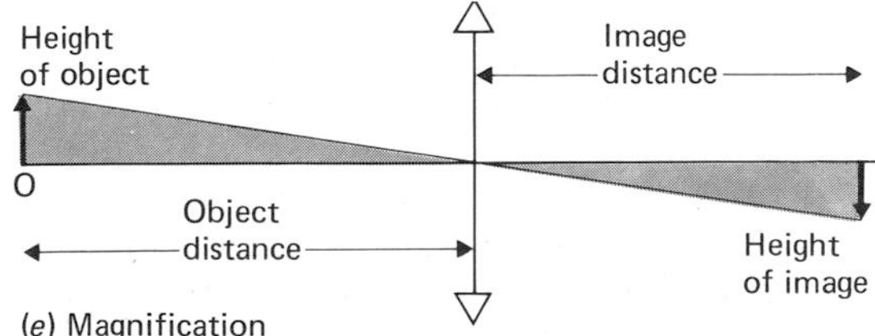

(*e*) Magnification

Fig. 6.28

Methods involving ray construction are important in a discussion of optical instruments, and the reader is advised to solve the problems in Question 6.17 graphically, as well as numerically. It is usual to represent thin lenses by straight lines, with hats to indicate whether they are converging or diverging lenses, as shown in Figs 6.28 and 6.29.

Fig. 6.28 shows that a converging lens produces a real, inverted image if the object distance is greater than the focal length, but a virtual, erect image if it is less. With a virtual object, caused by interposing the lens between a previous lens system and its image, the lens always forms a real erect image. An object at infinity produces a real inverted image in the focal plane.

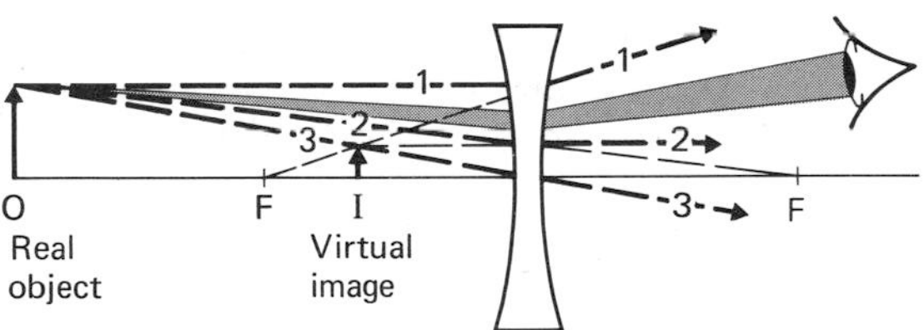

(*a*) A virtual image is formed from a real object

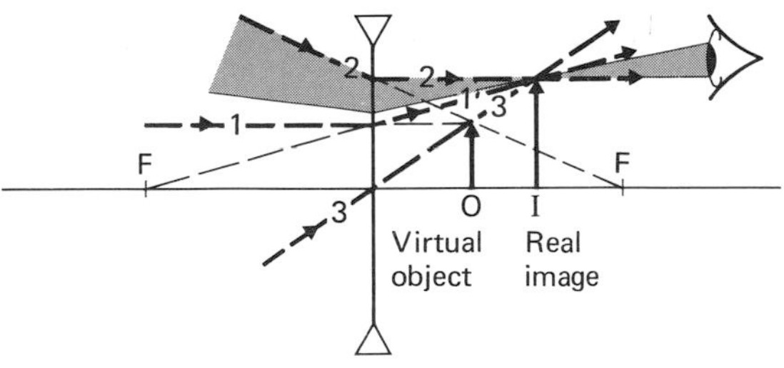

(*b*) A real image is formed from a virtual object

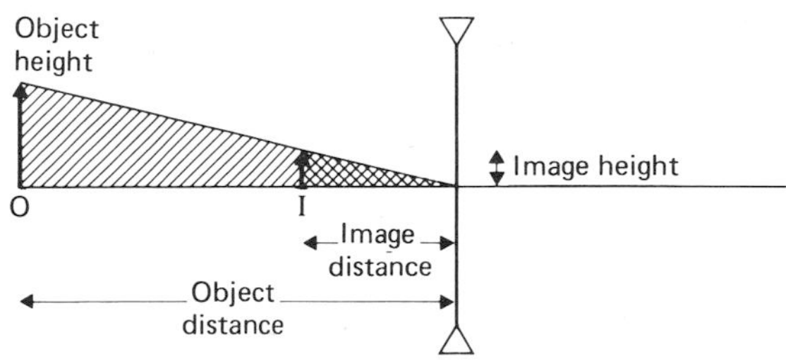

(*c*) Magnification

Fig. 6.29

In Fig. 6.29 we see that a diverging lens always produces a virtual, erect image of a real object, but can produce a real, erect image of a virtual object. Again, the image of an object at infinity lies in the focal plane, but this time it is virtual and erect.

The size of the final image in all cases is determined by the formula:

$$\text{magnification} = \frac{\text{image height}}{\text{object height}}$$

$$= \frac{\text{image distance}}{\text{object distance}}$$

since the two shaded triangles in Figs 6.28(*e*) and 6.29(*c*) are similar.

The power of a lens

Instead of describing a lens by its focal length, we sometimes use the *power of the lens, F,* defined by:

$$F = \frac{1}{f}$$

where the focal length is measured in *metres*. The unit of power is the *dioptre*, or *inverse metre*, named after the old-fashioned name for optics—*dioptrics*. Thus, a converging lens with a focal length of 0.2 m (200 mm) has a power of +5 dioptres: a diverging lens of the same focal length has a power of −5 dioptres. Ophthalmic opticians usually work in dioptres.

Defects of vision and their correction

In a normal eye the focal length of the eye lens can be adjusted so that objects at different distances can form images on the retina; this is done by changing the shape of the lens and is called *accommodation*. There are a wide variety of eye defects which prevent sharp images being formed; the two whose correction is easiest to understand are *short-sight* (*myopia*) and *long-sight* (*hypermetropia*).

A short-sighted person can see close objects clearly, but cannot see distant objects as the image is formed in front of the retina (see Fig. 6.30(*a*)). The most distant point at which an object can be clearly seen is called the *far-point*. To correct for short-sightedness, a diverging lens is used so that light from an object at infinity will appear to be coming from the far-point; the eye will then be able to focus it on the retina. The focal length of the spectacle lens, which is assumed to be adjacent to the eye, must be equal to the distance from the eye to the far-point. Thus, a person with a far-point at 4 m will require a diverging lens of power 0.25 dioptres ($f = 4$ m).

A long-sighted person, on the other hand, can see distant objects distinctly but cannot see close objects; the closest point that can be seen distinctly is called the *near-point* (see Fig. 6.30(*b*)). In a normal eye, the distance to the closest object that can be seen is called the *least distance of distinct vision* and is about 0.25 m; we will assume in our calculations that we want a person suffering from long-sight to be able to see an object at this distance. A converging lens is therefore used to make an object 0.25 m away have a virtual image at the wearer's near-point. Suppose that the person has a near-point at 1 m; the focal length of the spectacles required is given by:

$$\frac{1}{f} = \frac{1}{(0.25)} + \frac{1}{(-1)}$$

since the image is to be *virtual*. Thus

$$\frac{1}{f} = \frac{4-1}{1}$$

$$= \frac{3}{1}$$

i.e.

$$f = \tfrac{1}{3}\ \text{m}$$

$$= 0.33\ \text{m}$$

A converging lens of power 3 dioptres ($f = 0.33$ m) is required.

There are two other common defects. Whereas the defects above arise because the eyeball is too long or too short, it is also common for the eye to become unable to focus on near objects because of a lack of accommodation in old age; this defect is called *presbyopia*. Moreover, some people have corneas whose curvatures are not symmetrical; they find it more difficult to see horizontal lines than vertical lines, or vice versa, and are said to suffer from *astigmatism*. Some forms of astigmatism can be corrected with cylindrical lenses.

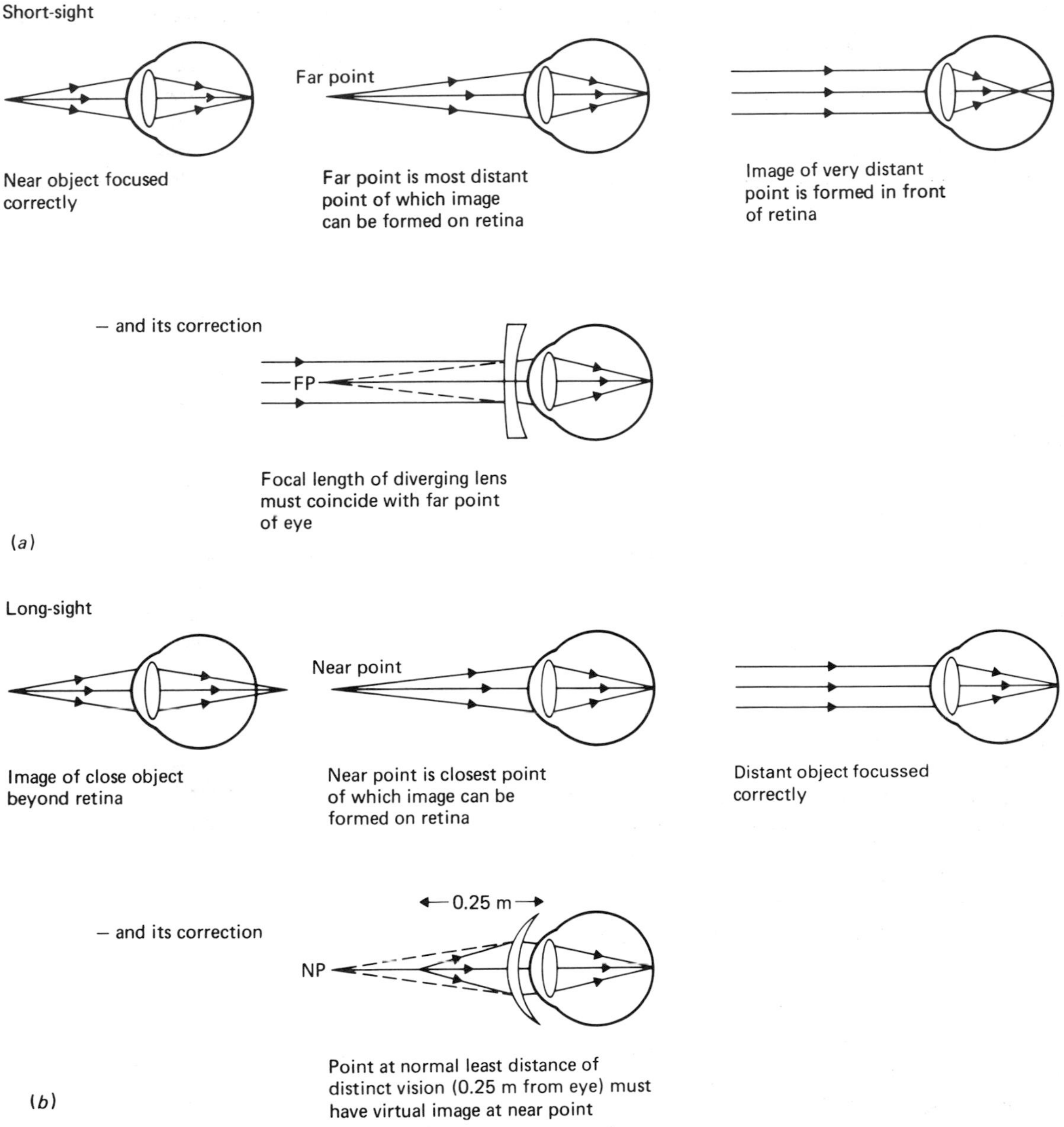

Short-sight

Near object focused correctly

Far point

Far point is most distant point of which image can be formed on retina

Image of very distant point is formed in front of retina

– and its correction

FP

Focal length of diverging lens must coincide with far point of eye

(*a*)

Long-sight

Image of close object beyond retina

Near point

Near point is closest point of which image can be formed on retina

Distant object focussed correctly

– and its correction

0.25 m

NP

Point at normal least distance of distinct vision (0.25 m from eye) must have virtual image at near point

(*b*)

Fig. 6.30 Two defects of the eye and their correction.

Q 6.18
What power, and type, of lens would be required for the following:

(*a*) a person suffering from long-sight whose near-point is at 0.5 m;
(*b*) a person suffering from short-sight whose far-point is at 1 m.

The lens maker's equation

For spherical mirrors we derived a simple equation connecting the focal length of the mirror with its radius of curvature; clearly there must be a similar equation for a lens, and the equation is sometimes called *the lens maker's equation*. The equation is more complicated than the corresponding one for a mirror since the focal length of a lens depends upon the radii of curvature of both surfaces and also upon the refractive indices of the lens and the surrounding medium.

Fig. 6.31 shows a typical ray passing through a thin lens; for the sake of clarity the refracting angle A of the prism section has been made to appear larger than it really is. The argument which follows applies only to paraxial rays, which make small angles with the axis, and therefore strike the surface of the lens at small angles of incidence.

The ray from the object O is refracted twice and undergoes a total deviation D, as shown; the real image I is formed where this deviated ray meets the undeviated ray, OXZI, which has passed through the optical centre of the lens.

Now $$\tan\alpha = \frac{h_1}{\text{OX}}$$

and $$\tan\beta = \frac{h_2}{\text{IZ}}$$

But as the lens is thin, P and Q are approximately the same height h from the axis:

$$h_1 \approx h_2 \approx h$$

and we also have

$$\text{OX} \approx \text{OC} = u$$
$$\text{IZ} \approx \text{IC} = v$$

Hence $$\tan\alpha \approx \frac{h}{u}$$

$$\tan\beta \approx \frac{h}{v}$$

and if α and β are both small angles:

$$\alpha \approx h/u$$
$$\beta \approx h/v$$

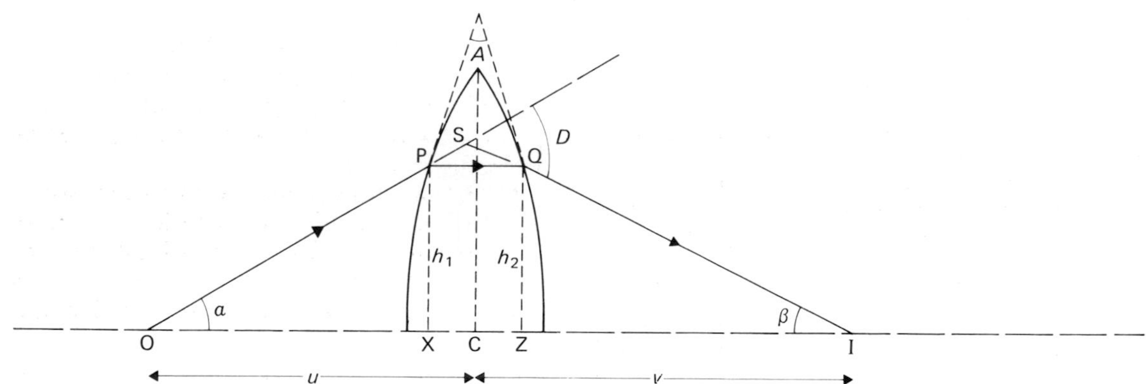

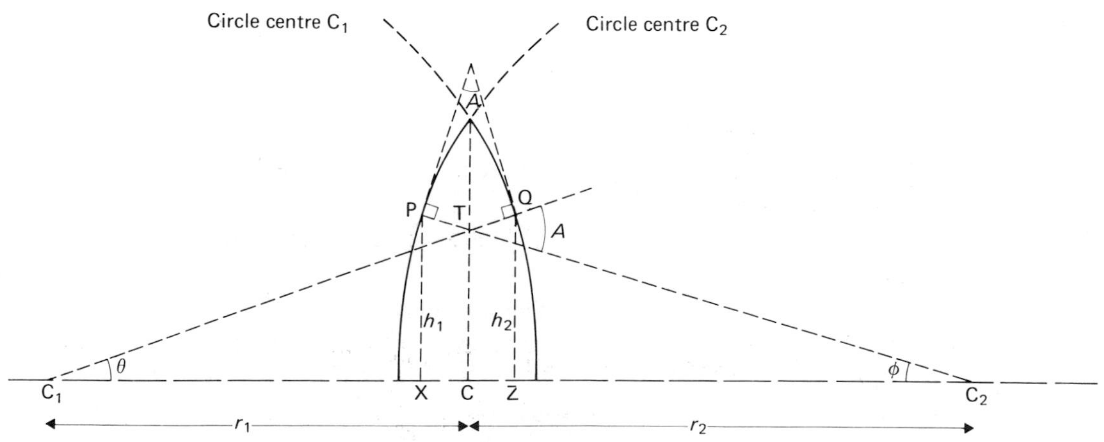

Fig. 6.31

Similarly, in the second diagram, where C_1Q and C_2P are normals to the lens surfaces:

$$\theta \approx h/r_1$$
$$\phi \approx h/r_2$$

Now, since D and A are external angles of triangles OSI and C_1TC_2, respectively:

$$D = \alpha + \beta$$

and
$$A = \theta + \phi$$

However, for a small-angle prism, from Equation (6.22):

$$D = (n_g - 1)A$$

Therefore
$$\alpha + \beta = (n_g - 1)A$$
$$= (n_g - 1)(\theta + \phi)$$

Hence
$$\frac{h}{u} + \frac{h}{v} = (n_g - 1)\left(\frac{h}{r_1} + \frac{h}{r_2}\right)$$

i.e.
$$\frac{1}{u} + \frac{1}{v} = (n_g - 1)\left(\frac{1}{r_1} + \frac{1}{r_2}\right)$$

Then, since
$$\frac{1}{u} + \frac{1}{v} = \frac{1}{f}$$

we have
$$\frac{1}{f} = (n_g - 1)\left(\frac{1}{r_1} + \frac{1}{r_2}\right) \qquad (6.25)$$

This is the equation sometimes called the *lens maker's equation*, or full lens formula.

Thus, for spherical lens surfaces, as the radii of curvature will be the same over the whole surface, the equation implies that the focal length of the lens will not depend on the point where the ray strikes the surface. However, as explained above, this prediction is only true for *paraxial* rays passing through a *thin* lens.

In order to obtain the same result for a diverging lens, it is necessary to adopt a sign convention for the radii of curvature:

The radius of curvature of a surface is taken as positive if the surface presents a convex face to the optically less dense medium; it is taken as negative if the surface presents a concave face to the optically less dense medium.

Q 6.19

Derive an identical result to that obtained above for a diverging lens.

Consider, for example, a biconvex lens made of glass of refractive index 1.5 and having surfaces with radii of curvature 0.1 m and 0.2 m respectively. Since both lens surfaces are convex to air, the radii of curvature are taken as positive.

Hence
$$\frac{1}{f} = (n_g - 1)\left(\frac{1}{r_1} + \frac{1}{r_2}\right)$$
$$= (1.5 - 1)\left(\frac{1}{+0.1} + \frac{1}{+0.2}\right)$$
$$= 0.5\left(\frac{2+1}{0.2}\right)$$
$$= \frac{1.5}{0.2}$$

i.e.
$$f = \frac{0.2}{1.5}$$
$$= 0.133 \text{ m}$$

When the lens is not placed in air, the quantity $(n_g - 1)$ becomes $\{(n_{\text{lens}}/n_{\text{medium}}) - 1\}$ and *is always taken to be a positive quantity*.

Q 6.20

(*a*) Calculate the focal length of a biconcave lens made of glass of refractive index 1.5 if the radii of curvature of the two faces are 0.15 m and 0.10 m respectively.
(*b*) What will be its focal length when immersed in water of refractive index 1.33?

Measurements on lenses

Measurements of the focal length of a converging lens

A quick estimate of the focal length of a converging lens can be obtained by forming an image of a distant object on a piece of paper; the distance from lens to paper is roughly equal to the focal length.

A more accurate method is to take measurements of the size of object and image, and of the image distance. The image is located by the method of no parallax. The magnification can be calculated from the size of object and image.

Now
$$\frac{1}{u} + \frac{1}{v} = \frac{1}{f}$$

Hence
$$\frac{v}{u} + \frac{v}{v} = \frac{v}{f}$$
$$\frac{v}{u} = \frac{v}{f} - 1$$

But
$$m = \frac{v}{u}$$

Therefore
$$m = \frac{1}{f} \cdot v - 1$$

This equation is of the form

$$y = kx + c$$

where k is the slope of a graph of y against x. Thus, if m is plotted against v, the slope will be $1/f$ and the intercept on the v-axis (when $m = 0$) will be f. Comparison of these two results will give an indication of the accuracy of the experiment.

Q 6.21
In a series of measurements to find the focal length of a converging lens, an object 2 cm high was used. The following table gives image distances, together with the height, h, of each image. Calculate the magnification in each case and plot a suitable graph to find the focal length of the lens:

v/cm	60.2	39.8	33.5	30.4	27.7	26.9
h/cm	4.0	2.0	1.4	1.0	0.8	0.7

The final method we shall consider is shown in Fig. 6.32. Suppose that an object 30 mm from a convex lens produces an image that is 70 mm from the lens; then, as light is reversible, the positions of object and image could be reversed. The new object distance will be 70 mm and the new image distance 30 mm; the object and image distances have been interchanged, but the separation of object and image is unchanged.

A similar situation is shown in the figure. In the first diagram u is the object distance and v the image distance. In the second diagram the new object distance u', is equal to the previous image distance, v; the new image distance v', is equal to the old object distance, u. The separation of object and image, d, has remained unchanged, and the lens has been moved a distance l, which can be measured. Points O and I are called *conjugate points* or *conjugate foci*, since they are interchangeable.

Now $$u = \frac{d-l}{2}$$

and $$v = l + \frac{d-l}{2}$$

$$= \frac{d+l}{2}$$

But $$\frac{1}{f} = \frac{1}{u} + \frac{1}{v}$$

and hence $$\frac{1}{f} = \frac{1}{\frac{d-l}{2}} + \frac{1}{\frac{d+l}{2}}$$

$$= \frac{2}{d-l} + \frac{2}{d+l}$$

$$= \frac{2(d+l)+2(d-l)}{(d-l)(d+l)}$$

i.e. $$\frac{1}{f} = \frac{4d}{d^2-l^2}$$

$$\therefore \quad \frac{d^2-l^2}{4d} = f$$

$$d - \frac{l^2}{d} = 4f$$

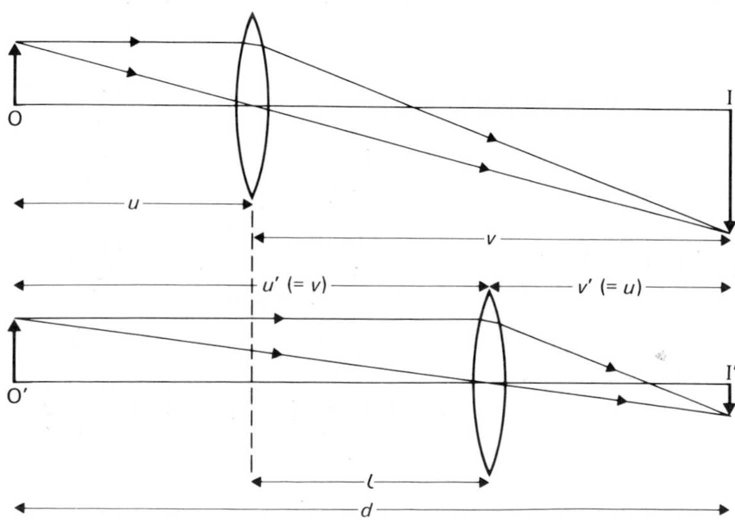

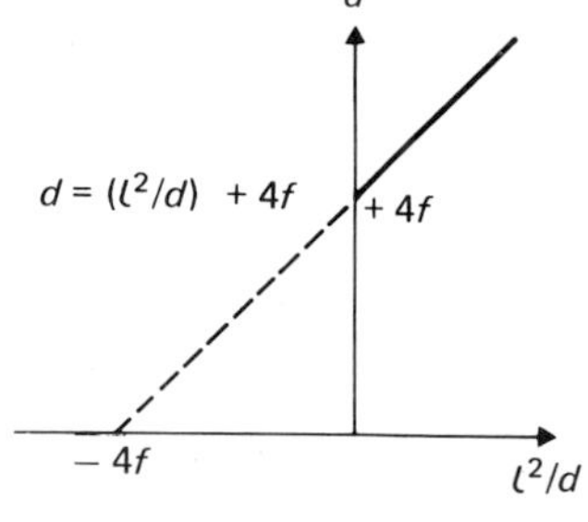

Fig. 6.32

i.e.
$$d = \frac{l^2}{d} + 4f$$

A graph of d against l^2/d will have an intercept $\pm 4f$ on both axes.

This method is especially useful when it is difficult to measure to the optical centre of a lens; it is, of course, always possible to measure the displacement of a lens from one position to another.

Note that the minimum separation of object and image, when $l = 0$, is 4f; this is a fact worth remembering when setting up an optical system.

Q 6.22

In an experiment to find the focal length of a converging lens, the object and screen were set a given distance, d, apart. The two positions of the lens for which a clear image could be obtained were then found, and the distance between them, l, was measured. The screen was then moved to a new position and the experiment repeated. Use the results quoted below to plot an appropriate graph and from it find the focal length.

l/cm	17.1	31.9	45.2	56.8	80.1
d/cm	123.1	127.8	135.1	142.9	160.1

Measurement of the focal length of a diverging lens

Measurements of the focal length of a diverging lens are more difficult because methods similar to those used for converging lenses would involve finding the position of a virtual image which cannot be shone on to a screen. However, one method which requires measurements on real images only is shown in Fig. 6.33(a); a converging lens, not in close contact, is used to provide a virtual object for the diverging lens.

The converging lens forms a real image, I′, which is located with a screen. The diverging lens is then interposed and the new position of the real image, I, ascertained in the same way. In the lens equation for the diverging lens, used to calculate the focal length, the object distance must be inserted as a *negative* number but the image distance as a *positive* number.

Q 6.23

In the above experiment, the initial real image was 200 mm from the converging lens. The diverging lens was then placed 150 mm from the converging lens, and the final image was 400 mm from the diverging lens. What was its focal length?

Boys' method for the radius of curvature of a surface of a converging lens

Finally, Fig. 6.33(b) shows an ingenious experiment for measuring the radius of curvature of a converging lens surface; this method was devised by Sir Charles Boys, and can be used for surfaces that are convex to the air. The surface whose radius of curvature is to be measured is used as a concave mirror, and to cut out extraneous light it is placed on a piece of black material.

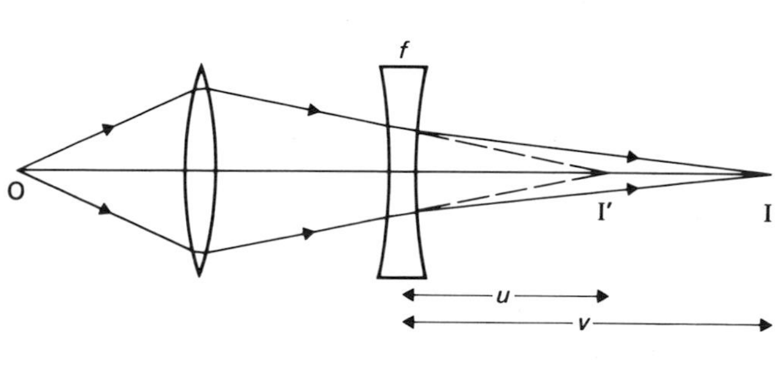

(a) A method for measuring the focal length of a diverging lens

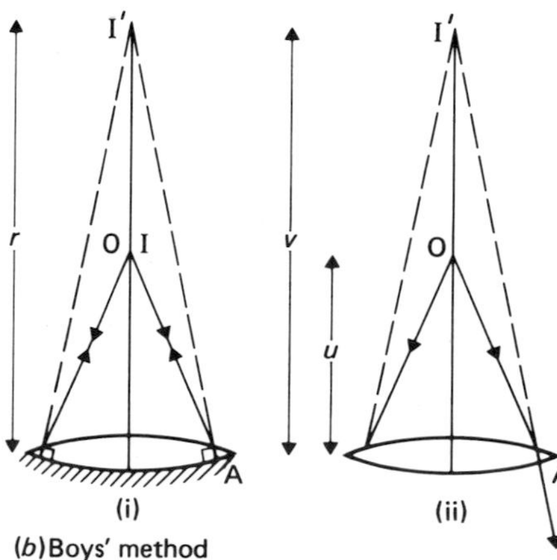

(b) Boys' method

Fig. 6.33

Most of the light incident on this reflecting surface will be transmitted and absorbed by the material, but a small proportion will be reflected.

An optics pin is mounted above the lens and moved up and down until it is found, by the method of no parallax, to be coincident with a real, inverted image; as the object and final image are at the same place, the object is said to be at a *self-conjugate point.* The light must have returned along its original path and must, therefore, have struck the reflecting surface at right angles. The line AI′ in the diagram is at right angles to the surface and thus I′ is at its centre of curvature. Now I′ is the virtual image formed by refraction in the first surface of the lens, and thus this virtual image distance, v, is numerically equal to the radius of curvature of the reflecting surface, r. Moreover, since there is no further refraction in the transmitted ray by the lower surface at A, as the ray is incident normally, I′ is also the virtual image of O formed by the lens as a whole, as shown in Fig. 6.33(b)(ii). Thus the distance from I′ to the optical centre of the lens, v, can be calculated from the object distance, u, and the focal length of the lens which is measured in some other way.

Thus
$$\frac{1}{u}+\frac{1}{v}=\frac{1}{f} \qquad (6.23)$$

$$\frac{1}{v}=\frac{1}{f}-\frac{1}{u}$$

The image distance will be a negative number since the image is virtual but, of course, the radius of curvature will be positive in accordance with our convention.

For a *diverging lens* the radius of curvature may be determined directly by reflecting light from its concave surface; when object and image are coincident, the object is at the centre of curvature of the nearer face.

Q 6.24

A biconvex lens has a focal length of 150 mm. It is laid on a piece of black velvet and an optics pin mounted above it. When the optics pin is 50 mm above the lens, a real, inverted image is coincident with it. Calculate the radius of curvature of the bottom surface of the lens.

6.6 Aberrations

When a point object does not produce a point image, we say that an aberration occurs; it may cause the image of an extended object to differ from the object in shape or sharpness, or make a white object appear coloured. Some aberrations occur only in lenses but others, notably spherical aberrations, occur both in lenses and in mirrors.

Spherical aberration

In our analysis of lenses and mirrors, we have frequently made small-angle approximations, as discussed on page 126. In practice, the rays furthest from the axis do not pass through exactly the same image point as those closest to it. This effect can be seen on careful examination of the photograph on page 146 and it is called *spherical aberration*; this was not evident in our simple theory because of the approximations made.

In Fig. 6.34 spherical aberration produced by a lens is shown. The optimum place to put a screen to observe the image is in the *plane of least confusion*; the reader should note the

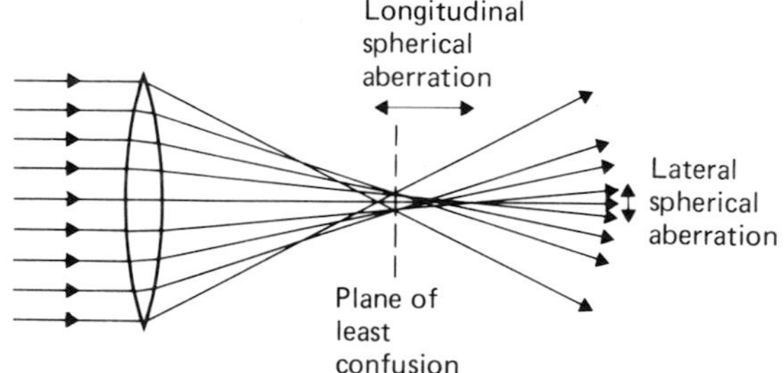

(a) Spherical aberration in a convex lens

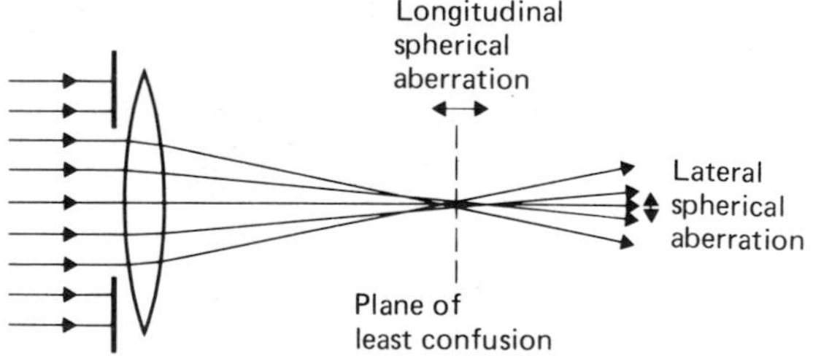

(b) Spherical aberration reduced by a lens-stop

Fig. 6.34

distinction between *lateral spherical aberration* and *longitudinal spherical aberration*. Spherical aberration can be corrected by grinding the lens to a carefully calculated aspherical shape; this is extremely difficult to do because in any lens grinding machine the shape produced most easily is a spherical one. The defect is minimized by dividing the refraction equally between the two surfaces of the lens; for instance, if the lens is plano-convex and a distant object is to be viewed, the rays parallel to the axis should be incident upon the convex side. We see from Fig. 6.34(b) that blurring of the image may be reduced by *stopping down* the lens—in other words reducing the effective diameter and eliminating the peripheral rays. This results in a less bright image and may make diffraction effects more important, as explained in Section 7.5. In the human eye, spherical aberration is corrected by a decrease in the refractive index of the eye lens away from the axis.

The effect of spherical aberration in a concave mirror is shown in Fig. 6.35(a). The bold curve is the envelope of the image-forming rays and is called the *caustic curve*; it may be seen when sunlight is reflected in a cup of tea. Spherical aberration in a curved mirror may be corrected for distant objects on the axis by making the surface *parabolic*; all rays from the object will then pass through the same image point, no matter at what point they strike the mirror. In theory, for a closer object, an ellipsoidal mirror should be used but, except for some cinema projectors, this is rarely done.

An alternative method, used in certain telescopes, is to place a piece of specially shaped plastic material in front of the mirror; this is known as a *Schmidt corrector* and is illustrated in Fig. 6.35(b). The convex central part slightly converges the beam so that the image is formed slightly inside the focal point; the concave outer section diverges the light so that it crosses the axis nearer to the focal point than would otherwise be the case. In this way, it is possible to bring all the rays to very nearly the same image point.

The final way of reducing spherical aberration is to stop down the mirror.

Other geometrical aberrations

There are a number of other geometrical aberrations which we shall consider briefly; these apply to both lenses and mirrors. Sometimes the image of an off-axis *point* becomes two *lines*; this is known as *astigmatism* and is illustrated in Fig. 6.36.

The second diagram shows that the image of a plane object at right angles to the axis is not always in a plane at right angles to the axis. In general, the image lies on a slightly curved surface; this is called *curvature of the field* and

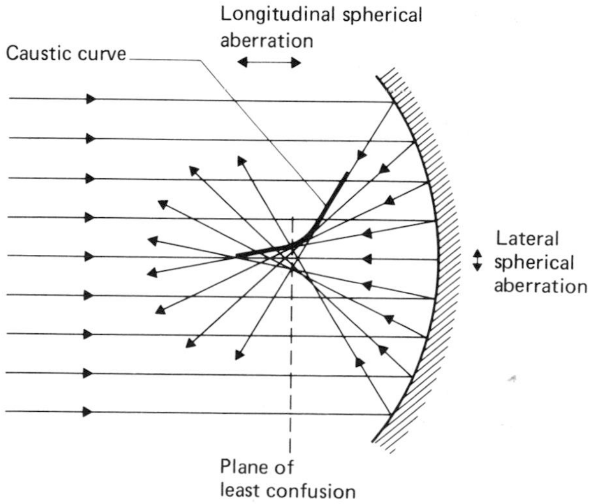

(a) Spherical aberration in a mirror

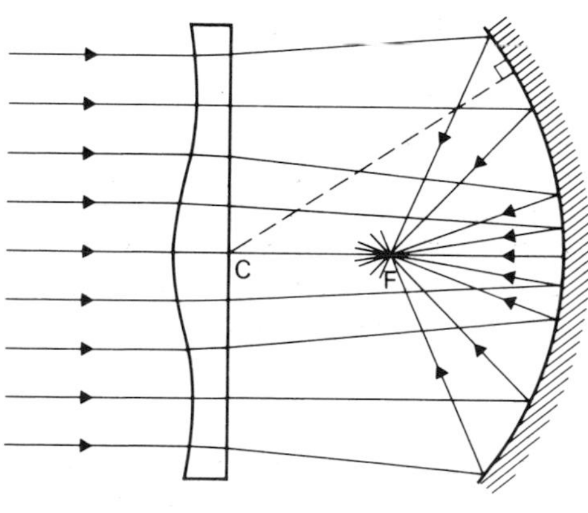

(b) Use of a Schmidt corrector

Fig. 6.35

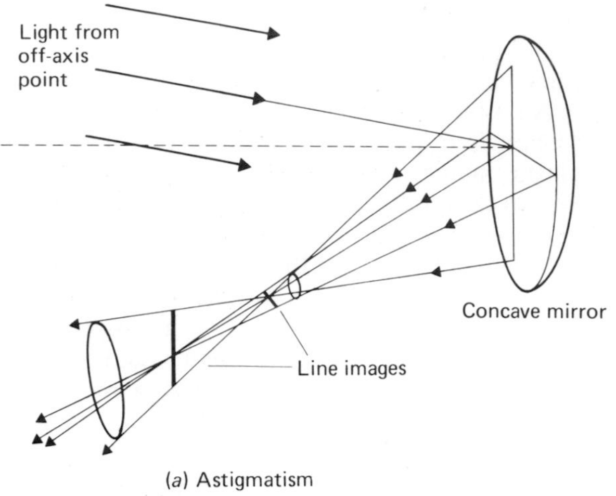

(*a*) Astigmatism

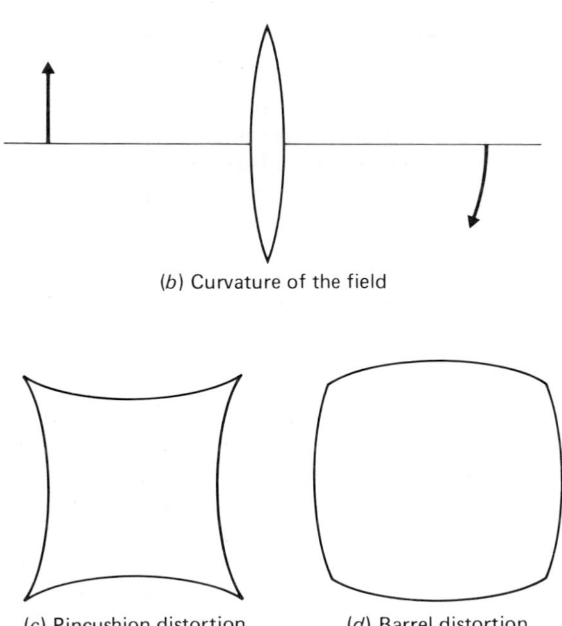
(*b*) Curvature of the field

(*c*) Pincushion distortion (*d*) Barrel distortion

Fig. 6.36 Other geometrical aberrations of lenses and mirrors.

is particularly important when a strong magnifying glass is used.

The final geometrical aberration we shall consider is called *distortion* and is caused by changes in magnification at different distances from the axis. If a postage stamp is viewed through a strong magnifying glass, the virtual image will not be rectangular but will appear with *pin-cushion distortion*; in this case, the magnification increases with distance from the axis. On the other hand, if the lens is held at arm's length so that the real inverted image of a distant window is seen, we observe *barrel distortion*; this is caused by a decrease in magnification as the distance from the axis is increased.

Chromatic aberrations

We have already shown how dispersion in prisms is caused by the variation in refractive index of glass with colour, as shown in Fig. 6.23. Dispersion also occurs in lenses for the same reason. Equation (6.25) tells us that:

$$\frac{1}{f}=(n_g-1)\left(\frac{1}{r_1}+\frac{1}{r_2}\right)$$

and as the refractive index of glass for blue light is greater than it is for red, we would expect the lens to have a greater focal length for red light than for blue. The effect of this in a lens is shown in Fig. 6.37(*a*) and is called *chromatic aberration*; the image becomes less sharp and is slightly coloured, especially at the edges.

The usual way of reducing chromatic aberration is to use an *achromatic doublet.* This consists of a converging and diverging lens, of different focal lengths and dispersive powers; two faces must have the same radius of curvature so that they can be cemented together with Canada balsam, as shown in Fig. 6.37(*b*). This is used because it has much the same refractive index as glass, and reflections at the glass/balsam surfaces are thus reduced.

The dispersion produced by the two lenses will be in opposite directions, since one is a converging lens and one diverging; if these dispersions can be made to be the same magnitude then they will cancel out. In fact, this can only be done for two wavelengths, usually red and blue light, and there will be slight dispersion at other wavelengths, as shown in the graph of Fig. 6.37(*c*). It is possible to show that *the ratios of the focal lengths must be numerically equal to the ratio of the dispersive powers.*

Now we require the achromatic doublet to act as a converging lens and therefore the convex lens must be more powerful than the concave lens; that is, the focal length of the convex lens must be less than that of the concave lens. It follows that the convex lens must be made

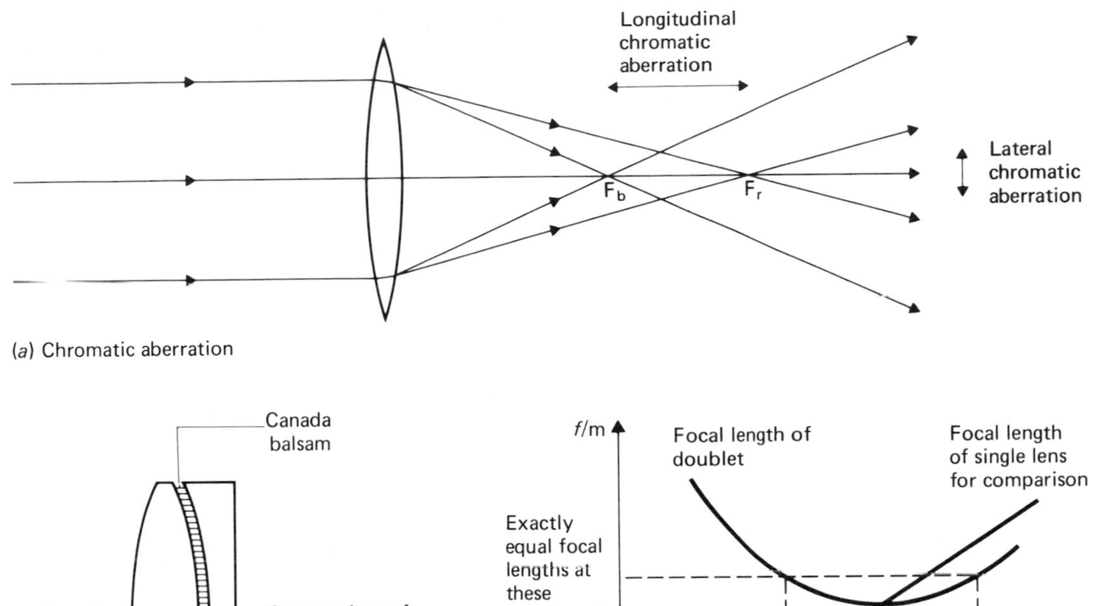

(*a*) Chromatic aberration

(*b*) An achromatic doublet

(*c*) Wavelengths for which the lens is made achromatic

Fig. 6.37

of a glass, such as a crown glass, which has a relatively low dispersive power while the diverging lens is made of a flint glass which has a higher dispersive power. The achromatic lens was developed by Dolland in 1753.

Bibliography

Books

Arnold, J.D. (1981) *Manual of Fibre Optic Communication.* Standard Telephone and Cables; a well-illustrated account of the physics of a fibre-optic communication system.

Bolton, W. (1986) *Patterns in Physics* (2nd Edn.). McGraw-Hill; Chapter 10 provides a good account of the most important material in this chapter.

Hecht, E. and Zajac, H. (1974) *Optics.* Addison-Wesley; an excellent, up-to-date textbook on a wide range of topics.

Jenkins, F.A. and White, H.E. (1981) *Fundamentals of Optics.* McGraw-Hill; one of the best undergraduate reference books on optics, this book can be used selectively by other students.

Overheim, R.D. and Wagner, D.L. (1982) *Light and Colour.* Wiley; provides a very well-illustrated account of geometrical and physical optics, and colour.

PSSC (1981) *Physics* (5th Edn). Heath; Chapters 17–19 are thorough and well-illustrated.

Rutherford, S.J. (1985) *Optical Fibres in School Physics.* SCSST (Experimenting with Industry).

Schawlow, A.L. (1969) *Lasers and Light.* Freeman; a Scientific American book covering a wide variety of topics.

Time Life (1972) *Light and Vision*; many fascinating examples of optics in unfamiliar situations.

Wright, G. (1973) *Elementary Experiments with Lasers.* Wykeham; Chapter 3 contains some interesting experiments to demonstrate lens aberration using a laser.

Software

Heinemann/Fiveways *Lenses*; simulates the effect of mirrors and lenses on light rays.

Hutchinson Software *Optics*; programmes on refraction, lens behaviour and defects of the eye provide diagrams in which parameters can be altered easily.

Longman Micro Software *Eye*; programmes showing the way in which the eye works with and without defects, including the effect on the depth of field of changing the aperture.

7
Wave optics

7.1 Superposition

In Chapter 9 we shall discuss the historical evidence for the wave nature of light and in this chapter we look at those properties of light which derive from the fact that it is a wave phenomenon; these properties are often referred to collectively as *physical optics*.

The first four chapters of this book reviewed the properties of oscillations and waves in general, showing how wave motions are propagated through media, how they can be reflected and refracted at interfaces, and what happens when two wavetrains meet. We saw that:

Wavetrains move through each other without undergoing any permanent alteration.

Moreover, in the region where they co-exist:

The total wave disturbance at a particular time is obtained by adding together at each point the displacements due to each wave passing through the medium.

This is the *Principle of Superposition*. Where two equal wave crests coincide, a crest of double amplitude is formed, and if a crest and trough of equal size superpose, a region of zero amplitude results.

In Section 4.3 we saw that when two plane wavetrains cross, regions of constructive and destructive superposition occur:

Constructive superposition (*or constructive interference*) *occurs whenever the two or more waves incident at that point are exactly in phase, so that the resultant amplitude of oscillation at the point is a maximum, equal to the sum of the amplitudes of the waves; these points are called antinodes, and lines along which all points are antinodes are called antinodal lines.*

Destructive superposition (*or destructive interference*) *between two waves occurs at a point whenever the two waves incident at the point are exactly out of phase so that the resultant amplitude is a minimum; if the amplitudes of the two waves are equal the resultant will be zero. These points are called nodes, and lines along which all points are nodes are termed nodal lines.*

As we explained in Section 4.3, where a fixed time-independent pattern of maxima and minima is produced we refer to the phenomenon by the usual name of *interference*.

Coherent light and conditions for interference

The above account may give the impression that any two sources of light, of identical frequency, and therefore colour, sending light into the same space, can produce an interference pattern; this is not so. To produce an interference pattern, the sources of light must bear a constant phase relationship to each other, so that there are places where there will always be maxima and places where there will always be minima; if, for example, the light from one source suddenly experienced a phase change of π radians, all the maxima would become minima and vice versa. If such random phase changes were to occur frequently, there would be no visible interference pattern at all.

An ordinary monochromatic light source emits a short packet of continuous waves lasting about 10^{-8} s; this packet will then end, and the new packet will have a random phase relative to the first. As explained in Chapter 10, each wave packet is a photon of light.

Fig. 7.1(*a*) shows why light from two separate sources will not form interference fringes. As the phase of each photon of light is determined randomly, the light from the two sources will sometimes be in phase and sometimes not; they are said to be non-coherent because there is no consistent phase relationship between them.

However, two sets of wavetrains derived from the same source can have a consistent phase relationship since both wavetrains undergo the same random phase changes. In Fig. 7.1(*b*) there is no path difference between the top two sets of wavetrains, and a path

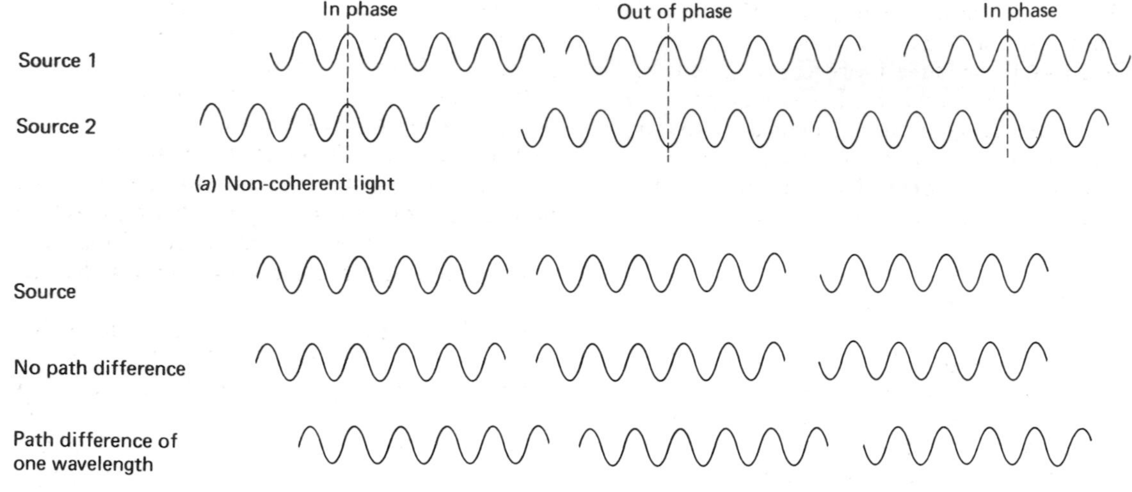

Fig. 7.1

difference of one wavelength between the first and third sets; in this case the waves are always one wavelength out of step. Interference will still occur if the path difference is increased, provided it is still much shorter than the length of one photon..

Thus the two wavetrains of light which produce interference must come from the same source. In arrangements such as Young's slits the wavefront is divided—by passing through two slits—to produce the two interfering wavetrains; this is called *division of wavefront.* In situations like that in an oil film—where the coloured fringes are produced by interference effects—part of the light energy is reflected at one surface and part at another; this is called *division of amplitude.*

In both methods a path difference is introduced between the two interfering wavetrains; since there is a random phase difference between successive photons, the path difference introduced must be much less than the 'length' of a photon. It takes about 10^{-8} s for a typical light photon to be emitted; the velocity of light is 3×10^8 m s^{-1}. Hence:

$$\text{'length of photon'} = \text{velocity} \times \text{time for emission}$$
$$= (3 \times 10^8) \times 10^{-8}$$
$$= 3\text{ m}$$

maximum path difference ≈ 0.3 m

A further restriction is that if the light is polarized the planes of polarization must be the same, otherwise the displacements will not superpose.

Q 7.1
The wavelength of green light is about 5×10^{-7} m. How many waves are emitted in 10^{-9} s? How many waves are there in 0.3 m?

Laser light sources

When light from a conventional source such as a light bulb is used to illuminate a pair of slits for an interference experiment, it is necessary first to pass it through a narrow slit, much reducing the energy available. This is because the photons across the width of the beam are not in phase with each other, and a broad beam will not, therefore, produce an interference pattern.

In a laser, each photon emitted in a pulse is stimulated by the light already in the beam, and thus the phases of the photons are not random; within the pulse all the photons have exactly the right phase to produce a coherent beam. The pulses last very much longer than the 10^{-9} s mentioned above and, during the time taken to emit a pulse, the light is coherent right across the width of the beam. Some types of laser produce a continuous wave and the light is continuously coherent.

Thus, lasers produce more coherent energy over a much longer period and can produce more intense fringe patterns. They are a useful, if expensive, tool for demonstrating phenomena in physical optics; some of the photographs in this Chapter were taken using laser light.

7.2 Young's slits

The concept of superposition had been applied to water waves in the seventeenth century by Isaac Newton, when explaining anomalous tides, but was not applied to light until the beginning of the nineteenth century. Thomas Young, who proposed that light was a transverse wave motion, revived interest in the wave theory of light and, in 1801, set up an experiment to observe interference using two pin holes in a blind as sources of light. We shall here describe the modern form of the experiment in which the pin holes are replaced by slits; this is now known as the Young's slits experiment. Young performed the experiment using sunlight, but we shall consider first the results obtained using monochromatic light.

The apparatus is set up as shown in Fig. 7.2; a sodium vapour lamp may be used as a convenient source since it produces light that is almost monochromatic. It illuminates a single slit which acts as the effective source of coherent light, as explained in the previous section. Huygen's principle has been used, in Section 3.3, to show how diffraction will cause light to spread out from a narrow slit, and thus the two slits in the second screen will be illuminated. They, too, act as sources of light and all points on the screen are illuminated by light from these two sources. Note that in all diagrams the wavelength of light is very much exaggerated! The diagram shows how the slits divide the wavefront of the light, as mentioned in the last section.

Consider a point P on the screen receiving light from the two slits S_1 and S_2. S_1N is a line drawn at right angles to S_2P. If the angle θ is small, S_1P and NP will be almost equal and hence the path difference will be S_2N. If the angles are small and the rays nearly parallel, it also follows that all the angles marked θ in the third diagram are equal. We also assume for the present that the distance from the source slit to each of the double slits is the same. Now in the triangle S_1S_2N:

$$\sin\theta = \frac{p}{s}$$

i.e. $$p = s\sin\theta \qquad (7.1)$$

The angle θ can be expressed in terms of the distance x and D, thus:

$$\sin\theta = \frac{x}{PQ} \qquad \text{in the triangle PRQ}$$

$$\approx \frac{x}{D} \qquad \text{as } \theta \text{ is small}$$

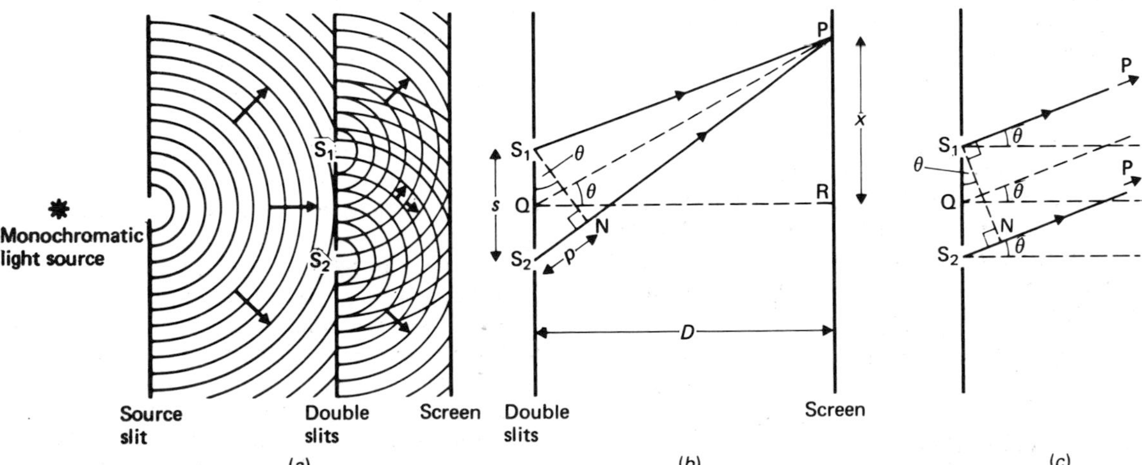

Fig. 7.2 The formation of an interference pattern using Young's slits.

Hence $$p = \frac{sx}{D} \tag{7.2}$$

Now, *constructive interference* at the screen will occur if the path difference is an integral number of whole wavelengths, as explained in Section 4.4. Hence, the condition for constructive interference is:

$$\text{path difference,} \quad p = m\lambda \tag{7.3}$$

where m can be any whole number including zero; thus, the path difference might be 0, λ, 9λ, 15λ, etc.

Now $$s \sin\theta = p \quad (7.1)$$
$$= m\lambda \quad \text{from (7.3)}$$

where $$\sin\theta = \frac{x}{D}$$

Thus, the condition for constructive interference may be written:

bright fringe
$$\sin\theta = \frac{m\lambda}{s} \tag{7.4}$$
$$\text{or } x = m\frac{\lambda D}{s} \tag{7.5}$$

We therefore expect there to be a series of bright fringes on the screen. The zero-order fringe ($m = 0$, $\sin\theta = 0$, $x = 0$) will be on the line QR and there will be further fringes arranged symmetrically to either side (at $m = \pm1, \pm2, \pm3 \ldots$; $\sin\theta = \pm\lambda/s, \pm2\lambda/s, \pm3\lambda/s \ldots$; $x = \pm\lambda D/s, \pm2\lambda D/s, \pm3\lambda D/s, \ldots$). The fringe with a value of m equal to one is called the *first-order fringe*. The distance between adjacent fringes, called the *fringe width*, is $\lambda D/s$.

The symbol m is used in optical interference work to stand for order number since n, which was used in Chapter 4, is also used to stand for refractive index.

Destructive interference will occur when a crest from one wave coincides with a trough from the other; in other words when the path difference is an odd number of half-wavelengths, as explained in Section 4.4. Hence the condition for destructive interference is:

$$\text{path difference,} \quad p = (m + \tfrac{1}{2})\lambda \tag{7.6}$$

where m is any whole number including zero; thus, the path difference might be $\frac{1}{2}\lambda$ ($m = 0$), $\frac{3}{2}\lambda$ ($m = 1$), . . . , $17\lambda/2$ ($m = 8$), etc.

Now $$s \sin\theta = p \quad (7.1)$$
$$= (m + \tfrac{1}{2})\lambda \quad \text{from (7.6)}$$

where $$\sin\theta = \frac{x}{D}$$

Thus, the condition for destructive interference may be written:

dark fringe
$$\sin\theta = (m + \tfrac{1}{2})\frac{\lambda}{s} \tag{7.7}$$
$$\text{or } x = (m + \tfrac{1}{2})\frac{\lambda D}{s} \tag{7.8}$$

Thus, spaced symmetrically between the bright fringes, there is a series of dark fringes; the spacing of the dark fringes is identical to the spacing of the bright fringes.

Q 7.2
Coherent light of wavelength 5.5×10^{-7} m is incident on two slits 1.1 mm apart. Calculate the angles at which the following lines will be found:

(*a*) the first-order bright fringe;
(*b*) the tenth-order bright fringe;
(*c*) the first dark fringe;
(*d*) the fourth dark fringe.

The fringes are observed on a screen 0.5 m from the slits. Calculate (answers (*e*) to (*h*)) how far from the zero-order fringe each of the above will be.

Fig. 7.3 shows a set of Young's slits fringes photographed with a thin beam of light from a laser. It also shows a graph of the variation of intensity with angle in the photograph; it can be proved that the intensity varies between maxima and minima according to the square of the cosine of the phase difference between the two components. The intensity of the light varies between maxima and minima throughout the region beyond the slits and thus, in theory, fringes may be observed anywhere there; they are known as *non-localized fringes*. In practice, the fringes are significant only where the intensity from each slit is large, and where the intensities are approximately equal.

Accurate measurements made on the pattern obtained enable us to calculate the wavelength of light used, using the equations above; the accuracy obtainable is remarkable considering

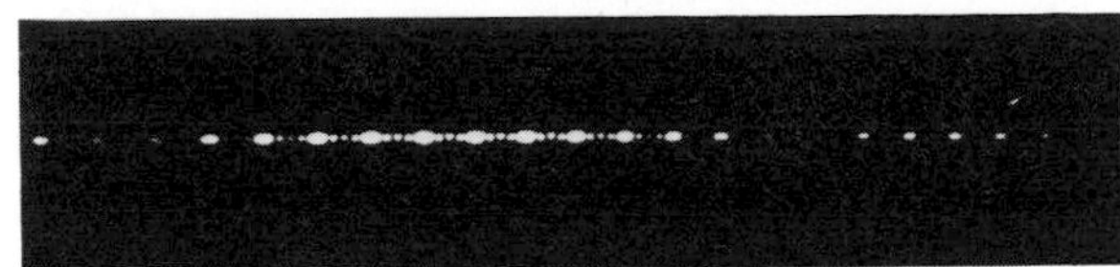

(*a*) Young's Slit fringes, observed with laser light. (Photograph by courtesy of Griffin and George.)

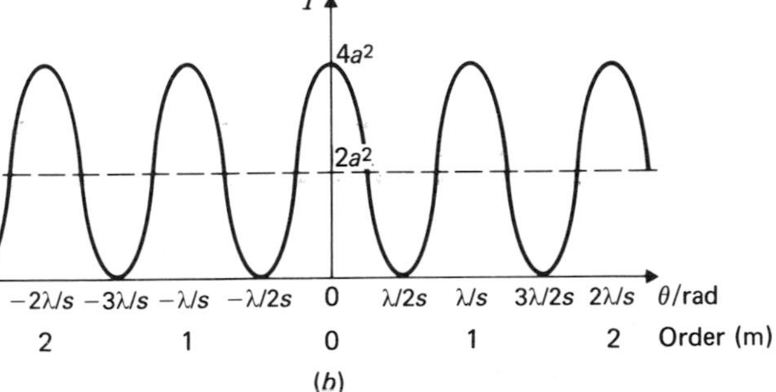

(*b*) The variation of intensity in a Young's Slits pattern

Fig. 7.3

that the wavelength is as small as 10^{-7} m. The separation of the slits is measured with a travelling microscope. When finding the fringe width, it is usual to measure, for example, ten fringe widths and divide by ten in order to obtain greater accuracy. The limiting factor on the accuracy of any determination of wavelength will always be the measurement of the separation of the slits, since if they are too close the separation will be difficult to measure, and if they are too far apart the fringes will be almost indistinguishable.

A simple method of observing the phenomenon of interference is to take a small undeveloped photographic plate and to scratch two lines across the emulsion with a penknife and straight edge; a separation of about 0.5 mm is convenient. The plate is held against the eye and a lamp with vertical filament, or a discharge tube, is viewed at a distance of about 3 metres. If the source is monochromatic, distinct fringes of the appropriate colour are observed. If a white light filament lamp is used, a number of white fringes are observed with coloured edges; these are termed *white-light fringes.*

Q 7.3
In a Young's slits experiment to measure the wavelength of light from a sodium vapour lamp, the following measurements were made:

slit separation	$=0.42\pm0.01$ mm
distance to screen	$=0.500\pm0.001$ m
ten fringe widths	$=7.21\pm0.01$ mm

Calculate the wavelength of the light and the maximum probable error in the answer.

White light fringes

Equation (7.4) shows that the separation of red fringes is about twice that of blue ones, since the wavelength is roughly twice as great. Thus, except for the central maximum, fringes of different colours will be in different places. If the slits are illuminated with white light, the first-order fringe will vary in colour from blue at the inside edge to red at the outside. The second-order fringe will be even broader and by the time the higher-order fringes are reached, the red from one fringe will overlap the blue from the fringe of next highest order; there will then be a general illumination that is almost white, as shown in Fig. 7.4.

In experiments using monochromatic light, it is sometimes necessary to determine which fringe is the central, zero-order fringe; this may be required where shifts of the whole system are to be detected by the motion of the central fringe. This identification may be done conveniently by passing white light into the system and observing the central white fringe.

Q 7.4
If the wavelength of red light is 6×10^{-7} m and of blue light 4×10^{-7} m, at what order of the red fringe system will fringes of these colours coincide in adjacent orders?

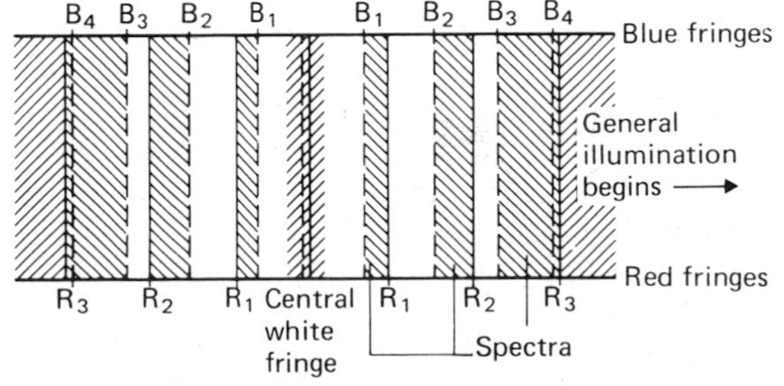

Fig. 7.4 White light fringes.

Optical path

We have seen in Section 6.3 that in a medium of absolute refractive index, n, the velocity of light, c_n, is equal to c/n. Thus, whereas the time taken to travel a distance l in a vacuum, t_0, is given by:

$$t_0 = l/c$$

the time to travel the same distance in the medium, t_n, is given by:

$$t_n = \frac{l}{c/n}$$

$$= \frac{nl}{c}$$

It is sometimes useful to regard the medium as having an *optical path length*, nl, (travelled at velocity c) rather than a geometrical path length, l, (travelled at velocity c/n).

Note that when light passes from a vacuum into the medium the frequency remains constant but the wavelength changes from λ_0 to λ_n:

$$\lambda_0 = \frac{c}{f}$$

$$\lambda_n = \frac{c_n}{f}$$

$$= \frac{c}{nf}$$

i.e. $$\lambda_n = \frac{\lambda_0}{n}$$

Therefore, the higher the refractive index, the shorter the wavelength and the greater the number of waves that will fit into a given length of medium; the number is determined by nl/λ_0. Thus, the optical path length, nl, determines the phase difference between the beginning and end of a length l of the medium.

Shifting a fringe system

In Fig. 7.5 a piece of glass of refractive index n and thickness l is placed in front of the lower slit in a Young's slits experiment, thus increasing the optical path length of all rays passing through the slit. The centre of the pattern will now be displaced downwards a distance x' to a point where:

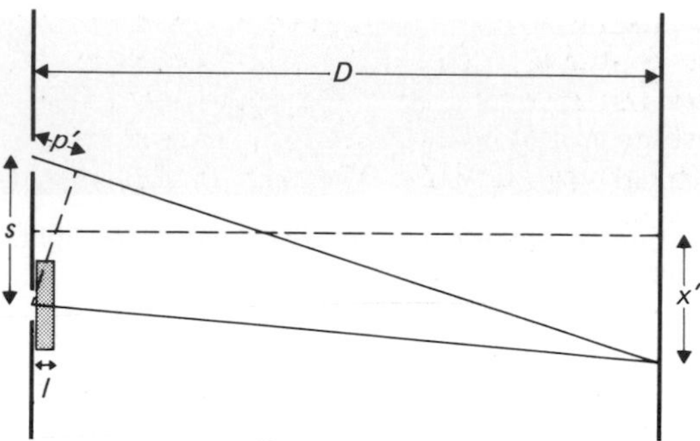

Fig. 7.5

extra geometric path length from upper slit = extra optical path length caused by glass

$$p' = nl - l$$

$$= (n-1)l$$

Now p' is given by Equation (7.2):

$$p' = \frac{sx'}{D}$$

Hence $$\frac{sx'}{D} = (n-1)l$$

$$x' = (n-1)\frac{lD}{s} \quad (7.9)$$

In general, the total optical path difference between the slits is given by:

$$\text{total optical path difference} = p' - (n-1)l$$

$$= \frac{sx'}{D} - (n-1)l$$

The condition for constructive interference will therefore be, from Equation (7.3):

$$\frac{sx'}{D} - (n-1)l = m\lambda$$

i.e. $$x' = m\frac{\lambda D}{s} + (n-1)\frac{lD}{s} \quad (7.10)$$

Without the piece of glass the condition for constructive interference would have been:

$$x = m\frac{\lambda D}{s} \quad (7.5)$$

Thus, the whole fringe system is displaced downwards a distance $(n-1)lD/s$.

This will also be the case if light arrives at the lower slit later in phase than at the upper slit; for instance, if the lower slit is further from the source slit. This will, in fact, often be the case but it results only in a shift of the fringe system as a whole; white light fringes can be utilized to identify the central fringe if necessary.

Q 7.5
A Young's slits experiment is set up using monochromatic light of wavelength 6×10^{-7} m. A slip of glass, 0.2 mm thick, is inserted in front of one slit and the fringe system moves a distance which is found to be equal to 200 fringe widths. Calculate the refractive index of the glass.

7.3 Other division of wavefront systems

Lloyd's mirror

Lloyd's mirror, and Fresnel's biprism which is described later, are alternative means of providing two coherent sources of light; both these arrangements utilize *virtual image sources* but the basic geometry is the same as in the case of Young's slits. The geometry of Lloyd's mirror is often important in problems involving the reflection and interference of radar beams.

Fig. 7.6 shows how a flat mirror or glass block is used to form a virtual image S_2 of the single slit S_1. Light can reach any point in the shaded region either direct from S_1 or by reflection from the mirror; in the latter case it appears to come from S_2. The light from S_1 and S_2 interferes in the shaded region and interference fringes can be seen on the screen. In the second diagram, the two paths taken by light to reach point P are shown; this diagram is almost identical to that for Young's slits and as before, the fringe width is equal to $\lambda D/s$, where s is the distance between S_1 and S_2.

One important point to note is that if the screen is placed so that it touches the end of the mirror, the path difference at the point of contact between the direct and reflected ray is zero, as shown in Fig. 7.6(*c*). However, a dark fringe is observed at the point of contact; this is because *a phase change of π radians occurs when a light ray is reflected from a medium of higher refractive index*.

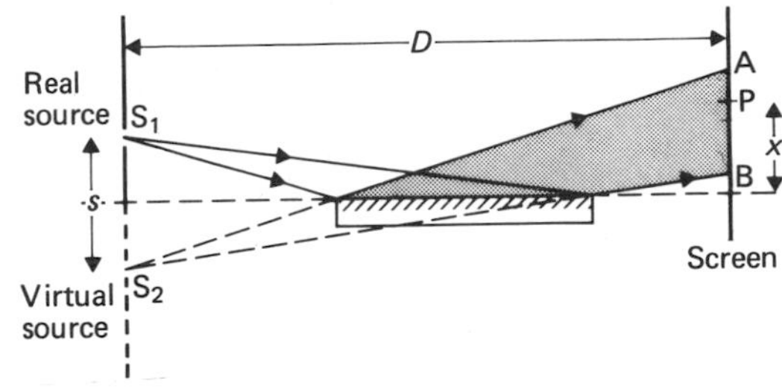

(*a*) Reflection in Lloyd's mirror

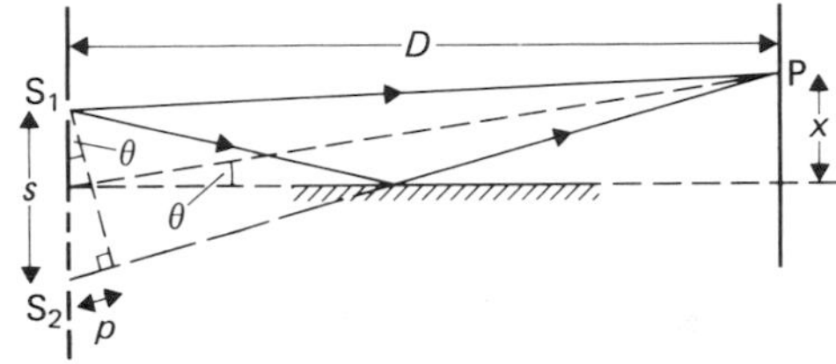

(*b*) The two rays producing interference at point P

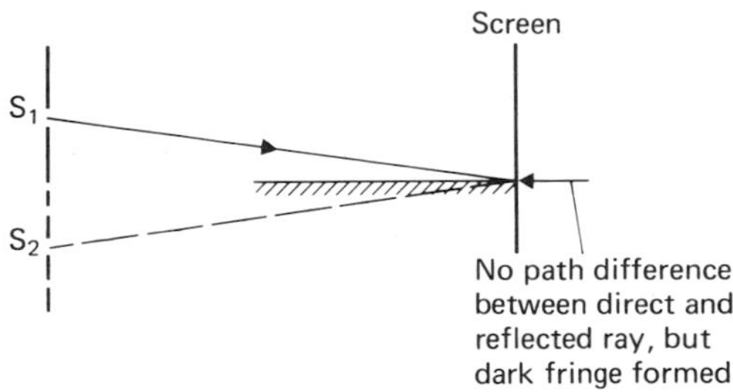

(*c*) The phase change at reflection

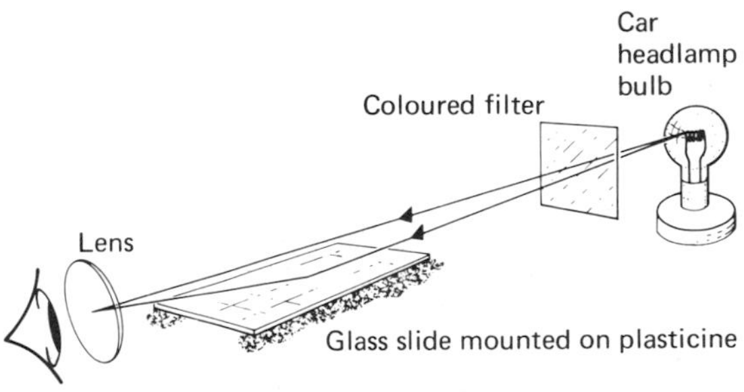

(*d*) An experiment to observe Lloyd's mirror fringes

Fig. 7.6

To set up a simple demonstration of Lloyd's mirror in the laboratory, place a bright lamp on the bench with its filament horizontal and mount a coloured filter in front of it, as shown. A microscope slide is then stuck to the bench with plasticine about 1 m away from the lamp, along a line at right angles to the lamp filament.

The slide is tilted until the image of the filament in the glass is as close as possible to the filament seen directly; a magnifying glass is used as shown in the diagram to look at the far end of the slide when fringes should be visible.

Q 7.6
A front-silvered mirror 0.1 m long is sunk into a bench so that its reflecting surface is flush with the bench top. A small monochromatic light source is set up 0.1 m from one end of the mirror, and Lloyd's mirror fringes are observed on a screen placed 0.2 m from the other end. The fringe width is 0.1 mm. If the light source is 1 mm above the bench surface, what is the wavelength of the light?

Fresnel's biprism

Fresnel's biprism produces brighter fringes than Young's slits because the system can admit a greater proportion of the wavefront. It is made from a plane sheet of glass with two facets ground to make a truncated prism which behaves like two small-angled prisms stuck together. One of the reasons why Fresnel proposed this arrangement was to counter suggestions that the bright fringes in Young's experiment were produced by some complicated effect at the edges of the slit and not by interference at all.

Fig. 7.7 shows how the biprism forms two virtual image sources, S_1 and S_2, of the real sources. The light which appears to come from these virtual sources interferes in the region shown and the distances x and D are measured in the usual way; the effective source separation, s, can be calculated as shown below. Thus the wavelength of the light may be determined.

The two diagrams in Fig. 7.7(*b*) show how the separation, s, is measured. S_1 and S_2 are regarded as the extreme points of a real object of size s, and a converging lens is used to form a real image of them on the screen. The separation s' is measured, and the object and image distances a and b noted. The lens is then placed in its conjugate position with a and b reversed and the new separation of the images s'' measured. In fact, of course, the source S and the biprism provide the object for the lens in both cases. Now by similar triangles:

$$\frac{s}{s'} = \frac{a}{b}$$

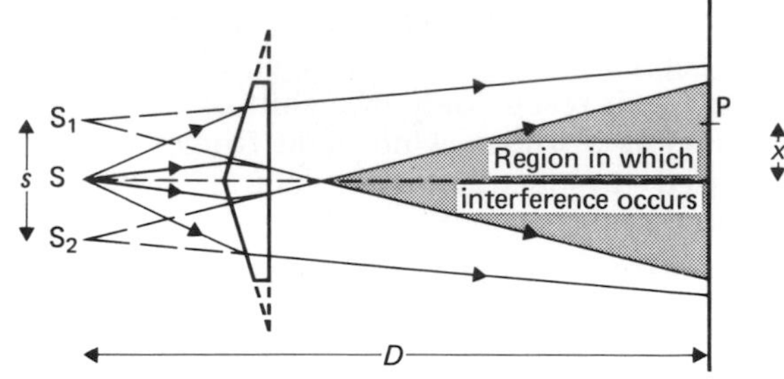

(*a*) Interference produced by Fresnel's biprism

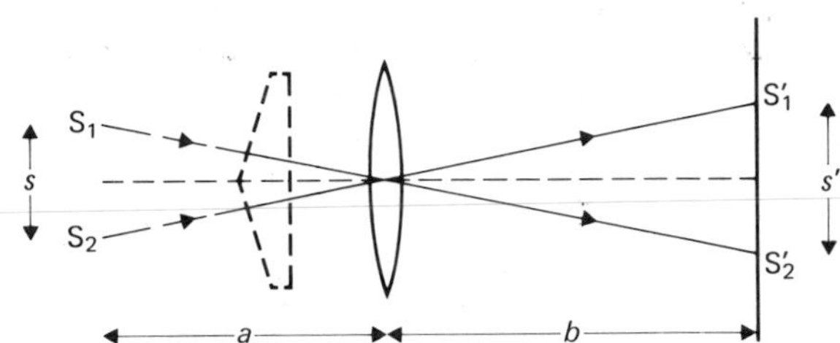

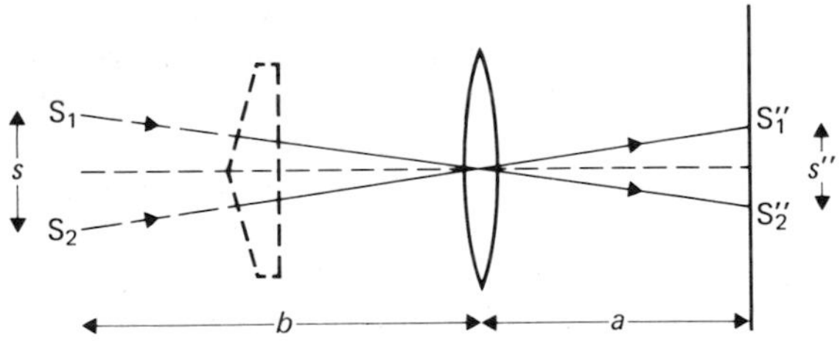

(*b*) Measurement of *s*

Fig. 7.7

and

$$\frac{s}{s''} = \frac{b}{a}$$

Multiplying we get

$$\frac{s^2}{s's''} = 1$$

i.e.

$$s = \sqrt{s's''}$$

Direct measurements of a and b are not used because of the difficulty of measuring to the centre of the source.

Q 7.7
In a determination of the wavelength of light using a Fresnel biprism, the following measurements

were made, as described above:

$$s' = 20 \text{ mm}$$
$$s'' = 2 \text{ mm}$$
$$D = 1.5 \text{ m}$$
$$\text{fringe width} = 0.12 \text{ mm}$$

Calculate the wavelength of the light used.

7.4 Interference by multiple reflections

In the examples we have considered so far, two coherent sources of light have been obtained from a single source by *division of wavefront*. For instance, in Young's slits experiment, part of the wavefront passes through S_1 and part through S_2; interference occurs between the light from these two sources.

The earliest observation of interference, however, was probably due to a different effect. Newton observed that if a convex lens is placed on a plane mirror, concentric coloured fringes are observed. Some of the incident light is reflected at the bottom surface of the lens and some at the plane mirror; in other words at the top and bottom surfaces of the air gap in between. Interference occurs between these rays because the light which travels to and from the mirror has travelled further than, and is thus out of phase with, the light reflected at the lens surface. The interference pattern produced is known as *Newton's rings* and is described in detail later. Similar colours may be observed when a patch of oil lies on a road and sunlight is reflected from top and bottom surfaces of the oil. In each case, at the upper surface of the gap or film, part of the light is reflected and part transmitted; the energy, and hence the amplitude, is divided and these types of interference are referred to collectively as *interference by division of amplitude*, or *thin-film interference*.

Interference conditions in thin films

Thin-film interference takes place under a variety of conditions; for instance the surfaces of the film may be parallel, as is the case in films used in certain types of interferometer, or inclined at a small angle, as in the formation of Newton's rings. However, the mathematical analysis is similar in all cases; we assume that if the surfaces are not parallel, the angle between them is so small that for light incident along the normal, refraction at the upper surface may be ignored. We shall restrict our analysis to these conditions.

In Fig. 7.8 we consider the repeated reflections that can occur when a single ray is incident on a film. As we are considering light at almost normal incidence on a small portion of the surface, the sides of the film are drawn parallel. A ray incident at the point A will be partially reflected and partially transmitted. The same process occurs at points B, C, D, E, F, G, ... Thus, in the diagram, rays 1, 2, 3, etc. are reflected and rays a, b, c, etc. are transmitted. In practice, instead of a finite number of such rays, there will be an infinite sequence of ever decreasing amplitude. The reflected rays emerge parallel to each other, as though from an object at infinity, and will interfere if brought to a focus at a point P by a lens, for instance as in a camera or eye.

The type of fringe formed at P will depend upon the angle θ_f, and hence upon the angle of incidence. Now, since $\angle YAB = \angle AYB$

$$AB = BY \qquad (7.11)$$

The geometrical path difference, p, between rays 1 and 2 is given by:

$$p = ABNC - AM$$

and thus the optical path difference, ϕ, is given by:

$$\phi = n(ABNC) - AM$$
$$= n(ABN) + n(NC) - AM$$

But, since AN and MC are wavefronts:

$$n(NC) = AM$$

Hence $\phi = n(ABN)$

$$= n(YB + BN) \qquad \text{from (7.11)}$$
$$= n(YN)$$

Now $\angle ANY = 90°$, and therefore:

$$YN = YA \cos \theta_f$$
$$= 2t \cos \theta_f$$

Hence $$\phi = 2nt \cos \theta_f \qquad (7.12)$$

As we noted on page 169 there is a phase change of π radians when a light beam is

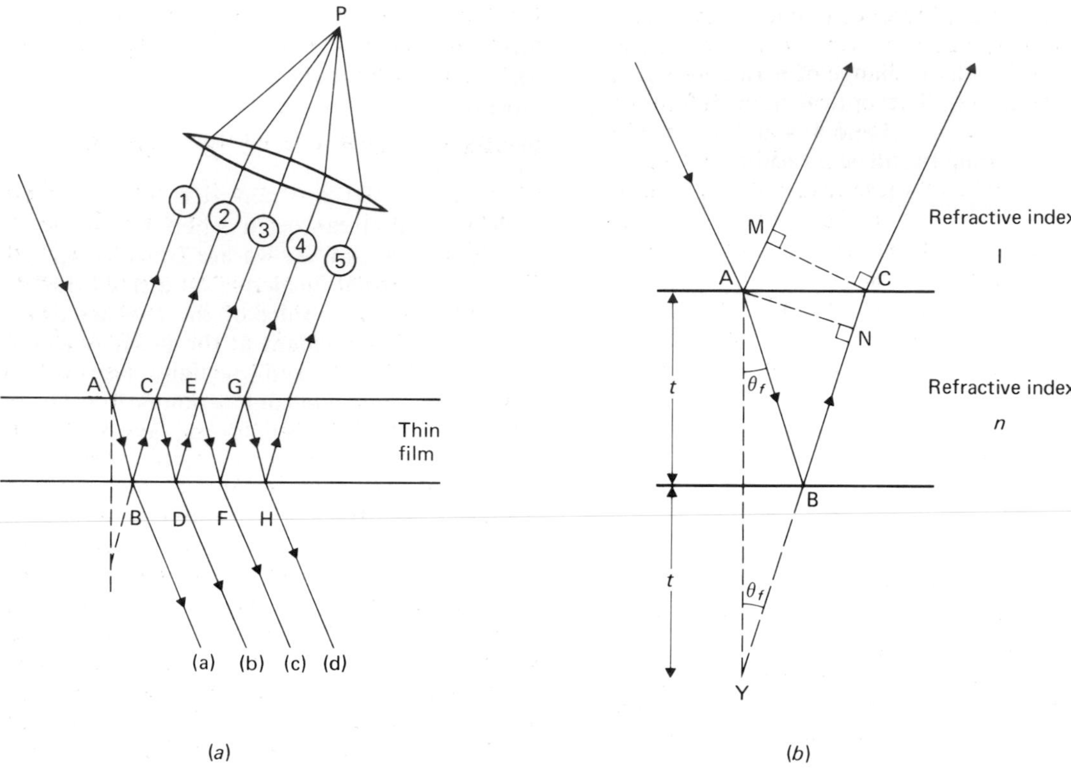

Fig. 7.8 Multiple reflections in thin films.

reflected at an air–glass surface and this is equivalent to a path difference of $\lambda/2$. There is no such phase changes at a glass–air surface. It is of no physical significance whether the sign of this is taken as positive or negative, and to produce equations comparable with earlier ones we shall take it to be positive or negative as required. Hence, the total effective optical path difference, Φ, between rays 1 and 2 is given by:

$$\Phi = (2nt \cos \theta_f) - \tfrac{1}{2}\lambda \qquad (7.13)$$

Let us consider cases in which the first two reflections are in phase:

$\Phi = 2nt \cos \theta_t - \tfrac{1}{2}\lambda = m\lambda$

where $m = 0, 1, 2, 3, \ldots$

i.e. $2nt \cos \theta_f = (m + \tfrac{1}{2})\lambda$ (7.14)

Now although rays 1 and 2 are in phase, rays 2 and 3 will be out of phase because the optical path difference between 2 and 3 is the same as between 1 and 2, but neither ray 2 nor ray 3 is reflected at an air–glass surface.

Hence, we can draw up a list of rays in phase with ray 1, and of those which are out of phase.

In phase with ray 1	Out of phase with ray 1
2	3
4	5
6	7
etc.	

Now, ray 3 has a smaller amplitude than ray 2 and thus the net effect of rays 2 and 3 is to reinforce ray 1; similarly the net effect of the pairs of rays 4 and 5, 6 and 7, etc. is to reinforce ray 1. Thus a maximum occurs for this particular value of θ_f.

Suppose, now, that the first two rays are out of phase. The condition for this is:

$\Phi = 2nt \cos \theta_f + \tfrac{1}{2}\lambda = (m + \tfrac{1}{2})\lambda$

where $m = 0, 1, 2, 3, \ldots$

i.e. $2nt \cos \theta_f = m\lambda$ (7.15)

The optical path difference between rays 1 and 2 is now $m\lambda$, but the rays are out of phase because of the phase change of π radians when ray 1 is reflected. The optical path difference between rays 2 and 3, 3 and 4, 4 and 5, etc. will also be $m\lambda$, but there will be no additional phase changes and thus all these rays will be in phase with each other, but out of phase with ray 1. It can be shown that the sum of the amplitudes of 2, 3, 4, 5, . . . would just cancel out ray 1 and thus a minimum of zero amplitude occurs.

Thus, we have the following conditions, where m is any whole number including zero:

$$\left.\begin{array}{r}\textit{constructive}\\ \textit{interference}\end{array}\right\} 2nt\cos\theta_{\mathrm{f}} = (m+\tfrac{1}{2})\lambda \qquad (7.14)$$

$$\left.\begin{array}{r}\textit{destructive}\\ \textit{interference}\end{array}\right\} 2nt\cos\theta_{\mathrm{f}} = m\lambda \qquad (7.15)$$

It will be noticed that these equations are different from those obtained for interference by division of wavefront, given on page 164. Sometimes we consider air films between glass surfaces; the phase change on reflection then occurs at the lower surface instead of the upper surface but the formulae are unaltered, except that n is now equal to unity for the air film.

Q 7.8

Carry out an analysis, similar to that above, to find under what conditions maxima and minima occur in the *transmitted* light. In scientific work it is more usual to use transmitted thin-film fringes than reflected ones.

We shall now consider a number of situations in which thin-film interference can be observed, and examine the differences between them.

Thin-film interference phenomena can be divided into three types:

(*a*) the thickness of the film varies but the angle θ_{f} does not; we include here wedge-shaped films and arrangements to observe Newton's rings;

(*b*) the thickness of the film is constant but θ_{f} varies; this is the case when viewing a parallel-sided film;

(*c*) the thickness of the film and the direction in which it is viewed both vary, which is likely to be the case when viewing oil and soap films.

It is helpful to remember that, in each case, the basic phenomena and mathematical analysis will be the same.

Fringes produced by a wedge-shaped film

Let us consider the fringes formed when light is incident on the air wedge formed between two almost-parallel glass surfaces; a pair of microscope slides, with a piece of paper separating them along one edge, is a convenient arrangement. The angle between the surfaces must be about 10^{-4} radian, and the light is assumed to be incident normally, so that refraction as the light passes through the glass can be ignored. The wedge could be illuminated by sodium light as shown in Fig. 7.9(*a*).

We consider the light reflected from the two surfaces on either side of the air film as shown, and interference occurs in the space immediately adjacent to the wedge; the angle between the wavefronts reflected from the two surfaces will be 2α. Now, interference can only be observed between *coherent* light rays, but the wedge is illuminated by a parallel beam of light which will have originated from different points in the source. The interference fringes we observe are formed when the light from *single rays* interferes after reflection at the two surfaces; we cannot observe interference between one ray reflected at the upper surface and a *different* ray reflected at the lower.

For light incident normally the fringe resulting from interference between the two reflections of one ray will be observed at the upper reflecting surface, as shown in Fig. 7.9(*c*). When light is incident at other angles, the interference fringes will be observed a distance, d, above the wedge, but still close to it, as shown. As the fringes can only be observed in one place they are called *localized fringes*; and the travelling microscope must be focused on the plane in which the fringes occur. As we have pointed out on page 166, fringes such as those produced by Young's slits or Lloyd's mirror can be observed anywhere throughout a region, and are therefore called *non-localized fringes*.

In the arrangement described above, a series of yellow and dark fringes is observed parallel to the edge of the wedge. As we shall see below, each fringe corresponds to a particular thickness of the wedge, and the fringes are known

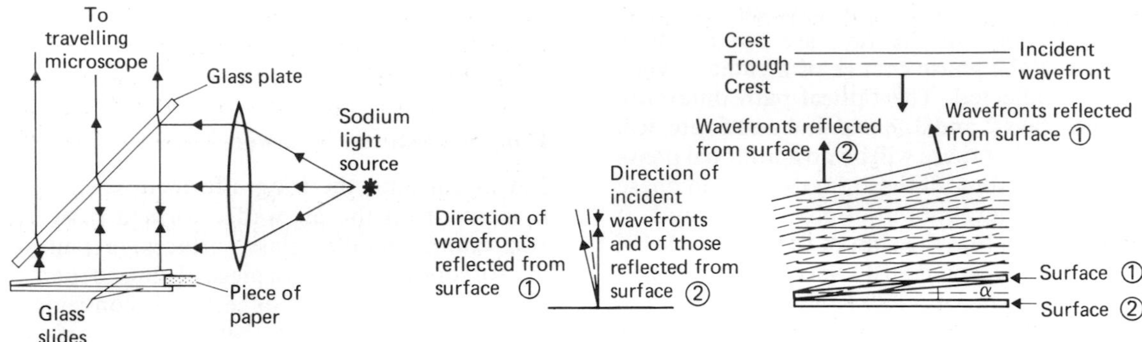

(*a*) An arrangement to observe wedge fringes

(*b*) Wavefronts reflected from both wedge surfaces

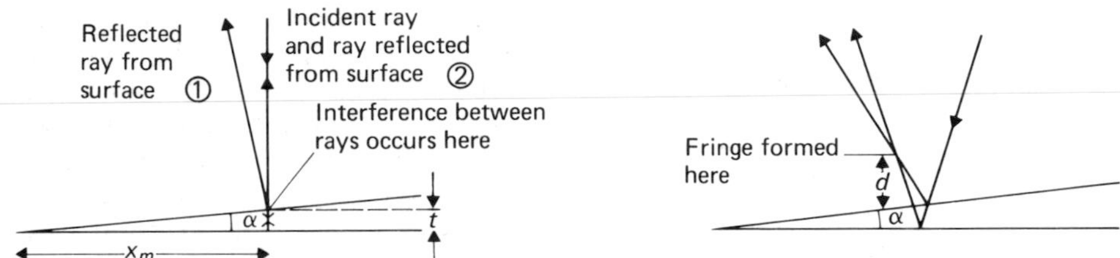

(*c*) The formation of localized fringes

NB In each of these diagrams the thickness of the air wedge is much exaggerated; in all cases it is much less than the thickness of the glass slide.

Fig. 7.9

as *fringes of equal thickness*. Because the wedge is illuminated with parallel light, the light rays by which the fringes are observed are all in the same direction, and thus θ_f is the same for all fringes.

If the air gap is increased in size any variation in the direction of the incident light, due to the extended nature of the source, becomes more important; at a point near the gap there may be constructive interference between two rays from one point on the source, and destructive interference between rays from another. Thus, the fringe pattern will disappear if the gap is more than a few hundred wavelengths wide, and hence interference cannot occur with light reflected from the *top* surface of the upper microscope slide.

The position of the light and dark fringes will depend upon the phase difference between the rays; if the angle of the wedge is small, and the light incident normally upon the top surface, the path difference will be $2t$, where t is the thickness of the wedge at that point. Now, if α is the angle between the plates and x_m the distance from the edge of the wedge:

$$t = \alpha x_m \qquad \text{since } \alpha \text{ is small}$$

i.e.
$$2t = 2\alpha x_m$$

Now, since there is a phase change of π upon reflection at the lower surface of the air wedge, corresponding to a change in the optical distance of $\lambda/2$, the condition for a bright fringe is:

$$2\alpha x_m - \tfrac{1}{2}\lambda = m\lambda$$

$$2\alpha x_m = (m + \tfrac{1}{2})\lambda \qquad (7.16)$$

$$\therefore \text{ fringe width} = \frac{\lambda}{2\alpha} \qquad (7.17)$$

These formulae can also be derived directly from that given in Equation (7.14) for the thin-film interference pattern:

bright fringe $\quad 2nt \cos \theta_f = (m + \tfrac{1}{2})\lambda \qquad (7.14)$

For an air wedge film, with normal incidence, we have:

$$\theta_f = 0$$

$$\therefore \cos\theta_f = 1$$

$$t = \alpha x_m$$

$$n = 1$$

bright fringe $$2\alpha x_m = (m + \tfrac{1}{2})\lambda \qquad (7.16)$$

This is the same equation as above.

If the wedge is formed from a medium of refractive index n, we have

bright fringe $$2\alpha x_m = \frac{(m + \frac{1}{2})\lambda}{n} \qquad (7.18)$$

If the refractive index of the wedge is greater than that of both media, as may be the case in a soap film, the phase change will occur upon reflection at its upper surface instead of its lower surface, but the formulae will be unchanged.

An important use of wedge fringes is to test whether two surfaces are exactly the same shape or not. If they are not, and the surfaces are pressed together, air wedges will be formed between them and fringes of equal thickness observed. The fringes will produce a *contour map* of the irregularities. Frequently we use this technique to test the flatness of a surface by testing it against an accurately flat block of glass; if the surface is sufficiently plane to show no fringes, it is said to be *optically flat.* Some fringes formed in this way are shown in Fig. 7.10.

At normal incidence the fraction of the incident light intensity that is reflected at an interface is

$$\left(\frac{n_1 - n_2}{n_1 + n_2}\right)^2$$

It is the same whether light is passing from medium 1 to medium 2, or vice versa. Thus, since only a small proportion—about 4% for air and glass—of the total intensity is reflected at the upper surface of the wedge, the amount of light reflected at the upper and lower surfaces will be very nearly equal; in consequence the amplitudes are very nearly equal and the dark fringes have almost zero amplitude and, therefore, intensity.

Q 7.9

In an experiment with monochromatic light of wavelength 5.4×10^{-7} m, an air wedge produces fringes with a spacing measured to be 5.4×10^{-4} m. What is the angle of the wedge?

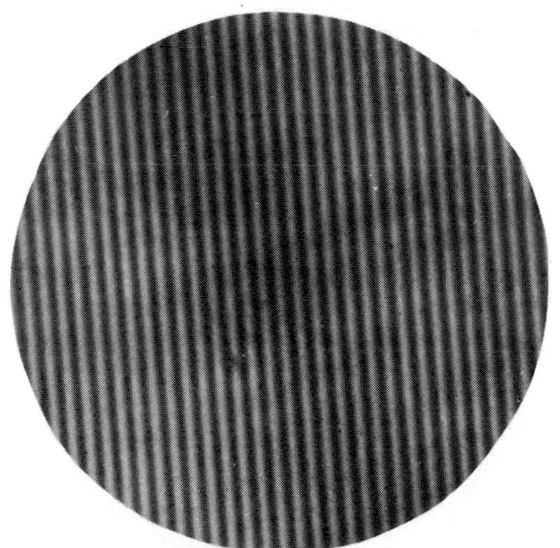

(a) Fringes viewed by reflection using an air wedge between two optical flats.

(b) An optical flat is used to test whether an aluminized sheet of mica is exactly flat.

Fig. 7.10

Newton's rings

Newton observed dark and bright rings when he placed a plano-convex lens on an optically flat glass surface and illuminated it with monochromatic light, as shown in Fig. 7.11; the arrangement became known as *Newton's rings*. The rings are contour fringes of equal thickness formed by the interference of light reflected from the bottom surface of the lens and the upper surface of the flat glass block. They are localized in the vicinity of the gap like wedge fringes. As before, we shall assume that the light crosses the surface at right angles so that refraction there can be ignored; this is justified because only the most central portion of the lens is used.

The thickness of the air film at a distance x from the point of contact can be calculated from Apollonius' theorem; r is the radius of the lower surface of the lens:

$$t = \frac{x^2}{2r}$$

The path difference between the rays reflected at the bottom surface of the lens, and at the top of the optically flat glass surface is thus x^2/r; in addition, there will be a phase change of $\pm\pi$ radian upon reflection at the surface of the flat. This is equivalent to a change in the optical path difference of $\pm\lambda/2$: the physical effect of a phase change of $+\lambda/2$ is the same as that of a change of $-\lambda/2$, and appropriate

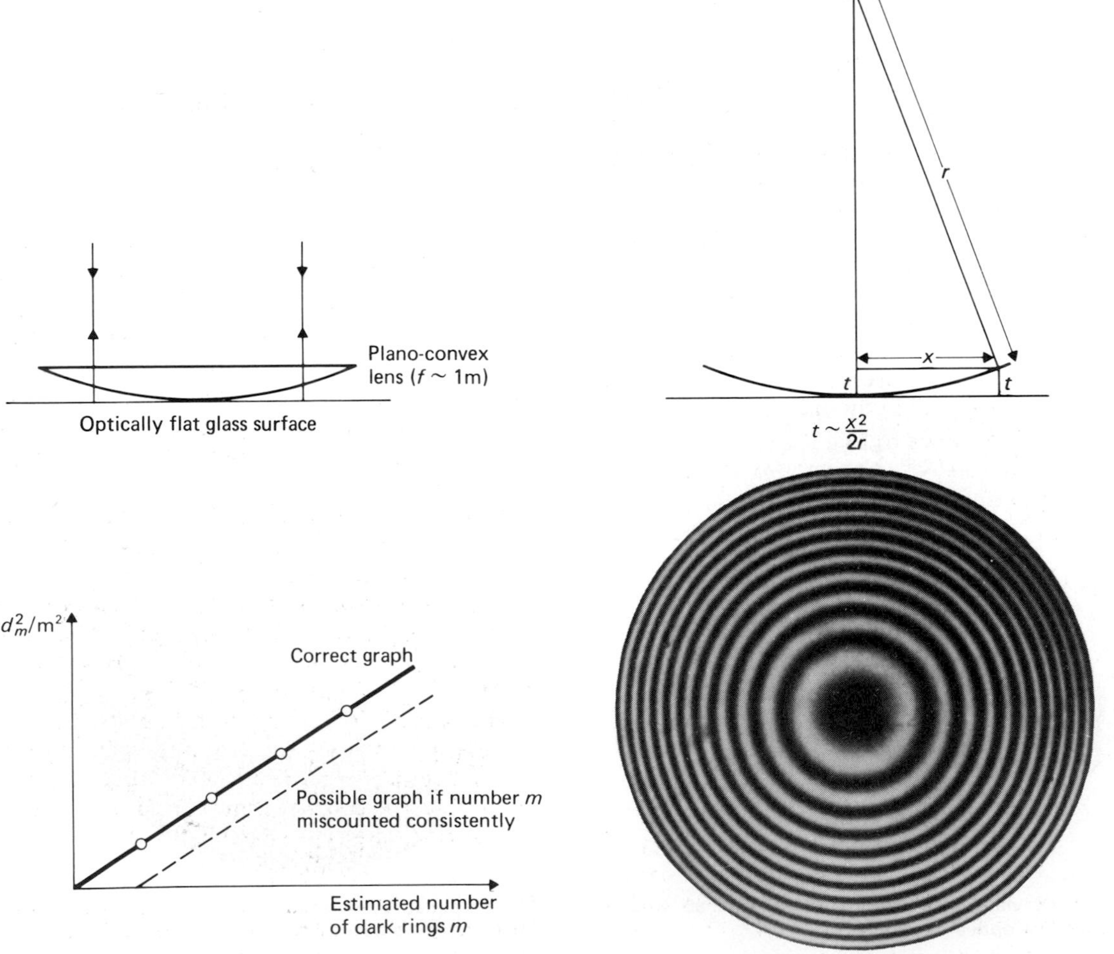

Fig. 7.11 The observation of Newton's Rings.

signs are chosen to give Equations (7.19) and (7.20) in the usual forms in terms of $(m+\frac{1}{2})\lambda$ and $m\lambda$. Hence, for bright rings:

$$\frac{x^2}{r}-\frac{\lambda}{2}=m\lambda$$

$$\frac{x^2}{r}=(m+\tfrac{1}{2})\lambda$$

i.e.

$$x^2=(m+\tfrac{1}{2})\lambda r \tag{7.19}$$

For dark rings:

$$\frac{x^2}{r}+\frac{\lambda}{2}=(m+\tfrac{1}{2})\lambda$$

i.e.

$$x^2=m\lambda r \tag{7.20}$$

The fringes will be alternate dark and bright concentric circles.

If there is good contact at the centre, and hence hardly any air gap, we observe a dark spot there. This is because of the phase change at the air–glass surface at the top of the flat block. However, the smallest speck of dirt can cause an air gap of much greater depth, in which case either a bright or dark spot can be formed. Young demonstrated the existence of the phase change by carrying out a modified experiment; he used a lens made up of crown glass and a plate of flint glass, and he replaced the air by carria oil, of intermediate refractive index. There is then a phase change at both surfaces and a bright spot is observed at the centre, even when the thickness of the oil film is very much less than one wavelength.

Equation (7.20) for the radius of the rings can be written

$$x=\sqrt{(mr\lambda)} \tag{7.21}$$

and we see that the radii of successive rings are proportional to $\sqrt{m}$ $(=\sqrt{1}, \sqrt{2}, \sqrt{3} \ldots$, etc.), with the result that the rings become closer together at greater distances from the centre.

Transmitted Newton's rings patterns can also be observed but show less contrast than in the reflected case.

Q 7.10
In our discussion of Newton's rings we ignored refraction where light crosses the convex surface of the lens, on the grounds that only the central portion of the lens was used. For a plano-convex lens of focal length 1 m, made of glass of refractive index 1.5, illuminated with light of wavelength 5×10^{-7} m, calculate:

(*a*) the radius of curvature of the convex surface;
(*b*) the radius of the tenth dark ring;
(*c*) the thickness of the gap for the tenth dark ring;
(*d*) the approximate angle between the ray forming the tenth ring and the normal to the surface.

Newton's rings may be used to find either the wavelength of the incident light, or the radius of curvature of the lens surface, if the other is known; we will look briefly at the former. It is usual to measure the diameter of the chosen dark ring rather than the radius; rings are narrower further from the centre and thus the position of an outer ring is better defined than the position of the central spot. A travelling microscope is used to measure the diameter of the mth ring, d_m.

Since $\quad x^2=m\lambda r \quad$ from Equation (7.20)

$$d_m^{\,2}=4m\lambda r$$
$$=(4\lambda r)m$$

Thus, a graph of $d_m^{\,2}$ against m should be a straight line through the origin and of slope $4\lambda r$. However, if there is poor contact at the centre, some of the central rings may not be present and the correct number of rings will not be counted; provided that this error is consistent, a straight line graph of slope $4\lambda r$ will still be obtained, but it will not pass through the origin.

Fizeau analysed the rings caused by sodium light, which he initially assumed to be monochromatic, so carefully that he was able to demonstrate the presence of two wavelengths of yellow light very close together; we refer to these as the sodium D lines of wavelength 5.890×10^{-7} m and 5.896×10^{-7} m.

Nowadays, it is more accurate to measure wavelengths with a spectrometer and grating, and it would be more usual to use Newton's rings to measure the radius of curvature of a lens surface than the wavelength of light.

Q 7.11
What graph has a slope equal to the radius of curvature of a lens surface in a Newton's rings experiment?

The apparatus can also be used to measure the absolute refractive index of a liquid placed in the space between the lens and optical flat. This method can be particularly useful if only a small quantity of liquid is available. Newton observed that, if this was done, the radius of the rings was reduced. If the refractive index of the liquid is n, the optical path difference is $2nt$, instead of $2t$.

Hence, instead of $x^2 = mr\lambda$

we have
$$x_n^2 = \frac{mr\lambda}{n} \qquad (7.22)$$

and the refractive index may be found from the formula:

$$n = \left(\frac{\text{radius of } m\text{th dark ring using air}}{\text{radius of } m\text{th dark ring using liquid}}\right)^2 \qquad (7.23)$$

We have, so far, assumed that a source of monochromatic light has been used to produce Newton's rings. If, however, the system is illuminated with white light a small number of 'rainbow-coloured' rings is observed. These are the result of overlapping rings of different colours and radii; the radius of a red ring will be about $\sqrt{2}$ times that of the blue ring of the same order. After a few rings the overlapping is so great that general illumination occurs.

Fringes of equal inclination

In our initial analysis of thin films we showed that for a bright fringe:

$$2nt \cos \theta_f = (m + \tfrac{1}{2})\lambda \qquad (7.14)$$

Thus, if a point source illuminates a parallel-sided film of constant refractive index, in order for different fringes to be seen the film must be viewed from different directions, as shown in Fig. 7.12(a); as the eye moves to positions in which θ_f and θ_a are different, bright and dark fringes will be seen.

However, if an extended source is used the eye can receive rays at different angles, such as θ_a and θ_a' simultaneously, as shown in Fig. 7.12(b). Thus, in some directions there will be a bright fringe and in other directions a dark fringe, depending upon the corresponding angles θ_f and θ_f'. An interference pattern can be observed *without moving the eye* and there will be bright fringes in directions where:

bright fringes
$$2nt \cos \theta_f = (m + \tfrac{1}{2})\lambda \qquad (7.14)$$

These fringes form arcs of circles and are known as *fringes of equal inclination.*

If the film is viewed at right angles to the surfaces all points with the same value of θ_a will show the same brightness; thus, the dark and bright fringes will be concentric circles and the fringes are known as *Haidinger fringes.* The centre may be dark or bright, depending upon the path difference. As the interfering rays entering the eye are parallel the fringes are said to be located at infinity. In order to produce good fringes, the plates must be parallel over the whole region involved to within a fraction of a wavelength, otherwise wedge fringes that are located in the film will be formed.

If the film is very thick, fringes will not be seen except at near-normal incidence. At other angles the reflected rays from the same incident ray will be so far apart that only one can enter the eye at a time and interference will not occur.

Fringes are also produced in the transmitted light and appear to be formed at infinity on the side of the film from which the light is incident. Such fringes, known as *Fabry–Perot fringes,* are widely used in interferometry for the measurement of small distances and the wavelength of light; we shall refer to them again when discussing the Michelson interferometer.

Colours in thin films

In the above analysis we have assumed that the incident light was monochromatic, but the most beautiful and frequently observed effects are those occurring with white light. If the film is thin, certain wavelengths, and therefore colours, will reinforce in directions along which other colours cancel. Hence, coloured patterns are observed which will be circular if the film thickness is constant. If the thickness changes, different colours will occur as can be seen in a soap film supported vertically; the colours move downwards as the soap film gradually drains. In this case, we have the third type of system mentioned at the beginning, in which both θ_f and the thickness of the film can change. The brilliant colours in the wings of some insects are the result of interference, in this case caused

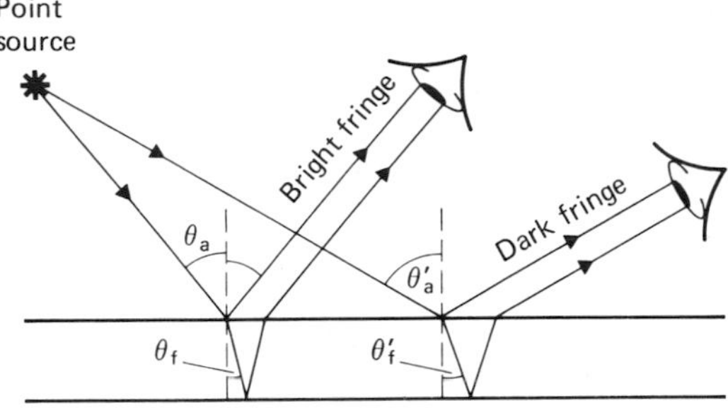

(*a*) Point source

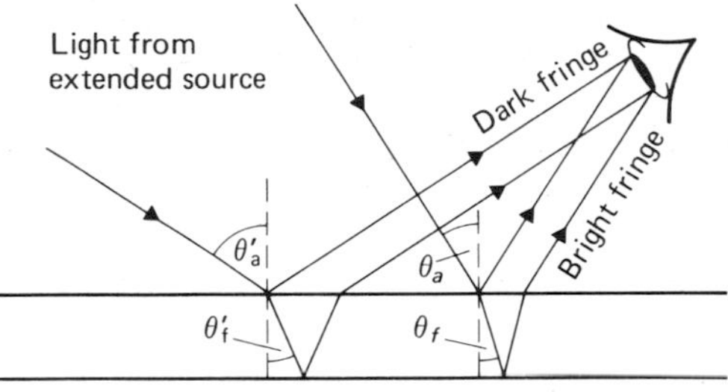

(*b*) Extended source – fringes of equal inclination

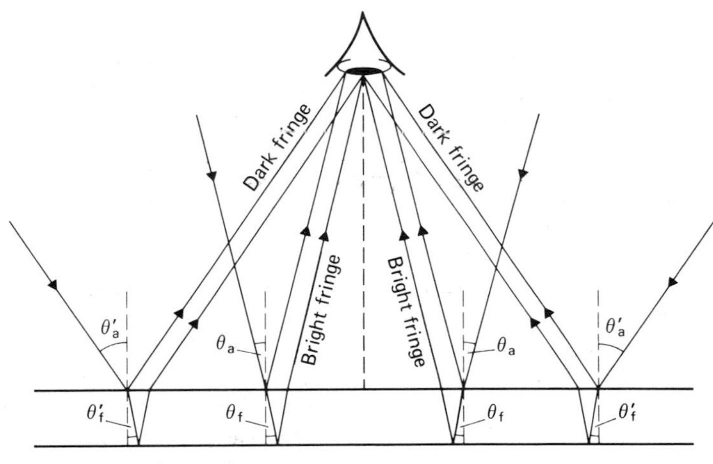

(*c*) Haidinger fringes

Fig. 7.12 The formation of fringes of equal inclination.

by parallel layers of equal thickness inside the wing.

Blooming

Undesirable reflections in lens systems are minimized by a process known as *blooming*; a thin coating of a substance intermediate in refractive index, n_c, between air and glass, such as magnesium fluoride, is applied to the surface of the lens. When light enters the lens reflections will occur at both surfaces of the coating with a phase change of π radian at each surface. If the optical thickness, $n_c t$, of the coating is made equal to $\lambda/4$ then the two reflected rays will be π radian out of phase for light of wavelength λ, and will interfere destructively, reducing the overall reflection, as shown in Fig. 7.13. Cancellation is complete only when the amplitudes of the two reflected rays are equal; this can be achieved by choosing a coating whose refractive index, n_c, is given by $\sqrt{n_a n_g}$. The thickness is calculated to give destructive interference for green light, in the middle of the spectrum, and no light of this wavelength is reflected; however, cancellation will not be perfect for red or blue light, at the extremities of the visible spectrum, and a small quantity of this light will be reflected—this is the reason why bloomed lenses appear magenta in colour.

The same principle is used in electrical transmission lines and waveguides to reduce the

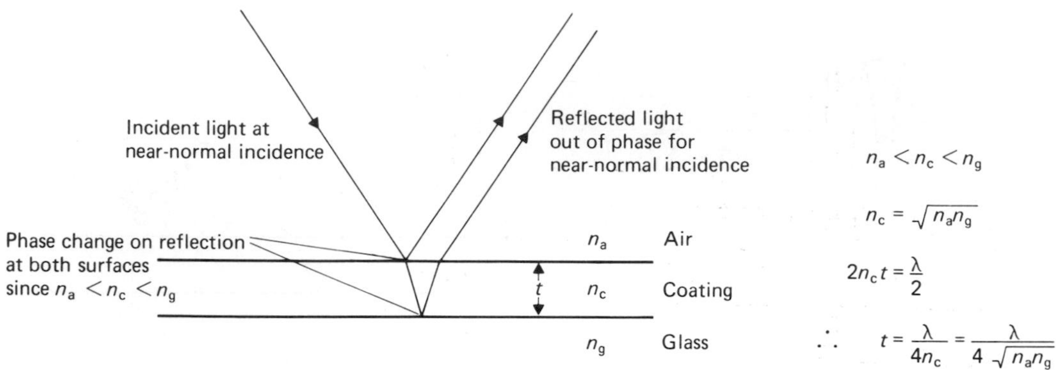

Fig. 7.13 Lens blooming.

reflection of power at a point where the geometry of the conductor or waveguide is changed; a short length of material of intermediate velocity of propagation is interposed. This is one example of the close interrelation between light and electromagnetism discussed further in Chapter 10.

Q 7.12
Calculate the required thickness for a lens coating to make a lens of refractive index 1.6 non-reflective for light of wavelength 5.5×10^{-7} m.

Interferometers

Interference fringes are widely used to measure small differences in length, and instruments using this principle are called *interferometers*. We have already discussed the simplest type of interferometer, on page 168, where we showed that when a slip of glass is placed in front of one of the slits in a Young's slits experiment, the fringe system is displaced. It follows that the refractive index of the glass slip could be measured by counting the number of fringes by which the pattern was displaced, using white light to identify the central fringe.

Lord Rayleigh developed this technique and Rayleigh's interferometer, illustrated in Fig. 7.14(a), is a modification of this method. A parallel beam of monochromatic light passes through two slits, S_1 and S_2, and then through two enclosed glass tubes; the beams are brought to a focus to form interference fringes, which are observed through an eyepiece. If both tubes are initially filled with the same gas, of absolute refractive index n, and one tube is then evacuated, the fringes will shift across the field of view; suppose that m fringes pass the crosswires during this process. The refractive index of the medium in the tube has changed from n to 1 and, if the tube is of length l, the optical path has changed by $(n-1)l$.

Hence $$(n-1)l = m\lambda$$

from which the refractive index of the gas may be determined. Rayleigh used this method to find the refractive indices of the then recently discovered gases, argon and helium.

Albert Michelson was an American physicist and Nobel prize winner who devoted almost the whole of his life to the experimental study of light; and in particular to the measurement of its velocity. He devised the Michelson interferometer, based on the division of amplitude, and used it in the famous Michelson–Morley experiment in 1881. The significance of this experiment is discussed in Chapter 9, but we shall here consider the operation of the interferometer shown in Fig. 7.14(b).

Light from an extended source is passed through a lens to illuminate uniformly a glass block, G_1, which is usually half-silvered on the rear side as shown. The beam is split into two, half being reflected from mirror M_1 and half from M_2. The beams are re-combined by G_1 and enter the observer's eye as shown. Although the glass block and mirrors are uniformly illuminated, for clarity only the central ray is shown passing through the system. The compensating plate, which is identical to the reflecting plate, is used to make the total path

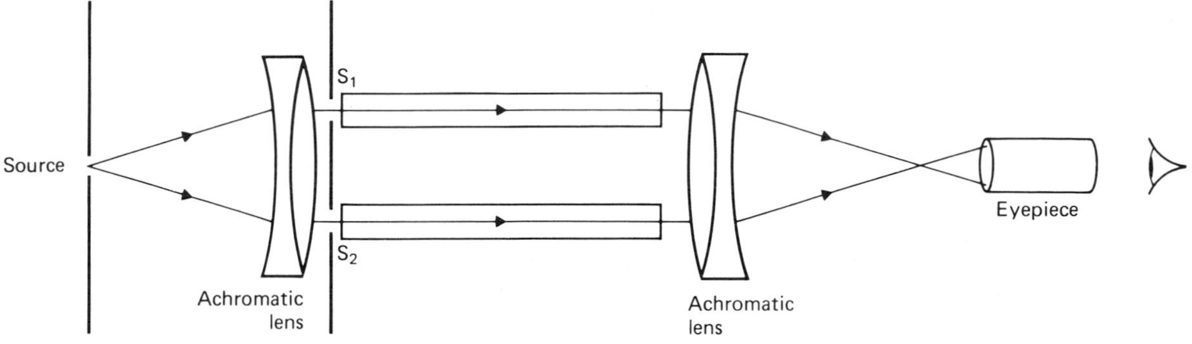

(*a*) Rayleigh interferometer

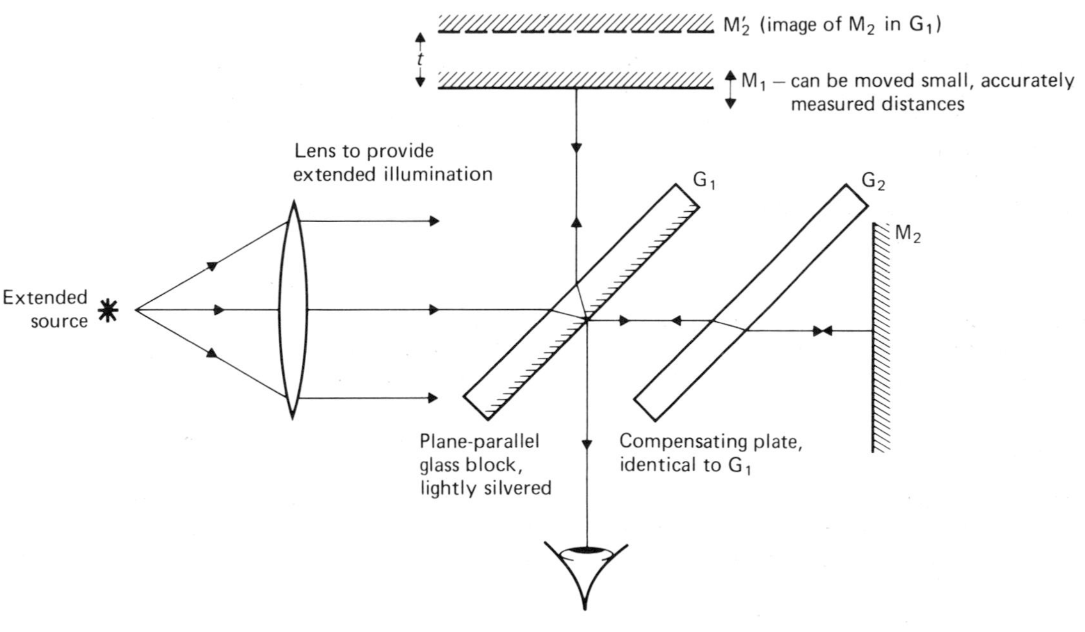

(*b*) Michelson interferometer

Fig. 7.14

through glass identical for rays reflected both at M_1 and at M_2, and thus to correct for possible dispersion effects. It is most important that the mirrors are accurately perpendicular, and that the glass blocks are at exactly 45° to them.

Michelson in fact made the beams travel backwards and forwards several times and thus achieved a total path of about 11 m. If the path lengths are different, interference fringes will be formed whose brightness will depend upon the phase difference between the re-combining beams; the arrangement is thus a sensitive test of the equality of the two optical path lengths. The formation of the fringes is best understood by considering the reflection M_2' of the mirror M_2 in the glass block G_1. The arrangement is then seen to be equivalent to a parallel-sided film in which the surfaces are at M_1 and M_2'; circular Haidinger fringes are thus observed.

If the mirrors are not exactly at right angles, and thus M_1 and M_2' are not exactly parallel, thin localized air wedge fringes are seen.

In the case of the formation of circular fringes, if the separation M_1M_2' is t, and the absolute refractive index of air n_a, the optical path difference between the interfering rays is $2n_at \cos \theta_f$. Both rays undergo a phase change of π radian upon reflection at the silvered surface and the condition for a *bright fringe* is then given by:

$$2n_at \cos \theta_f = m\lambda \qquad (7.15)$$

In the earlier analysis, Equation (7.15) referred to a dark fringe, since the phase change of π radians occurred only in one of the two rays. If the separation of the mirrors is increased, new fringes will keep appearing at the centre of the pattern and these can be counted; this procedure can be used to measure either the movement of the mirror or the wavelength of the light.

Michelson used his interferometer to compare the wavelength of various lines in the cadmium spectrum with the standard metre, and also to investigate the fine structure of spectral lines. In the Michelson–Morley experiment it was used to compare the velocity of light in two directions at right angles, as explained in Section 9.9.

7.5 Diffraction

In Section 3.3, we stated that when a wavefront passes through an opening it will spread out beyond it in a way that can be explained by Huygens' principle; this phenomenon is called *diffraction*. When we observe diffraction more carefully, we find that the amplitudes of the spreading wavefronts are not the same in all directions; there is a pattern of maxima and minima similar to that already described for cases of interference. These maxima and minima are produced by the superposition of wavelets originating from different points in the opening; it is an interference pattern produced by an infinite number of sources infinitesimally close together, and a series of bright and dark fringes is seen.

Diffraction occurs when waves pass through an opening, or when they meet an obstacle and diffract into its 'shadow'. The smaller the opening or obstacle the more noticeable is the effect of diffraction, and the greater the separation between maxima and minima.

Diffraction at an aperture can be observed under a wide variety of conditions, and for convenience these are divided into two types—Fresnel diffraction and Fraunhofer diffraction; Augustin Fresnel was the first person to complete a mathematical analysis of the phenomenon. He did this in 1815.

In theory, to observe a *Fraunhofer diffraction pattern* the point source used must be so distant that the incident wavefronts are effectively plane, and the pattern must be observed on a screen placed at infinity as shown in Fig. 7.15(*a*). In practice, the parallel light is produced by placing an illuminated slit at the focal point of a converging lens, and the fringes are observed on a screen placed in the focal plane of a second lens (Fig. 7.15(*b*) and (*c*)).

If the second lens is not used, and the screen is placed at a finite distance, a different pattern of fringes is seen which is called a *Fresnel diffraction pattern*. Another type of Fresnel pattern will be observed if the incident waves are not plane, wherever the screen is placed beyond the slit.

Thus Fraunhofer diffraction patterns are merely special cases where diffraction is observed under carefully prescribed conditions; it happens to be very much easier to analyse mathematically and most of this section will, therefore, be devoted to it. First, however, we will look at one or two examples of Fresnel diffraction.

Fresnel diffraction

The Fresnel diffraction patterns for slits and circular apertures are rather different from the corresponding Fraunhofer patterns, and are shown in Fig. 7.16. A particularly interesting case is the pattern produced by a circular obstacle. Fresnel predicted that there would be a bright spot in the middle of the shadow and to other physicists' great surprise, this was indeed found to be the case; it was an important piece of evidence for the wave nature of light since it could not be explained by the corpuscular theory. The pattern produced by any shape of aperture depends upon the position of the screen on which the pattern is observed; as the screen is moved further away from the

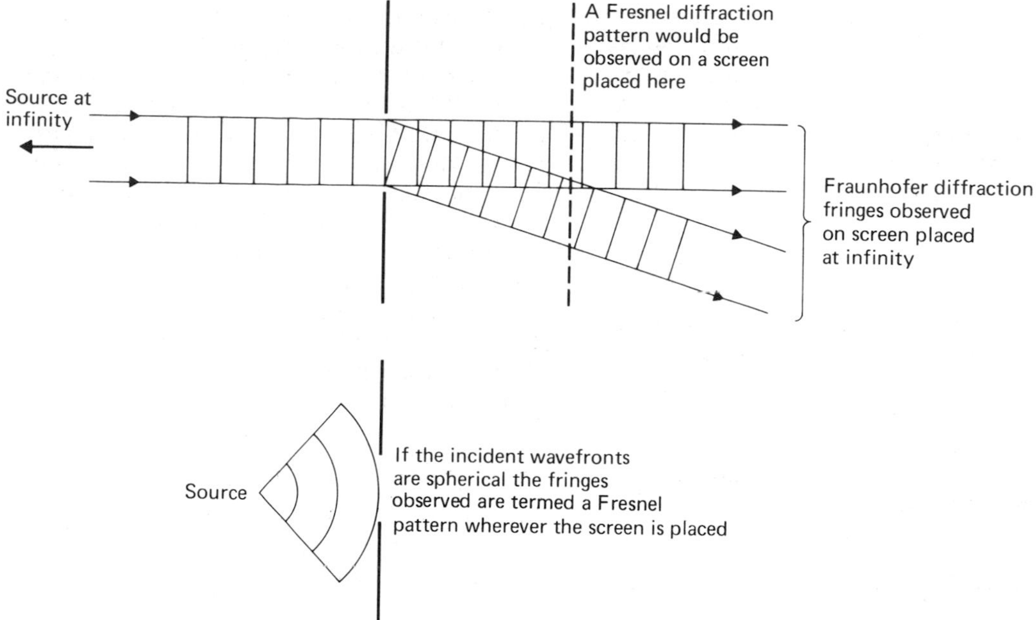

(*a*) Fresnel and Fraunhofer diffraction patterns compared

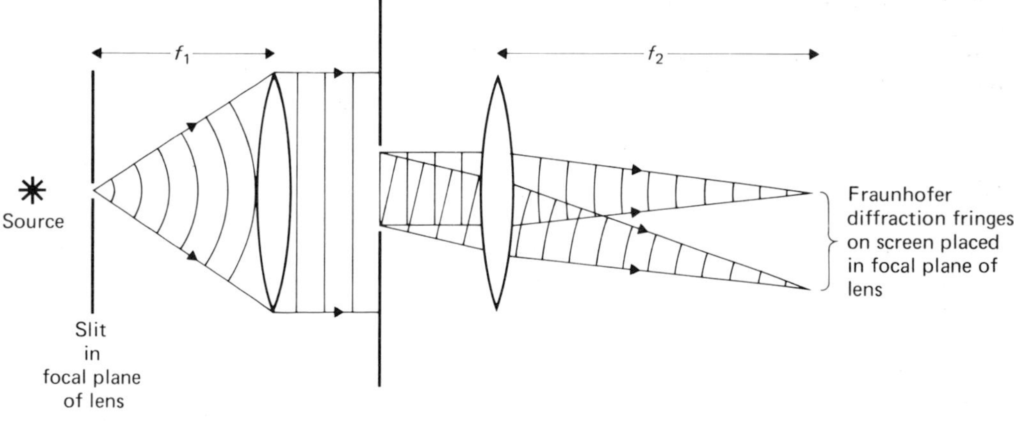

(*b*) The usual arrangement for observing Fraunhofer diffraction

(*c*) Fraunhofer diffraction due to a slit source and a single narrow slit aperture

Fig. 7.15

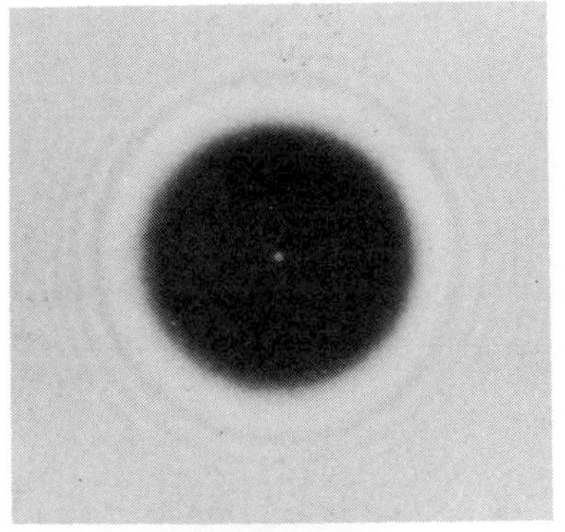

(*a*) Fresnel diffraction due to a point source and circular obstacle

(*b*) A point source and small circular aperture

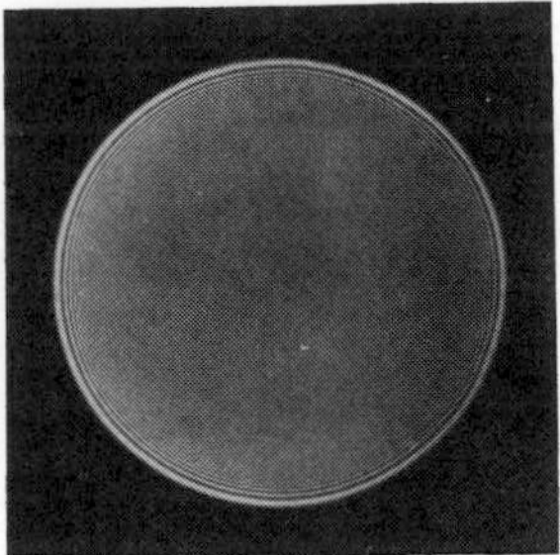

(*c*) A point source and large circular aperture

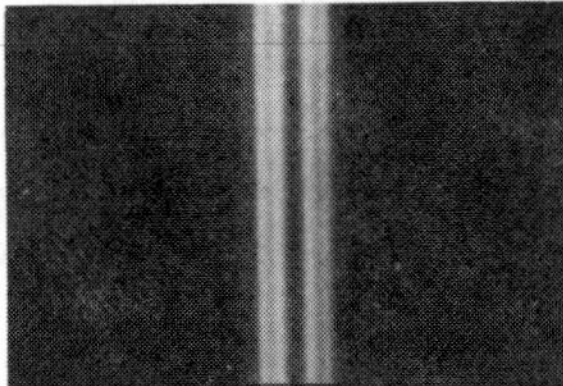

(*d*) A slit source with a slightly wider slit aperture

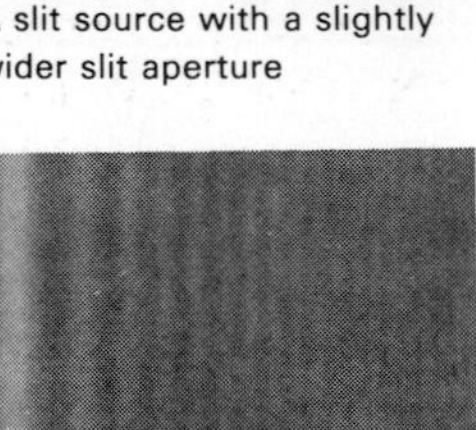

(*e*) A slit source and straight edge

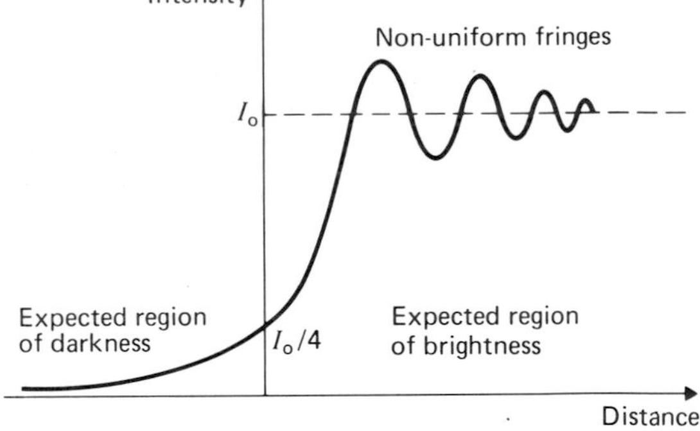

(*f*) The distribution of intensity in a Fresnel edge diffraction pattern

Fig. 7.16 Fresnel diffraction patterns.

aperture, the pattern becomes more and more like that observed under Fraunhofer diffraction conditions.

An interesting case of Fresnel diffraction can be observed at the edge of a rain drop. When looking at a distant light through a rain drop on a window pane we would expect to find, at the edge of the drop, a sharp boundary between light and dark. Careful examination, however, reveals that the light region is crossed by fine dark lines, which are part of an edge diffraction pattern, and there is some brightness in the dark region; a similar pattern is shown in Fig. 7.16(*e*) and (*f*). Try the experiment some time, especially if you are sceptical!

Many other patterns can be observed looking through easily made slits or small holes at bright lamps in your home.

These patterns of maxima and minima may also be demonstrated using 3 cm radio waves, obstacles or slits a few centimetres in size, and a small sensitive detector.

Fraunhofer diffraction at a single slit

We have already learnt that a Fraunhofer pattern will be observed when two conditions are satisfied:

(*a*) the aperture is illuminated by parallel

light, usually provided by a source placed in the focal plane of a converging lens;

(*b*) the resulting fringe pattern is observed in the focal plane of a second converging lens, or at infinity.

Let us consider the Fraunhofer diffraction pattern produced by normal incidence at a single slit. The interference occurs between rays originating at different points in the slit, and then travelling different distances to the screen, as shown in Fig. 7.17. The lens, used to focus the image on to the screen, does not introduce any further optical path differences and therefore the path differences to be considered are those such as NA in Fig. 7.17. It will be seen that the rays leaving the slit along the normal have no such path difference between them and at the centre of the pattern there will therefore be a maximum.

We will now determine in which directions the intensity is a minimum. We have just stated that interference occurs between light from different points in the slit, and in Fig. 7.18(*a*) we have divided the slit into two halves; each half is subdivided further into small sections

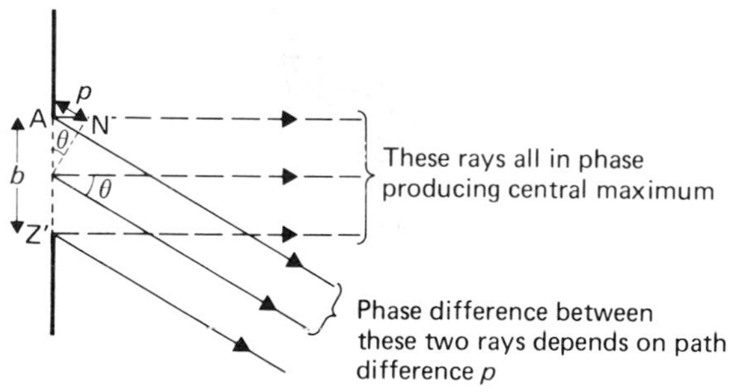

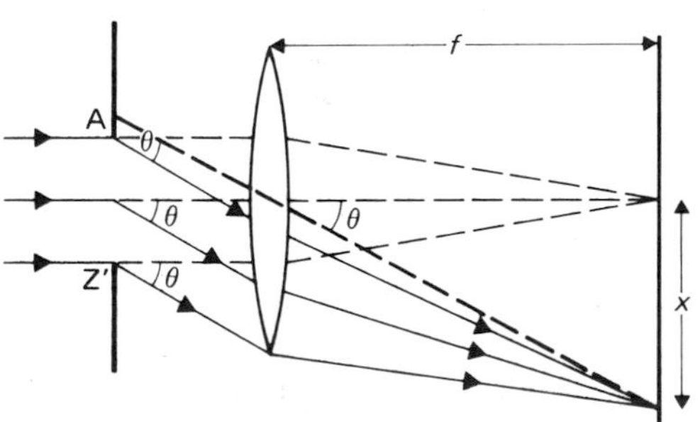

Fig. 7.17 The observation of a Fraunhofer diffraction pattern.

labelled A to Z in the top half, and A′ to Z′ in the lower. Now consider rays from each section that have been diffracted through an angle θ, as shown. The path difference between the rays from sections A and A′ is p, where p is given by:

$$p = \tfrac{1}{2}b \sin\theta \qquad (7.24)$$

As we are considering parallel rays, the path differences between the rays B and B′, C and C′ . . . Z and Z′ will be the same as between A and A′. Thus, if the rays from A and A′ cancel, because there is a path difference of $\lambda/2$ between them, rays from B and B′, C and C′ . . . Z and Z′ will also cancel; light from the top half of the slit interferes destructively with light from the lower half of the slit and there will be darkness at the angle θ. The condition for destructive interference at the first minimum is therefore:

$$\tfrac{1}{2}b \sin\theta = \lambda/2$$

$$b \sin\theta = \lambda \qquad (7.25)$$

The next minimum will occur at approximately twice the angle as shown in Fig. 7.18(*b*), and this can be proved by dividing the slit into four sections, each further subdivided as before. There will be a path difference of $\lambda/2$ between corresponding points in the top and second sections, and also between corresponding points in the bottom two sections; thus there will again be destructive interference and a dark fringe is produced. A similar argument can be made for increased multiples of θ and we see that the general condition for destructive interference is given by:

dark fringes
$$b \sin\theta_m = m\lambda \qquad (7.26)$$

In between each of these minima there must be a *secondary maximum*, in addition to the central *primary maximum*. We might expect that these secondary maxima would be situated midway between the minima but a rigorous analysis shows that this is not quite true; they are situated at a slightly smaller angle, and their intensity decreases rapidly, as shown in the graph. The central maximum, which spreads through $2\lambda/b$ from $-\lambda/b$ to $+\lambda/b$, is seen to be twice the width of the subsidiary maxima, which only spread through λ/b (Fig. 7.18(*c*)). A full analysis requires an integration across the width of the slit and readers are referred

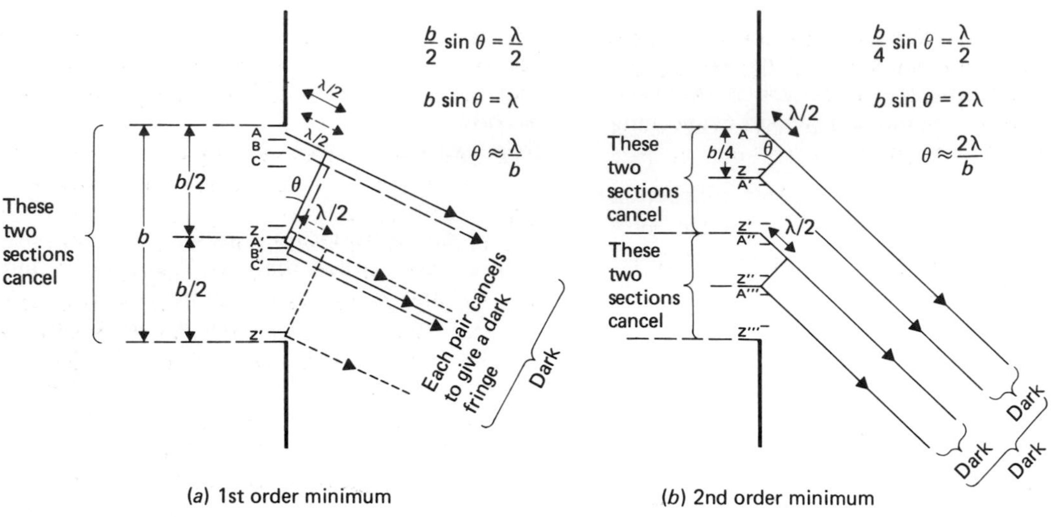

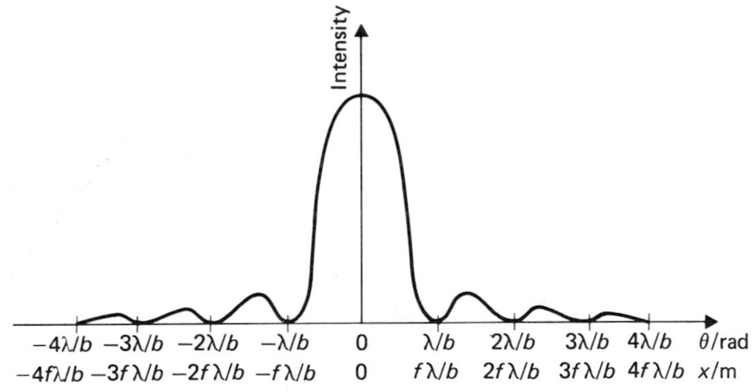

(c) Intensity distribution in Fraunhofer diffraction

Fig. 7.18 Fraunhofer diffraction at a single slit.

to texts such as Jenkins and White, *Fundamentals of Optics*, for a more rigorous treatment. The less rigorous argument used above does, however, help us to see the physical origin of the fringes observed.

The positions of the minima are given by the equation:

$$b \sin \theta_m = m\lambda \qquad (7.26)$$

as stated above. If θ_m is small this can be rewritten in the form

$$\theta_m = m\frac{\lambda}{b} \qquad (7.27)$$

As the distance from lens to screen in Fig. 7.17 must be the focal length of the lens, f, the linear distance from the centre of the pattern to a minimum is given approximately by:

$$x = f\theta_m$$

i.e.

$$x = m\frac{f\lambda}{b} \qquad (7.28)$$

Thus, for small angles, the minima are equally spaced and the spacing is roughly twice as great for red light as for blue, since the wavelength is twice as large. Moreover, the narrower the slit the broader the diffraction pattern. If the slit could be made the same width as the wavelength of light the first minimum would be at 90° and the whole screen would be occupied by the central maximum; we shall discuss the relationship of this result to the Young's slits experiment and diffraction grating in the next Section. On the other hand, if the slit is very wide, the spacing of the fringes may be so small that the pattern becomes difficult to see. A single-slit diffraction pattern is shown in Fig. 7.19.

Q 7.13
In each of the following questions assume that the wavelength of the light is 5×10^{-7} m.
Calculate:

(*a*) the angular fringe width of a Young's slits pattern if the slits are 0.5 mm apart;
(*b*) the angular fringe width between successive diffraction minima from a slit of width 0.5 mm.

In a pair of Young's slits the width of each slit is likely to be perhaps one-tenth of the slit separation. Calculate:

(*c*) the angular fringe width of the diffraction pattern produced by a slit of width 0.05 mm.

By *angular fringe width* we mean the difference in angle between one maximum and the next, or one minimum and the next.

Fraunhofer diffraction at double slits

When we discussed the interference of light in a Young's slits experiment, we considered only interference between light from *different slits* and not interference between light from *different parts of the same slit*; the phenomenon we have called diffraction. In fact both must occur simultaneously. Young's analysis predicts that there will be a series of dark and bright fringes across the field of view. The brightness of these fringes is determined by the diffraction pattern which is superimposed upon them—we might say that the fringes are *modulated* by the diffraction pattern. This effect could be ignored if the central maximum extended to 90° on either side; as noted above this is the case when the slits are the same width as the wavelength of light. We will now consider what happens if this is not the case; we assume that the separation of the slits is considerably larger than their width.

Because a lens focuses all parallel light on to the same point on a screen, the diffraction patterns formed by two adjacent slits will, in the usual arrangement for the observation of Fraunhofer diffraction, superpose. When Fraunhofer discovered this fact, he was surprised to see that the pattern was crossed by a number of finely spaced fringes, which we would recognize as the Young's slits pattern. The lenses shown in the arrangement in Fig. 7.20 are to ensure that the diffraction is of the Fraunhofer type; they are not, of course, necessary for the observation of the Young's slits pattern and do not introduce any additional optical path differences.

Fig. 7.20 also shows the observed pattern; the dotted line represents the Fraunhofer diffraction curve which determines the intensity of the Young's slits fringes. If a minimum in the diffraction pattern coincides with a bright fringe, the bright fringe will be missing and two such *missing orders* on each side are marked; whether, and where, these will occur will depend upon the spacing and width of the slits. The variation in brightness of the fringes is seen in the photograph. The equation for *maxima* in the Young's slits pattern is:

$$s \sin \theta = m\lambda$$

i.e.
$$\sin \theta = m\lambda/s \qquad (7.4)$$

and the equation for a *minimum* in the diffraction pattern is

$$b \sin \theta = m'\lambda \qquad (7.26)$$

i.e.
$$\sin \theta = m'\lambda/b$$

If a minimum in the diffraction pattern coincides with a maximum in the double-slit

Fig. 7.19 Fraunhofer diffraction at a single slit.

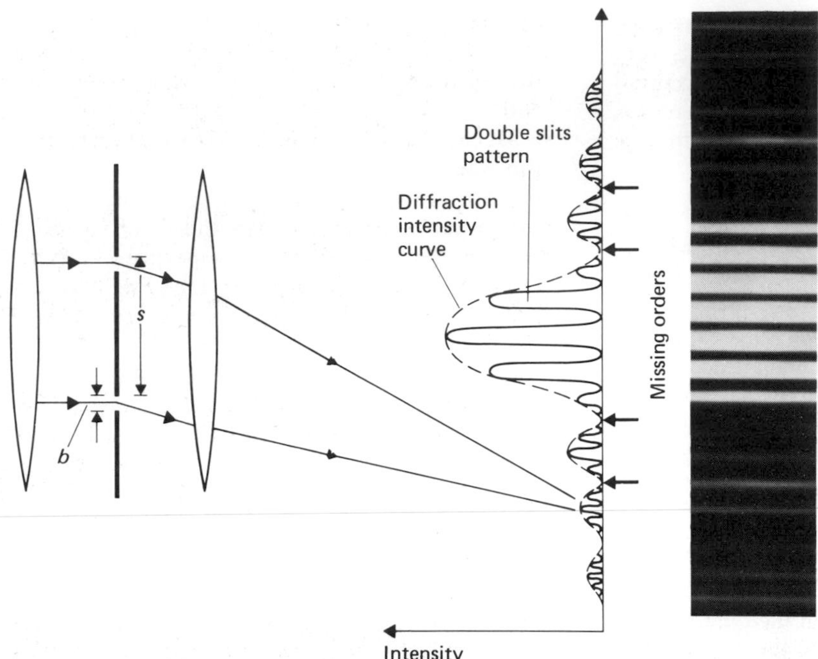

Fig. 7.20 Missing orders in an interference pattern.

pattern there is a missing order; the condition for this to occur is:

$$m\frac{\lambda}{s} = m'\frac{\lambda}{b}$$

i.e.

$$m = \frac{s}{b}m'$$

Thus, if the slit separation is ten times the slit width, the missing orders in the Young's slits pattern will be the tenth ($m' = 1$, $m = 10$), twentieth, thirtieth and so on.

Q 7.14
A Young's slits experiment is set up with slits 1 mm apart. In the resulting fringe pattern it is observed that one in every twenty fringes is missing. What is the width of the slits?

7.6 The diffraction grating

In Section 7.2 we showed that two slits illuminated by parallel light gave a series of dark and bright interference fringes, which were caused by the superposition of light from the two slits. If we increase the number of slits an interesting thing happens, as shown in Fig. 7.21; the brightness and sharpness of the fringes increases but their separation remains unchanged. A diffraction grating consists of a very large number of slits; try using a feather as a simple grating.

Interesting patterns can also be obtained by looking at a light through a piece of nylon material in which the perpendicular sets of threads behave like two gratings at right angles; a good opportunity for this is a dark and rainy night when street lamps can be viewed through the fabric of an umbrella.

The theory of the diffraction grating

In Fig. 7.22 each slit in a diffraction grating acts as a source of secondary wavelets; Huygens' principle enables us to predict the possible directions in which continuous wavefronts will travel beyond the slits by drawing envelopes to the secondary wavefronts; in all other directions the wavefronts will superpose destructively. One wavefront—the zero order—is propagated in the original direction, and there is also a

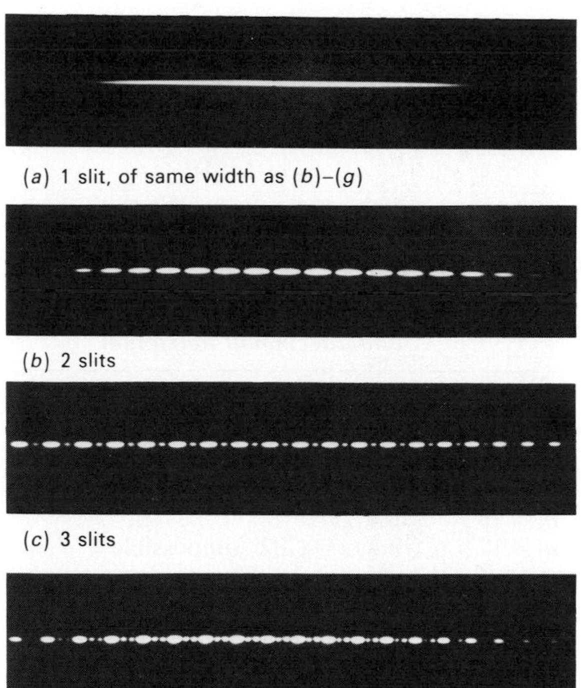

(*a*) 1 slit, of same width as (*b*)–(*g*)

(*b*) 2 slits

(*c*) 3 slits

(*d*) 4 slits

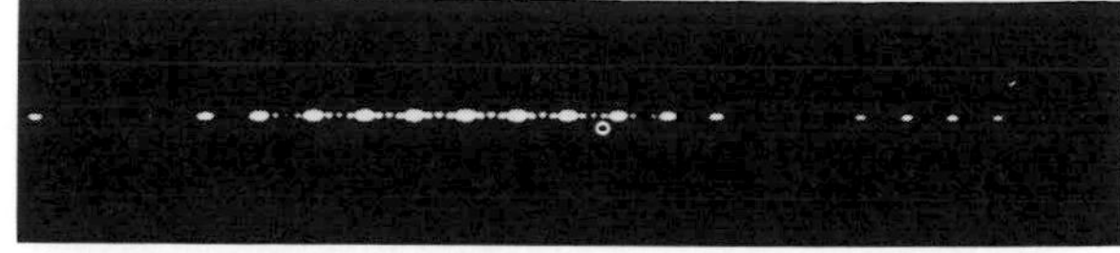

(*e*) 5 slits

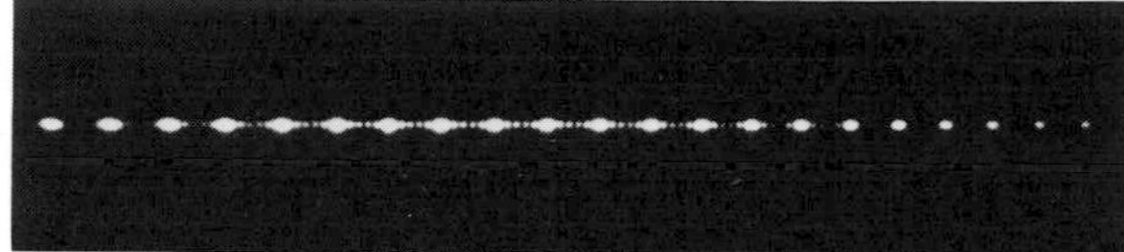

(*f*) 6 slits

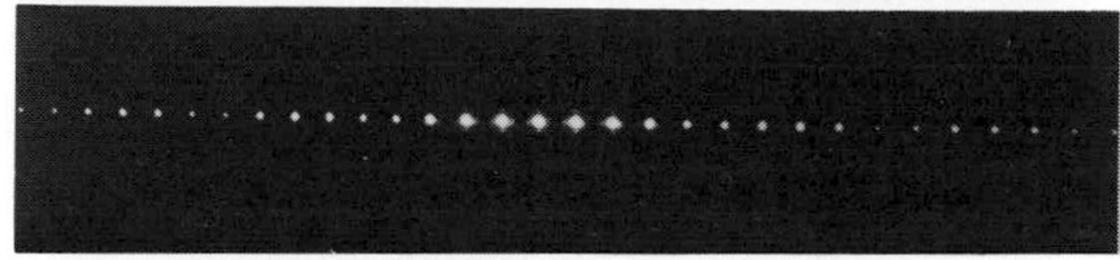

(*g*) A coarse grating with the same slit separation and slit width as in photographs (*b*)–(*f*); the secondary maxima are so faint as to be almost invisible

(*h*) A grating with smaller slit separation and slit width has a greater fringe separation, and broader central region

(Photographs by courtesy of *Griffin & George*)

Fig. 7.21 Interference patterns produced by differing numbers of slits.

series of other possible wavefronts starting with the first order and the second order marked in Fig. 7.22(*a*).

In Fig. 7.22(*b*) we show the situation from the point of view of rays incident normally on the grating. Consider those rays which emerge from the grating at an angle θ to the incident direction. The path difference, p, between rays from the top two slits is given by:

$$p = s \sin \theta$$

and these two rays will therefore be in phase if:

$$s \sin \theta = m\lambda$$

i.e.

$$\sin \theta = m \frac{\lambda}{s} \tag{7.4}$$

This is the same relationship as was derived for Young's slits. Now the path difference for the lower two rays is also p and therefore these two rays will also be in phase if:

$$\sin \theta = m \frac{\lambda}{s}$$

This argument can be extended to any number of slits and the light coming from a grating will be in phase if:

bright fringes

$$\sin \theta = m \frac{\lambda}{s} \tag{7.4}$$

$$\theta = m \frac{\lambda}{s} \quad \text{if } \theta \text{ is small} \tag{7.29}$$

Thus, the condition for constructive superposition is the same as that for Young's slits.

A more detailed analysis shows that the fringe system produced by gratings contains extra subsidiary maxima, as shown in Fig. 7.24; we will, in this simplified treatment, ignore these.

The most important feature of the fringes produced by a diffraction grating is that they are very much sharper than those produced by Young's slits. Similar intensity graphs are plotted for the two systems in Fig. 7.24, and the difference is clearly seen; it is the sharpness of the maxima that makes the grating such a powerful tool in the analysis of spectra. An argument showing why the maxima are so sharp is given on page 192.

As in the case of Young's slits, there is a broader pattern for light of longer wavelengths, or for smaller slit separations. Thus, a grating can be used for the production of spectra and the smaller the line spacing the greater will be the dispersion. The directions in which maxima are to be found is determined from the wavelength of light, as explained above.

For example, suppose a grating has 10^5 lines per metre and is illuminated with light of wavelength 6×10^{-7} m.

Now $$s = \frac{1}{10^5} = 10^{-5}\ \text{m}$$

For maxima: $$\sin\theta = m\frac{\lambda}{s} \qquad (7.4)$$

$$= m\frac{6\times10^{-7}}{10^{-5}}$$

$$= m(6\times10^{-2})$$

The directions in which maxima of different orders will occur are tabulated below:

Order	*Angle between incident beam and direction of maximum*
$m=0$	$\theta=0$
$m=1$	$\theta=3.4°$
$m=2$	$\theta=6.9°$
$m=10$	$\theta=36.9°$
$m=16$	$\theta=73.7°$
$m=17$	$m\lambda/s=1.02$ impossible

Note that the number of orders is limited by the restriction of θ to angles less than 90°, and thus there are no maxima beyond $m=16$ since $\sin\theta$ cannot be greater than one.

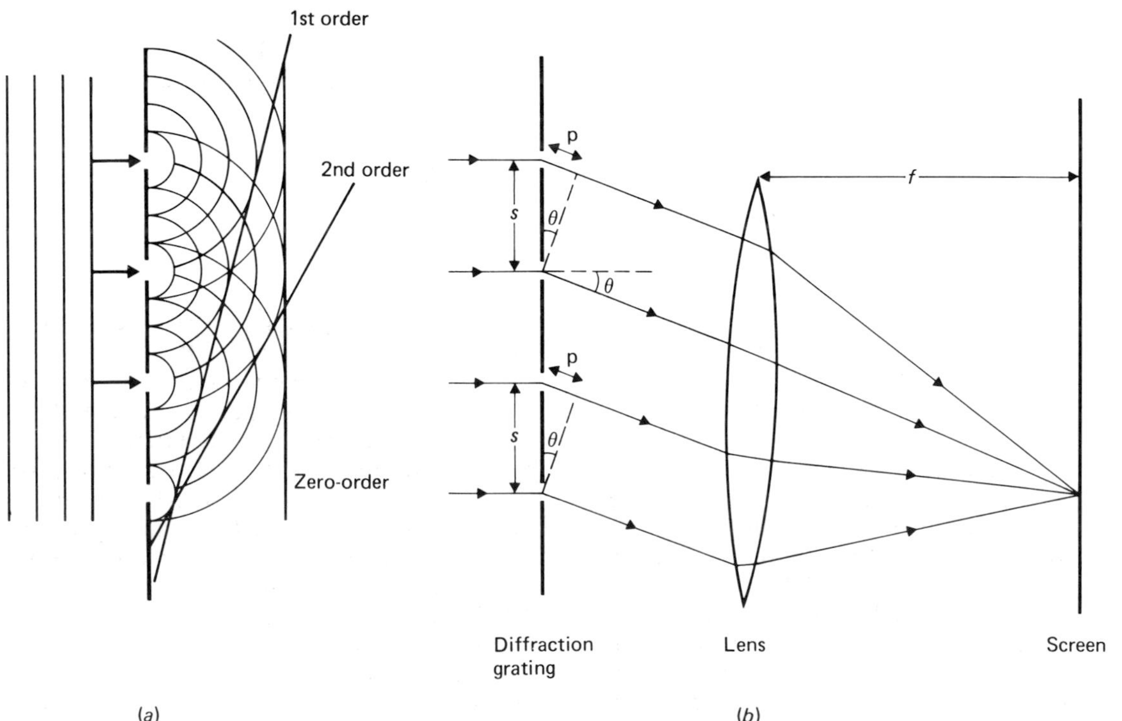

Fig. 7.22 The theory of a diffraction grating.

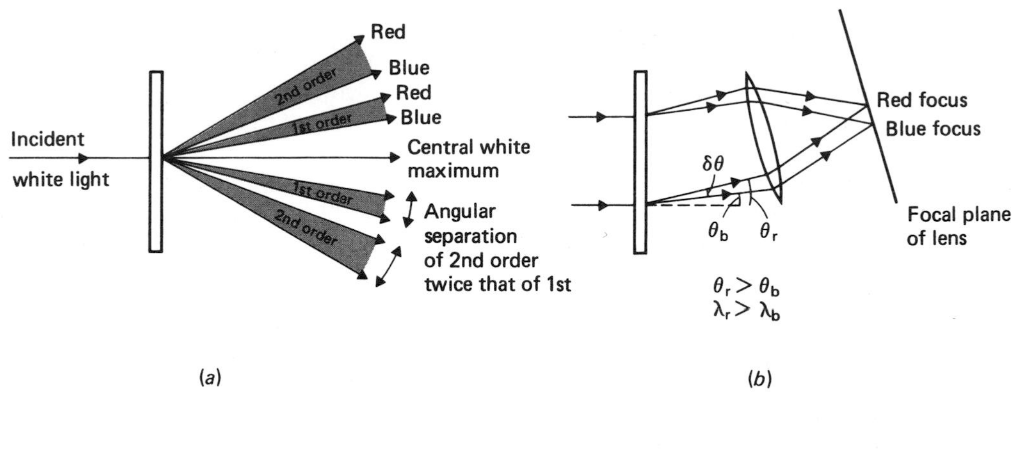

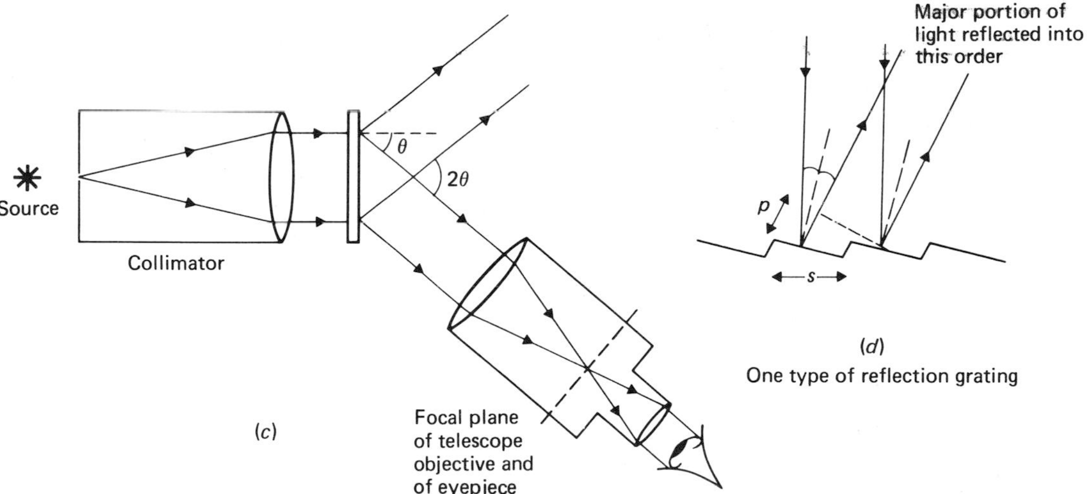

Fig. 7.23 The formation of spectra by gratings.

Q 7.15
The same grating as that above produces a second-order maximum at an angle of 5.7°. What is the wavelength of the incident light?

The formation of spectra by gratings is shown in Fig. 7.23. If the incident light is white, the central maximum will also be white and the higher orders may overlap as in the production of white light fringes. Fig. 7.23(a) shows the arrangement of the different orders; note that the red light is diffracted through a greater angle than the blue, which is contrary to *refraction* by a prism. In Fig. 7.23(b), we see how a lens may be used to produce a focused spectrum in one order. Fig. 7.23(c) shows the arrangement of a transmission grating spectrometer; its lens system is described in Section 8.6.

Grating spectrometers have a number of advantages over prism spectrometers and are almost always preferred for accurate measurements. Using a grating it is easier to distinguish between lines of almost identical wavelength; for a grating with about 6000 lines per centimetre the visible spectrum in the first order will be spread over about 10°, whereas a typical 60° high-dispersion prism gives a dispersion of only about 5°.

Many gratings work by reflecting light from the flat areas between lines ruled on a surface, instead of transmitting light through slits. They are then called *reflection gratings*, and the theory is similar to that given above for transmission gratings. A portion of one type of reflection grating is shown in Fig. 7.23(*d*); this grating is blazed for the reason explained below.

Types of grating

In about 1820, Fraunhofer constructed the first gratings of fine silver wires stretched between parallel screws; there were only thirty wires to the centimetre, but he was able to make surprisingly accurate determinations of wavelength with them. Later he made gratings by the 'modern' method of ruling lines with a diamond point on glass, producing about 500 lines to the centimetre; the ridges between the rulings act as the slits in the grating.

This method was improved by H.A. Rowland in about 1880; he made a very accurate screw to advance the diamond point between each line and ruled gratings on soft metal with as many as 17 000 lines to the centimetre. Similar methods are still used, but much more sophisticated control systems have been developed to ensure that correctly shaped grooves are ruled in the right places.

Replica transmission gratings can be made from a reflection grating by pouring a layer of transparent plastic called collodion over it and stripping it off when dry. Semi-opaque ridges are formed, separated by clear strips. In the modern manufacture of replica gratings, the original lines are ruled in a spiral on a soft metal cylinder around which the replica grating is moulded; cheaper gratings are merely stamped out.

In the most recently developed method of producing gratings, a metal surface is first deposited onto a glass plate. A laser beam is then used to form intense interference patterns, which can be used to vaporize strips of metal, leaving transparent strips behind.

Gratings used in spectroscopy are usually 'blazed' reflection gratings where the reflecting surface is at a carefully controlled angle to the plane of the grating, as illustrated in Fig. 7.23(*d*); about 85% of the incident light of a particular wavelength will go into the order for which the grating was designed. Some reflection gratings are concave and thus no further lenses are needed to form an image; this enables us to investigate wavelengths in the ultraviolet to which glass would be opaque, and to avoid the chromatic aberrations which are produced in a lens.

The sharpness of maxima

We have already seen that the importance of a diffraction grating is that it gives sharp, bright fringes that are very suitable for measurements of wavelength. The distribution of the intensity of light in the pattern produced by Young's slits, and that produced by a grating, were shown in Fig. 7.24.

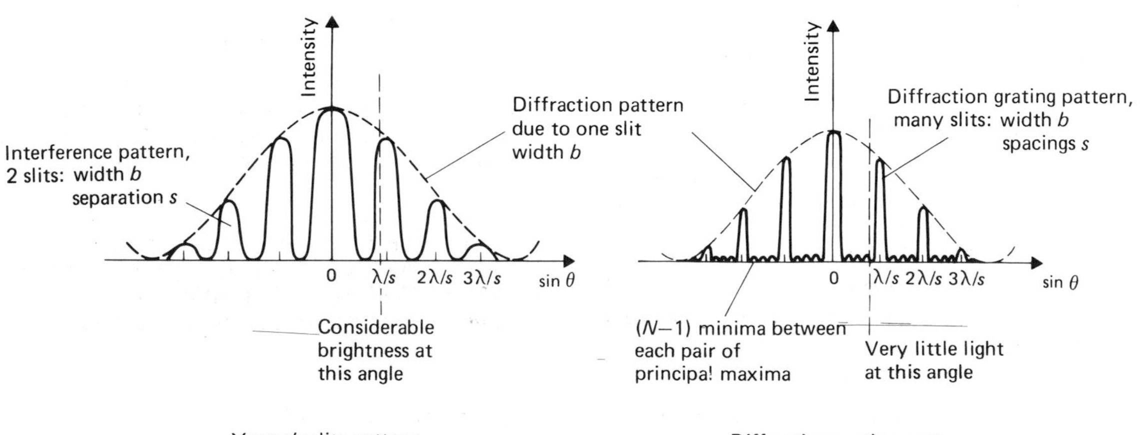

Fig. 7.24

Readers will recall from the last section that the brightness of Young's fringes is modulated by the diffraction pattern due to one slit, and thus their brightness varies as shown. The maxima are at λ/s, $2\lambda/s, \ldots$ but we see that there is still considerable brightness at angles to either side of $m\lambda/s$.

The pattern produced by a diffraction grating, with the same slit spacing as a pair of Young's slits, has maxima at the same angles with their brightness modulated by diffraction in an identical fashion. However, the maxima are very much sharper and at angles to either side of $m\lambda/s$ there is almost zero intensity. As there are more slits to transmit the light, and the principal maxima are sharper, the brightness of the maxima must, by the conservation of energy, be much greater than before; in consequence, the intensity scales for the two graphs are very different. In the analysis which follows we shall show why the principal maxima are sharper than when produced by Young's slits.

Let us consider a grating with N slits, with a spacing such that at an angle θ there is a first-order maximum, as shown in Fig. 7.25(*a*). The path difference between the top slit in the top half and the top slit in the bottom half of the grating will be $N\lambda/2$. Suppose now that we consider the superposition of light emerging at a slightly greater angle $(\theta+\Delta\theta)$; the path difference between successive slits is increased by a small amount to $(\lambda+\delta)$ and thus the path difference between light from the top slits in each half must now be $(\frac{1}{2}N\lambda+\frac{1}{2}N\delta)$ as shown in Fig. 7.25(*b*).

Suppose that the light from these two slits is now just half a wavelength out of phase, and thus destructive interference occurs. Then:

$$\text{increase in path difference} = \frac{\lambda}{2}$$

i.e.

$$\frac{N}{2}\delta = \frac{\lambda}{2}$$

$$\delta = \lambda/N \qquad (7.30)$$

Now if the light from the top slits in each half of the grating is exactly out of phase, so will be the light from the second slits in each half.

In this way, there will be destructive interference between successive pairs of slits $(1, \frac{1}{2}N+1)$, $(2, \frac{1}{2}N+2)$, $(3, \frac{1}{2}N+3), \ldots$ despite the fact that light from adjacent slits is very nearly in phase; thus the overall intensity will be zero, and the maxima will be very narrow. The greater the number of slits, the smaller the increase in path difference between light from adjacent slits needs to be for this to occur, and the narrower are the maxima.

The path difference, p, between adjacent slits is given by

$$p = s\sin\theta \qquad (7.1)$$

when the angle is exactly θ. Hence for a first order maximum to occur:

$$\lambda = s\sin\theta$$

Now, when the path difference is increased just sufficiently for the intensity to fall to zero we have:

$$\lambda+\delta = s\sin(\theta+\Delta\theta)$$

instead of $\lambda = s\sin\theta$

However $\sin(A+B) = \sin A\cos B + \cos A\sin B$

Hence

$$\lambda+\delta = s\sin\theta\cos\Delta\theta + s\cos\theta\sin\Delta\theta$$

The angle $\Delta\theta$ is small and we therefore have:

$$\cos\Delta\theta \approx 1$$

$$\sin\Delta\theta \approx \Delta\theta$$

giving $\lambda+\delta = s\sin\theta + s\cos\theta\,\Delta\theta$

But $\lambda = s\sin\theta$

and therefore $\delta = s\cos\theta\,\Delta\theta$

i.e.

$$\Delta\theta = \frac{\delta}{s\cos\theta}$$

or

$$\Delta\theta = \frac{\lambda}{Ns\cos\theta} \qquad \text{from (7.30)}$$

Fig. 7.25(*c*) shows that $(Ns\cos\theta)$ is the projected size of the grating, D', seen from the position of the relevant maximum. Thus the ability of the grating to produce fine lines depends on its effective size; the grating can be regarded as a 'large slit' through which the waves pass. The larger the apparent size, the smaller the increase or decrease in angle has to be for the intensity to fall to zero, and the sharper are the fringes. The result may be written:

$$\Delta\theta = \frac{\lambda}{D'} \qquad (7.31)$$

For example, a grating with 10 000 lines per centimetre and a total width of 2 cm produces a first-order maximum at an angle of 30° with light of wavelength 5×10^{-7} m.

$$D' = (Ns)\cos\theta$$
$$= (\text{grating width})\cos\theta$$
$$= (2\times10^{-2})\times0.866$$
$$= 1.73\times10^{-2}\text{ m}$$

But $\Delta\theta = \lambda/D'$.

Hence

$$\Delta\theta = \frac{5\times10^{-7}}{1.73\times10^{-2}}$$
$$= 2.89\times10^{-5}\text{ rad}$$

i.e. $\Delta\theta = 1.65\times10^{-3}$ degrees

The total angular width of the maximum will be 3.3×10^{-3} degrees; in practice, inaccuracies in the ruling will prevent the definition being quite as good as this.

Q 7.16

A grating with 5000 lines per centimetre is 1.5 cm wide. It is illuminated with light of wavelength 6×10^{-7} m. Find the angle at which the second-order maximum occurs, and its total angular width.

The production of spectra

As a diffraction grating will produce very narrow maxima in directions that are determined by the wavelength of light, it can be used most successfully to produce spectra; in particular, it can separate two spectral lines that are close together.

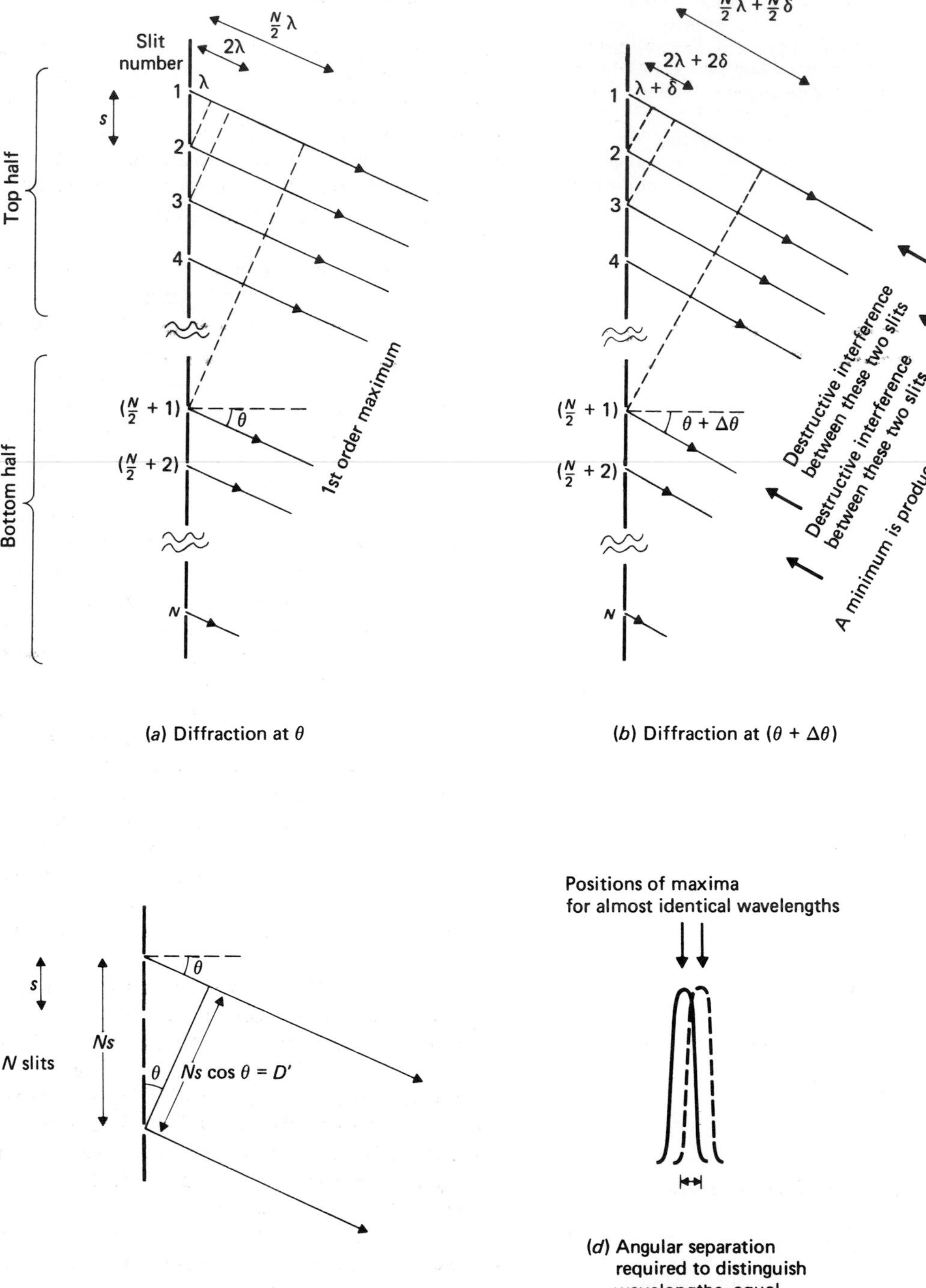

(*a*) Diffraction at θ

(*b*) Diffraction at $(\theta + \Delta\theta)$

(*c*) Projected size of grating

(*d*) Angular separation required to distinguish wavelengths–equal to half-width of maximum

Fig. 7.25 The sharpness of maxima formed by a diffraction grating.

We showed above that for the grating specified the half-widths of a first-order maximum at $\theta = 30°$ for light of wavelengths 5×10^{-7} m is 2.89×10^{-5} rad. For a maximum of a different colour to be clearly discernible it would have to be farther from the original maximum than 2.89×10^{-5} rad, as shown in Fig. 7.25(*d*). Now, for a first-order maximum:

$$\lambda = s \sin \theta \tag{7.4}$$

Therefore

$$\frac{d\lambda}{d\theta} = s \cos \theta$$

i.e.

$$\Delta\lambda = s \cos \theta \Delta\theta \tag{7.32}$$

The grating has 10^4 lines per centimetre and therefore the spacing, *s*, must be 10^{-6} m. So, in the region of 30°, the change in wavelength required to produce a change in angle of 2.89×10^{-5} rad, is given by

$$\Delta\lambda = 10^{-6} \times 0.87 \times (2.89 \times 10^{-5})$$

i.e.

$$\Delta\lambda \approx 2.5 \times 10^{-11} \text{ m}$$

The wavelength of the light considered was 5×10^{-7} m and thus the grating will separate wavelengths whose fractional difference in wavelength is given by:

$$\frac{\Delta\lambda}{\lambda} = \frac{2.5 \times 10^{-11}}{5 \times 10^{-7}}$$
$$= 5 \times 10^{-5}$$

It will, in theory, detect a change in wavelength of five parts in one hundred-thousand.

Q 7.17

Use the algebraic equations and method above to show that for a grating of *N* slits the fractional difference in wavelength that can be detected in the *m*th order is given by:

$$\frac{d\lambda}{\lambda} = \frac{1}{mN} \tag{7.33}$$

Copy this proof into your notes.

The ratio $d\theta/d\lambda$ is called the *dispersive power* of the grating. Equation (7.32) shows that for a first-order maximum it can be expressed as:

$$\frac{d\theta}{d\lambda} = \frac{1}{s \cos \theta}$$

and for an *m*th order maximum we can derive:

$$\frac{d\theta}{d\lambda} = \frac{m}{s \cos \theta} \tag{7.34}$$

The dispersive power is, therefore, increased in higher orders, and also if the slit separation is reduced.

If θ is small, variation in $\cos \theta$ it is not important and the *angular separation of the colours*, $\Delta\theta$, *is proportional to the difference in their wavelengths*, $\Delta\lambda$; under these conditions a *normal spectrum* is said to be produced.

7.7 Polarization

The final physical 'property' of light that we shall discuss is one of the most important, for it provides evidence that light is a transverse wave motion rather than a longitudinal one. Most of the phenomena described earlier in this chapter could be observed with a longitudinal wave of suitable wavelength, but this is not true of polarization.

We shall discuss the nature of light more in later chapters, but the reader must understand now what is meant when we say that it is an *electromagnetic wave*. There are electric and magnetic fields at right angles to each other, and their magnitudes vary with both time and position in the wave; thus, their magnitudes at a particular instant may be compared with the physical displacement of a wave on a string. Because the electric vector is most closely connected with the interaction of light with matter, it is the direction of the electric vector that we shall discuss; the plane containing the direction of propagation and the electric vector is called the *plane of polarization* (Fig. 7.26).

In *unpolarized light* the direction of the electric vector changes at random; this is because different atoms in the source produce light with random planes of polarization. This is like changing the direction in which the end of a rope oscillates when sending waves along it. Light that is *plane* or *linearly polarized* has a consistent plane of polarization, so that the electric vectors are always parallel.

Radio and radar waves, which are produced by the oscillation of charges in aerials, are polarized with their electric vectors parallel to the direction of oscillation of a charge. Many receiving aerials can only respond to one plane of polarization; thus, if a receiving aerial is rotated with respect to a transmitting aerial two maxima and two minima will be detected per revolution. Television and radio aerials should be aligned with the detecting rod in the plane of polarization of the signal.

The same effect may be observed with light. Polaroid is a material which prevents the transmission of light with one particular plane of polarization. If a light source is viewed through two sheets of polaroid, as shown in Fig. 7.27, and the second sheet of polaroid is rotated, then there will be total darkness twice per revo-

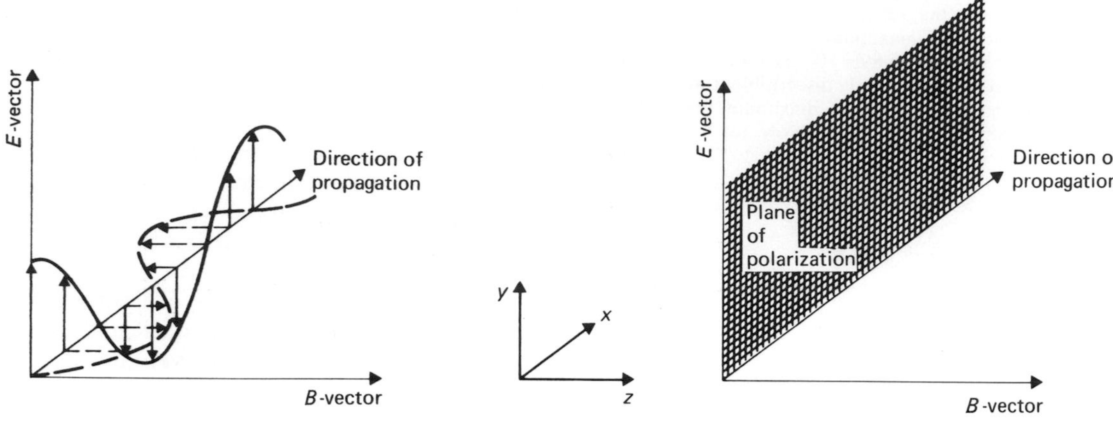

(a) Plane polarized light

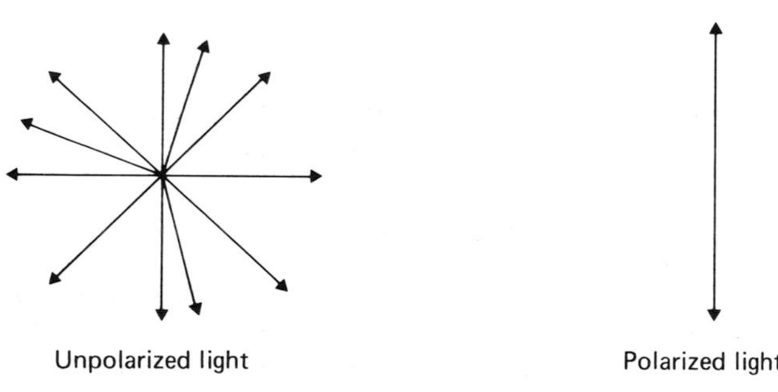

(b) Direction of E-vector

Fig. 7.26

lution. This occurs when the planes of polarization that are removed from the beam are at right angles; the sheets of polaroids are said to be *crossed*. If the polaroids are aligned so that they both inhibit E-vectors in the same direction, they are said to be *parallel*.

Methods of producing polarized light

We shall next consider a number of methods by which polarized light can be obtained.

(1) *Polarization by selective absorption*

This is the method we have met already in the use of polaroid to produce polarized light. Selective absorption can also be illustrated by an experiment using 3 cm radar waves. These are radiated from an aerial in a plane polarized beam, as described above, and passed through a metal grill to the detector, as shown in Fig. 7.28. When the metal grill is parallel to the electric vector the energy is absorbed because

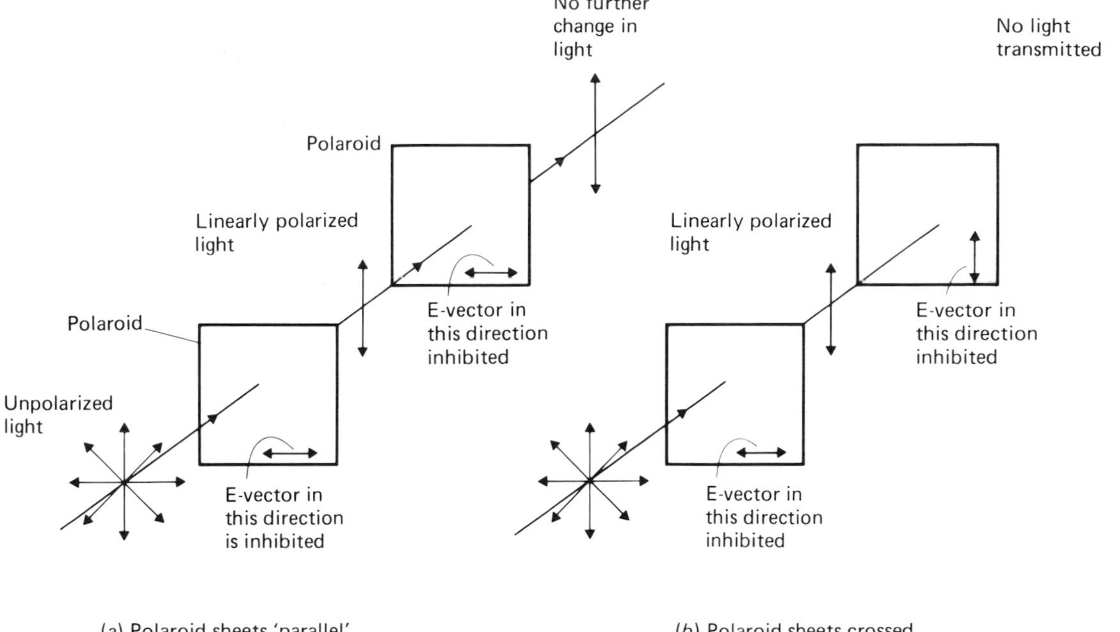

(*a*) Polaroid sheets 'parallel' (*b*) Polaroid sheets crossed

Fig. 7.27

it can be used to set electrons in the grill into oscillation, and there is no wave beyond the grill; when the grill is at right angles to the electric vector the energy cannot be absorbed in this way and the wave is able to continue to the receiver.

Crystals of some minerals and some organic compounds selectively absorb one of the components of unpolarized light; substances which have this property are said to be *dichroic.* Attempts were made in the nineteenth century to produce large dichroic crystals, so that wider polarized beams could be produced, but in 1932, an American undergraduate called Edwin Land discovered a better method; he lined up lots of small organic crystals in sheets of nitro-cellulose. The crystals were of herapathite, a sulphate of iodo-quinine. The sheets were larger and more robust than the single crystals and have become known as *polaroid.* There are now a number of types of polaroid on the market, but the principle of their manufacture is very similar to that used by Land. As the absorption of light is fairly selective as to wavelength, the transmitted beam is coloured and polaroid, for example, often appears dark green.

(2) *Polarization by double refraction*

When a beam of light is passed through certain single crystals, such as calcite or quartz, it may be split into two parts; thus a double image is produced as shown in Fig. 7.29(*a*). One ray, known as the *ordinary* ray or O-ray, obeys Snell's law, whereas the other, known as the *extraordinary ray* or E-ray, does not; for instance, when the incident ray is normal to a surface, the extraordinary ray is still refracted so that its angle of refraction is not zero. The ordinary ray will always lie in the plane of incidence, whereas the extraordinary ray will only do so in special cases. Typical paths are shown in Fig. 7.29(*b*).

If the two rays are passed through a second crystal, each will form two more rays, an ordinary and an extraordinary one. As the second crystal is rotated relative to the first, the intensities of these four rays increase and decrease in turn; we conclude that the ordinary

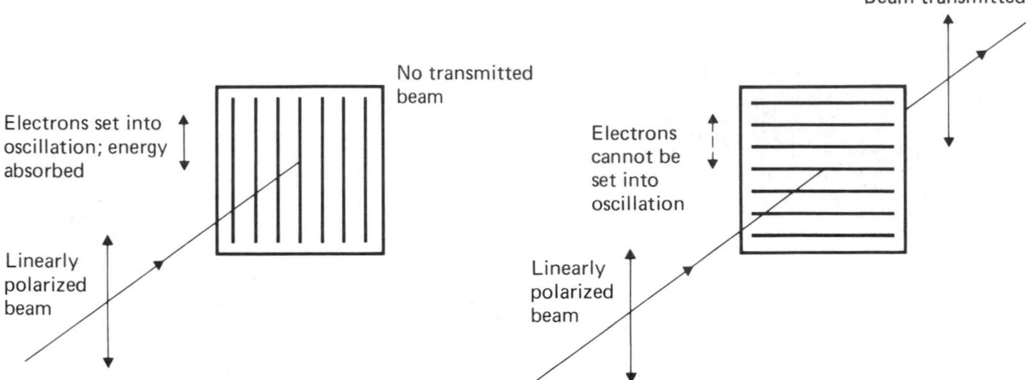

Fig. 7.28 Polarization by selective absorption.

and extraordinary rays produced in the first crystal are plane-polarized at right angles; Fig. 7.30(*b*) shows the propagation of the E- and O-wavefronts in a crystal where the extraordinary wave moves more quickly than the ordinary wave.

The anomalous behaviour of the E-ray is caused by the fact that its velocity through the crystal is different in different directions. Any line parallel to the axis of symmetry of the crystal is called an *optic axis*, as shown in Fig. 7.30, and in this direction the two rays have the same velocity. In other directions, the extraordinary ray may move faster or slower depending upon the type of crystal. Crystals like this, which have different optical properties in different directions, are said to be *anisotropic*, whereas those with the same optical properties in all directions are said to be *isotropic*.

In the special case shown in Fig. 7.30(*a*), the incident ray is normal to the surface ABCD, and lies in the principal section BB′DD′. A *principal section* is a plane which contains an optic axis and is perpendicular to one or more crystal surfaces; in this case, to ABCD and A′B′C′D′. The ordinary ray is undeviated and the extraordinary ray, although deviated, still lies within the principal section, which is also a plane of incidence. If the crystal is rotated about the ordinary ray, the extraordinary ray will describe a circle about it as well. Both rays are plane-polarized, the E-ray with its plane of polarization parallel to the principal section, the O-ray with its plane of polarization perpendicular to the principal section.

(3) *Polarization using a Nicol prism*

A Nicol prism (see Fig. 7.31) consists of a double refracting crystal such as calcite; the crystal is cut in half as shown and then stuck together again with Canada balsam. As this material has a lower refractive index than calcite for the O-ray, total internal reflection can occur at the glass–balsam interface; it has a higher refractive index than calcite for the E-ray and thus total internal reflection is not possible here.

ABCD is a principal section of the crystal, containing the optic axis and at right angles to a crystal face. The incident ray is therefore split into two plane-polarized components. The O-ray is polarized at right angles to the paper and the E-ray is polarized in the plane of the paper; thus the transmitted light is plane-polarized as shown.

(4) *Polarization by reflection*

So far we have looked at three methods of producing polarized light by transmission; this final method depends upon the reflection of light from an interface. A piece of polaroid can be used to show that at a certain angle of incidence, usually about 57°, the light reflected from a glass surface is completely plane-polarized; the plane of polarization is the same as the plane of the interface, as shown in Fig. 7.32. Further experiments will show that the angle at which this phenomenon occurs depends upon the refractive indices of the media involved.

(*a*)

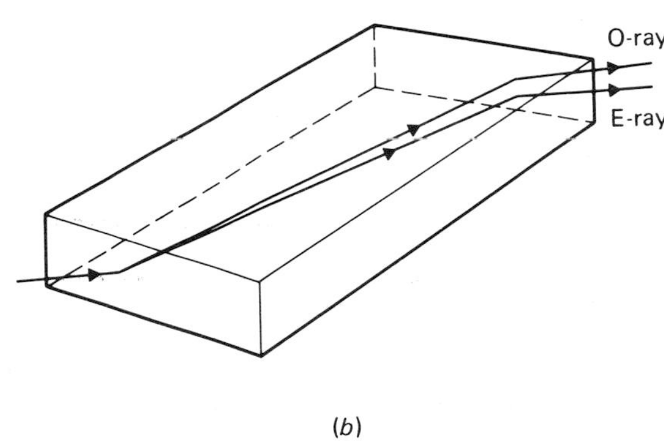

(*b*)

Fig. 7.29 Double refraction.

Now, when light strikes a glass block at an angle of incidence θ_a, some of the light will be reflected, at an angle of reflection also equal to θ_a; the remainder of the light will be refracted, at an angle of refraction, θ_g, given by:

$$n_g \sin \theta_g = n_a \sin \theta_a \quad \text{from (6.11)}$$

Malus found that when $n_a = 1$, and $n_g = 1.55$, the angle of incidence at which the reflected ray was plane-polarized was 57.17°. Now

$$1.55 \sin \theta_g = 1 \sin 57.17$$

$$\sin \theta_g = \frac{0.8403}{1.55}$$

$$= 0.5421$$

i.e. $$\theta_g = 32.83°$$

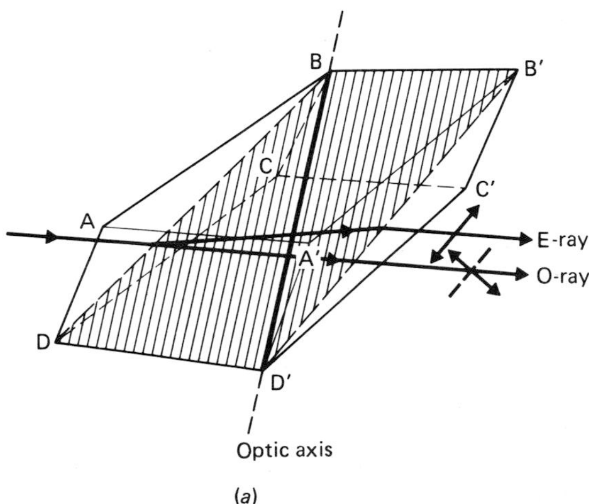

(*a*)

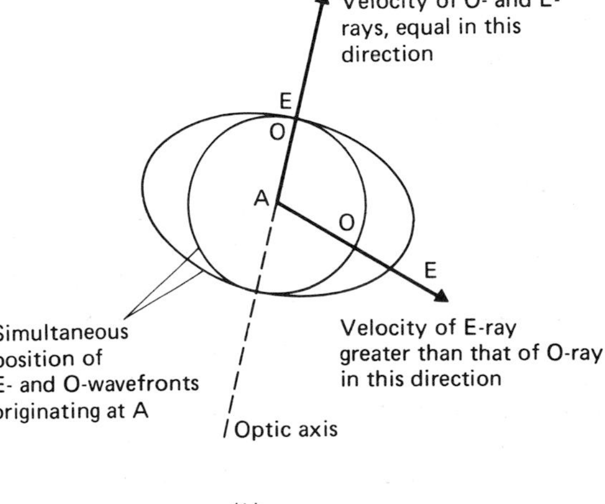

(*b*)

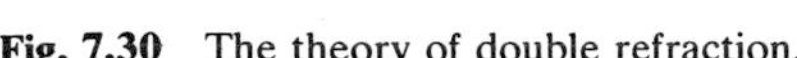

Fig. 7.30 The theory of double refraction.

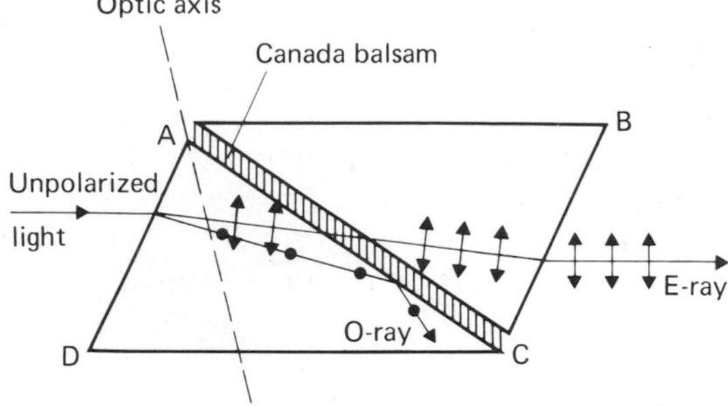

Fig. 7.31 A Nicol prism.

For the experiment described:

$$\theta_a + \theta_g = 57.17° + 32.83°$$

i.e.
$$\theta_a + \theta_g = 90° \tag{7.35}$$

It follows from the figure that the reflected and refracted rays are at right angles; this is always the case whenever completely plane-polarized light is reflected from an interface, although the angles θ_a and θ_g will depend upon the refractive indices of the media.

A plausible explanation of polarization by reflection is as follows. The incident ray sets the electrons in the atoms of the glass into oscillation in the direction of the electric vector, and the energy is re-radiated as the transmitted and reflected beams. If the latter is at 90° to the refracted ray, then only those vibrations perpendicular to the plane of incidence can be reflected; those in the plane of incidence have no component perpendicular to the direction of reflection, since they are at 90° to the refracted ray. Since the reflected ray is plane-polarized, the transmitted ray must also be partly plane-polarized, by the conservation of energy.

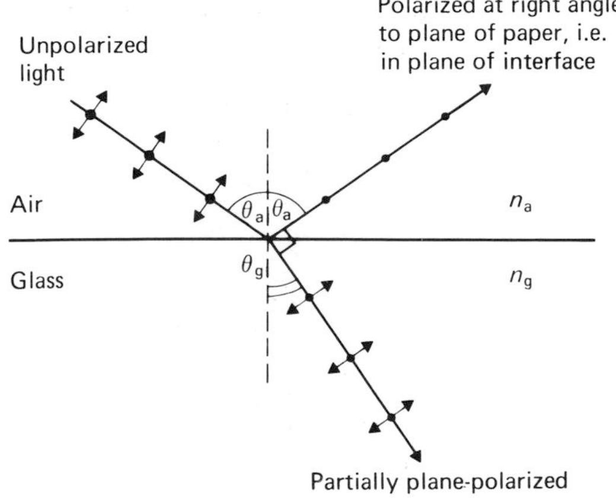

Fig. 7.32 Polarization by reflection.

The size of the angle of incidence can be calculated directly from the refractive indices of the two media. In the case we have discussed:

$$\theta_a + \theta_g = 90° \tag{7.35}$$

and
$$n_a \sin\theta_a = n_g \sin\theta_g$$
$$= n_g \sin(90 - \theta_a) \quad \text{from (7.35)}$$

i.e.
$$n_a \sin\theta_a = n_g \cos\theta_a$$

Hence
$$\tan\theta_a = \frac{n_g}{n_a}$$

This analysis can be applied to any two media, of refractive indices n_1 and n_2, and the angle of incidence at which the reflected ray is plane-polarized, θ_p, is always given by:

$$\tan\theta_p = \frac{n_2}{n_1} \tag{7.36}$$

This is known as *Brewster's law*; if medium 1 is air we have:

$$\tan\theta_p = n_2 \tag{7.37}$$

and θ_p is known as *Brewster's angle.*

When light is reflected from a metal surface, such as the silvered surface of a mirror, the effective refractive index of the metal is so high that $\theta_p \approx 90°$; light reflected at smaller angles is not polarized.

On the other hand, light from car headlights reflected off a road surface is partially plane-polarized. A good way of cutting down glare is to wear polarid sun glasses with a suitable plane of polarization.

Q 7.18
Calculate:

(*a*) the effective refractive index of a medium if, when light is incident in air on its surface, Brewster's angle is 89.9°;
(*b*) Brewster's angle for an air–water surface ($n_w = 1.33$).

Malus' law

When a light is viewed through two pieces of polaroid the first piece, which polarizes the light, is called the *polarizer* and the second, which detects the presence and direction of polarization, is called the *analyser*. Any such arrangement of polarizer and analyser is called a *polariscope*.

We have described a number of ways of producing polarized light, any of which could be used in the polarizer and analyser of a polariscope. It is comparatively awkward to use polarization by reflection and this is rarely used now, though it was once the main experimental method.

Earlier in the Section we considered what happens when polarizer and analyser are parallel and what happens when they are crossed. Malus' law enables us to predict the transmitted intensity when polarizer and analyser are inclined at an intermediate angle, θ, as shown in Fig. 7.33.

Suppose that E_0 is the amplitude of the electric vector incident on the analyser, after plane polarization in the polarizer. Now this vector can be resolved in two directions; in the plane of polarization of the analyser and at right angles to it. The component in the plane of polarization of the analyser is $E_0 \cos\theta$, and this will be the amplitude transmitted by the analyser; if $\theta = 0$ it will be equal to E_0, and if $\theta = 90°$ it will be zero.

Now $\quad \text{intensity} \propto E^2$

$$\propto E_0^2 \cos^2\theta$$

Therefore $\quad I = I_0 \cos^2\theta$

which is known as *Malus' law*; it can be used as the basis of a method of comparing two light intensities by reducing the brighter intensity as described above until the intensities appear to be equal, and then measuring the angle θ.

It is important to point out that the polarizer has also reduced the intensity of the original light, independently of the relative orientation of the analyser. By restricting the transmitted light to one plane of polarization, it reduces the intensity by one half, and thus the amplitude of the transmitted electric vector, E_0, is less than the incident vector by the factor $1/\sqrt{2}$.

Q 7.19
What is the angle between the planes of polarization of a polarizer and analyser when the final intensity is half the incident intensity? What will be the ratio of final and incident intensities when $\theta = 0°$ and $\theta = 90°$?

Applications of polarization

(1) *The wave nature of light*

A most important academic application of the theory of polarization is to show that light must be a transverse wave motion, rather than a longitudinal one; longitudinal waves cannot be polarized. However, the phenomenon also shows the importance of coherence in superposition experiments. An unpolarized beam of light can be regarded as consisting of two beams, plane-polarized at right angles, and with random phase differences between them. Suppose that these two components are now separated, for instance by double refraction, and one plane of polarization is rotated through 90°, as explained in (3), below; the two beams now have identical planes of polarization and we might expect to be able to perform interference experiments with them. This is not, in fact,

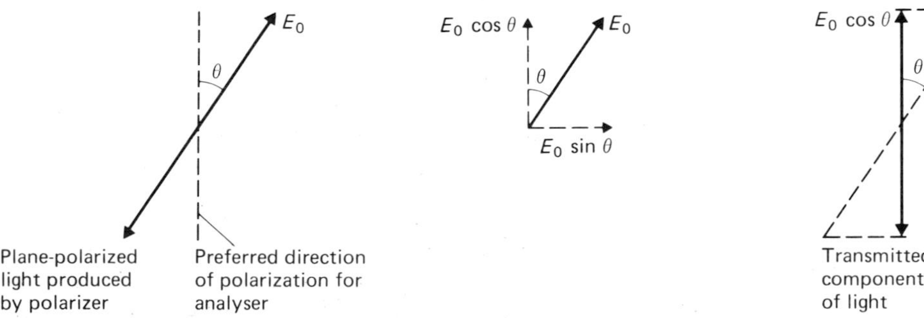

Fig. 7.33 Malus' law.

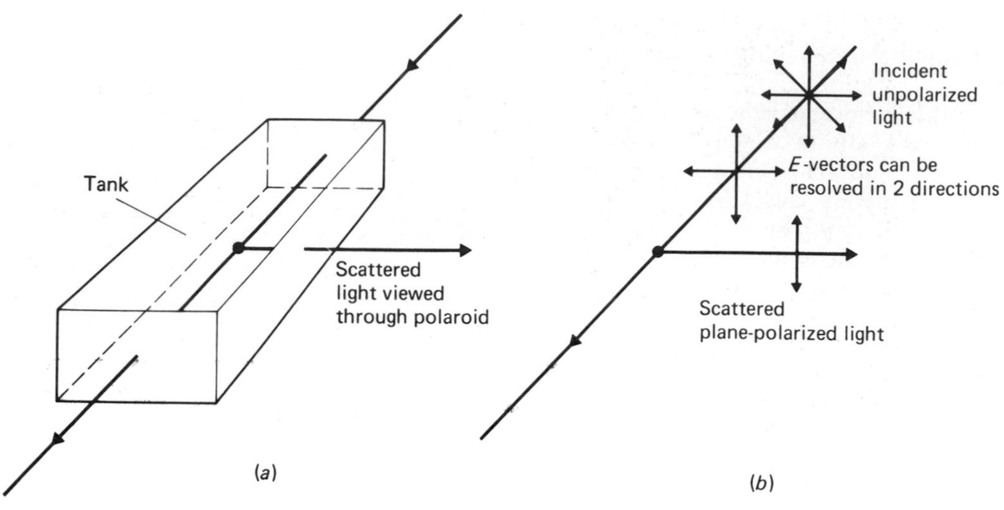

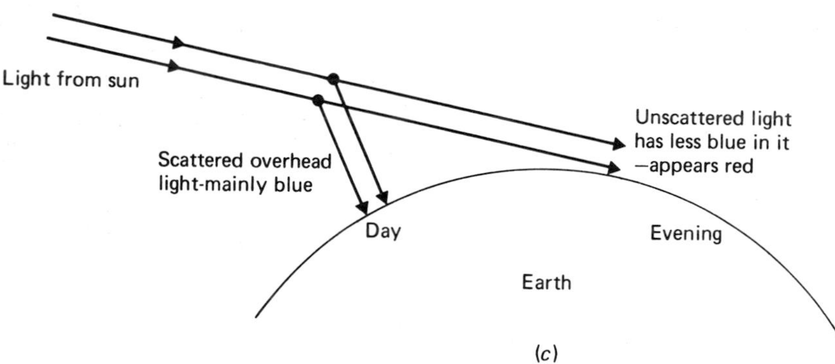

Fig. 7.34 Polarization by scattering.

possible because the two beams were initially incoherent.

(2) *Scattering*

If a beam of light is passed through a tank of water containing just one drop of milk, as shown in Fig. 7.34(a), the light scattered from the milk molecules can be shown, by looking through polaroid, to be polarized. The reason is similar to that given earlier to explain the polarization of light by reflection and is illustrated in Fig. 7.34(b). The electrons are made to oscillate when the light hits them and the oscillation can be resolved into two components—oscillation along the direction of the *scattered ray* and oscillation at right angles to it. The plane of polarization of the scattered E-vector is determined by the direction of this latter oscillation, and the scattered light is thus plane-polarized; there is no component of the E-field in the direction of propagation of the incident ray, since light is a transverse wave motion.

The amount of scattering by very small molecules is proportional to $1/\lambda^4$, so we would expect more blue light to be scattered than red; it is for this reason that the sky looks blue, as shown in Fig. 7.34(c). The light that is left in the original beam will thus contain a higher proportion of red and make a sunset appear

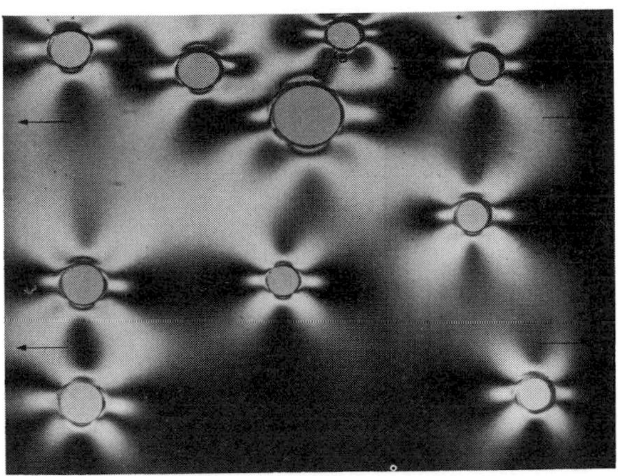

Fig. 7.35 The application of photo-elasticity to the study of crack propagation in an aircraft wing, at the Royal Aircraft Establishment, Farnborough. (*Crown Copyright.*)

red. This variation with frequency does not occur to such a large extent with larger particles such as water droplets and the light scattered from clouds therefore, looks white. Bees have cells in their eyes which are actually sensitive to the polarization of light and they utilize them to provide a sense of direction; a similar principle is used in a type of *solar compass* for use in the Arctic when the sun is below the horizon.

(3) *Optical activity*

Some solids, such as quartz crystals, and certain liquids, such as sugar solution, can rotate the plane of polarization of polarized light. The extent of the rotation depends on the 'length' of the sample and on the nature of the substance and its concentration, if in solution. Measurements of the angle of rotation can be used to determine the concentration of sugar solutions.

In an analogous experiment, we find that if spirals of wire are inserted into polystyrene spheres, they will rotate the plane of polarization of 3 cm microwaves; this occurs whether the spheres are arranged in a lattice or jumbled up. If the spirals are right-handed, the plane of polarization will rotate to the right and vice versa. We conclude that the atoms in a sugar molecule, or quartz crystal, must also have an asymmetric 'spiral' shape, and this is true of most organic substances in living things; the spiral shape is usually left-handed!

(4) *Photo-elasticity*

Glass and some plastics become double refracting when they are stressed, because the stress tends to line up the molecules. The O-ray and E-ray will move at different speeds and thus phase differences will arise between them; if the sample is illuminated by polarized light and viewed through an analyser, coloured interference patterns can then be observed. This phenomenon is used by engineers to observe stresses in models of bridges, etc.; Fig. 7.35 shows a stress concentration in the region of a tear in strained polyurethane rubber, used to represent an aircraft wing skin.

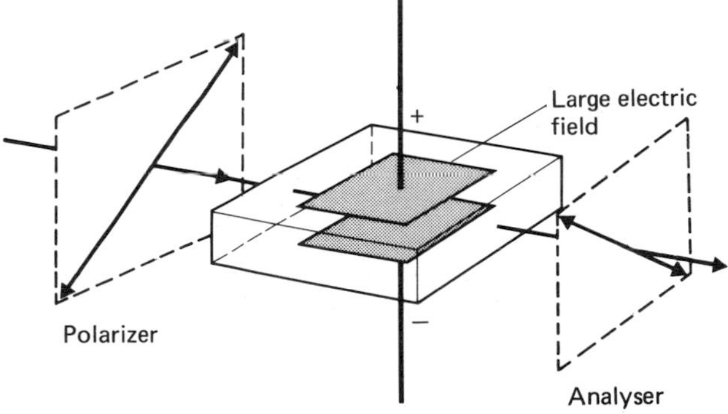

Fig. 7.36 A Kerr cell.

(5) *Kerr cell*

In 1875, Kerr discovered that glass becomes double refracting when subjected to an intense electric field, and this is true also of some gases and liquids, such as nitrobenzene. A Kerr cell is shown in Fig. 7.36 mounted between crossed polaroids; the liquid behaves as though the optical axis were in the direction of the electric field. When the cell becomes double refracting it tends to alter the plane of polarization and some of the light can therefore pass through the analyser. With suitable liquids, the time lag before the effect takes place may be as little as 10^{-9} s, and the cell therefore acts as a very rapid electro-optical shutter. It has been used in accurate measurements of the velocity of light.

Bibliography

Books

Boulind, H.F. (1972) *Waves or Particles?* Longman; this short and straightforward book includes some excellent colour illustrations on physical optics.

Braddick, H.J.J. (1965) *Vibration, Waves and Diffraction.* McGraw-Hill.

French, A.P. (1968) *Electromagnetic Waves and Optics.* Nelson; an MIT introductory science text.

Hecht, E. and Zajaz, A. (1974) *Optics.* Addison-Wesley; advanced in places but relates the performance of optical systems clearly to the wave properties of light.

Jenkins, F.A. and White, H.E. (1981) *Fundamentals of Optics.* McGraw-Hill; standard university book.

Longhurst, R.S. (1973) *Geometrical and Physical Optics.* Longman; this university text contains an excellent selection of fringe photographs.

Revised Nuffield Advanced Physics (1986) *Particles, Imaging and Nuclei.* Longman; contains a well-illustrated article on 'Imaging' which shows how the wave nature of light affects the image that we see.

Revised Nuffield Advanced Physics (1986) *Teachers' and Students' Guides.* Longman; for unit J in volume 2, the Teachers' book contains details of some suitable experiments together with the underlying theory and the Students' book contains some particularly useful structured questions.

PSSC (1981) *Physics* (5th Edn.). Heath; Chapter 25 gives a brief, well-illustrated account of the more important phenomena.

Shurcliff, W.A. and Ballard, S.S. (1964) *Polarized Light.* van Nostrand; a fairly mathematical introduction to this topic.

Tolansky, S. (1968) *Revolution in Optics.* Penguin, Chapters 7, 9; two chapters on modern interferometers and some recent applications of interference and diffraction.

van Heel, A.C.S. and Velzel, C.H.F. (1968) *What is Light?* Weidenfeld and Nickolson Chapters 3–5; a non-mathematical discussion of the most important topics.

Walker, J. (1980) *Light and its Uses.* Freeman; a Scientific American book applying some of the ideas in this chapter and the next.

Wright, G. (1973) *Elementary Experiments with Lasers.* Wykeham, Chapters 1, 4 and 6; details of basic experiments to demonstrate the phenomena of physical optics using a laser.

Software

Edward Arnold, Chelsea Science Simulations *Interference and Diffraction of Waves*; enables the intensity distribution to be calculated for point or slit sources over a range of parameters, displaying the result as a table or a graph—in a more complex model the detailed application of Huygens' Principle is considered.

G'SN Educational Software *Physics Suite* 'Multiple Slit Interference'; enables intensity distribution to be plotted quickly for a wide variety of parameters.

Longman Micro Software *Dynamic Modelling System*; 'SLIT', '1 SLIT' and '2 SLIT' provide an opportunity to look at the intensity distribution and the model on which it is based for one, two or more slits.

8
Optical instruments

8.1 Introduction

Optical instruments have a number of functions; they may help us to see an object more clearly, by making it appear larger; they may collect as much light coming from the object as possible, and thus make it appear brighter; they may enable us to record for posterity what an object looked like; and so on. One of the most difficult functions to understand is that of magnification, and we shall spend a little time discussing it in general before we consider specific types of instrument.

Linear magnification and angular magnification

If a converging lens forms a real image of a real object on a screen, we can compare the size of the image and object, shown in Fig. 8.1(*a*):

$$\text{linear magnification} = \frac{\text{image height}}{\text{object height}} \quad (8.1)$$

The term *linear magnification* is used to distinguish this from *angular magnification*, which we shall consider below.

When we look at an object or image its apparent size will in fact depend, not only upon its size, but also upon its distance from one's eye, as shown in Fig. 8.1(*b*). The size of the image on one's retina is determined by the angle subtended by the object at one's eye, θ; if θ is small we may write:

$$\theta = \frac{\text{object height}}{\text{distance from eye to object}} \quad (8.2)$$

where θ is measured in radians.

If we say that an optical instrument makes an object appear larger, we mean that the image subtends a larger angle at the eye than the object did, and thus a larger final image is produced on the retina. We define:

$$\text{angular magnification, } M = \frac{\text{angle subtended at eye by image}}{\text{angle subtended at eye by object}} \quad (8.3)$$

Angular magnification, sometimes called magnifying power, is calculated for an instrument adjusted in the way in which it is usually used; it is then said to be in *normal adjustment.* For a microscope it is most convenient to have the final image at the *least distance of distinct vision*, or *near point*, since this is the closest that the image can be seen clearly; if the observer is drawing what he sees, the final image will be roughly the same distance from the eye as the bench on which he is drawing. Astronomers, however, usually use telescopes adjusted so that the final image is at infinity since the observer's eye can be completely relaxed; the telescope is then said to be in *normal adjustment.* Slightly greater angular magnification will be obtained with the final image at the near point and the telescope is then in *near-point adjustment.*

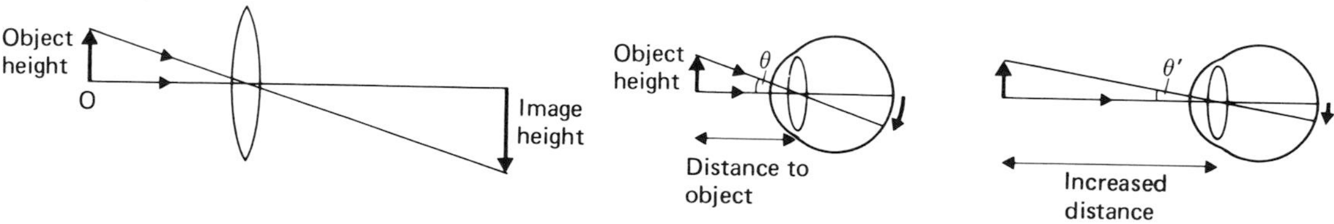

(*a*) Magnification produced by a lens

(*b*) The variation in the apparent size of an object

Fig. 8.1

Usually, in optical instrument calculations, we assume that the near point is 250 mm from the eye.

The angle subtended by the object without the instrument also depends on how far the object is from the observer's eye. When carrying out calculations for telescopes, of course, the angle is that subtended by the object *in the position in which it happens to be*. In a microscope we are interested in comparing the apparent size of the image with the *largest* possible apparent size obtainable without it; to achieve this *the object is considered to be held at the near point*. Thus microscope and telescope calculations will be carried out in slightly different ways.

Field of view and the brightness of the image

The *field of view* of an instrument, such as a telescope or microscope, is the angle subtended at the eye by the largest object which can be seen through the instrument. It is determined by the size of the eye lens.

The *brightness* of the image is determined by the amount of light entering the instrument, and thus partly by the size of the objective lens.

In any instrument there is an optimum position in which to place the eye which is called the *exit pupil*; a metal cap with circular hole is usually placed there so that the pupil of the eye is automatically in the right place. If the size of the exit pupil is equal to the size of the pupil of the eye, maximum field of view, brightness and resolving power are obtained.

These ideas are applied to individual instruments in the sections following.

Depth of field and depth of focus

A lens, or optical instrument, forms an image in a particular place only from objects at one distance from the lens; objects at other distances will be out of focus, as shown in Fig. 8.2. However, an object at a different distance may *appear* to be satisfactorily in focus, especially if the lens aperture is reduced, as shown. The variation in object distance for which the image is adequately focused is called the *depth of field*.

The maximum distance the film on which an image is recorded can be moved, without the image appearing out of focus, is called the *depth of focus*. These terms are widely used in photography and further discussed in Section 8.7 on the camera.

Resolving power

The ability of an optical system to distinguish two adjacent objects is called its *resolving power*; it determines how much fine detail may

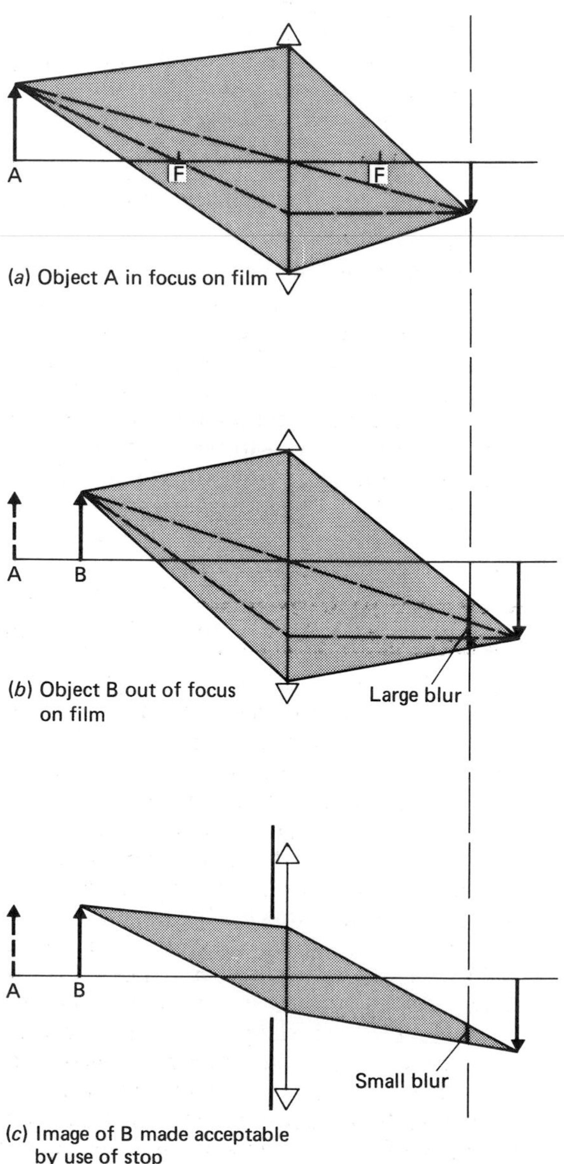

Fig. 8.2

be seen on an extended object. If we look at a distant approaching car at night, we cannot distinguish its headlamps as separate sources if it is more than about 2 km away; when it is closer, and their angular separation greater, they can be seen as separate lights. Similarly, if we look through a hole in a piece of card at a multiple light source, as shown in Fig. 8.3(a) we find that as we move away from it the lamps' filaments become blurred together and indistinguishable. The *angular resolution*, θ, of a system is defined as the minimum angular separation at which two objects can be resolved by the system. *Resolving power* is then defined by:

$$\text{resolving power} = \frac{1}{\text{angular resolution}} \qquad (8.4)$$

and is measured in $(\text{rad})^{-1}$.

We find that the lamps are easier to resolve if the hole we look through is larger, and we also find that they are more distinct if the wavelength of the light is shorter. This dependence upon wavelength and aperture suggests that the blurring of the separate sources is caused by diffraction at the hole. Although optical instruments usually have circular apertures, we will just consider blurring caused by diffraction at a slit; this is simpler than, but similar to, diffraction at a circular aperture.

If there were no diffraction at the slit, each source would form a sharp image. However, as diffraction occurs, there is a central maximum, of width determined by the wavelength and slit width, and to each side a succession of minima and secondary maxima. Each source will produce an adjacent identical pattern; these are shown projected on a screen in Fig. 8.3(b). In each case, the angle between the central maximum and the first-order minimum is λ/b.

When both sources illuminate the one slit, instead of two line images being formed, two diffraction patterns will be seen. Under what condition can we still see that there are two images? This requires us to resolve the two central maxima. Rayleigh made the arbitrary assumption which has become known as *Rayleigh's Criterion*; he said that the two maxima ought to be resolvable if the first-order minimum of one coincided with the central maximum of the other, as shown in Fig. 8.3(c). If the maxima are resolved, then we can see that there are two distinct sources. Note that as the sources are self-luminous, they must be *incoherent* so *interference cannot occur.*

If this criterion is accepted, then Fig. 8.3 shows that the minimum angular separation at which the sources can be resolved is λ/b. Hence

$$\text{angular resolution, } \theta = \lambda/b \qquad (8.5)$$

$$\text{resolving power} = 1/\theta$$

i.e.

$$\text{resolving power} = b/\lambda \qquad (8.6)$$

Thus, the angular resolution of a system will be decreased if the aperture is made larger and the wavelength made shorter; the resolving power will be increased.

Whenever there is an aperture in an optical system we must consider the extent to which diffraction effects will prevent us resolving two close points on an object; this problem is of the utmost importance in the design of optical systems as it determines how much detail we can see. For a circular aperture, of diameter D, it can be shown that for the Rayleigh criterion to be satisfied, the angular separation, θ, must be greater than $1.22\ \lambda/D$.

i.e.

$$\text{angular resolution, } \theta = 1.22\frac{\lambda}{D} \qquad (8.7)$$

$$\text{resolving power} = \frac{D}{1.22\lambda} \qquad (8.8)$$

The requirement to achieve high resolution suggests that telescopes and microscopes should have large apertures; a compromise must be found between this and the problem of aberrations introduced by large aperture lenses and mirrors.

Q 8.1.
What is the resolving power and angular resolution of a lens of diameter 30 mm when illuminated with light of wavelength 6×10^{-7} m?

The pupil of the eye is, of course, an aperture in an optical system, of diameter about 3 mm in daytime.

Hence

$$\theta = 1.22\frac{\lambda}{D}$$

and for light of wavelength 5×10^{-7} m

$$\theta = 1.22\times\frac{5\times10^{-7}}{3\times10^{-3}}$$

$$= 2\times10^{-4}\ \text{rad}$$

i.e.

$$\theta = 42 \text{ seconds of arc}$$

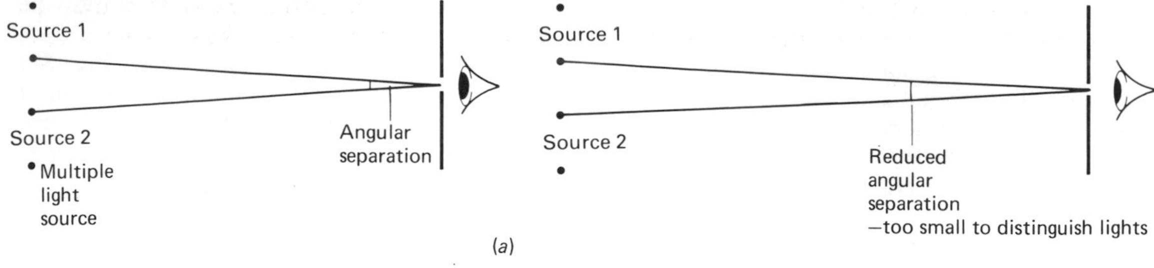

(*a*)

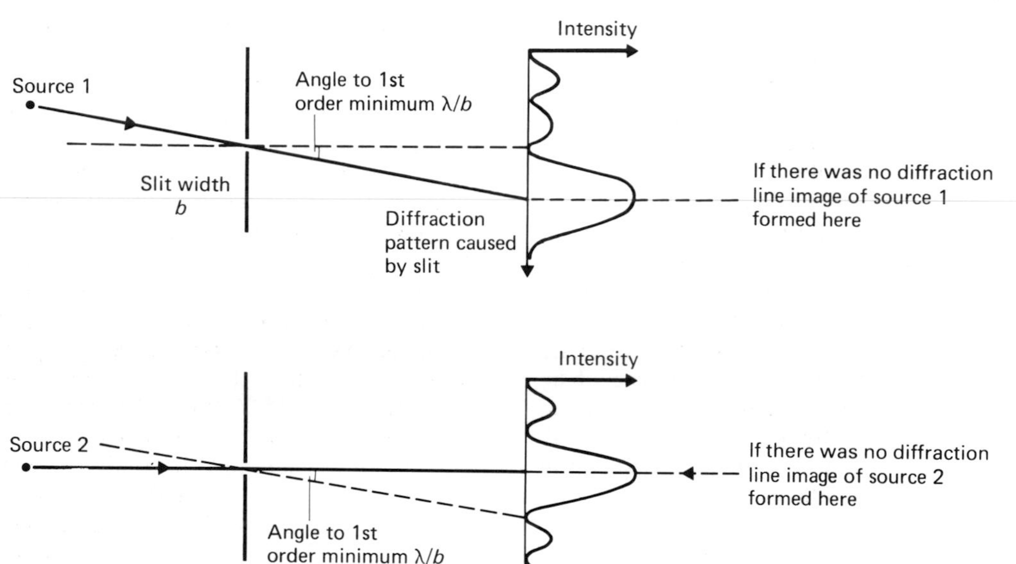

(*b*) Diffraction pattern produced by separate line sources at slit

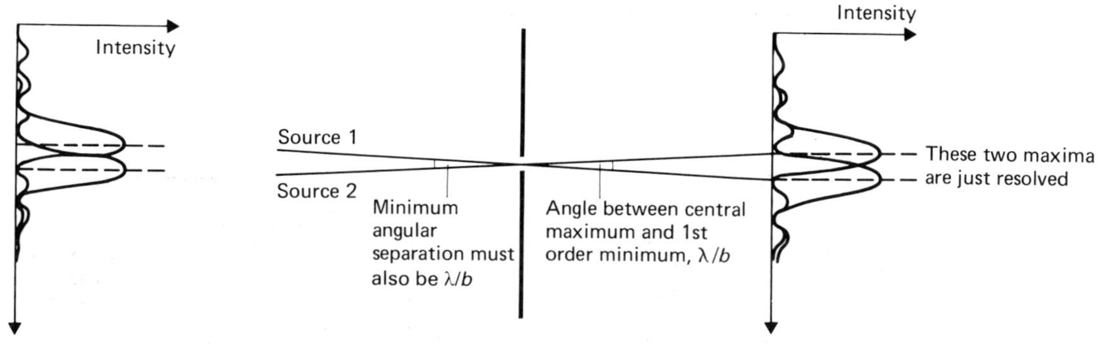

(*c*) *Rayleigh's criterion*—these two central maxima will be resolved if the first order minimum of one coincides with the maximum of the other

(*d*) Rayleigh's criterion applied to two adjacent sources

Fig. 8.3 Factors affecting the resolution of a pair of light sources.

Thus, at a distance of 5 m we would expect to distinguish objects a distance x apart, where:

$$\begin{aligned} x &= 5 \times \theta \\ &= 5 \times 2 \times 10^{-4} \\ &= 10 \times 10^{-4}\ \text{m} \\ &= 1.0\ \text{mm} \end{aligned}$$

We should, therefore, be able to distinguish two points about one millimetre apart on a wallpaper at the far end of a fair-sized room; readers are invited to compare their own performance with this estimate.

It is likely that the reader will find, in fact, that his powers of resolution are not quite as good as this; the average person can only resolve objects provided that they are more than 1 minute of arc apart, or 3×10^{-4} rad, because the definition attainable by the eye is further restricted by the spacing of the light-sensitive cells on the retina.

Telescopes and microscopes increase the apparent size of the object, and thus the angle subtended at the eye by adjacent points on it. As the resolving power of the telescope itself will be much larger than that of the eye, the eye will be able to resolve finer detail than it would be able to unaided.

Radio telescopes

Before leaving this topic we will consider briefly why radio telescopes are so large.

The parabolic mirror of a radio telescope acts as an aperture, which in the case of Jodrell Bank is 80 m in diameter. The radio waves that it receives from a star may have a wavelength of 0.2 m.

Hence
$$\begin{aligned} \theta &= 1.22\lambda/D \\ &= 1.22 \times \frac{0.2}{80} \\ &= 3 \times 10^{-3}\ \text{rad} \end{aligned}$$

We saw above that the human eye can distinguish objects down to 3×10^{-4} rad which is, of course, considerably better than the Jodrell Bank radio telescope.

Thus, the resolving power of Jodrell Bank is considerably less than that of our own eyes, at the wavelengths to which it is sensitive. This is because the wavelength of the radio waves from stars is very large compared with that of visible wavelengths, and the size of the dish does not make up for this. To get some idea of the resolution 'seen' by Jodrell Bank, make a hole in a piece of tissue paper with 36 swg copper wire and look through it; this hole has the same angular resolution as the radio telescope.

To improve the resolution at Jodrell Bank to equal that of our eyes, it would be necessary to make a dish ten times the size, and that is clearly absurd, especially for a steerable telescope. Instead, astronomers use a radio interferometer. The principle of this is shown by the analogous arrangement in Fig. 8.4; 3 cm radio waves are used in this experiment.

The transmitter is placed at one end of the room and the other items are placed on a table

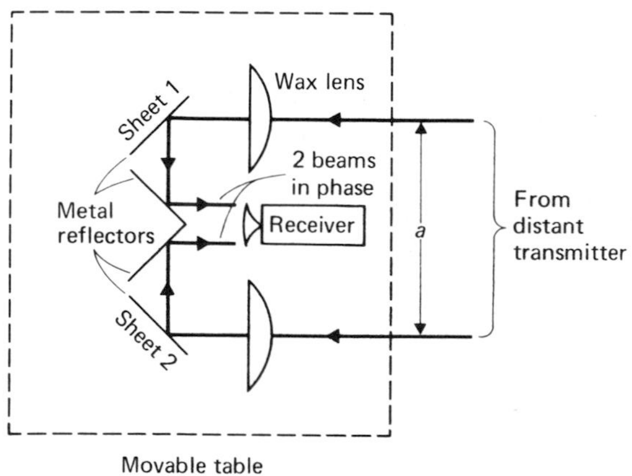

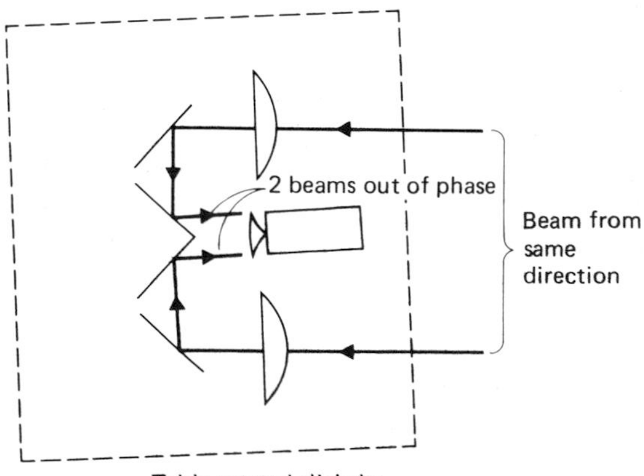

Fig. 8.4 An analogue interferometer.

at the other end, so that they can be rotated on it as a unit. Signals from the distant transmitter are collected by the lenses and reflected at two metal sheets into the receiver. If they arrive in phase at the receiver, there is a large response, and if they arrive out of phase, the response is zero. The reflectors are first adjusted so that each path, on its own, gives the maximum response; both reflectors are then put in place so that the superposed signal is detected at the receiver. We now find that the apparatus can be adjusted much more sensitively than before to locate the direction of the source more accurately.

Thus, if the table is rotated slightly, as shown in the second diagram, the beams of light reflected from sheets 1 and 2 travel slightly different distances to the receiver; they become out of phase. The effective aperture of the device is the distance, *a*, in the diagram.

The radio interferometer at Cambridge can be set up with a pair of relatively small dish aerials located up to 1.6 km (1 mile) apart; the rotation of the earth is used to enable the interferometer to build up slowly the information that could be collected by one very large dish. It has a much higher resolving power and is more suitable for some purposes than the 80 m Jodrell Bank telescope.

Q 8.2
Calculate the angular resolution and resolving power of the '1 mile' radio interferometer, for radio waves of wavelength 0.2 m. Assume that the 'aperture' is circular.

The use of cameras with optical instruments

In the sections on telescopes, microscopes and spectrometers, we shall show an observer viewing a virtual image; the lens in his eye then forms a real image on his retina. In practice, the observer's eye is frequently replaced by a camera and the real image formed on the film enables a permanent record of events to be kept.

The permanence of the record is not the only advantage of using a camera. The longer a camera shutter is open, the more light enters the camera and the more exposed is the film; the brightness of the image on the film depends on the sum total of all the light entering the camera during the exposure time, and the camera is therefore an *integrating device*. The brightness of an image on the retina, on the other hand, depends on the instantaneous intensity of the light from the source. Thus, a camera may form an observable image of an object too faint to be seen by the naked eye; this is particularly important in astronomy, where the intensity of light from a star in a distant galaxy may be very low. Fig. 8.5 shows a photograph of the Orion nebula.

There are situations, of course, where the use of a camera is difficult. Objects viewed through a microscope may be moving and so too are objects in the sky. Telescopes have complex driving systems to ensure that the same stars or planets are kept in view for prolonged periods.

Q 8.3
The image on the retina is retained for about 1/16 s. A photographic exposure of a star might last 4 hours. How many times more light would enter a camera in 4 hours than the eye in 1/16 s, assuming the apertures are the same? The eye pupil is about 6 mm in diameter at night; how wide would it have to be to receive as much light in 1/16 s as a camera of aperture 6 mm does in 4 hours?

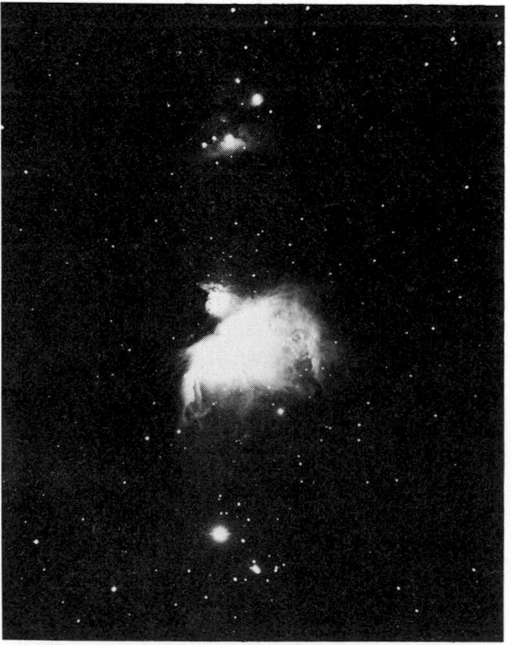

Fig. 8.5 Orion Nebula: stars in this photograph are at distances ranging from 8 to 400 light years away.

8.2 Refracting telescopes

Telescopes are used to provide angular magnification of distant objects and increased resolution of fine detail, and to collect as much light as possible to form a bright image. They are used to observe distant objects on the earth but an important application is in astronomy. As we shall see, the simplest form of telescope produces an inverted imagc but this does not matter except when viewing terrestrial objects; at the end of this Section we shall look at one type of telescope that will form an erect image, for use on earth.

The astronomical telescope

A simple astronomical telescope is shown in Fig. 8.6. The objective produces a real inverted image of a distant object and this image is then viewed through an eyepiece lens, as shown in Fig. 8.6(*a*); this eyepiece lens is a converging lens used as a magnifying glass, forming an enlarged virtual image as explained in Section 8.4. The intermediate image is often arranged to be in the focal plane of the eyepiece lens and the final, virtual image is at infinity; the telescope is then said to be in *normal adjustment*. With this arrangement, the observer's eye is more relaxed and, as this is more comfortable for long periods of viewing, this is how telescopes are usually used. Fig. 8.6(*b*) shows a telescope in normal adjustment forming an image at infinity of an object at infinity.

A simple telescope may be made in the laboratory to illustrate the above description. A convex lens of focal length 500 mm is held in a clip attached to one end of a metre rule, as shown in Fig. 8.6(*c*). This lens is then pointed towards a bright lamp several metres away, and a ground-glass screen moved up and down the ruler; when it is about 500 mm from the lens a small, inverted image of the lamp is observed on the screen. A second convex lens, of focal length 50 mm, is now mounted in a similar clip and used to view the image; at any distance from the screen less than or equal to 50 mm, an inverted, magnified image will be seen through the lens. When the distance from lens to screen is exactly 50 mm, the image can be viewed with a relaxed eye, for it is at infinity. If the screen is now removed, without refocusing the eye, the final image can still be seen; the intermediate image is formed whether or not it is cast on a screen.

The angle subtended at the telescope objective by the object is θ_o, as marked in Fig. 8.6(*b*). This is the angle between light rays from the top of the object and those from the bottom; the bottom of the object is assumed to be on the axis. When calculating angular magnification, we assume that the angle subtended at the eye by the object is the same as the angle subtended at the telescope objective; in other words, we assume that the length of the telescope is negligible compared with the distance from telescope to object. For an object at infinity the intermediate image will be in the focal plane of the objective. Therefore, the angle θ_o is given by:

$$\theta_o = \frac{h}{f_o} \qquad \text{if } \theta_o \text{ is small}$$

where h is the height of the intermediate image.

The angle subtended at the eye by the image is shown in the diagram as θ_i. For a telescope in normal adjustment, the intermediate image must lie in the focal plane of the eyepiece lens, as well as of the objective, and we then have:

$$\theta_i = \frac{h}{f_e} \qquad \text{if } \theta_i \text{ is small}$$

Thus, the angular magnification of a distant object produced by a telescope in normal adjustment is calculated as:

$$\text{angular magnification, } M = \frac{\theta_i}{\theta_o} = \frac{h/f_e}{h/f_o}$$

i.e.

$$M = \frac{f_o}{f_e} \qquad (8.9)$$

Q 8.4
A laboratory telescope similar to that described earlier has an objective lens of focal length 0.8 m, and an eyepiece lens of focal length 0.2 m. What is its angular magnification and length?

Thus, to obtain a large angular magnification, we require an objective of long focal length and

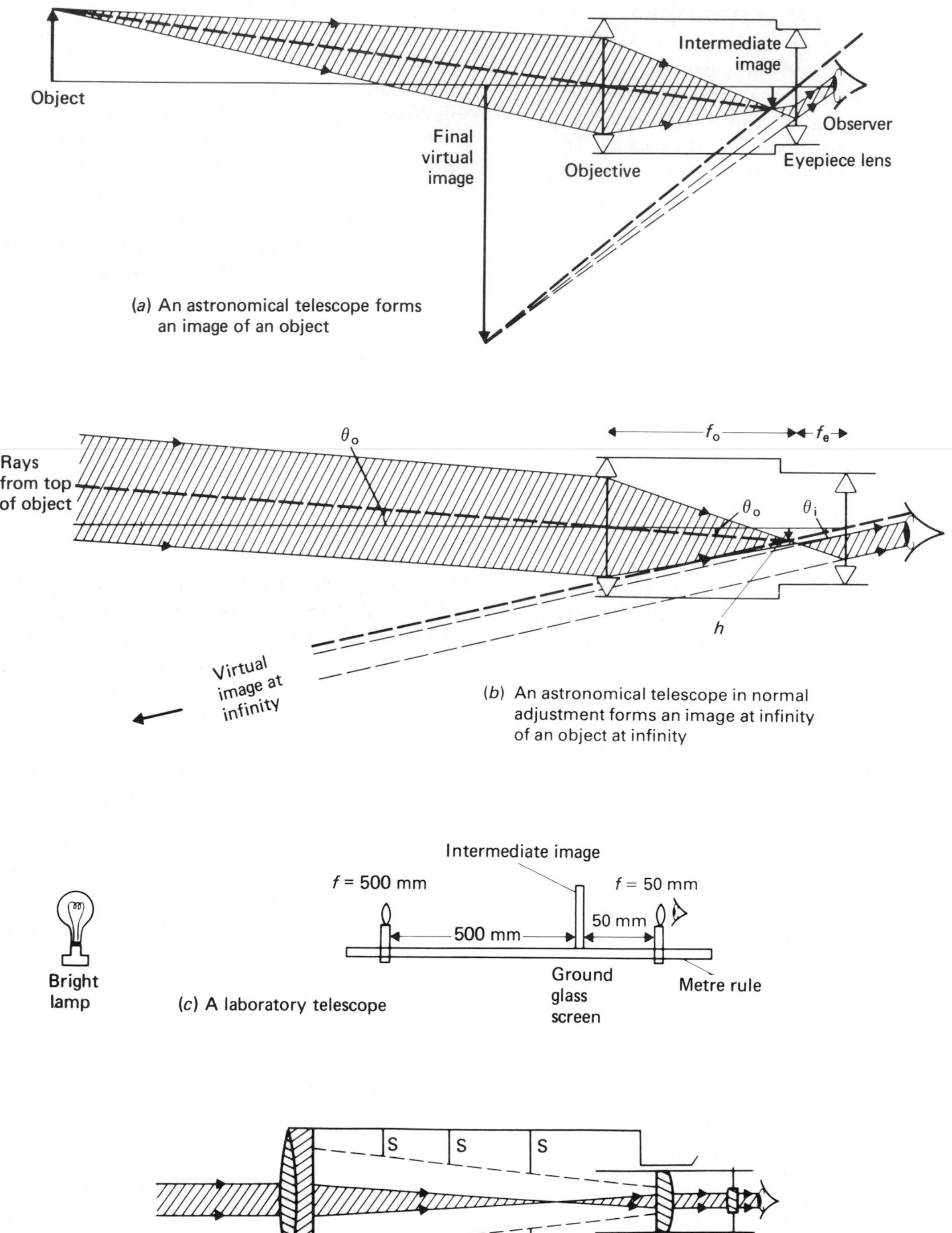

(*a*) An astronomical telescope forms an image of an object

(*b*) An astronomical telescope in normal adjustment forms an image at infinity of an object at infinity

(*c*) A laboratory telescope

(*d*) A simple refracting telescope

Fig. 8.6

an eyepiece lens of short focal length. When telescopes were first invented, it was not possible to produce very short focal length eyepiece lenses without introducing severe chromatic aberrations. To gain significant angular magnification, therefore, the focal length of the objective had to be very long indeed, resulting in telescopes of great length since the separation of the lenses must be (f_o+f_e). One example of such a telescope is illustrated in Fig. 8.7. The problem of chromatic aberration in the lenses of a telescope can be reduced by using more complicated lens systems. Nowadays telescope eyepieces almost always consist of multiple lens systems as shown, for example, in Fig. 8.6(*d*); objectives consist of achromatic doublets. The stops, S, prevent stray incident light, at an angle to the axis, from being reflected from the telescope walls; the latter are painted matt black for the same reason.

A slightly greater angular magnification is produced if the final image is at the near point, instead of at infinity; in this case, the lens spacing is not equal to (f_o+f_e) and in most telescopes this spacing can be altered by a focusing arrangement.

Q 8.5
Draw a ray diagram to show how a telescope can produce a virtual, inverted image 0.25 m from the eyepiece, of an object that is at infinity. Take $f_o = 0.8$ m and $f_e = 0.1$ m and measure the angular magnification from the diagram.

Resolving power

The resolving power of the telescope will be determined by the diameter of the objective lens; the wider this is, the finer the detail that can be resolved. It is, however, difficult to manufacture large, perfectly-shaped, homogeneous pieces of glass; once made, they are also difficult to support without sagging, which produces unacceptable changes in the radii of curvature and thus the focal length of the lens. One of the largest telescope objectives is at Yerkes Observatory in America and has a diameter of 1 m. For light of wavelength 5.6×10^{-7} m, the angular resolution will be:

$$\text{angular resolution} = 1.22\frac{\lambda}{D} \qquad (8.7)$$

$$= 6.8\times10^{-7}\ \text{rad}$$

This is several orders of magnitude better than the angular resolution of the eye. Because of the importance of objective diameters, they are normally quoted when referring to a telescope; we speak of the 1 m (40 inch) refracting telescope at Yerkes. We shall see in the next section that reflecting telescopes can have larger objective diameters and thus achieve much better resolution.

A wide objective lens also admits more light into the system, and hence increases the brightness of the image.

Entrance and exit pupils

As stated above, the brightness of an image is determined partly by the amount of light entering the system. In general, the aperture which limits the amount of light entering the system is known as the *entrance pupil*: thus the entrance pupil of a telescope is the objective, as shown in Fig. 8.8(*a*).

Light from the objective, passing through the eyepiece, will form an image of the objective as shown in Fig. 8.8(*b*); the diameter of this image is called the *exit pupil.* If the eye is placed here, with a pupil at least equal in diameter to the exit pupil, it will receive *all* the rays from the objective which subsequently pass through the eyepiece as shown in Fig. 8.8(*c*); thus, it will observe the largest possible field of view. The exit pupil is usually a few millimetres beyond the last element in the eyepiece, and frequently the eyepiece is so constructed that the eye cannot be held any closer to the element than this. The position of the exit pupil is found by applying the lens equation to the eyepiece. Hence, by similar triangles:

$$\frac{\text{entrance pupil, } D}{\text{exit pupil, } d} = \frac{f_o}{f_e}$$

and hence, from Equation (8.9), we have

$$\text{angular magnification } M = \frac{D}{d} \qquad (8.10)$$

The angular magnification of a telescope may be measured by illuminating the objective and casting an image of it onto a screen placed where the exit pupil will be; the sizes of the entrance pupil and the exit pupil can then be measured directly and the angular magnification calculated. The sharply defined

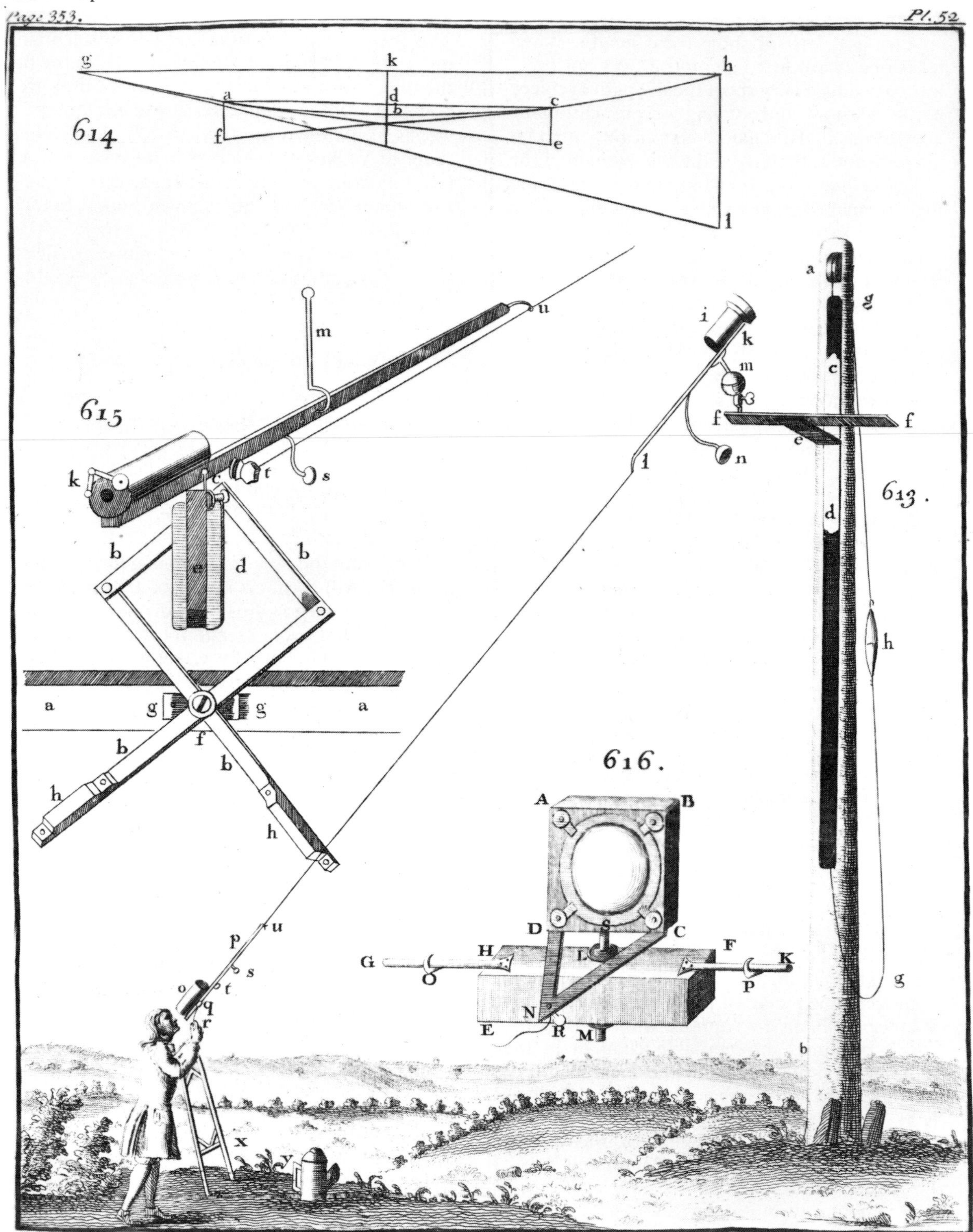

Fig. 8.7 Huygens' astronomical telescope. Smith (1738) *Optics.*

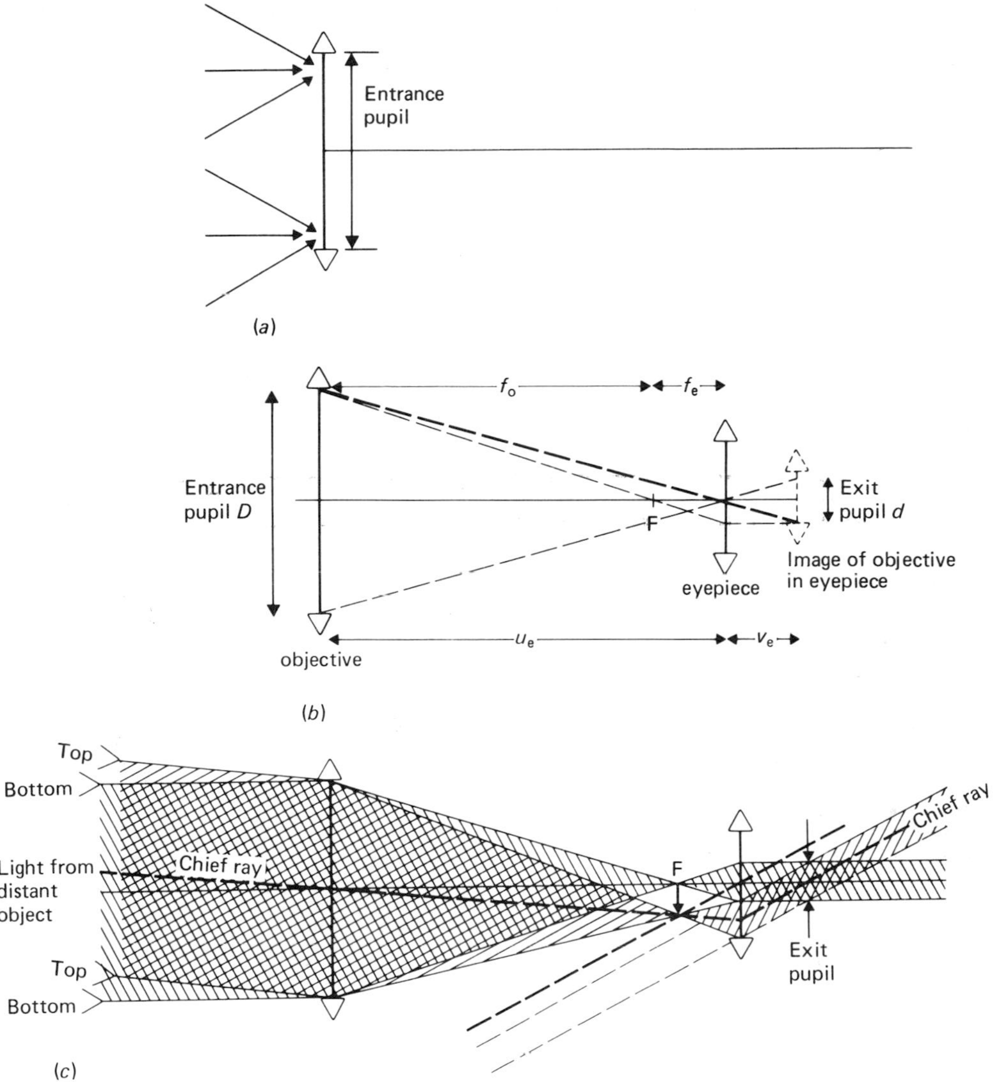

Fig. 8.8 Entrance and exit pupils.

disc of light that is the exit pupil is sometimes called the *eye ring*.

The ray passing through the centre of the entrance pupil is called the *chief ray*; it will pass through the centre of the exit pupil, as shown in Fig. 8.8(*c*).

Similar results can be derived for the reflecting telescope and compound microscope, described in later sections.

Q 8.6
Calculate, for a telescope in normal adjustment, with $f_o = 0.8$ m and $f_e = 0.02$ m, how far from the eyepiece the exit pupil will be. If the objective lens is 50 mm in diameter, calculate the size of the exit pupil.

Galilean telescope

Galileo was the first person to construct a telescope which produced an erect final image, and so could be used for observing distant terrestrial objects. A long focal length objective lens is used, as before, but the eyepiece consists of a short focal length *diverging* lens. It is placed between the objective and the intermediate image, which then acts as a virtual object for the eyepiece, as shown in Fig. 8.9. If the virtual object is in the focal plane of the eyepiece lens, the final image will be at infinity.

Galilean telescopes have a very small field of view because the eye cannot be placed at the exit pupil, and are nowadays only used in opera glasses; in modern binoculars totally reflecting prisms are used to invert the image, as explained in Section 6.3.

Q 8.7
Draw a ray diagram of an alternative form of terrestrial telescope, which has an additional converging lens to give a final erect image. Show that this telescope has a greater field of view than a Galilean telescope; what will be its disadvantage and how is this usually overcome?

8.3 Reflecting telescopes

The advantages of reflecting telescopes

We have already seen the need for large objectives in optical systems, in order to give good resolution and increased brightness; however, a large objective lens for a refracting telescope is difficult to manufacture, difficult to support and suffers from chromatic aberrations. Although the largest refracting telescope has a 1 m (40 inch) objective it is, in fact, unusual to have lenses of diameters much larger than 0.2 m (8 inch), and to obtain larger objectives reflecting telescopes are used. The largest reflecting telescope, which is in Russia, has an objective of diameter 6 m (240 inch); the largest Western telescope is the 5 m (200 inch) instrument at Mt. Palomar, in America.

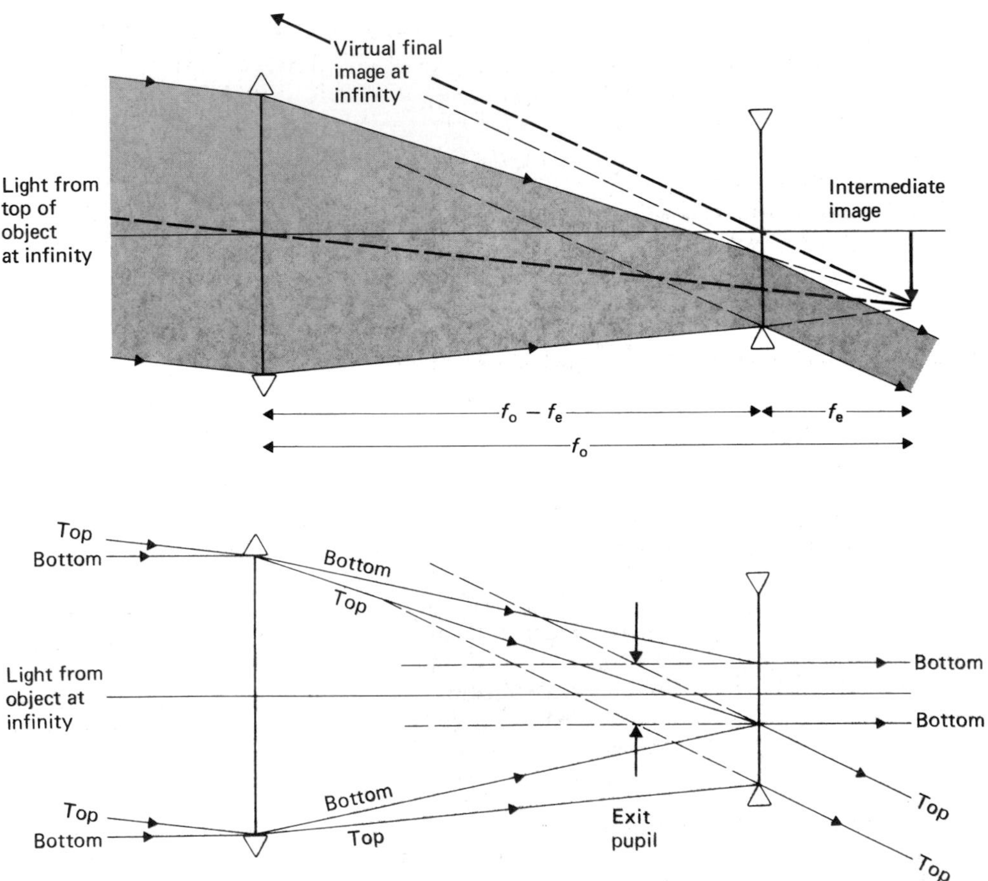

Fig. 8.9 The Galilean telescope.

One of the difficulties in the manufacture of large lenses is that stresses are set up during the cooling process of any large block of glass, which cause the refractive index to vary from place to place; it is interesting to note that the glass for the mirror at Mt. Palomar was gradually cooled over a whole year. Stresses in the glass used for mirrors matter less, of course, because the light does not pass through the glass. It is also possible to consider alternative substances to glass for the construction of large mirrors; glass ceramics are particularly suitable since they have a low expansivity and therefore the heat generated in grinding, and the temperature changes during use, are less of a problem.

The cost of grinding a mirror will be less than for a lens since there is only one surface to be ground instead of the four required in an achromatic doublet. It will also be less likely to flex than a lens since it can be supported over the whole of its rear surface.

The most important optical advantage is that, as there is no refraction of the light, there can be no chromatic aberration introduced by the objective. The fact that the light does not pass through any glass also means that there will be less loss of brightness in the system; as explained in Section 6.2, the light is reflected at the front surface of the glass with an efficiency of up to 85%.

The optical system of a reflecting telescope

In a reflecting telescope, the intermediate image is formed by reflection in a large concave mirror, as shown in Fig. 8.10(*a*). In principle, this image could then be viewed through an eyepiece by an observer in exactly the same way as in a refracting telescope; this is shown in Fig. 8.10(*b*). We see from the two diagrams that for a reflecting telescope in normal adjustment:

$$\text{angular magnification, } M = \frac{\theta_i}{\theta_o} \qquad (8.3)$$

i.e.

$$M = \frac{f_o}{f_e} \qquad (8.9)$$

However, as is clear from the diagram, the observer and eyepiece are likely to obstruct too large a fraction of the incoming light, unless the concave mirror is extremely large.

It is more usual to deflect the light to form a second intermediate image out of the incident beam, and it can then be magnified without cutting out any further light. There are many ways of doing this, and the two most common are shown in Figs 8.10(*c*) and (*d*); Cassegrain's system was designed one year after Newton's. Newton's system does not alter the angular magnification at all, whereas the Cassegrain system can increase it by up to five times by increasing the size of the second intermediate image; diagram (*d*) shows the size of the second intermediate image being reduced by this system. With these arrangements the obstruction caused by the small mirrors can be negligible; major telescopes are usually provided with a number of systems for use in different circumstances.

On page 159 we discussed the use of a Schmidt plate to correct spherical aberrations in a concave mirror; this is used in conjunction with a spherical mirror which is easier to construct than a paraboloidal mirror. The system forms sharp images of points further from the axis and is of use where a large field of view is required.

The greater aperture of the reflecting telescope gives it a much greater resolving power than a refracting instrument. The 5 m (200 inch) Mt. Palomar telescope has an angular resolution given by:

$$\begin{aligned} \text{angular resolution, } \theta &= 1.22\frac{\lambda}{D} \\ &= 1.22 \times \frac{5.6 \times 10^{-7}}{5} \\ &= 1.4 \times 10^{-7}\ \text{rad} \\ \text{resolving power} &= 1/\theta \\ &= 7.3 \times 10^{6} \end{aligned}$$

This is five times the resolving power of the Yerkes telescope, which is one of the largest refractors. Fig. 8.11 shows the largest British telescope, the 2.5 m Isaac Newton telescope belonging to the Royal Greenwich Observatory.

Q 8.8
The Mt. Palomar reflector has a focal length of 16.6 m. What would its magnification be if used in Newtonian form with an eyepiece of focal length 100 mm?

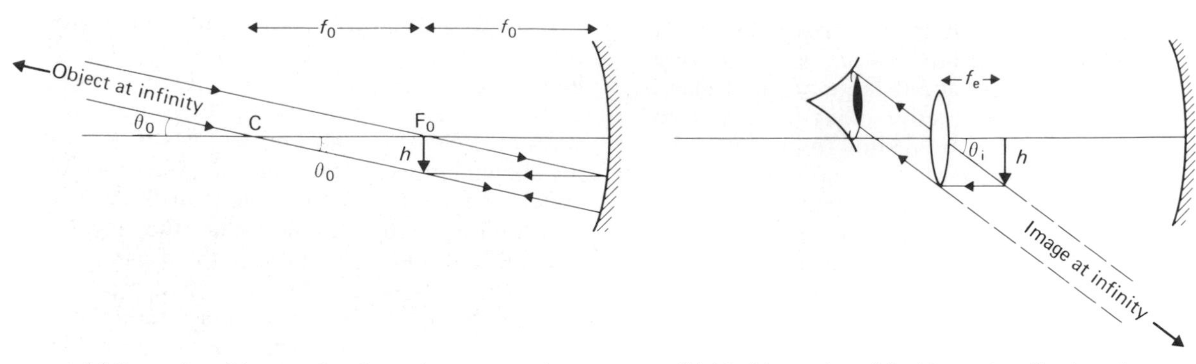

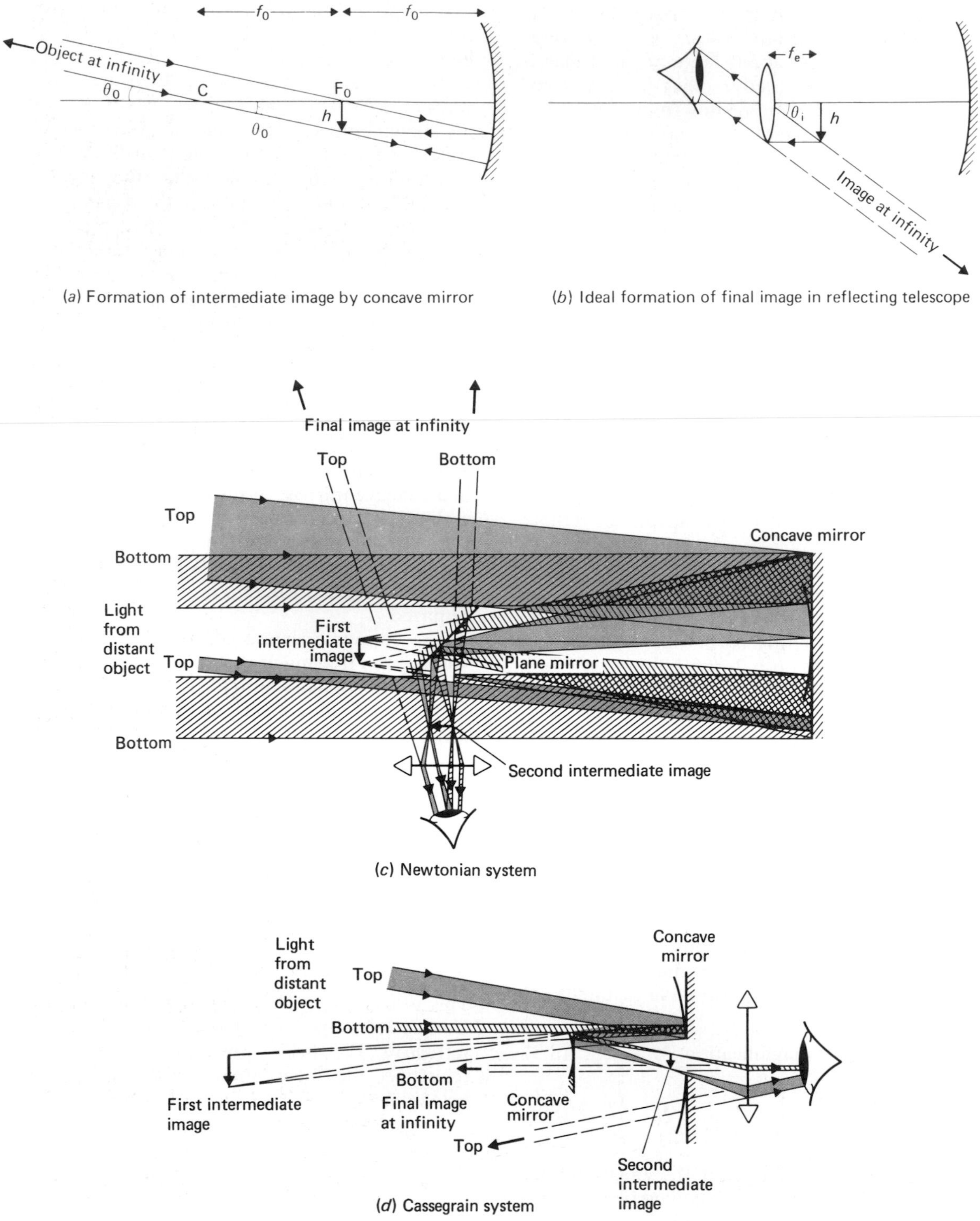

f_0
f_0
Object at infinity
θ_0
C
F_0
h
θ_0
(a) Formation of intermediate image by concave mirror
f_e
θ_i
h
Image at infinity
(b) Ideal formation of final image in reflecting telescope
Final image at infinity
Top
Bottom
Top
Concave mirror
Bottom
Light from distant object
First intermediate image
Top
Plane mirror
Bottom
Second intermediate image
(c) Newtonian system
Light from distant object
Top
Concave mirror
Bottom
Bottom
First intermediate image
Final image at infinity
Concave mirror
Top
Second intermediate image
(d) Cassegrain system

Fig. 8.10 The reflecting telescope.

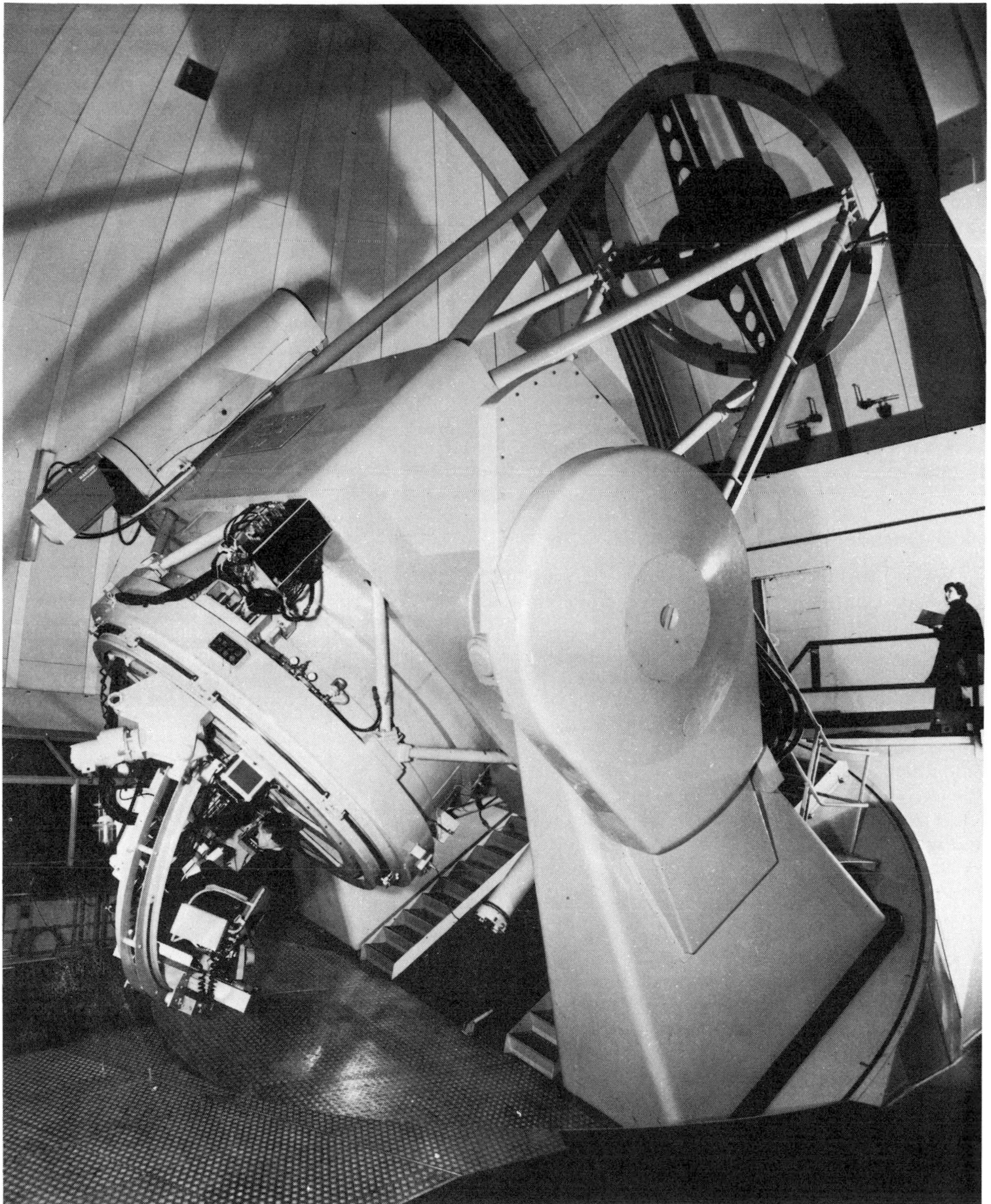

Fig. 8.11 The 2.5 m Isaac Newton reflecting telescope when it was at the Royal Greenwich Observatory, Herstmonceux; to give improved conditions for observation, this is now on the island of La Palma, in the Canaries.

8.4 The magnifying glass

The simplest optical instrument giving an enlarged image of a small object is the magnifying glass, which is merely a converging lens of short focal length. It is probable that crude magnifying glasses were used by the Romans and certain that they were in use by AD 1000; early examples were made from glass beads.

An object that is to be viewed with a magnifying glass must be held between the lens and its focal point; an erect, virtual image is formed as shown in Fig. 8.12(*b*). In the analysis which follows, we shall make the simplifying assumption that the eye is immediately behind the lens. The angular magnification produced by a magnifying glass held some distance from the eye, as is often the case, is smaller and more difficult to analyse.

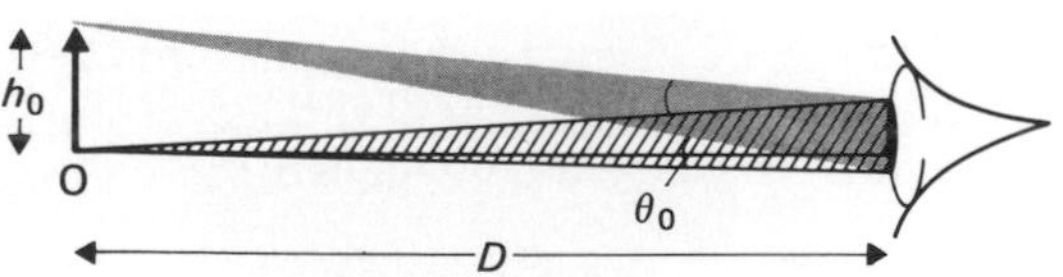

(*a*) Without lens, object must be a distance *D* from observer's eye

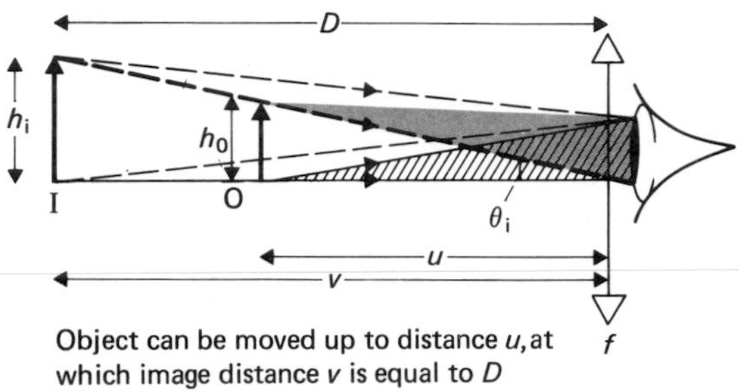

Object can be moved up to distance *u*, at which image distance *v* is equal to *D*

(*b*) Magnifying glass in near-point adjustment

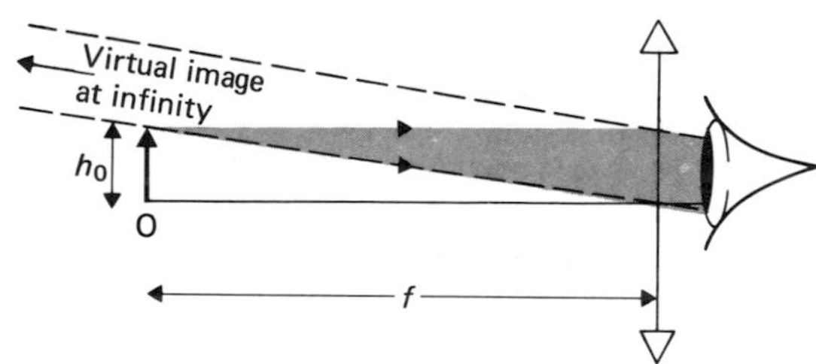

(*c*) Magnifying glass with image at infinity

Fig. 8.12

Near-point adjustment

In Fig. 8.12(*b*) the image is formed at the least distance of distinct vision, *D*, from the eye; in this position the magnifying power will be as large as possible and the magnifying glass is in near-point adjustment. The object is being held closer to the eye than the near-point but it can be seen distinctly because the magnifying glass forms an image at the near-point; the angle subtended at the eye by the image is θ_i. Without the lens the object must be moved further from the eye until it is at the near-point, as shown in Fig. 8.12(*a*), and the angle subtended at the eye decreases to θ_o. Thus, *a magnifying glass produces angular magnification because it enables you to hold the object closer to the eye than would otherwise be possible.*

Now, from Fig. 8.12(*b*), if the angles are small:

$$\theta_i = \frac{h_i}{D} \quad \text{by definition}$$

therefore $$\theta_i = \frac{h_o}{u} \quad \text{by similar triangles}$$

Note that the object could not be seen at O without the aid of the lens because it is nearer to the eye than the least distance of distinct vision. If the object is to be seen distinctly without the lens we must have

$$\theta_o = \frac{h_o}{D} \quad \text{as in Fig. 8.12}(a)$$

Hence, angular magnification

$$M = \frac{\theta_i}{\theta_o} = \frac{h_o/u}{h_o/D}$$

i.e. $$M = \frac{D}{u} \tag{8.11}$$

The distance from the lens, *u*, at which the object must be held to produce an image at the least distance of distinct vision, can be calculated from the lens equation; the image distance will be negative since the image is virtual:

$$\frac{1}{u}+\frac{1}{v}=\frac{1}{f}$$

$$\therefore \quad \frac{1}{u}=\frac{1}{f}-\frac{1}{v}$$

$$=\frac{1}{f}-\frac{1}{(-D)}$$

Hence $$\frac{D}{u}=\frac{D}{f}+\frac{D}{D}$$

i.e. $M=\dfrac{D}{f}+1$ *in near-point adjustment* (8.12)

from Equation (8.11). If the observer is using a magnifying glass to view a detailed object which he is drawing on a piece of paper, it is convenient to have the image at the near point; the image and the drawing will be the same distance from his eye. The magnifying glass is said to be in *normal adjustment.*

Far-point adjustment

The magnifying glass could be used in what is termed *far-point adjustment,* with the image at infinity; this situation is shown in Fig. 8.12(*c*). We have again:

$$M=\frac{\theta_i}{\theta_o}$$

$$=\frac{D}{u}$$

In this case, where the image distance is equal to $(-\infty)$, the lens formula gives us:

$$\frac{1}{u}+\frac{1}{v}=\frac{1}{f}$$

$$\therefore \quad \frac{1}{u}=\frac{1}{f}-\frac{1}{(-v)}$$

$$=\frac{1}{f}+\frac{1}{\infty}$$

i.e. $u=f$ since $1/\infty$ is zero

Hence $M=\dfrac{D}{f}$ *in far-point adjustment* (8.13)

Thus, to obtain a high angular magnification a lens of short focal length, and therefore small radius of curvature, must be used. In fact, the angular magnification is limited to about ×10 by aberrations, and to obtain more magnification and greater resolving power a compound microscope must be used.

Q 8.9
A man, with a least distance of distinct vision of 250 mm, uses a lens of focal length 100 mm as a magnifying glass. What will be the angular magnification:

(*a*) in near-point adjustment?
(*b*) in far-point adjustment?

8.5 The compound microscope

The compound microscope was developed in about 1610 to enable a more detailed examination to be made of accessible objects than was possible with either the eye or a magnifying glass. The arrangement of lenses is similar to that in a telescope, as we see in Fig. 8.13, but because the object is to be close to the objective lens, the theory is rather different.

Each of the lenses shown in Fig. 8.13 will, in fact, be compound to reduce aberrations, but they are shown as single lenses for simplicity. The angular magnification can be made very large but it is, in practice, limited by diffraction effects; the oil-immersion objective, described later, can be used to increase the resolving power of the instrument.

The objective lens forms a real, inverted image as in a telescope. However, to increase the angular magnification the object distance must be small with respect to the image distance; the intermediate image is enlarged, not diminished, as shown. The image distance is restricted by the length of the microscope, and thus an objective of very short focal length is required. The eyepiece, used as a magnifying glass, further increases the angular magnification although, unlike a telescope, most of the magnification in a microscope arises at the objective. In Fig. 8.13(*a*), the final image is formed at the near-point, and the instrument is in normal, or near-point, adjustment; in Fig. 8.13(*b*) it is formed at infinity, and the

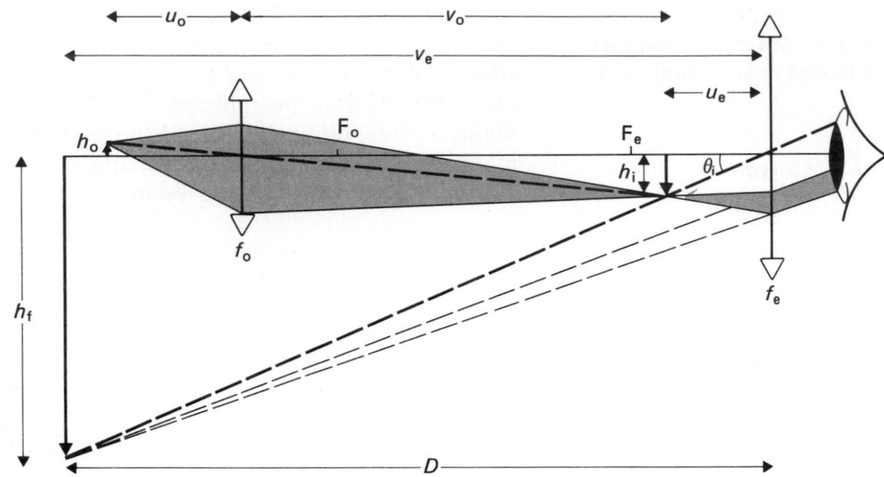

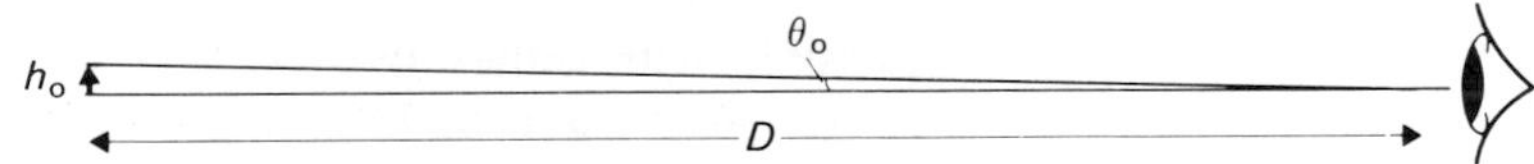

(*a*) A microscope in normal (near-point) adjustment

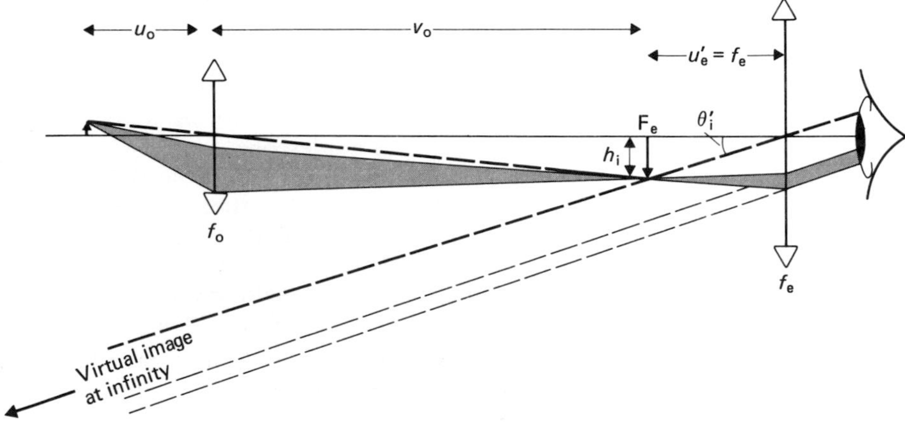

(*b*) A microscope in far-point adjustment

Fig. 8.13 The compound microscope.

instrument is in far-point adjustment. We assume, in our analysis, that the eye is placed close to the eyepiece, as in our discussion of the magnifying glass.

Microscope in near-point adjustment

In Fig. 8.13(a) we see that the angle subtended at the eye by the image, θ_i, can be expressed as:

$$\theta_i = \frac{h_f}{D} \qquad \text{if } \theta_i \text{ is small}$$

i.e.
$$\theta_i = \frac{h_i}{u_e} \qquad \text{by similar triangles}$$

Now, the intermediate image is acting as a real object for the eyepiece. Therefore

$$\frac{1}{u_e} + \frac{1}{v_e} = \frac{1}{f_e}$$

$$\frac{1}{u_e} = \frac{1}{f_e} - \frac{1}{v_e}$$

$$= \frac{1}{f_e} - \frac{1}{(-D)}$$

i.e.
$$\frac{D}{u_e} = \frac{D}{f_e} + 1 \qquad (8.14)$$

The size of the intermediate image, h_i, can be calculated from the object and image distances for the objective:

$$\frac{h_i}{h_o} = \frac{v_o}{u_o} \qquad (8.15)$$

Now, the largest angle which the object can subtend at the naked eye and still be clearly seen, θ_o, must be when it is held at the least distance of distinct vision, D. Thus, angular magnification

$$M = \frac{\theta_i}{\theta_o}$$

$$= \frac{h_i/u_e}{h_o/D}$$

i.e.
$$M = \left(\frac{h_i}{h_o}\right)\left(\frac{D}{u_e}\right)$$

Hence, from Equations (8.14) and (8.15), we have:

$$M = \frac{v_o}{u_o}\left(\frac{D}{f_e} + 1\right) \qquad (8.16)$$

This equation should be compared with Equation (8.12), which gives the angular magnification for a magnifying glass in near-point adjustment. We see that the angular magnification of a microscope could be written:

total angular magnification

$$= \left(\begin{array}{c}\textit{linear magnification}\\ \textit{produced by objective}\end{array}\right) \times \left(\begin{array}{c}\textit{angular magnification}\\ \textit{produced by eyepiece}\end{array}\right)$$

Microscopes are usually used in near-point adjustment when we wish to draw what we see; the image viewed and the drawing are likely to be the same distance from the eye. Hence, for a *microscope*, this is termed *normal adjustment.*

Microscope in far-point adjustment

Fig. 8.13(b) shows a microscope in far-point adjustment, with the final image at infinity; the eyepiece has been moved further from the objective, so that the intermediate image is at the focal point of the eyepiece. We see that the angle subtended at the eye by the final image, at infinity, is again the same as that subtended by the intermediate image:

$$\theta_i' = \frac{h_i}{u_e'} \qquad \text{if } \theta_i' \text{ is small}$$

But
$$u_e' = f_e$$

and hence
$$\theta_i' = \frac{h_i}{f_e} \qquad (8.17)$$

The position of the intermediate image with respect to the objective is unchanged, and hence its height is still h_i.

We have, as before:

$$\frac{h_i}{h_o} = \frac{v_o}{u_o} \qquad (8.15)$$

Now, the largest angle subtended by the object in a position where it can be clearly seen, θ_o, is given by:

$$\theta_o = \frac{h_o}{D} \qquad (8.18)$$

where D is the least distance of distinct vision.

Thus, the angular magnification is given by:

$$M = \frac{\theta_i'}{\theta_o}$$

$$= \frac{h_i/f_e}{h_o/D} \quad \text{from Equations (8.17) and (8.18)}$$

i.e. $$M = \left(\frac{h_i}{h_o}\right)\left(\frac{D}{f_e}\right)$$

Therefore $$M = \left(\frac{v_o}{u_o}\right)\left(\frac{D}{f_e}\right) \qquad (8.19)$$

Again, comparison with Equation (8.13) shows that this formula may be written:

total angular magnification

$$= \left(\begin{array}{c}\textit{linear magnification}\\ \textit{produced by objective}\end{array}\right) \times \left(\begin{array}{c}\textit{angular magnification}\\ \textit{produced by eyepeice}\end{array}\right)$$

The angular magnification of the microscope in far-point adjustment is slightly smaller than in near-point adjustment. However, it is less tiring to use the instrument like this for long periods since the eye can be completely relaxed; to examine some particular detail, the instrument can always be changed into near-point adjustment by adjusting the focusing, which alters the spacing between the lenses.

Microscope calculations

near-point adjustment $$M = \frac{v_o}{u_o}\left(\frac{D}{f_e}+1\right) \qquad (8.16)$$

far-point adjustment $$M = \frac{v_o}{u_o}\left(\frac{D}{f_e}\right) \qquad (8.19)$$

In order to calculate the angular magnification of a microscope it is necessary to calculate first the object and image distances for the objective.

Consider a microscope with an objective of focal length 3 mm and an eyepiece of focal length 12 mm. Suppose the object distance for the objective is 3.1 mm.

Then $$\frac{1}{u_o}+\frac{1}{v_o}=\frac{1}{f_o} \quad \text{for the objective lens}$$

$$\therefore \quad \frac{1}{v_o}=\frac{1}{f_o}-\frac{1}{u_o}$$

$$= \frac{1}{+3}-\frac{1}{+3.1}$$

$$= \frac{0.1}{+9.3}$$

i.e. $$v_o = 93 \text{ mm}$$

Hence $$\frac{v_o}{u_o} = \frac{93}{3.1}$$

$$= 30$$

In near-point adjustment we have:

$$M = \frac{v_o}{u_o}\left(\frac{D}{f_e}+1\right) \quad \text{from (8.16)}$$

$$= 30\left(\frac{250}{12}+1\right) \quad \text{if } D = 250 \text{ mm}$$

i.e. $$M = 655$$

In far-point adjustment the angular magnification will be given by:

$$M = \frac{v_o}{u_o}\left(\frac{D}{f_e}\right) \quad \text{from (8.19)}$$

$$= 30\left(\frac{250}{12}\right)$$

i.e. $$M = 625$$

In more complicated calculations u_o and v_o can be calculated, for instance, by knowing the position of the intermediate image and the spacing between the lenses.

Q 8.10
A microscope has an objective of focal length 3.5 mm and an eyepiece of focal length 10 mm. If the microscope views an object 5 mm below the objective lens, what will be the angular magnification:

(*a*) in near-point adjustment?
(*b*) in far-point adjustment?

What will be the spacing between the lenses in each case?

The resolving power of a microscope

Calculations of resolving power depend on more factors for a microscope than for a telescope. In a microscope the object studied is illuminated by a small source and so the light scattered from two adjacent points on it may be *coherent* and interfere; the lack of resolution caused by this effect is in addition to that caused by diffraction at the objective. It is as though two adjacent points act like the slits in a Young's slits experiment.

The maximum useful angular magnification of a microscope is set by the limit of resolution; there is no point in further magnification if the object is merely seen to be indistinct because of lack of resolution in the image. Similarly, good resolution at the objective is wasted unless there is sufficient angular magnification to enable those points made distinct by the objective to be seen as distinct by the eye or camera.

Consider two points on an object O_1 and O_2, a distance x apart, as shown in Fig. 8.14(a). If the objective subtends an angle 2α at the object, the path difference between rays from O_1 and O_2 striking the edge of the objective is $x \sin \alpha$ as shown in diagram (b); if the space between object and lens is filled with a medium of refractive index n the optical path difference will be $nx \sin \alpha$. Thus rays from O_1 will travel an optical distance $nx \sin \alpha$ *further* than rays from O_2 to the bottom of the lens, and $nx \sin \alpha$ *less* to the top.

Ernst Abbe showed that the objects would be resolved if:

$$2nx \sin \alpha = \lambda$$

$$x = \frac{\lambda}{2n \sin \alpha}$$

$n \sin \alpha$ is called the *numerical aperture* of the system.

Theoretically, for a lens in air ($n = 1$), if the aperture of the objective were very much greater than the distance from it to the object, so that α became equal to 90°, a limit of resolution of $\lambda/2$ could be attained; two points a half wavelength apart on the object could be distinguished. To approach this standard of resolution objective lenses are required with a focal length as small as 0.3 mm, and very oblique light will be required to enter the objective.

To obtain oblique light, wide-angle illumination by a source and *condenser lens* is required, as shown in

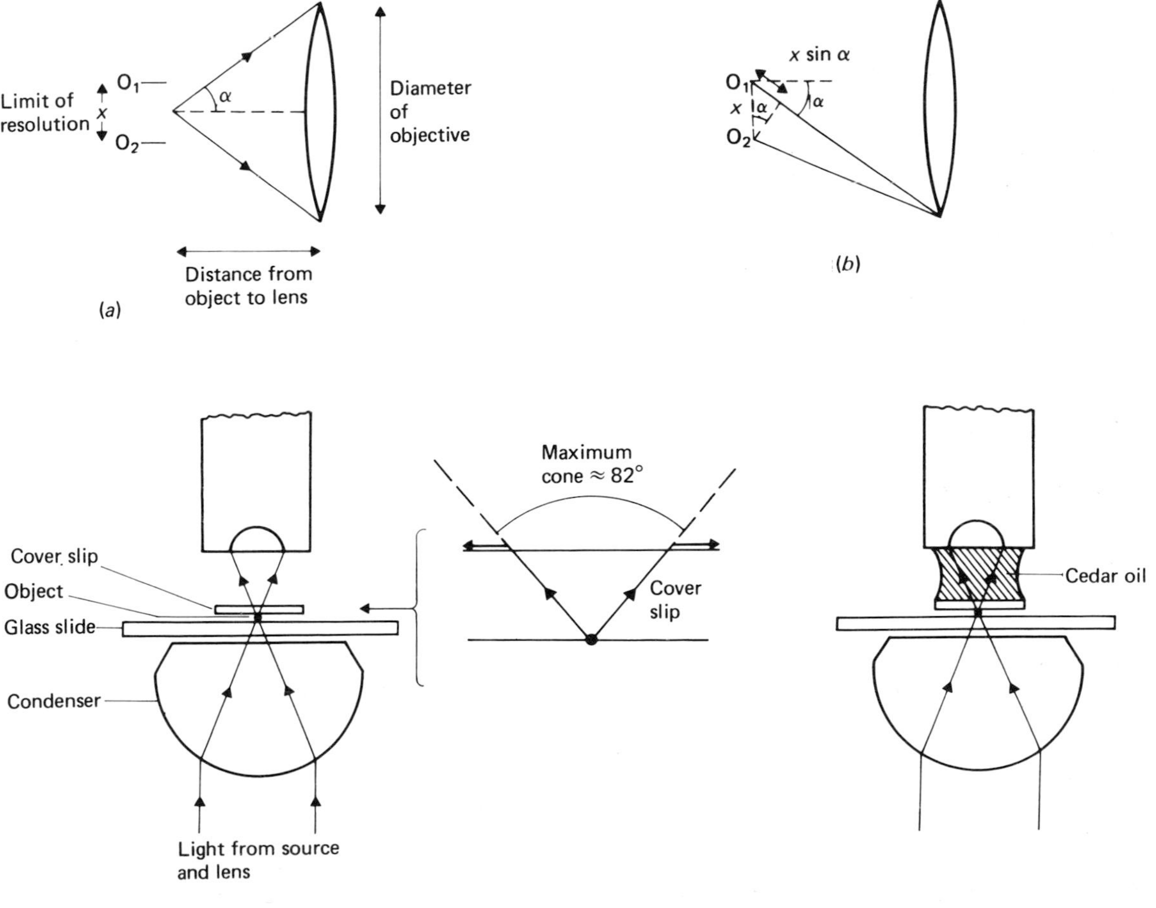

(c) A conventional microscope system

(d) An oil-immersion objective

Fig. 8.14

Fig. 8.14(*c*); good illumination is also essential to maintain the brightness of the image despite very high magnification.

The brightness of the image is also limited by refraction at the upper surface of the cover slip, and for glass of refractive index 1.5 the useful cone of rays from the subject is restricted to 82°, as shown. If a liquid of refractive index approaching that of the cover slip is placed between it and the objective, as shown in diagram (*d*), the cone angle can be increased considerably. *Oil-immersion objectives* also decrease the limit of resolution by increasing the numerical aperture of the system, as shown by Abbe's equation above.

Finally, the resolution of a microscope can be increased by decreasing the wavelength of the light; if ultraviolet light is used, the limit of resolution may be halved, depending on the wavelength used. In *ultraviolet microscopy* quartz lenses are required as glass is opaque to the wavelengths used. In *electron microscopes* the fact that electrons can have a wave nature, discussed in Chapter 11, enables them to form images; because their wavelength is very much less than that of light, the resolution can theoretically be increased by a factor as large as 100 000, though present techniques cannot take full advantage of the shortness of electron wavelengths.

8.6 The spectrometer

A *spectrometer* is an optical instrument used for measuring accurately the directions of light rays, after they have been refracted by prisms or have been diffracted at gratings. Student instruments can measure angles to an accuracy of about one minute of arc, and good industrial spectrometers achieve accuracies of less than a second of arc.

When used with a prism on the table, a spectrometer can be used for measuring the angle of the prism and its angle of minimum deviation; hence the refractive index of the glass can be determined using Equation (6.20). It can also be used for observing and measuring spectral lines.

Nowadays, *grating spectrometers* are more common than *prism spectrometers*. Accurate determinations of wavelength can be made from measurements of the angles at which the spectral lines are produced, provided that the spacing of the grating is also known; this can be determined, if necessary, by using other light of accurately known wavelength. Once the spectra of different elements and compounds have been determined, spectrometers can be used to analyse substances; this is their most common industrial application.

In many applications, it is more convenient to replace the eyepiece with a camera; this will be especially useful when the spectral lines are very faint, because cameras are integrating devices and the brightness of the picture can be increased by lengthening the exposure. Sometimes the spectrum is scanned with a photo-sensitive detector, whose output can be fed into a pen recorder or computing system. As photocells have a narrow wavelength range to which they are sensitive, thermopiles are sometimes used instead, detecting the heating effect of the spectrum over a much wider wavelength range. In each of these modifications, the instrument is known as a *spectrograph*, because it keeps a permanent record.

A less expensive way of viewing a spectrum is through a direct-vision spectroscope. By combining two crown glass prisms, of lower refractive index and dispersive power, with one flint glass prism it is possible to arrange that the deviations cancel at one wavelength, usually yellow, but that there is still some dispersion; a virtual image of a spectrum at infinity is observed by the eye, sometimes against a calibrated wavelength scale.

Applications of spectroscopy are discussed further in Chapter 11.

The optical system of a spectrometer

The optical system of a spectrometer is shown in Fig. 8.15. The function of the collimator is to illuminate gratings and prisms with parallel light; the length from slit to achromatic lens can be carefully adjusted so that it is equal to the focal length of the lens. The slit is of adjustable width, so that the amount of light admitted to the system can be varied; the spectral line is an image of the slit, so for accurate work the slit width should be reduced until the lines to be observed are just visible. The edges of the slits are usually knife edges and care must be taken not to touch them together. The source should be broad enough to illuminate the full width of the collimator lens evenly; sometimes a small prism enables light from a standard source to be admitted to the bottom half of the slit so that the two spectra, which are formed above each other, can be compared.

The function of the table is to support the grating or prism, and clamps are supplied to

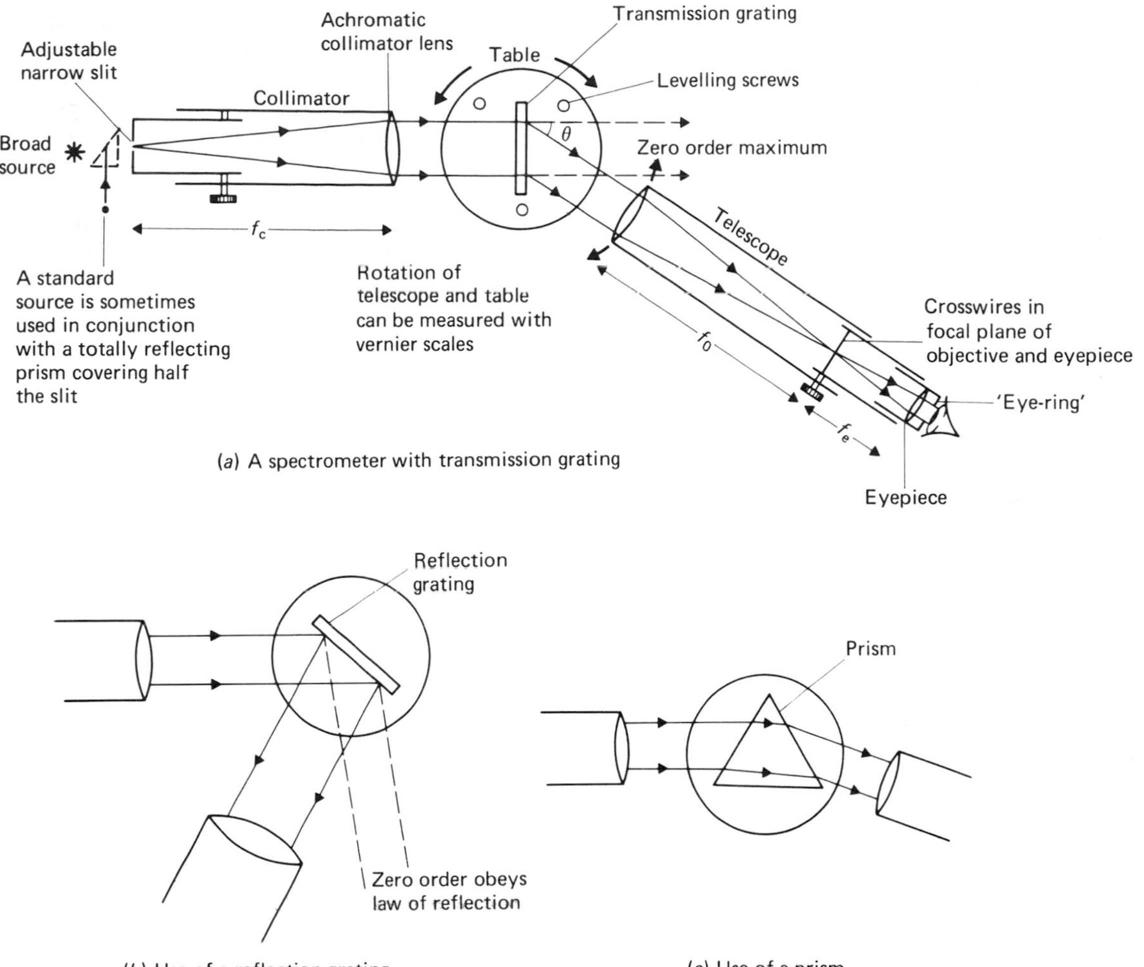

(*a*) A spectrometer with transmission grating

(*b*) Use of a reflection grating

(*c*) Use of a prism

Fig. 8.15

attach them to its top; it can rotate about a vertical axis, and then be clamped in position. Levelling screws are provided so that the refracting edge of the prism, or the grating lines, are always parallel to the axis of rotation.

The light due to one particular spectral line entering the telescope is parallel, as though from a distant line object, and the objective forms an image in its focal plane; there is also a set of crosswires in that plane. The intermediate image, and the crosswires, are then viewed through a multiple lens eyepiece; for simplicity, only one lens is shown in Fig. 8.15. The distance from eyepiece to crosswires is made equal to the eyepiece focal length, and the distance from crosswires to objective is adjusted to be equal to the focal length of the objective.

Both the collimator and telescope can be rotated round the same vertical axis as the table, and their positions can be adjusted by fine screws. Once clamped, their angular positions are measured accurately with vernier scales.

The optical system of an industrial spectrometer will be the same in principle, but the detailed arrangement may be different. In chemical analysis, the substance being analysed must be made to emit or absorb radiation. Some methods of doing this are shown in Fig. 8.16.

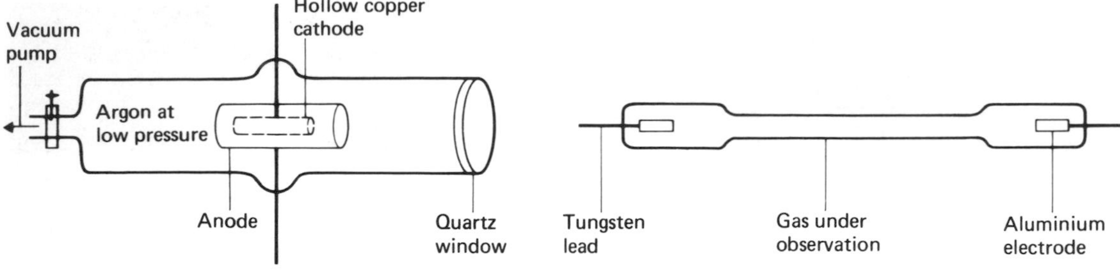

(a) Hollow cathode lamp

(b) Gas discharge tube

Fig. 8.16 The production of industrial spectra.

8.7 The camera

In a camera a converging lens system forms a real, inverted image on light-sensitive emulsion; the paper can subsequently be treated in such a way that a permanent image of the object is formed upon it. We will consider each optical component of the camera in turn, and then apply the particular problem of depth of focus to photography.

Camera components

As will be evident from any photograph, the field of view of a camera is large, frequently about 30°; this is very much larger, for instance, than the 1° field of view of a typical microscope. Thus, comparatively oblique rays must pass through the lens and this makes the reduction of aberrations more difficult. The cheaper type of fixed-focus camera usually has a convex meniscus lens, as shown in Fig. 8.17(*a*); this is *stopped down* so that only the central portion is used and quite good results can then be obtained. The shape of the lens reduces curvature of the field and use of the *stop* reduces spherical aberrations; the *aperture*, or diameter, of the lens that is used is usually no more than one-eleventh of the focal length. Frequently a cemented doublet is used instead of a single meniscus lens to reduce chromatic aberrations. In more expensive cameras combinations of lenses are used which are too complex to discuss here in detail; the Zeiss 'Tessar' system is shown in Fig. 8.17(*b*).

The aperture of the lens is determined by an iris diaphragm, so that the amount of light entering the camera under different conditions may be controlled. This diaphragm can be seen in the early camera shown in Fig. 8.17(*c*); it consists of metal plates that slide over each other to reduce or enlarge the opening. Instead of measuring aperture as a distance, it is usual to use the *relative aperture* instead:

$$\text{relative aperture} = \frac{\text{diameter of aperture}}{\text{focal length}}$$

The area of the aperture determines the amount of light entering the camera, and an increase in the focal length causes an increase in the size of the image; hence the relative aperture determines the brightness of the image. When discussing camera adjustments we often talk about the *f-number*, or stop-number:

$$f\text{-number} = \frac{1}{\text{relative aperture}} = \frac{\text{focal length}}{\text{diameter of aperture}}$$

Thus an f-number of 2, written $f/2$, means that the aperture diameter is half the focal length; an f-number of 4 means that the aperture diameter is one-quarter the focal length.

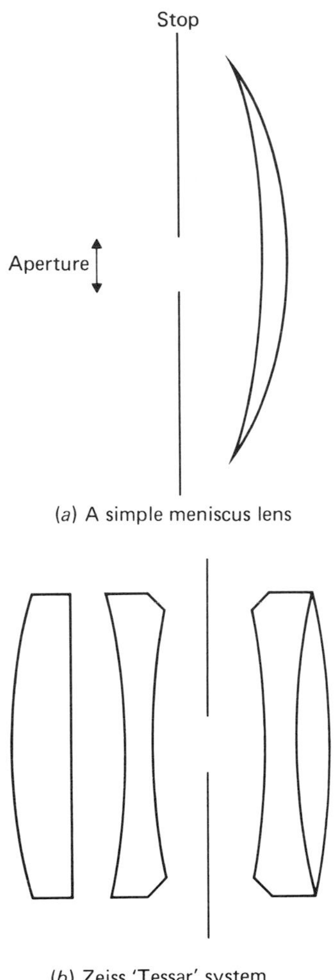

(*a*) A simple meniscus lens

(*b*) Zeiss 'Tessar' system

exposure-time adjustment
adjustment to focus image
iris diaphragm
aperture adjustment

(*c*) A Six-20 Kodak Junior camera dating from about 1935

Fig. 8.17

For a lens of given focal length, the increase in f-number from 2 to 4 corresponds to a reduction of the aperture diameter by a factor of 2, and of the aperture area by a factor of 4. The aperture area, controlling the amount of light entering the camera, is thus inversely proportional to the square of the f-number.

A sufficient choice of aperture-stops is provided when successive stops change the aperture area by a factor of 2. The following table gives the appropriate values for f^2, starting from 4, and hence the full range of f-numbers available on a good camera; some of the square roots are approximations.

f^2	4	8	16	32	64	128	256	512	1024
f	2	2.8	4	5.6	8	11	16	22	32

← larger aperture   higher f-number →

The f-numbers available on the camera shown in Fig. 8.17(*c*) are marked under the lens.

The time for which light enters the camera, or *exposure time*, is controlled by a *shutter*. This may be an iris diaphragm, like that used to control the aperture; in a multiple lens system it is placed between two lenses and can give exposure times as short as 1/500 s. A more sophisticated system, the *focal plane shutter*,

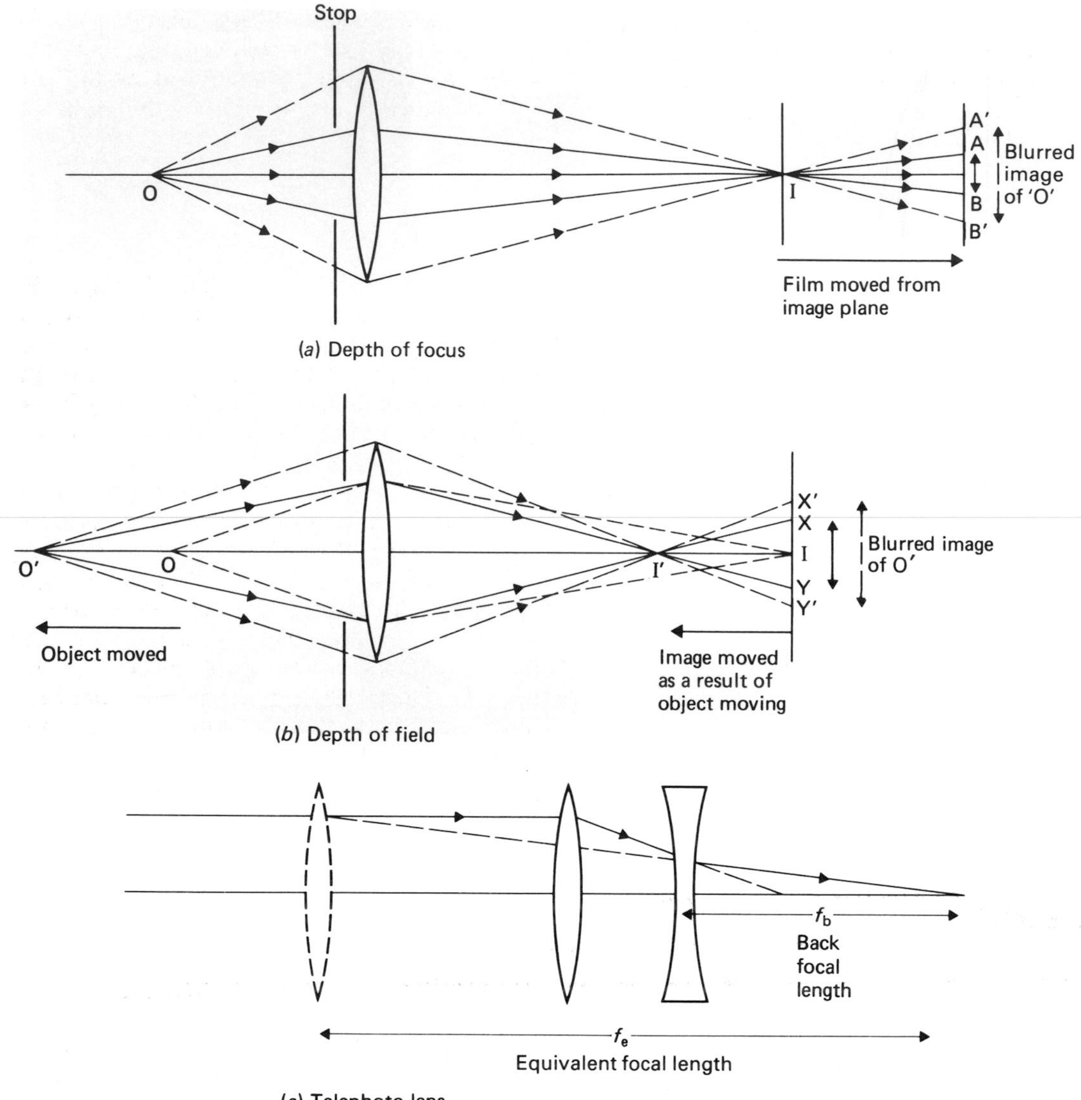

Fig. 8.18

uses an opaque blind immediately in front of the film; when the picture is taken a slit, of variable width, in the blind moves across the film admitting light to each part of the film as it does so. Exposure times as short as 1/1000 s can be obtained by this method.

To focus the image onto the film the lens can be moved backwards and forwards. Older cameras had the type of bellows slide shown in Fig. 8.17(*c*), but more modern cameras have a lens which can be screwed in and out. There are four ways in which the operator may be able to check that the object is in focus:

(1) by estimating how far away it is and using figures marked on the lens casing;
(2) by using a coupled range-finder which utilizes a 'stereoscopic vision' effect;
(3) there may be a ground glass screen which can be inserted instead of the film, and on which the image can be observed. This is more common in large 'plate' cameras used by professional photographers;

(4) in a single lens reflex camera there is a mirror in front of the film, inclined at 45° to it, which reflects the image on to a viewfinder on top of the camera; the optical path to the viewfinder is identical to that to the film. When the shutter is operated the mirror is moved simultaneously out of the way.

The distance from lens to film is always very near to the focal length, unless objects are very close indeed to the camera.

Depth of focus

The *depth of focus* of a camera is the distance the film can be moved without making the image on it indistinct. The *depth of field* is the distance that an object can be moved to or from the camera without the image on the film becoming indistinct; in practice this distance is also often referred to as the depth of focus. These distances are shown in Fig. 8.18.

In both cases an extended image, A′B′ or X′Y′ is formed on the screen; this will be a patch of light instead of a point. The diagrams show that if the lens is stopped down the size of these patches is reduced in proportion to the reduction in aperture. When a photograph is taken we frequently require objects at different distances from the camera to be in focus at the same time; this is strictly speaking not possible, but if the lens is stopped down the blurring may not be noticeable. If the f-number is increased in this way, however, less light will enter the camera and the exposure will need to be longer; this may prove unsatisfactory if the object is moving. Selecting the correct f-number for a photograph is always a matter of compromise.

Finally we must ask what degree of blurring is acceptable. We have seen, in Section 8.1, that the eye can resolve two objects if their angular separation is more than 1 minute of arc (3×10^{-4} rad). At a distance of 0.25 m two objects on a photograph will have an angular separation of 1 minute, if the distance between them, x, is given by:

$$\begin{aligned} x &= 0.25\times\theta \\ &= 0.25\times(3\times10^{-4}) \\ &\approx 7\times10^{-5}\text{ m} \\ &\approx 0.07\text{ mm} \end{aligned}$$

Thus, as two objects 0.07 mm apart on a photograph can only just be resolved, blurring of up to this amount will be unnoticed by the eye. The resolution is also limited by the grain size in the film.

Telephoto lenses

When a convex lens forms a real image of a distant object the size of the image is proportional to the focal length, as the image is formed in the focal plane. Hence, for photographing distant objects, lenses of long focal lengths are required, which are cumbersome to use and require a long camera.

A telephoto lens provides the same focal length but with a more compact arrangement, as shown in Fig. 8.18(c). The length of camera required is determined by the *back focal length*, but the equivalent *focal length* can be a great deal longer. Telephoto lenses are usually of fixed focal length and a variety of interchangeable telephoto lenses is required for maximum flexibility; nowadays, more complex zoom lenses are used instead.

The *telephoto magnification* is defined as the ratio of equivalent focal length to back focal length.

Bibliography

Fuller, G.R. (1970) *Optical Systems.* Heinemann: a brief but thorough introduction to different types of instruments.

Jenkins, F.A. and White, H.E. (1981) *Fundamentals of Optics.* McGraw-Hill, Chapter 10; a thorough treatment of optical instrument theory.

PSSC (1968) *College Physics.* Raytheon, Chapter 10; a well-illustrated account of the physics of optical instruments.

Readhead, A.C.S. Radio astronomy, by very long baseline interferometry; *Scientific American* Volume 246 (b), June 1982; examines the achievements of radio telescopes using interferometry techniques when the baseline is thousands of kilometres long.

Part 3

Light, Waves and Matter

9

Light: waves or particles?

9.1 The birth of geometrical optics

In the next two chapters, we shall examine the way in which our understanding of the nature of light has developed over the years, and the experimental basis for that knowledge. At the same time, there is a good opportunity to see something of the methods of scientists, always testing existing theory against new facts, and always facing the possibility that one crucial experiment could challenge the basis of all their thinking to date. This chapter will look mainly at the wave–particle debate around the time of Newton, and its supposed resolution in the nineteenth century; in the next chapter, we shall discover that even as the debate seemed to be resolved, fresh facts re-opened the whole question.

Although it is usual to think of the study of light as dating from the time of Newton, a great deal was in fact discovered by the ancient Greeks. Euclid, in the third century BC, used the angle law of reflection to construct theorems about mirrors.

The fact that light rays change direction upon passing from one medium to another had been noted even earlier than the time of Euclid, in this case by Aristotle in the fourth century BC. He observed that an oar placed partly in air and partly in water appeared to be bent at the surface in a similar way to that shown in Fig. 9.1.

Fig. 9.1 The author reflects on waves.

The earliest quantitative study of refraction was undertaken in the first half of the second century AD by Ptolemy. In addition to stating the laws of reflection in the form in which we now know them, he also produced some well-documented experimental results on the refraction of light. He graduated a circle by dividing it into 360 degrees and immersed it partly in water as shown in Fig. 9.2. He used two

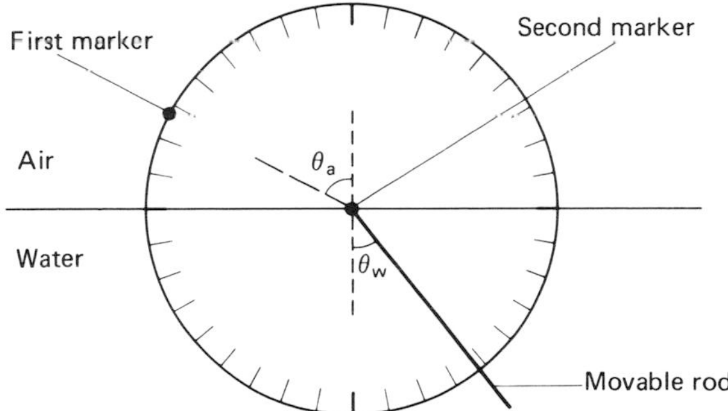

Fig. 9.2 Ptolemy's experiment to investigate refraction.

markers; the first placed at a point on the circumference, and the second at the centre of the circle. The movable rod was then adjusted so that it appeared to be in line with the two markers. He was then able to measure the angles θ_a and θ_w shown in Fig. 9.2. His results are set out in Table 9.1:

Table 9.1

θ_a/degree	10	20	30	40	50	60	70	80
θ_w/degree	8	$15\frac{1}{2}$	$22\frac{1}{2}$	29	35	$40\frac{1}{2}$	$45\frac{1}{2}$	50

Ptolemy, *Optics*, Book V.

Ptolemy apparently made little attempt to analyse the results and, in any event, his results were not accurate enough to give the law now accepted; he appeared satisfied with having obtained a series of readings.

The unnatural feature of the results quoted is that for equal changes in θ_a the *difference* between values of θ_w decrease by $\frac{1}{2}°$ each time. This seems rather contrived, being also completely fallacious, and raises the interesting question as to whether the results were 'fudged' to give the numerical simplicity that was fashionable at the time!

Further research on refraction was carried out by Alhazen, an eleventh century Arabian scholar who wrote *Optical Thesaurus*, a text which remained a standard work until the seventeenth century. He discovered that the ratio θ_a/θ_w was approximately constant for small angles of incidence.

The laws of refraction, as we know them, are traditionally ascribed to Snell, but were probably originally discovered experimentally by Hariot (1560–1621). He placed a prism ABC on a piece of paper with the face AB perpendicular to a line NOR drawn on the paper, as shown in Fig. 9.3. He looked along the line RO and placed a pin P in such a position that it appeared to be in line with RO. From his knowledge of the geometry of the prism, he could calculate the angles θ_a and θ_g. Repeating the experiment with prisms of different geometry, he was able to obtain a series of results and he found that:

$$\frac{\sin \theta_a}{\sin \theta_g} = \frac{1}{0.63} = 1.59.$$

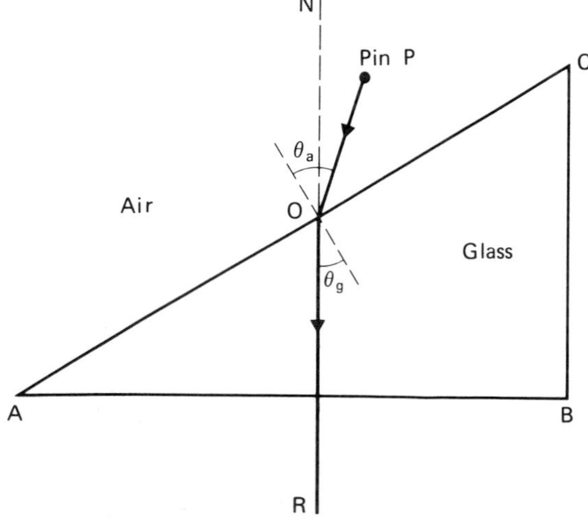

Fig. 9.3 Hariot's experiment to investigate refraction.

This result shows good agreement with modern determinations for most types of glass, which have refractive indices between 1.5 and 1.9.

Willebrord Snell (1591–1626), who was Professor of Mathematics at Leiden, derived the same result in 1621, also from experimental results. Descartes, who re-stated and published the law in 1637, is responsible for attaching Snell's name to it. His own derivation, however, was a theoretical one using the same ideas that Newton used later in his corpuscular theory.

The Rainbow

As early as the fourteenth century the rainbow was investigated in terms of multiple reflections and refractions of light. It was, however, Descartes in his treatise, *Les Meteores*, who provided an explanation for the position of the first and second bows; Fig. 9.4 shows an illustration from this treatise which was published in 1637 as a supplement to the *Discours de la Methode*.

Although primarily a philosopher and mathematician, Descartes showed in his analysis of the problem a willingness both to observe and to experiment. Observing that a rainbow occurred whenever drops of water were illuminated by the sun, he concluded that it must be caused by the way in which the rays of light fell upon, and passed through, the drops. He also observed that the rainbow's appearance did not depend upon the size of the raindrops, of which there were a great variety, and he therefore experimented with a large glass sphere filled with water.

Descartes' investigation of the glass sphere provided an explanation of the rainbow in terms of internal reflection within the raindrop, as shown in Fig. 9.4. He was unable to explain the formation of colours which, as Newton was later able to show, depend upon the variation in the refractive index of water for different colours of light.

9.2 Fermat's least-time principle

Snell's law provided physicists with a means of predicting how light behaved when it passed

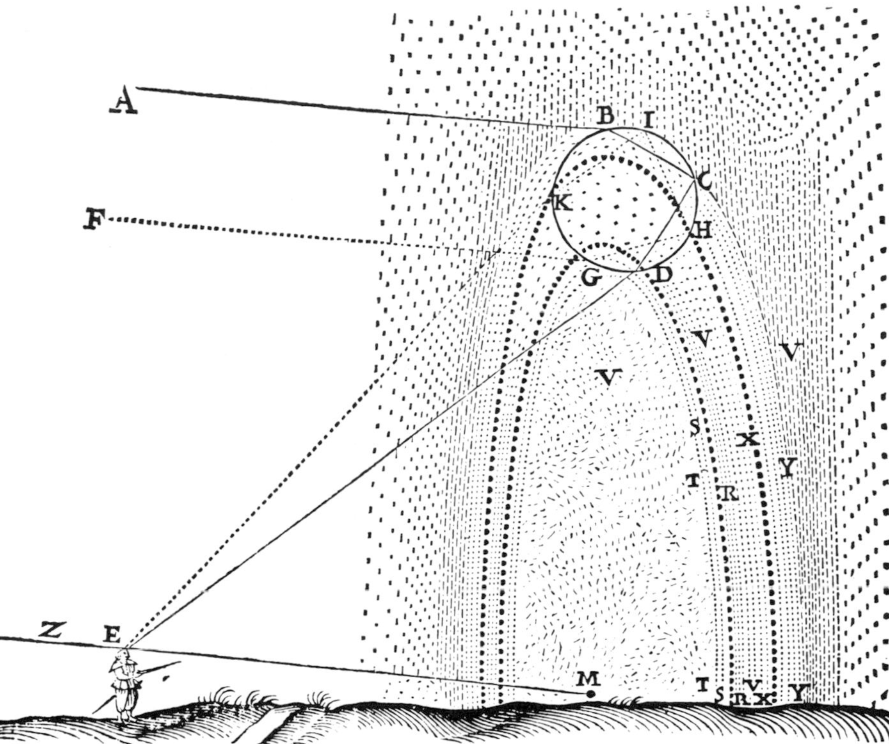

Fig. 9.4 Descartes' theory of the formation of a rainbow. Descartes (1637) *Les Meteores.*

from one medium to another, but it gave little indication as to why this relationship was obeyed. We shall see later how this question can be related to the question of the nature of light, but here we take a brief look at a different approach developed by Fermat.

Hero of Alexandra, who probably lived in the second century AD, originally advanced the hypothesis that light travels by that path which involves the shortest distance. This hypothesis can be applied successfully to the problem of the reflection of light in a mirror, and thus the statement that *the angle of reflection is equal to the angle of incidence* is equivalent to the statement that *the light travels from P to Q via the mirror in the shortest possible distance.* It is easy to see that this result cannot be true in the case of refraction and, in about 1650, Fermat modified Hero's principle to say that *light travels by the path involving the shortest time* instead of the shortest distance.

If we are to consider the time taken for a ray of light to pass from one point to another, we must make allowance for the medium through which the light passes as we have seen, for example, that light travels more slowly in glass than in air.

Now, although the shortest time between two points on a plane situated in one medium must necessarily be that for a straight line path, this is not the case where two media are involved. If you are standing near the bank of a lake and someone is drowning in the water a little way out from the shore much further down the lake, you will not, or should not, swim in a straight line from where you are to the drowning person. Instead, you should run along the bank until you are almost opposite him and then swim out, since you swim much more slowly than you run.

In the same way, light will travel further in the less dense medium, than it would on a 'straight line route', in order to travel less far in the more dense medium.

A number of consequences follow from this principle. First, we can justify the principle of reversibility. If a path takes the least time in one direction, then it will take the least time in the opposite direction also.

Second, we have seen in a lens system how all rays from an object striking a lens surface will pass through one image point; there are thus a number of possible paths from object to image. Fermat's principle suggests, and lens theory confirms, that this can only happen if the time taken along all paths is the same; this fact can be used to design lens systems.

Third, it should be pointed out that in some optical systems the path followed is that corresponding to a maximum, rather than a minimum, time; it is better to say that *paths immediately adjacent to the correct one will have almost identical times.*

In conclusion, Fermat's principle requires that light travels more slowly in glass than in air, and this prediction was to be of great importance in the development of optics.

Q 9.1

A pin is placed 5 cm from the side of a parallel-sided glass block of thickness 10 cm and of refractive index 1.5. A ray of light makes an angle of incidence of 60° with the first side and passes through the block to a second pin, placed 5 cm from the other side of the block. Calculate:

(*a*) the optical path for this ray;
(*b*) the time taken for a light ray to pass between the two pins;
(*c*) the optical path in a straight line from one pin to the other;
(*d*) the time a ray would take to traverse this path. Note that this time is greater than for the actual path in (*b*).

Take the velocity of light in air to be 3×10^8 m s^{-1}.

9.3 The transmission of light

In about 1650, whilst carrying out his researches into the nature of vacua, Torricelli observed that light could be transmitted without a material medium. Aristotle, in the fourth century BC, had also rejected atomism, holding that light is not made up of particles but is some kind of action occurring in a transparent medium filling space. The question of the nature of light and the 'nature of the medium' in a vacuum was to remain in the minds of physicists until it led finally, through the Michelson–Morley experiment, to the theory of relativity.

It was seen, even in the seventeenth century, that the investigation of the speed of light was an important issue and the first problem was whether it had a finite speed or whether light travelled instantaneously. Galileo, in his book, *Two New Sciences* (1638) stated that 'any theory of light must take into account whether light travels with finite or infinite speed' and went on to suggest a means of establishing which of these was true;

> 'I devised a method by which one might accurately ascertain whether illumination, i.e. the propagation of light, is really instantaneous. The fact that the speed of sound is as high as it is, assures us that the motion of light cannot fail to be extraordinarily swift. . . . Let each of two persons take a light contained in a lantern, or other receptable, such that by the interposition of the hand, the one can shut off or admit the light to the vision of the other. Next, let them stand opposite each other at a distance of a few cubits and practise until they acquire such skill in uncovering and occulting their light that the instant one sees the light of his companion he will uncover his own. After a few trials the response will be so prompt that without sensible error the uncovering of one light is immediately followed by the uncovering of the other. Having acquired skill at this short distance, let the two experimenters, equipped as before, take up positions separated by a distance of two or three miles and let them perform the same experiment at night, noting carefully whether the exposures and occultations occur in the same manner as at short distances: if they do, we may safely conclude that the propagation of light is instantaneous; but if time is required at a distance of three miles which, considering the going of one light and the coming of the other, really amounts to six, then the delay ought to be easily observable. . . . In fact, I have tried the experiment only at a short distance, less than a mile, from which I have not been able to ascertain with certainty whether the appearance of the opposite light was instantaneous or not, but if not instantaneous, it is extraordinarily rapid.'

Unfortunately Galileo's experiment could only show that the velocity was either infinite or very large, for the accuracy was clearly not great enough to enable a measurement to be made.

Q 9.2

How long does a ray of light take to travel 6 miles (9.5 km)?

The first piece of evidence on the matter was put forward in 1676 by the Danish astronomer Olaf Roemer. His method was based on the observation of the first satellite of Jupiter, and is shown in Fig. 9.5.

The sun, S, sends out light in all directions, some of it towards Jupiter, J. The first satellite of Jupiter, whose orbit is marked, passes into Jupiter's shadow at X and emerges at Y. The earth's orbit is marked and points A, B, C, D, E and F lie on it. Roemer measured the period of the orbit of the satellite by observing the time between successive disappearances at X, from points such as B and C, and by observing successive reappearances at Y, from points such as E and F.

Suppose that when the earth is at the point B, Jupiter's moon is just disappearing into shadow and suppose that the satellite next disappears about $42\frac{1}{2}$ hours later, when the earth is at C. Now, if the velocity of light is finite, the time taken for the earth to travel from B to C is not exactly equal to the time taken for one revolution of the satellite; because the earth is further from Jupiter when it is at B, the observed time lapse will be decreased by the time taken for light to travel from C to B.

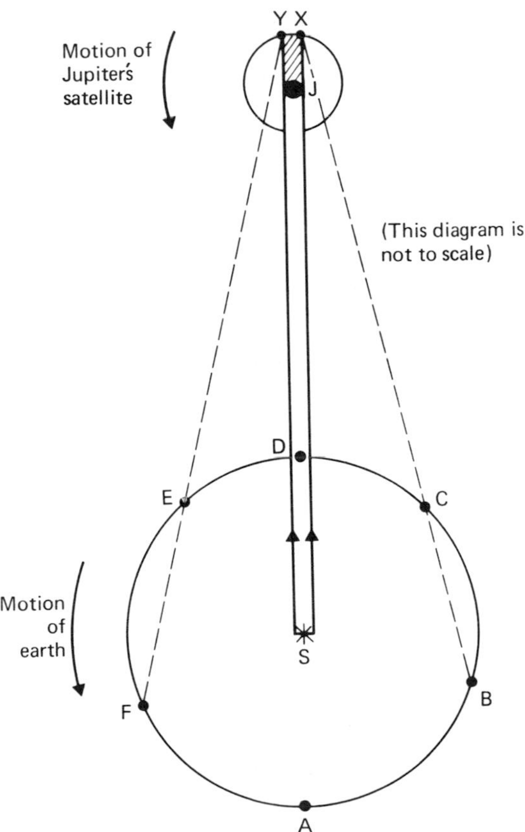

Fig. 9.5 Roemer's measurement of the velocity of light. Figure adapted from Huygens' (1678) *Treatise on Light* (reprinted by Dover).

Similarly, let E be a point from where the satellite is seen to emerge from shadow and F the point where this is next observed. Then the time lapse between E and F will be more than the period of the orbit by the time taken for light to travel from E to F. Thus the evidence for the finite velocity of light would be provided if the period measured when the earth was moving away from Jupiter was greater than that measured when it was moving towards it.

When Roemer first analysed his results on the lines described above, he found no such difference, suggesting that the velocity of light was infinite. However, he improved the accuracy of his observations by allowing the satellite to carry out 40 revolutions between the observations at B and C, E and F. This gave an observable time difference from which he was able to calculate that light would take 22 minutes to travel right across the diameter of the earth's orbit from D to A.

Roemer's work, carried out while he was resident in Paris, was mainly based on observations at the Royal Academy in Paris, of which he was a member. Given that the diameter of the earth's orbit is roughly 2×150 million km, we can calculate the velocity of light from his data.

$$\text{Velocity} = \frac{300 \times 10^6 \text{ km}}{22 \text{ min}}$$
$$= \frac{3 \times 10^{11} \text{ m}}{1320 \text{ s}}$$
$$= 2.3 \times 10^8 \text{ m s}^{-1}$$

The modern value is close to 3×10^8 m s^{-1}—a considerable tribute to Roemer's work.

It was to be almost 200 years before techniques for the accurate measurement of the velocity of light would be developed, but in the meantime Roemer's work showed beyond doubt that it was finite but very high. This also explained why earlier attempts to measure it using light signals flashed between hills had been unsuccessful.

9.4 Early evidence for a wave theory

From what has been said so far, it would appear that by the middle of the seventeenth century, optics had been reduced to a fairly well documented subject. It was known that light travelled in straight lines at high speed until it encountered a different medium. It could then be reflected or refracted and laws had been produced to predict the precise behaviour in both cases. In addition, Fermat had produced a theory which predicted both these laws and other phenomena. The development of the subject hereafter should have been a case of merely applying these laws and principles to the relevant practical problems.

It was in 1669 that Bartholinus, looking through a calcite crystal, observed two images instead of one. Calcite is a form of calcium carbonate sometimes called Iceland spar and the phenomenon is known as double refraction. As explained in Section 7.7, the rays forming one image lie in the plane of incidence and obey Snell's law whilst the rays forming the other image do not. Clearly these observations could not be explained by the simple 'ray treatment' of light that we have discussed so far.

Some doubt had, in fact, already been expressed about this simple ray picture. Grimaldi, who had been Professor of Mathematics at Bologna and was an ardent experimenter, had carried out some interesting light experiments which appeared in his book *Physico-mathesis de Luine, Coloribus et Iride*, published posthumously in 1665.

Grimaldi examined the sharpness of shadows, using a hole in a blind as his source of light. He observed that as the source was made smaller, so the shadow became sharper, as a ray picture would predict; however, if the process was continued until the source was minute, the shadow became diffuse again. He also noticed that very small obstacles gave shadows that had fringes. We now call this phenomenon 'diffraction' and it was discussed in detail in Section 7.5.

These experiments showed that the simple picture of absolutely straight light rays was inadequate. There must be circumstances other than those of simple reflection and refraction, where light can change its direction of travel. The peculiar significance of Grimaldi's experiments is that the change in direction occurred whilst travelling in one medium and not at an interface between two media.

Grimaldi observed that the effects which he had described occurred only when the hole or obstacle was itself extremely small. He thought that diffraction implied a periodic quality in light, rather similar to that seen in water waves. Robert Hooke, who re-discovered diffraction in 1672, was to write that light spreads out from a source 'like the rings of waves on the surface of the water'.

9.5 Colour and the spectrum

As has been explained earlier, the rainbow had provided early evidence that coloured light could be produced from white light. However, although Descartes had shown how the position of rainbows could be explained, no satisfactory theory of their colour could be formulated.

Newton, who was to perform conclusive experiments on this subject, became interested in it when he attempted to construct an astronomical telescope as an undergraduate. He found that the main image was surrounded by a coloured border and it was perhaps in an attempt to solve this problem that he set out to investigate colour.

To do this, he bought a prism at Stourbridge Fair in 1666 with which he carried out the following series of experiments shown in Fig. 9.6. The extract is quoted from his paper, 'A New Theory About Light and Colours' published in *Philosophical Transactions of the Royal Society*, February 1672:

> 'In the year 1666 (at which time I applied myself to the grinding of optic glasses of other figures than spherical), I procured me a triangular glass prism, to try therewith the celebrated phenomena of colours. And in order thereto, having darkened my chamber, and made a small hole in my window-shuts, to let in a convenient quantity of the sun's light, I placed my prism at its entrance, that it might be thereby refracted to the opposite wall. It was at first a very pleasing divertisement, to view the vivid and intense colours produced thereby; [Fig. 9.6(*a*)]. . . . Then I suspected, whether by an unevenness in the glass or other contingent irregularity, these colours might be thus dilated. And to try this, I took another prism like the former, and so placed it, that the light passing through them both might be refracted contrary ways,

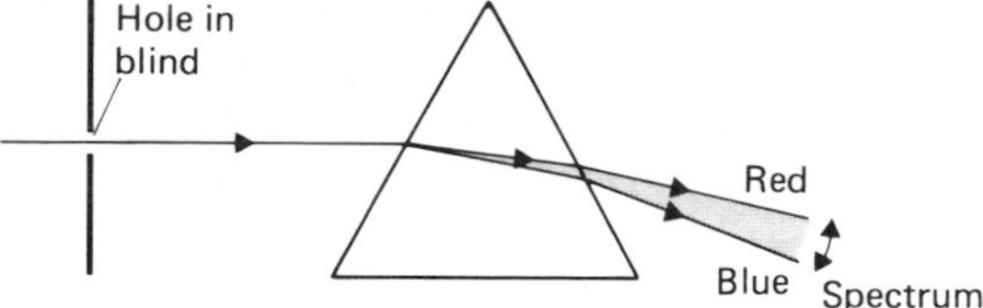

(*a*) One prism produces a spectrum

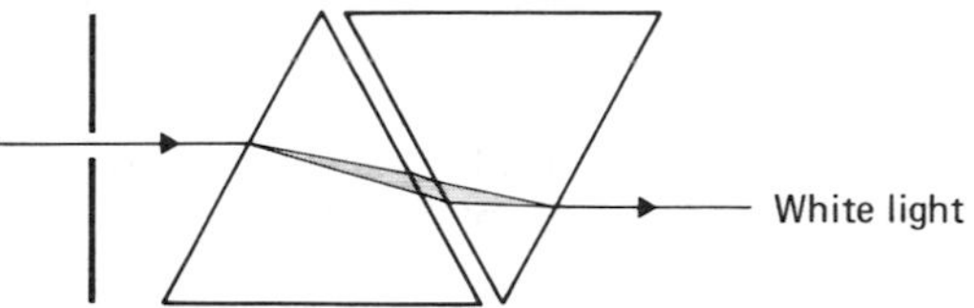

(*b*)Second prism can convert spectrum back into white light

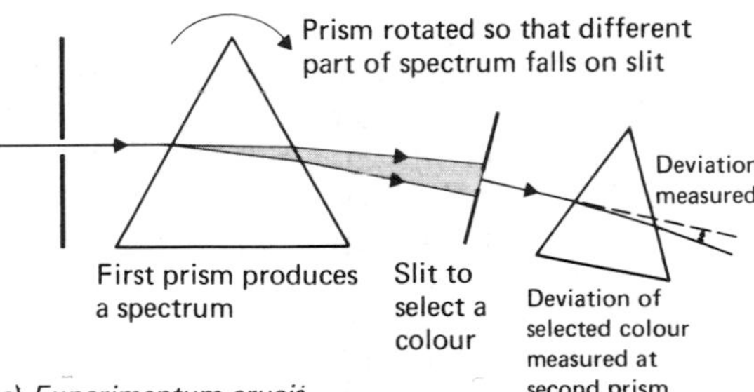

(*c*) *Experimentum crucis*

Fig. 9.6

and so by the latter returned into that course from which the former had diverted it. For by this means I thought the regular effects of the first prism would be destroyed by the second prism, but the irregular ones more augmented, by the multiplicity of refractions. The event was that the light, which by the first prism was diffused into an oblong form, was by the second reduced into an orbicular one, with as much regularity as when it did not at all pass through them [Fig. 9.6(*b*)]. . . . The gradual removal of these suspicions at length led me to the Experimentum Crucis, which was this: I took two boards and placed one of them close behind the prism at the window, so that the light might pass through a small hole, made in it for the purpose, and fall on the other board, which I placed at about 12 feet distance, having first made a small hole in it also, for some of the incident light to pass through. Then I placed another prism behind this second board, so that the light trajected through both the boards might pass through that also, and be again refracted before it arrived at the wall. This done, I took the first prism in my hand, and turned it to and fro slowly about its axis, so much as to make the several parts of the image cast, on the second board, successively pass through the hole in it, that I might observe to what places on the wall the second prism would refract them. And I saw by the variation of those places, that the light, tending to that end of the image, towards which the refraction of the first prism was made, did in the second prism suffer a refraction considerably greater than the light tending to the other end. And so the true cause of the length of that image was detected to be no other than that light is not similar or homogenial, but consists of Difform Rays, some of which are more Refrangible than others; and therefore that, according to their particular Degrees of Refrangibility, they were transmitted through the prism to divers parts of the opposite wall [Fig. 9.6(*c*)].

In the first experiment, Newton showed that a beam of light passed through a prism, produces an elongated patch of light, varying in colour from red to blue, which he termed a *spectrum* (Fig. 9.6(*a*)).

To test whether the colours arose through any unevenness in the glass, he passed the coloured light through a second identical prism, arranged as shown in Fig. 9.6(*b*). If the mere presence of the prism were responsible, then the effect would be increased by the second prism. Instead, the coloured light was combined together to produce white light again, suggesting that the colours are not 'added' by the prism but are part of the white light to begin with.

To test his hypothesis that white light consists of a mixture of different colours of light each of which is refracted differently by the prism, Newton performed his '*experimentum crucis*' shown in Fig. 9.6(*c*). Using the first prism to produce rays of light of different colours and placing the slit so as to select one particular colour, he then examined the refraction of coloured light by the second prism. He found that, in fact, the blue light was deviated through a larger angle than the red and he therefore concluded that:

> 'light consists of Difform rays, some of which are more Refrangible than others'.

We would say that glass has different refractive indices for different colours and that the angle of deviation caused by a prism depends on the refractive index. Detailed results, based on the geometry of the prism, have been discussed earlier in Chapter 6, but we are here concerned mainly with Newton's contribution to the understanding of the nature of light.

Although Newton's work was to be of great importance in the development of optics, it was at first badly received and strongly criticized. Robert Hooke, in particular, who was Secretary of the Royal Society, argued that light consisted of waves in an aether and that colours arose when the light was refracted at an angle to the 'direction of thrust' of the aether particles. His caustic criticism and the ensuing debate caused Newton to delay publication of his greatest optical work, *Opticks*, until 1704, the year after Hooke's death.

Unlike Hooke, Newton believed in a corpuscular theory of light in which a ray consisted of a stream of minute particles, capable of reflection and refraction, which produced images by their action on the retina. Reflection was viewed in terms of balls bouncing off a table, as shown in Fig. 9.7(a). It is easy to show, if the collision is elastic, both that the speed of the particles is constant and that the angles of incidence and reflection are equal.

(a) A time exposure of a ball being reflected from a steel plate. The angle of reflection equals the angle of incidence.

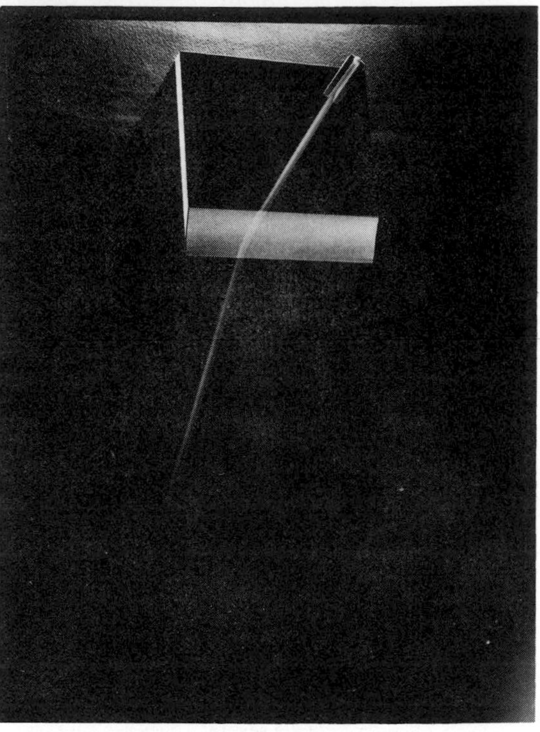

(b) A time exposure of a ball bearing rolling from a higher to a lower surface. The change in direction of the ball shows "refraction".

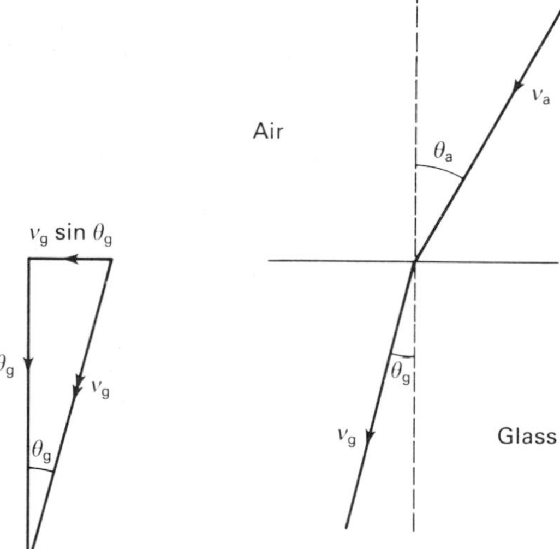

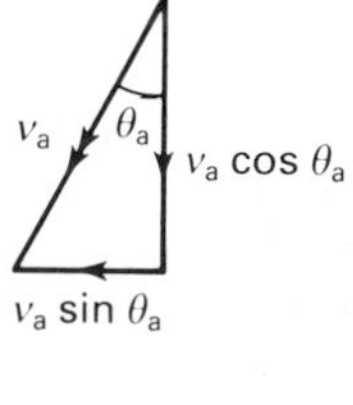

(c) The geometry of refraction

Fig. 9.7 The corpuscular model of reflection and refraction.

Newton explained refraction by arguing that the corpuscles travelled more quickly in glass than in air, being attracted into the glass surface as they travelled towards it. He assumed that the component of velocity parallel to the surface remained unchanged; again, an analogue using rolling balls is shown in Fig. 9.7. From Fig. 9.7(c), it can be seen that:

$$v_a \sin \theta_a = v_g \sin \theta_g$$

$$\frac{\sin \theta_a}{\sin \theta_g} = \frac{v_g}{v_a} = n$$

Since $\theta_a > \theta_g$, Newton predicted that the velocity of light in glass was greater than that in air. As we shall see in the next section, the rival wave theory of light made just the opposite prediction.

The formation of a spectrum can be explained by requiring light of different colours to have different velocities in glass, though it is interesting to note that Newton had to allow for a periodic character in light to identify the different colours:

> 'If by any means those [vibrations] of unequal bigness [wavelength] be separated from one another, the largest beget a sensation of a Red colour, the least or shortest a deep Violet, and the intermediate ones, of intermediate colours; much after the manner that bodies, according to their several sizes, shapes and motions, excite vibrations in the Air of various bignesses, which according to those bignesses, make several Tones in Sound.'
> § *Words in square brackets are not original.*

Newton (1704) *Opticks*

9.6 Huygens' wave theory

We have already described how Newton ascribed the properties of light to a corpuscular theory, and how he was able to use this theory to explain phenomena such as reflection and refraction. In this section, we shall see how another school of thought, led by the Dutch scientist Christian Huygens, attributed the behaviour of light not to a corpuscular theory, but to a wave theory.

Huygens, observing that sound spread out from a source like water waves, believed it to consist of some sort of wave motion; he also believed that it was caused by the oscillation of the surface of the source, and transmitted by collisions between the 'small bodies' in air.

In 1665, Hooke had argued that light, too, spread out 'like the rings of waves on the surface of the sea'. Huygens also observed the parallel between the behaviour of light and the behaviour of sound and water, and he therefore set forth a new wave theory in his *Treatise on Light.* This was presented to the French Academy of Science in 1678, but not published until 1690.

He discussed two phenomena which seemed to him to be in contradiction with Newton's corpuscular theory. These were, firstly, that light has a high but finite velocity and, secondly, that two light rays can pass through each other without alteration. He argued that corpuscles would be unable to travel fast enough to fit in with Roemer's then recent measurement of the velocity of light, and that they would collide when two beams crossed.

He also assumed that the transmission of light must consist of the motion of some kind of matter, for it was given out by hot bodies which must themselves contain moving particles. For:

> 'this is assuredly the mark of motion, at least in the true Philosophy, in which one conceives the causes of all natural effects in terms of mechanical motion.'

Huygens (1690) *Treatise on Light*

Thus light, like sound, consisted of wave-like motions in a medium, emitted by the vibrations of particles in the source. Huygens could see that the vibrations must be much more violent for the emission of light than for the emission of sound. He also knew that Torricelli had shown that light could pass through a vacuum which raised the difficult question as to the nature of the medium through which the light was passing.

Huygens termed this medium the *aether*. He discussed the way in which a row of hard steel balls, placed so as to touch each other, will transmit an impact from a ball hitting one end, with the result that the ball at the other end will move; this behaviour is today familiar in a toy device known as a 'Newton's cradle'. Huygens reasoned that this motion depended upon the balls yielding elastically to a collision and that the impact was not transmitted instantaneously. He assumed that the speed of transmission depended only upon the elasticity of the balls and not upon the size of the impact.

He supposed the aether to consist of particles, presumably massless, which were very hard and very elastic. The transmission of light through the aether involved the motion and collision of aether particles, and their high elasticity accounted for the high value of the velocity of light. As the elasticity is constant throughout the aether, the velocity of light is a constant and does not decrease as the distance from the source increases. It was not, in fact, necessary either for the particles to be in straight lines, or for them to be stationary, for their motion due to the passage of light was merely superimposed upon their ordinary motion.

Huygens suggested that each point on a body acts as a source of spherical waves, and is at their centre. Thus, point A in Fig. 9.8 is seen to be giving out spherical waves. As should be clear from the above discussion, these were assumed to be longitudinal waves rather than transverse and were not thought to be generated regularly. Thus the equally spaced semicircles do not represent different waves sent out at regular intervals all viewed at the same time, but rather the same wave viewed at different, regularly spaced, time intervals. We would now refer to one of these semicircles as a *wavefront*, being defined as a line along which all the particles are in phase.

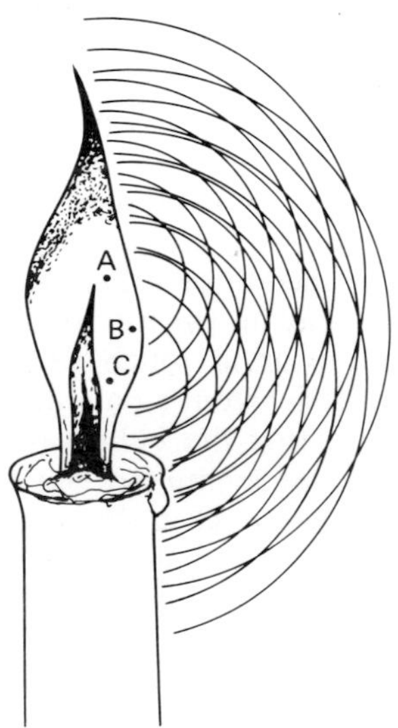

Fig. 9.8 Wavefronts given out by a candle. Huygens (1678) *Treatise on Light* (reprinted by Dover).

Each point on the flame is acting as a source of waves and thus points B and C are also at the centre of spherical wave patterns. These waves spread out into the same space that is affected by the waves from A. Thus one particular aether particle will be affected by many wave motions all at the same time, through impacts from all the other particles around it. Huygens did not see this as a problem:

> 'it being certain that one and the same particle of matter can serve for many waves coming from different sides or even from contrary directions, not only if it is struck by blows which follow one another closely but even for those which act on it at the same instant'.

Huygens (1690) *Treatise on Light*

In the same way, a ball bearing receiving a number of simultaneous impacts from other ball bearings will accelerate according to the resultant of those impacts. This is probably the earliest statement of the *Principle of Superposition*, though it was not seen as such at the time.

Now, the particle motion that is caused by the arrival of a wave at a point will affect not only the next particle in a line leading away from the source, but also all the other particles that touch it. Thus, each point on a wave disturbs the medium all around it and acts as a secondary source of waves; this is the first postulate of *Huygens' Principle*. The second postulate is that the envelope of all these secondary wavelets at a later time gives the new position of the wave. In Chapter 3, we saw that this construction could help us to understand phenomena such as the rectilinear propagation of light, refraction and diffraction; it is also used in the theory of double refraction.

Huygens' treatment of refraction depended upon light having a lower velocity in glass than in air, and he did attempt to provide an explanation for this, although experimental techniques were not good enough at that time to provide a direct measurement:

> 'and, moreover, one may believe that the progression of these waves ought to be a little slower in the interior of bodies, by reason of the small detours which the same (i.e. body) particles cause. In which different velocity of light I shall show the cause of refraction to consist'.

Huygens (1690) *Treatise on Light*

9.7 Newton's Opticks

We have already explained that Newton and his followers believed in a corpuscular theory of light whereas Huygens and his school believed that light was an irregular, longitudinal wave motion. Both theories could be applied to the phenomenon of refraction but with exactly opposite predictions; the corpuscular theory predicted that light would travel more quickly in glass than in air, whilst the wave theory predicted that it would travel more slowly. Unfortunately, the velocity of light being very large, it was almost two hundred years before experimental techniques had advanced sufficiently to enable this to be used as a test of the two theories. In the meantime, the greater influence of Newton amongst men of science enabled his views to become generally accepted despite the fact that, to our eyes at least, the weight of evidence came down in favour of Huygens.

Q 9.3
Newton thought that the velocity of light in glass was 1.5 times larger than in air; Huygens thought that the velocity in air was 1.5 times larger than in glass. The velocity in air was known to be about $3 \times 10^8\ \mathrm{m\,s^{-1}}$. What would need to be the length of the apparatus to observe a predicted difference between the time in glass and the time in air of 0.1 s according to Huygens' theory? What would be the predicted time difference according to Newton's theory for the same length experiment?

Newton was not, however, without his critics. Hooke was one of his most bitter opponents and, as has already been pointed out, Newton delayed the publication of *Opticks* until after Hooke's death. We have also seen, in Section 9.5, that Newton had to allow that the corpuscles produced waves in the aether in order to explain the phenomenon of colour. He was not without his doubts, which he expressed in a letter written to Hooke:

> 'Tis true, that from my theory I argue the corporcity of light; but I do it without any absolute positiveness . . . I know, that the properties, which I declar'd of light, were in some measure capable of being explicated not only by that but by many other mechanical hypotheses',

Newton (1672)

On the other hand, he was unable to completely accept a wave theory either, for he was concerned that there was little evidence to show that light could go round corners. In the following quotation from *Opticks*, he used the word 'motion' to mean 'wave':

> 'Are not all Hypotheses erroneous, in which Light is supposed to consist in Pressure, or Motion, propagated through a fluid Medium? . . . And if it consisted in Pressure or Motion, propagated either in an instant or in time, it would bend into a Shadow. For Pressure or Motion cannot be propagated in a Fluid in right Lines, beyond an Obstacle which stops part of the Motion, but will bend and spread every way into the quiescent Medium which lies beyond the Obstacle. . . . The Waves on the Surface of stagnating Water, passing by the sides of a broad Obstacle which stops part of them, bend afterwards and dilate themselves gradually into the quiet Water behind the Obstacle. . . . But Light is never known to follow crooked Passages nor to bend into the Shadow. The Rays which pass very near to the edges of any Body are bent a little by the action of the Body, as we shew'd above; but this bending is not towards, but from the Shadow, and is perform'd only in the passage of the Ray by the Body, and at a very small distance from it. So soon as the Ray is past the Body, it goes right on'.

Newton (1704) *Opticks*

It is interesting to note that even in the above passage, where Newton cites a lack of evidence for the bending of light round obstacles as an argument against the wave theory, he admits that the effect is present but in small degree. He was aware of diffraction but at the same time was worried that it was not an obvious enough effect. The clue in fact lies in the phrase 'a very small distance', for what Newton just failed to put his finger on was that this is a small effect visible only when the obstacle, or slit, is very small. Water waves, of wavelength several feet, display the phenomenon around objects that are the same order of size as their wavelength; most objects or slits are many orders of magnitude larger than the wavelength of light.

Even so, it seems likely that others read into Newton's reluctance to accept the wave theory more than he intended. Certainly it was not until the beginning of the nineteenth century that the corpuscular theory was to be seriously questioned, and the intervening lack of interest suggests that most physicists believed that Newton's theory was the key to optics.

Newton himself was less sure. The last section of *Opticks* is devoted to a series of thirty-one questions such as that quoted above; many of these revealed his own doubts about the inability of the corpuscular theory to explain the facts before him. Despite the awe in which he was widely held, he wrote towards the end of his life:

> 'I do not know what I may appear to the world; but to myself I seem to have been only like a boy, playing on the seashore and diverting myself in now and then finding a smoother pebble or a prettier shell than ordinary, while the great ocean of truth lay all undiscovered before me',

Newton (1727).

9.8 Young and Fresnel

There were few developments in the theory of light between the publication of *Opticks* in 1704 and the beginning of the nineteenth century. Despite some criticism of his experimental methods by German philosophers such as Hegel, Newton's authority was such that his corpuscular theory was widely accepted, and the crucial test of comparing the velocities of light in glass and air was not yet technically possible. The debate about the nature of light was not re-opened until another great scientist, with a breadth of interest almost as wide as Newton's, turned his attention to the problem.

Thomas Young was born in Somerset in 1773 and as a schoolboy showed exceptional talents. By the age of fourteen he could speak a wide variety of classical and modern languages and he was later to be a pioneer in the interpretation of Egyptian hieroglyphs. His many authoritative contributions to the *Encyclopedia Britannica* included articles on Annuities, Carpentry, Egypt, Road Making and Tides; his epitaph in Westminster Abbey reads 'a man alike eminent in almost every department of human learning'.

At the age of nineteen, he went to London to study medicine and his early research was into the mechanism of the eye and the nature of colour. His PhD was awarded for work on the human voice and initiated his interest in the physics of sound waves. Newton had applied the idea of the interference of waves to the explanation of tides but, as we have seen, had not extended the idea to the study of light. Young based many of his ideas on observations of waves on a lake and revived the hypothesis that light was a wave motion. He believed that if light consisted of particles, they would be likely to be given out at a range of speeds and hence would not form a well-defined image; however, he knew that the velocity of a wave was a property of the medium and hence that all waves from the source travelled at the same speed.

Newton had observed that when white light was reflected from a thin film, coloured patterns were often formed; these patterns are familiar to us in soap bubbles, or films of oil on the road. He examined the effect in some detail by placing a lens on a reflecting surface and observing the concentric series of coloured rings that is formed; we now refer to these as *Newton's rings* and they were discussed in detail in Section 7.4. He discovered the relationship between the colour of the light reflected at a point and the thickness of the air film there and suggested that the corpuscles of light, when they passed through the bottom surface of the glass, excited vibrations in the air film. He believed that these vibrations reached the glass block before the corpuscle and excited it into 'fits' of easy reflection which alternated with 'fits' of easy transmission; whether a bright ring would be reflected would depend on the 'fit' of the glass when the corpuscle arrived, and this would depend in turn upon the thickness of the film. The theory is unsatisfactory mainly because the mechanism is invented only to explain this phenomenon and is of no relevance elsewhere.

Young observed the superposition of waves occurring on lakes and realised that the resultant effect depended on the phase difference between the waves. In the arrangement to observe Newton's rings, some of the light is reflected at the lower surface of the lens and some at the upper surface of the glass block; the phase difference between these two rays will give rise to the coloured rings as explained in Section 7.4.

Young also showed that interference patterns could be formed when light from two slits superposed on a screen. The theory of his double-slit experiment has already been given for waves in general, in Section 4.4; and it was applied to the interference of light in Section 7.2. The original description of how he used this experiment to obtain an estimate of the wavelength of light is given in this extract from a paper read to the Royal Society; modern experiments would give a mean wavelength of about one 46-thousandth of an inch, and a frequency of about 540×10^{12} Hz:

> 'Supposing the light of any given colour to consist of undulations, of a given breadth, or of a given frequency, it follows that these undulations must be liable to those effects which we have already examined in the case of the waves of water, and the pulses of sound. It has been shown that two equal series of waves, proceeding from centres near each other, may be seen to destroy each other's effects at certain points, and at other points to redouble them. . . . We are now to apply the same principles to the alternate union and extinction of colours.
>
> In order that the effects of two portions of light may be thus combined, it is necessary that they be derived from the same origin, and that they arrive at the same point by different paths, in directions not much deviating from each other . . . the simplest case appears to be, when a beam of homogeneous light falls on a screen in which there are two very small holes or slits, which may be considered as centres of divergence, from whence the light is diffracted in every direction. In this case, when the two newly formed beams are received on a surface placed so as to intercept them, their light is divided by dark stripes into portions nearly equal. . . . The middle of the two portions is always light, and the bright stripes on each side are at such distances, that the light, coming to them from one of the apertures, must have passed through a longer space than that which comes from the other, by an interval which is equal to the breadth of one, two, three, or more of the supposed undulations, while the intervening dark spaces correspond to a difference of half a supposed undulation, of one and a half, of two and a half, or more.
>
> From a comparison of various experiments, it appears that the breadth of the undulations constituting the extreme red light must be supposed to be, in air, about one 36 thousandth of an inch, and those of the extreme violet about one 60 thousandth; the mean of the whole spectrum, with respect to the intensity of light, being about one 45 thousandth. From these dimensions it follows, calculating upon the known velocity of light, that almost 500 millions of millions of the slowest of such undulations must enter the eye in a single second.
>
> (1803) *Philosophical Transactions of the Royal Society*

In his paper Young also discussed the observations of diffraction seen by Grimaldi. It now seems surprising that his work was received at first with a great deal of scepticism and hostility. Support for Young came in 1815 from Fresnel, who made similar proposals about the nature of light, based on his independent study of diffraction. Fresnel was a French bridge and road engineer who worked on optics in his spare time; his lens system designed for lighthouses is still in use today.

Newton knew that light spreads into a shadow, blurring its edge and demonstrating a slight deviation from rectilinear propagation. In Section 9.4 we discussed the work carried out on this effect by Grimaldi, and in Fig. 9.9 we see a modern photograph showing the alternate dark and bright lines occurring in the shadow of a pinhead. Fresnel wrote several papers on diffraction and in 1818 he won a prize offered by the French Academy for the best discussion to date.

Ironically the widespread acceptance of the wave theory came about because of the opposition of Poisson, who was one of the judges in the French Academy competition. He proved, using Fresnel's equations, that at the centre of a circular shadow, there should be a small bright spot; this was clearly absurd and the theory must be wrong. Fortunately Fresnel was an experimental physicist as well as a mathematician and he arranged to test this prediction by experiment. A modern photograph is shown in Fig. 9.10 and we see that both Fresnel and Poisson were right!

Thus, support for the corpuscular theory began to crumble and it was no surprise when, in 1850, Fizeau and Foucault found that the velocity of light in water was indeed less than in air; Foucalt's method is described in the next section.

It is important to make some distinctions between Huygens' wave theory and that of Young. Huygens had imagined light to consist of isolated longitudinal pulses whereas Young and Fresnel regarded it as a periodic transverse

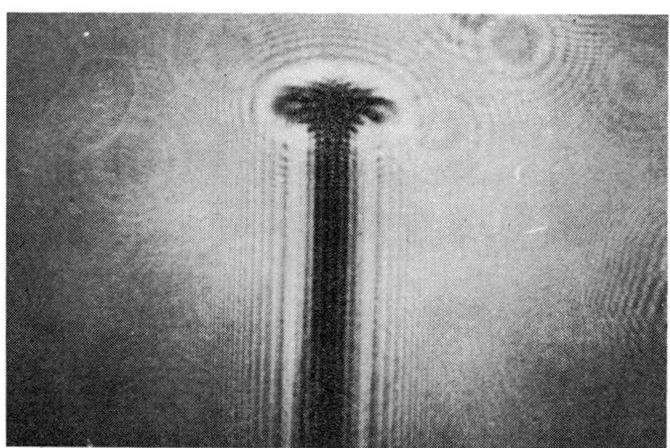

Fig. 9.9 Diffraction fringes formed in the shadow of a pin-head.

(Photograph by courtesy of *Griffin & George*)

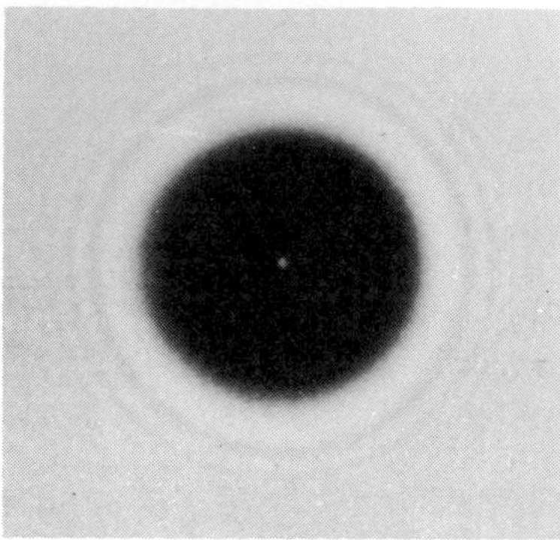

Fig. 9.10 Poisson's spot.

wave motion; this arose from their understanding of double refraction and of polarization.

In 1817, Young proposed that light waves must be asymmetrical about their direction of propagation, and therefore be transverse rather than longitudinal. However, this raised a new problem because transverse waves had only previously been observed in solid media which offered resistance to twisting; liquids and gases transmit only longitudinal waves. Thus 'space', which transmits light, must be a 'solid' and the aether (ether) is seen to be a transparent, solid, all-pervading substance. It was known that the velocity of both transverse and longitudinal waves in a solid was given by equations of the form $c = \sqrt{(E/\rho)}$, as shown in Section 2.5; since the velocity of light is very high, it follows by analogy that the aether must be of high elasticity and low density. The elasticity, for instance, must be higher than that of steel.

We have seen already, in Chapter 7, and will see again in Chapter 10, that light is now thought of as an electromagnetic, rather than a mechanical, wave and some of these problems were peculiar to the mechanical viewpoint. However, the research stimulated on waves in solids was of wide application, for instance to the study of earthquakes, and the concept of the aether was to be of importance in the development of the theory of relativity. In the nineteenth century, scientists could not conceive of a wave motion without a medium and thus the existence of the aether was a conceptual necessity.

9.9 The velocity of light

It has already become evident how important an accurate measurement of the velocity of light was to be to the development of an understanding of light; the final confirmation of the truth of Young's wave theory would only come when it was shown experimentally that light travelled more slowly in water or glass than in air.

Foucault's method

The first accurate terrestrial determination of the velocity of light was carried out in 1849 by a wealthy Frenchman called Fizeau; the value he obtained of 3.13×10^8 m s^{-1} is now regarded as being rather high compared with the currently accepted value of 2.998×10^8 m s^{-1} but nevertheless it represented, at the time, an enormous improvement in accuracy over previous measurements. Details of Fizeau's method may be found in books such as Jenkins and White, *Fundamentals of Optics*.

The following year, a different method was developed by a former collaborator of Fizeau's, called Foucault. He obtained a value of 2.98×10^8 m s^{-1}, and also showed that light travels more slowly in water than in air. We choose to discuss this method in some detail, because it is readily available for use in laboratories; the apparatus described differs slightly from that used by Foucault, to conform with present practice.

The apparatus is shown in Fig. 9.11. A light source illuminates a set of crosswires, which act as an object for the small concave mirror M_r a distance s away; M_r can be rotated by a motor, and the large fixed concave mirror M_f is so placed that the axis of rotation of M_r is at the centre of curvature of M_f. The position of the crosswires is adjusted until an image of them is formed by M_r at the centre of M_f.

Now, the light falling on M_f and forming an image of O at I_1 will itself be reflected by M_f; since all optical systems are reversible, M_r will form a further image of the image I_1, and it will be formed at O. In other words, O can be the object and I_1 the image, or I_1 can be the object and O the image. Most of the light forming the image at O is reflected by a glass plate to form an image at I_2 which can be viewed through an

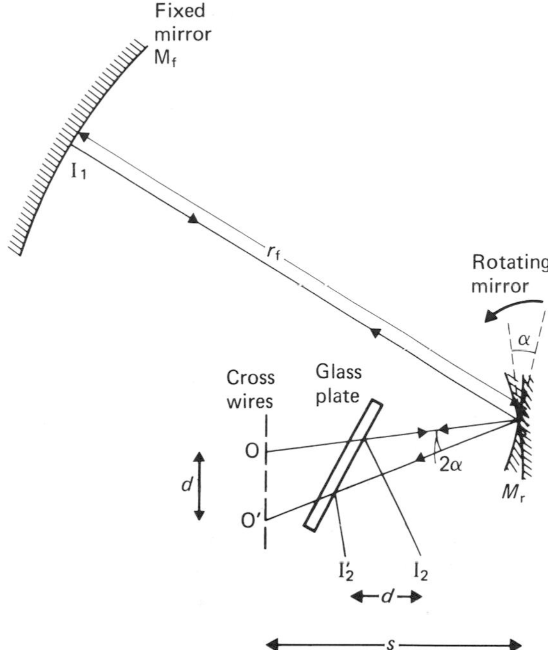

Fig. 9.11 Foucault's method to measure the velocity of light.

eyepiece. Thus, I_2 is an image of O formed after two reflections in M_r, and one in M_f.

Suppose that the mirror M_r is now rotating in an anticlockwise direction. When the light returns to M_r from M_f, the mirror M_r has turned through an angle α. The image of I_1 will now be formed at O′ instead of O, and by reflection in the glass plate, at I_2' instead of I_2. The image I_2' will also be fainter because light can only pass through the system for a small part of each revolution of the mirror M_r.

Consider the ray from O striking M_r at its mid-point. Since the distance from M_f to M_r is r_f, the radius of curvature of M_f, the ray will then strike M_f along a normal, no matter at what point it meets the mirror's surface. It will therefore be reflected along its own path. The time taken, t, before it returns to the centre point of M_r is given by:

$$t = \frac{2r_f}{c} \qquad (9.1)$$

where c is the velocity of light. If the mirror is rotating with a frequency f, the angle, α, through which the mirror will have turned in this time will be given by:

$$\alpha = 2\pi ft$$

$$= \frac{4\pi f r_f}{c} \qquad \text{from (9.1)}$$

If the mirror is rotated through an angle α then the reflected ray will be rotated through an angle 2α from the original incident direction, where:

$$2\alpha = \frac{8\pi f r_f}{c} \qquad (9.2)$$

The image of O, instead of being reflected back to O, will therefore be displaced a distance d to O′, where:

$$d = 2\alpha s$$

i.e. $$d = \frac{8\pi f r_f s}{c} \qquad \text{from (9.2)}$$

OO′ is reflected in the glass plate and d is therefore also the displacement of the image viewed in the eyepiece, I_2I_2'. It is usual to measure $2d$, the displacement of the image when the direction of rotation of the motor is reversed. Now since

$$d = \frac{8\pi f r_f s}{c}$$

we have $$c = \frac{8\pi f r_f s}{d} \qquad (9.3)$$

The frequency of rotation, f, can be measured by reflecting the light beam into a photo-diode once per cycle and counting the number of pulses in a given time with a scaler.

Q 9.4

In an experiment to measure the velocity of light using Foucault's method the following data were obtained:

radius of fixed mirror	(2.05 ± 0.02) m
distance from crosswires to rotating mirror	(0.6 ± 0.005) m
frequency of rotating mirror	(485 ± 15) Hz
distance moved by spot when rotation reversed	$(1.2 \pm 0.3) \times 10^{-4}$ m

Calculate the probable value of the velocity of light from this information; do the limits of error include the commonly accepted value?

As well as making an accurate determination of the velocity of light in air, Foucault placed a tube of water, of length about 3 m, between the two mirrors and showed conclusively that the velocity of light in water was less than that in air. This was, for the time being, the already heralded death knell of the corpuscular theory of light.

Michelson–Morley experiment

It would not be right to leave a discussion of the measurement of the velocity of light without mentioning the Michelson–Morley experiment, which was crucial to the development of the theory of relativity.

Light travels at nearly 3×10^8 m s^{-1}—relative to what? In the middle of the nineteenth century, physicists argued that the velocity was relative to the aether—the all-pervading elastic medium discussed earlier. If this was the case, and if the earth moved through the aether in its orbit around the sun, then the velocity of light relative to the earth would depend upon the direction in which it was transmitted relative to the aether. If the light was transmitted in the same direction as the aether moving past the earth's surface, then its speed relative to the earth would be increased; if it were transmitted in the opposite direction then the speed would be decreased. Michelson and Morley set out to test this deduction.

A simplified version of their experiment is shown in Fig. 9.12 and utilizes the Michelson interferometer described in Section 7.4. A beam of light, travelling in the same direction as the earth, and therefore into the 'aether wind', is split into two parts by a partially reflecting glass block G. Each beam then travels an identical distance to the mirrors M_1 or M_2, where it is reflected. The two beams are recombined at G and viewed by an observer at O. With an extended source, circular patterns of maxima or minima are formed, as explained in Section 7.4.

Suppose that the paths are identical in length. Classical theory predicts that light would travel slowest from G to M_2 fastest from M_2 to G, and at an intermediate speed along GM_1 and M_1G. As we see in Question 9.5, we expect the ray travelling to and from mirror M_1 to take a different time from the ray reflected from M_2,

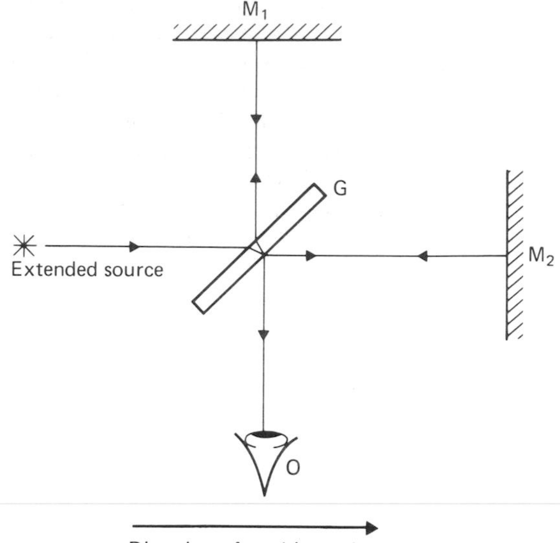

Fig. 9.12 The Michelson–Morley experiment.

introducing phase differences. Michelson and Morley tested this conclusion by rotating the whole apparatus, so that the directions of the two rays relative to the aether wind were changed; they were surprised to observe no change in the interference pattern.

Q 9.5
A wind blows from the west at 50 km hour^{-1} and an aeroplane is capable of an air speed of 100 km hour^{-1}. Calculate:

(*a*) the time taken for the aeroplane to fly 50 km west and return to its starting point;
(*b*) the time taken for the aeroplane to fly 50 km north and return to its starting point.

Note that these times are different.

Thus, Michelson and Morley showed that the velocity of light relative to the earth is the same regardless of the earth's motion through the aether. From this experiment, Einstein developed the idea that the velocity of light *in vacuo* is the same relative to all objects; there is no basic reference frame relative to which velocities are measured. Thus, if a rocket passes me at 2×10^8 m s^{-1} and sends a light signal forwards at a speed relative to itself of $3\times$

10^8 m s^{-1}, the speed of that light relative to me is only 3×10^8 m s^{-1}. Clearly such strange concepts require an entirely new way of thinking and this was incorporated in the Special Theory of Relativity, published in 1905. This provided a new system of mechanics which was identical to that of Newton at low relative velocities, but entirely different when velocities approached the speed of light itself; in particular the idea of a stationary reference frame for the measurement of velocities was discarded.

Bibliography

The Bibliographies to Chapters 9 and 10 appear together at the end of Chapter 10.

10
Light: waves and particles

10.1 Electromagnetic waves

In Chapter 9, we saw how the corpuscular theory of light was replaced by the wave theory, which could be used satisfactorily to explain the various phenomena known as physical optics. The next development was the realization that light was a form of electromagnetic wave and that the previously separate sciences of electromagnetism and light could be united through the theories of Maxwell.

James Clerk Maxwell was a Scottish physicist who contributed important theories to diverse fields of physics during his comparatively short life, from 1831 to 1879. He produced significant works on the kinetic theory of gases and on thermodynamics, but his most important achievement was the synthesis of light and electromagnetism into a single science. Although, as we shall see, his theories have their limitations in some circumstances, they are nevertheless fundamental to our understanding of light and have many practical applications; in the same way, Newton's laws of motion are still relevant today, despite the development of the theory of relativity.

In the 1860's, Maxwell showed theoretically that an oscillating charge will lose energy by radiating it into the surrounding space. The radiation was shown to be in the form of a wave motion, with electric and magnetic fields as the variable quantities. If this radiation reaches a wire, an alternating voltage is induced by the oscillating fields; this is the basis of radio reception. Maxwell predicted that the radiation would obey the laws of reflection and refraction, and that in a vacuum the velocity of propagation would be nearly 3×10^8 m s^{-1}, the same as that of light. He concluded that light itself must be an *electromagnetic wave*:

> 'This velocity is so nearly that of light, that it seems we have strong reason to conclude that light itself (including radiant heat and other radiations if any) is an electromagnetic disturbance in the form of waves propagated through the electromagnetic field according to electromagnetic laws.'

(1865) *Philosophical Transactions of the Royal Society*

Hertz's 'electric waves'

The experimental verification of Maxwell's work did not occur until 1886 when Hertz, a German physicist, showed that an oscillating current emitted electromagnetic waves with similar properties to light. There is a most readable account of his original experiment in *Electric Waves* by Hertz, now reprinted by Dover.

In a modern version of his experiment, a voltage of about 5000 V is applied to a gap between two polished rods, as shown in Fig. 10.1(*a*). A second pair of rods is connected to a sensitive galvanometer; the diode enables alternating currents in the right-hand circuit to be detected. When sparks are produced between the two rods in the transmitting circuit, a signal is detected in the receiving circuit at distances up to several metres away.

When the voltage is first applied to the rods, opposite charges will be stored on each; a simplified shape for the electric field set up in the gap is shown in Fig. 10.1(*b*). The stored charge, and the resulting field, increase until ionization of the air occurs and a spark jumps across the gap; conduction through the air is easy once ionization has occurred. At this instant, there is a current down both rods and through the gap, and a magnetic field is formed around the rods, as shown in Fig. 10.1(*c*); the charge stored on the rods is decreasing and so is the electric field. When the current, and hence the magnetic field, are decreasing a voltage is induced in the rod which will tend to maintain both current and field. As a result, the rods become oppositely charged and an electric field is set up in the opposite direction as shown in Fig. 10.1(*d*). A current now begins to flow in this new direction (Fig. 10.1(*e*)) and the charges reverse again. The cycle of oscillation of electric and magnetic field will continue, but with decreasing amplitude, until the electric field is too small to cause ionization; the rods will then charge up to form another spark. The oscillations in the gap are analogous to those in an organ pipe, and the rods and the space between them act like a resonator.

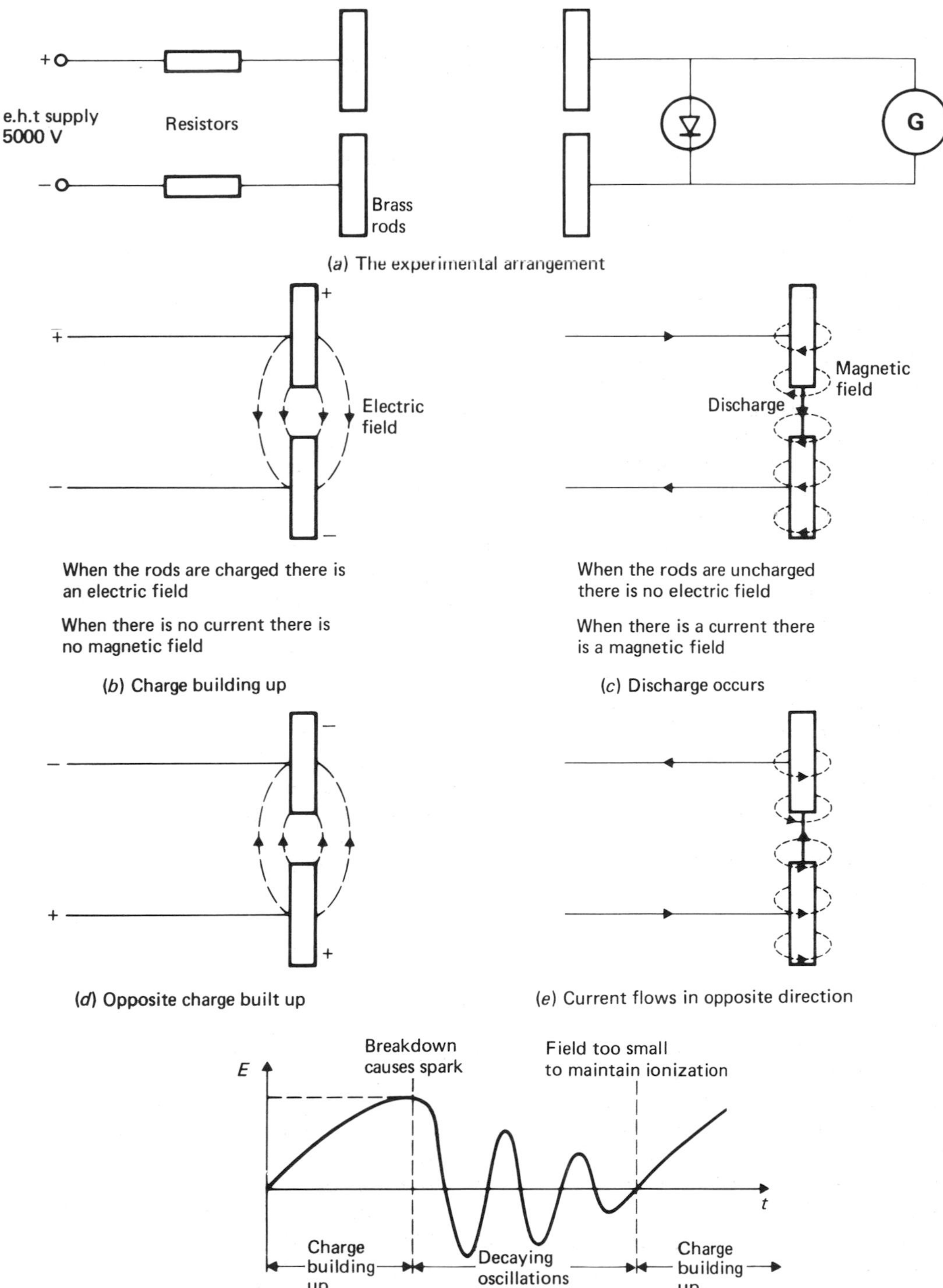

Fig. 10.1 The spark transmitter.

Thus, there is in the region of the gap an alternating electric and magnetic field; this is, of course, more complicated than the simple picture presented above. These oscillations cause the propagation of a wave consisting of an alternating electric and alternating magnetic field, and it is this *electromagnetic wave* that induces an alternating voltage and current in the receiving circuit. Experiments, similar to those carried out by Hertz, to demonstrate the wave nature of *electromagnetic radiation* are described in Section 10.2.

Marconi and radio

Hertz's experiment showed that electromagnetic waves could be transmitted, at least in a laboratory, as Maxwell had predicted. The possibilities of their use for communication over greater distances was immediately appreciated by an Italian called Marconi. Despite widespread scepticism, he gradually extended the range of his transmissions until, fifteen years later in 1901, he claimed the first successful trans-Atlantic transmission. Although there is now some doubt about his earliest results, we know that radio signals can be transmitted around the world by reflection from the ionosphere. At the time, his achievement was perhaps as unlikely as an ant being able to see round to the other side of a ball-bearing; Marconi, however, persevered at a task which everyone else thought to be impossible, and in due course this perseverance brought him a considerable fortune.

An aerial for the reception of short wavelength, very high frequency (VHF), radio transmissions is usually a metal rod about 1 metre long. The alternating electric field in the wave induces an alternating voltage in the rod which is then amplified. For long-wavelength transmissions, where the wavelength may be several hundred metres or more, the receiving aerial is often a coil wound on a ferrite rod which acts like a search coil in an alternating magnetic field. Thus, one method depends on the alternating electric field and the other on the alternating magnetic field, both of which are associated with electromagnetic waves.

Q 10.1
The diameter of a ball-bearing is 0.5 cm and of the earth about 6×10^3 km; the wavelength of light is about 5×10^{-7} m, and of radio waves about 500 m. Calculate the ratio wavelength/diameter for each situation described above.

10.2 The electromagnetic spectrum

Common properties of electromagnetic waves

Maxwell predicted that light was an electromagnetic wave, and both Hertz and Marconi showed that other electromagnetic waves, with wavelengths 10^6 to 10^9 times as great, could be produced and transmitted. We now know that electromagnetic waves can have wavelengths varying from about 10^{-13} m to 10^5 m, and this enormous range is called the *electromagnetic spectrum*; that range of it to which our eyes are sensitive, from about 4×10^{-7} m to 7×10^{-7} m, is called the *visible spectrum* or *light*. The only difference between that part which lies within the visible spectrum and the rest of the electromagnetic spectrum is the wavelength, and hence frequency, of the waves. The term *light* is sometimes taken to mean that which comes from the sun, which is a slightly wider range of wavelengths than the visible spectrum, including ultraviolet and infrared waves as well.

As all types of electromagnetic radiation are waves, they ought to show wave properties. Various experiments involving wave superposition have been described in earlier chapters to show the wave nature of light; similar experiments must be possible with other electromagnetic waves. Lloyd's mirror is a convenient experiment to set up and in Fig. 10.2 three arrangements of Lloyd's mirror are shown to demonstrate the interference of light, 3 cm radio waves and 30 cm radio waves. The 30 cm waves are transmitted using the same rods as were used to demonstrate Hertz's experiment; to demonstrate interference, a constant-frequency oscillator must be used instead of the high voltage spark described earlier. Fig. 10.2(*d*) shows how stationary waves can be set up using either VHF radio or ultra high frequency (UHF) television signals; in this case the wavelength would be of the order of 3 m or 0.5 m respectively. It is not easy to demonstrate the wave properties of X-rays and γ-rays in a laboratory, but the reader may be familiar with the principles of X-ray diffraction experiments which are widely used to determine the structure of substances.

Q 10.2
In a Lloyd's mirror arrangement for microwaves, the distance from transmitter to receiver is 200 cm and the transmitted wavelength is 3 cm. A maximum response is obtained when the distance from transmitter to the point at which reflection occurs on the reflecting plate is 106 cm. The plate is moved further away from the direct line; what will be the distance from transmitter to plate when the next maximum occurs? There is no phase change upon reflection at the metal plate.

Q 10.3
An experiment is set up as shown in Fig. 10.2(*d*) to measure the wavelength of a UHF television transmission. It is found that the detected signal fluctuates between two maxima when the plate is moved 15 cm. What is the wavelength of the waves?

We have already discussed the polarization of light in Section 7.7. If the receiving pair of rods in Hertz's experiment is rotated until it is at right angles to the transmitting pair no signal will be detected; the planes of polarization of transmitter and receiver are at right angles and correspond to crossed polaroids. This may also be demonstrated using 3 cm microwaves, and radio and television signals; thus it is essential to set up an aerial appropriately for the plane of polarization of the signals transmitted.

Maxwell was impressed by the fact that his electromagnetic waves had a theoretical velocity close to that of light; since then it has been possible to measure the velocity of electromagnetic waves *in vacuo* over a wide range of wavelengths and some typical results are quoted in Table 10.1; the close similarity between these results suggests strongly that all the waves are identical, apart from their wavelengths and therefore frequencies. In fact, since 1983, a metre has been *defined* as the distance travelled in 1/299 792 458 of a second by light in a vacuum, and thus the velocity of light is

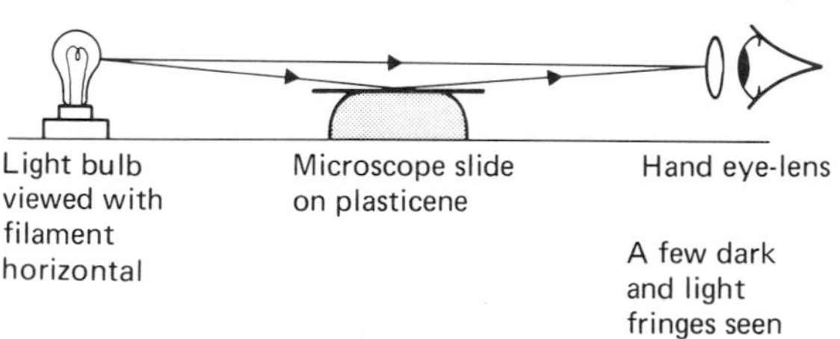

(*a*) Lloyd's mirror interference for light, $\lambda \sim 10^{-6}$ m

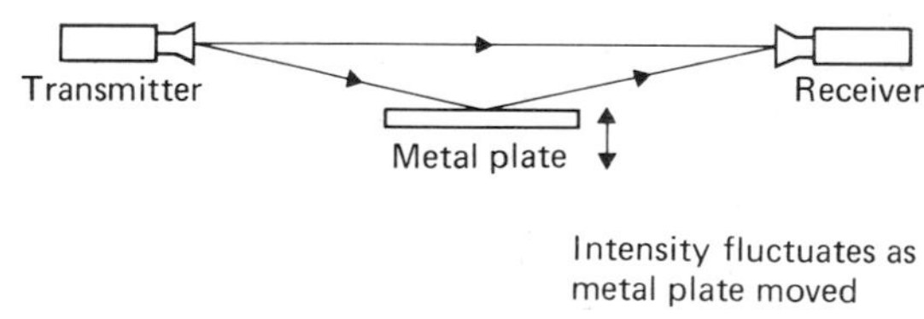

(*b*) Lloyd's mirror interference for microwaves, $\lambda \sim 0.03$ m

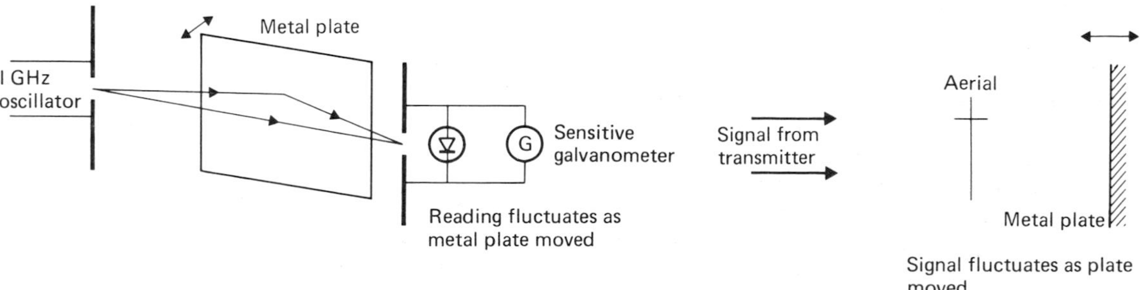

(*c*) Lloyd's mirror interference for I GHz waves, $\lambda \sim 0.3$ m

(*d*) Stationary wave interference for radio and u.h.f. television signals ($\lambda \sim 3$ m, 0.5 m respectively)

Fig. 10.2

299 792 458 metres per second exactly. We will consider below how some of these measurements at different wavelengths may be made.

Hertz knew that the frequency of the oscillations produced by his apparatus depended upon the capacitance and inductance of the complete transmitting circuit. He measured the wavelength of his electromagnetic waves by interference of the direct wave with those reflected from a metal plate using the method shown in Fig. 10.2(*d*); the receiver gave a null reading at the position of the nodes formed. In one measurement, he found that when the frequency of oscillations was 3×10^7 Hz, the wavelength of the emitted wave was 9.6 m.

Hence
$$\begin{aligned} c &= f\lambda \\ &= (3\times10^7)\times9.6 \\ &= 2.88\times10^8\ \mathrm{m\,s^{-1}} \end{aligned}$$

This was close enough to the accepted value for the velocity of light to justify Maxwell's claim that light was an electromagnetic wave.

Q 10.4
How far apart were the nodes in Hertz's experiment?

Another indirect method was used in 1946 by Essen and Gordon–Smith. They measured the frequency of the stationary electromagnetic waves that could be set up inside a metal cylinder, or resonator. It can be shown that the frequency, *f*, of waves set up in a right-circular cylinder is given by:

$$f = c\sqrt{\left[\left(\frac{r}{\pi D}\right)^2 + \left(\frac{n}{2L}\right)^2\right]}\left[\left(1-\frac{1}{2Q}\right)\right]$$

where D and L are the diameter and length of the cylinder, n is the number of half-wavelengths in the resonator, r a constant depending upon the mode and Q the quality factor. For electromagnetic waves of wavelength about 0.1 m they calculated that the velocity of light must be $2.997\,92\times10^8\ \mathrm{m\,s^{-1}}$; this value is quoted in Table 10.1.

The velocity of γ-rays

The value for the velocity of γ-rays was obtained by the experiment shown in Fig. 10.3(*a*). The coincidence counter can distinguish separate events if there is a time of at least 10^{-9} s between them. Positrons emitted by the ^{64}Cu source are stopped and annihilated in its metal container, and two γ-rays are emitted simultaneously as a result. The length of cable between counter 2 and the coincidence counter is adjusted until the pulses from counter 1 and counter 2 arrive simultaneously. The time taken by γ-ray 1 to travel the extra distance in air to counter 1 is now equal to the time taken for the signal from counter 2 to travel the extra length of the cable. If the speed of signals in coaxial cable is known, the velocity of γ-rays in air may be calculated. The value obtained was $2.983\times10^8\ \mathrm{m\,s^{-1}}$ using γ-rays of wavelength 2.5×10^{-12} m.

The velocity of microwaves

The velocity of 3 cm radio waves, or microwaves, can be measured using the arrangement in Fig. 10.3(*b*). The output of the microwave transmitter is modulated by a rapid switching circuit so that it produces short pulses at 200 kHz. These pulses are sent through a wax lens to a distant reflector and returned to a detector; after amplification they are displayed on a double-beam oscilloscope along with the transmitted pulses. The oscilloscope timebase is set to have a sweep speed of $1\ \mu\mathrm{s\,cm^{-1}}$, and from the spacing of the two pulses the time delay can be estimated. In the upper diagram, there is only a small path in air and the delay is much less; the reflected signal has, however, still been changed in shape.

Great care is required to align the reflectors, and even so the reflected signal is small and must be amplified; the time delay that is observed is also small. However, although the result of the experiment is only likely to fall within about 25% of the value for the velocity of light, it does suggest that with more accurate methods, they could be found to be the same.

Q 10.5
In a typical experiment the oscilloscope trace was set at a sweep speed of $1\ \mu\mathrm{s\,cm^{-1}}$, and the spacing of the pulses on the screen was 3 mm. The total path was 100 m. Calculate the velocity of the microwaves.

Table 10.1

Type of wave	Wavelength/m	Velocity/$10^8\ \mathrm{m\,s^{-1}}$
Radio	6.4	2.997 8 ± 0.000 3
Radar	0.1	2.997 92 ± 0.000 09
Infrared	4.2×10^{-3}	2.997 925 ± 0.000 001
Light	5.6×10^{-7}	2.997 931 ± 0.000 003
γ-rays	2.5×10^{-12}	2.983 ± 0.015

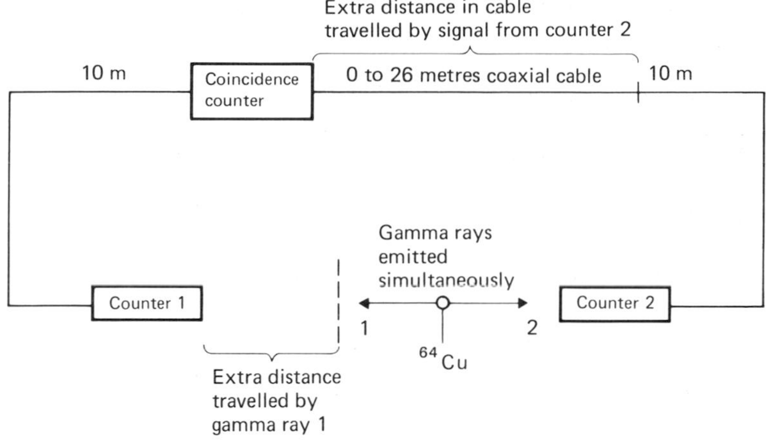

(*a*) Velocity of gamma rays

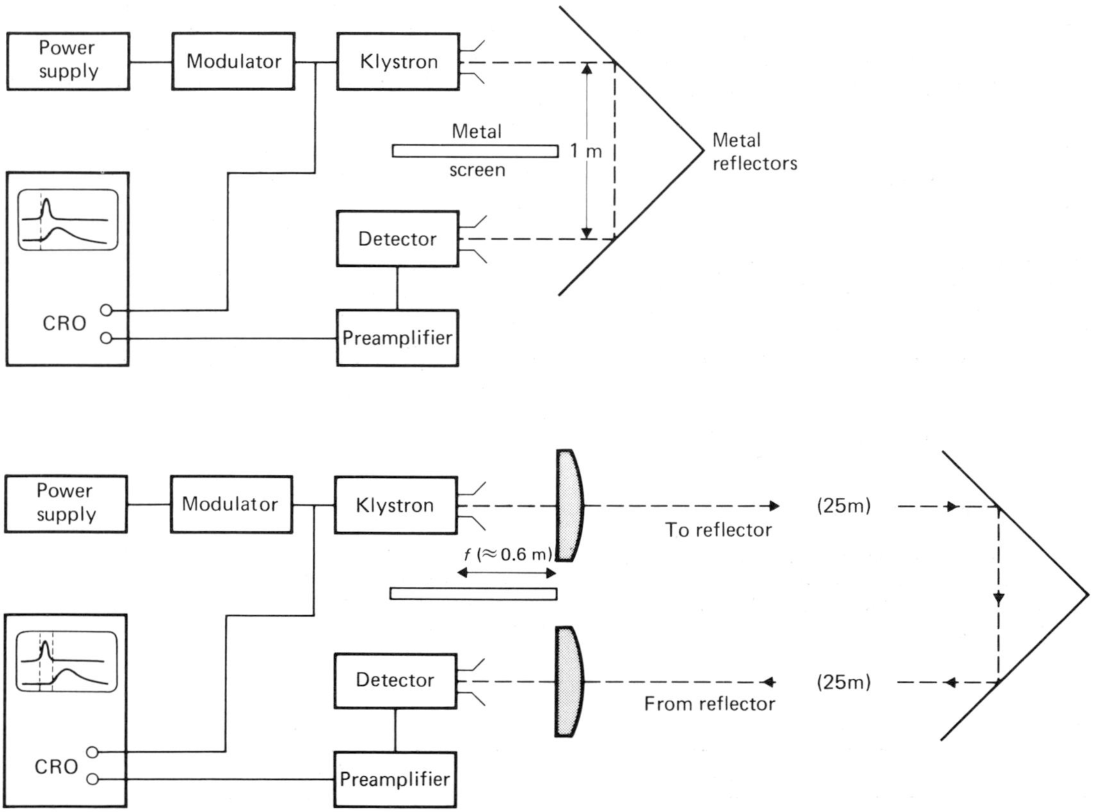

(*b*) Velocity of microwaves

Fig. 10.3

It must be emphasized that Maxwell proved that the velocity of light and of other electromagnetic waves are the same *in vacuo*; in other media dispersion will occur as the velocity of electromagnetic waves depends on their wavelength. In practice, the difference in velocities between *air* and a vacuum is so small that it can usually be ignored. The refractive index of air is about 1.000 277 at room temperature, and is almost independent of wavelength.

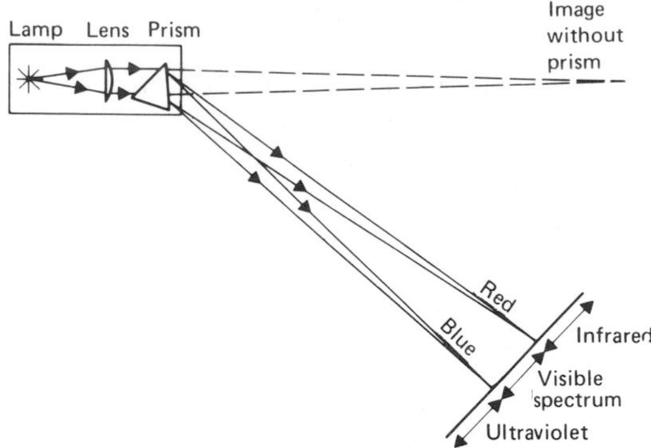

Fig. 10.4

Ultraviolet and infrared radiation

The wavelength of an electromagnetic wave is determined by its frequency, and this in turn is determined by the source of the energy. In this section, we shall see that certain ranges of the electromagnetic spectrum, such as radio waves or X-rays, are given names that are connected with the conditions of emission or detection.

As stated earlier, the visible spectrum contains those wavelengths to which our eyes are sensitive. The most important source of light in man's evolution is the sun, and it is not surprising then that the middle of the visible range coincides with that wavelength at which the sun gives out most energy. The radiation from the sun does not, however, contain only visible light; it also contains radiation of longer and shorter wavelength. It is the radiation of shorter wavelength, known as *ultraviolet* radiation, that makes us go brown in hot weather, whereas that of longer wavelength, known as *infrared* radiation, keeps us warm.

It is easiest to demonstrate the existence of ultraviolet and infrared radiation using a projector lamp; a similar experiment could be performed in bright sunlight. The parallel beam projector is set up to illuminate a high-dispersion prism, as shown in Fig. 10.4. Without the prism in place, the lens is adjusted so that it forms an image of the lamp the same distance away as the screen, and the prism is then placed in the position of minimum deviation. A visible spectrum is formed upon the screen.

A detector can be made of a photo-transistor, dry cell and milliammeter, connected in series. Although it gives a response when placed in the visible region, the response increases towards the red end of the spectrum and there is a peak *beyond* the visible red region. This is, therefore, the region of the spectrum where the most energy is arriving and it is termed the *infrared*; the wavelength is longer than for red light and the frequency lower. Hot objects give out more infrared radiation than cooler ones and infrared radiation is associated with the emission of heat. Infrared photography from satellites is used to survey vegetation, as some types of ground cover give out more heat than others.

Fluorescent paper gives out visible light when ultraviolet radiation falls upon it, so it may be used to enable us to observe the presence of ultraviolet in the spectrum. The paper is pinned to the screen and some fluorescence can then be seen beyond the blue end of the visible spectrum. It is helpful also to pin totally non-fluorescent paper to the screen for contrast, but this is hard to obtain as nowadays all paper has a certain amount of fluorescence to make it appear whiter. A more extended ultraviolet spectrum is seen if the light is allowed to fall on daylight printing paper; when it is developed the blackened area goes considerably beyond the blue end of the spectrum. The region beyond the blue is termed the ultraviolet and this light has a shorter wavelength but higher frequency. *Larger amounts of ultraviolet radiation are dangerous to the eyes and care must be taken with ultraviolet lamps*; fortunately, a considerable proportion of any ultraviolet present can be absorbed by a glass screen. In fact, to get the best results in the ultra-violet region of the spectrum, a quartz prism and 'quartz window' in the lamp should be used.

Electromagnetic spectrum chart

The electromagnetic spectrum is shown in the chart of Fig. 10.5; the column headed 'energy of a photon' may be ignored until the end of the chapter. The frequencies and wavelengths of the main types of radiation are shown; the boundaries are marked with wavy lines to indicate that they are approximate and, in many cases, arbitrary. Some wavelengths can be classified as two types of wave depending upon their origin; for instance, waves of wavelength 10^{-11} m are usually termed X-rays if they come from electron transitions outside the nucleus but γ-rays if their origin is in energy transitions inside the nucleus, as shown on the chart. *There is, of course, no physical difference between these waves at all, other than in their frequency and wavelength.*

The possible origin of each type of wave is shown to the left side of the chart, some of the possible detection methods are shown in the middle and a selection of uses to which the waves are put, on the right. We shall see later that different frequencies of electromagnetic radiation come in lumps of energy of different size; this explains both why they are produced in different ways and why different methods of detection are available for different frequencies. Further reference to this chart will be made and it should be carefully studied as required.

In our discussion of the nature of light so far the evidence requires a wave interpretation. The phenomena of physical and geometrical optics have been explained, important connections established with the subject of electromagnetism, and a coherent picture of the electromagnetic wave spectrum presented. It was perhaps the zenith of classical physics. A.A. Michelson, in 1894, even said that there were no fundamental discoveries left to be made, but only increasingly refined determinations of the physical constants to six or seven decimal places. However, as we shall see in the next section, the problems that remained led to a complete revolution in physical thinking equivalent only to that which occurred at the time of Newton. Just as Newton's laws of motion and gravitation enabled the purely empirical laws, such as those of Kepler, to be explained in terms of a smaller number of more fundamental laws, so quantum mechanics provided a basis for matters as diverse as magnetism, spectra, atomic bonds and the structure of matter itself.

Q 10.6
The chart of the electromagnetic spectrum is on a logarithmic scale. Suppose that a larger chart is to be made with a linear scale. The scale is to be such that the wavelength range from 10^{-6} m to 10^{-7} m—roughly the visible spectrum—occupies 1 cm. How long, roughly, will the chart be if it is to include wavelengths from 10^{-13} m to 10^{5} m? Try to relate this to a well known distance. How long would a ray of light take to travel along the chart?

10.3 The photo-electric effect

We have placed considerable emphasis in earlier sections on the similar properties of electromagnetic radiation of different wavelengths. However, we must also be able to explain why it is, for instance, that light affects ordinary photographic film, whereas radio waves do not, however intense the radiation or however long the exposure. It would appear that it is not only the energy density that is important in these effects, but also the frequency or wavelength of the radiation concerned.

There were other problems at the end of the nineteenth century which required explanation. In his experiments, Hertz had applied a high voltage across two metal spheres separated by a small gap and he noticed that the sparks jumped the gap more readily if it was illuminated by ultra-violet radiation.

The following year, Hallwachs showed that if a metal plate is charged negatively and placed in a beam of ultraviolet radiation, the charge is rapidly lost; on the other hand if it is charged positively, the charge is retained. This experiment can be repeated conveniently using a gold-leaf electroscope with a zinc cap placed a few centimetres from an earthed mesh; ultraviolet light releases negative charges from the surface of the zinc, but not positive charges.

In 1899 Lenard showed that these negative charges were electrons, particles which had only recently been identified by J.J. Thompson. The emission of electrons, stimulated by ultraviolet light, explained Hertz's observations; the

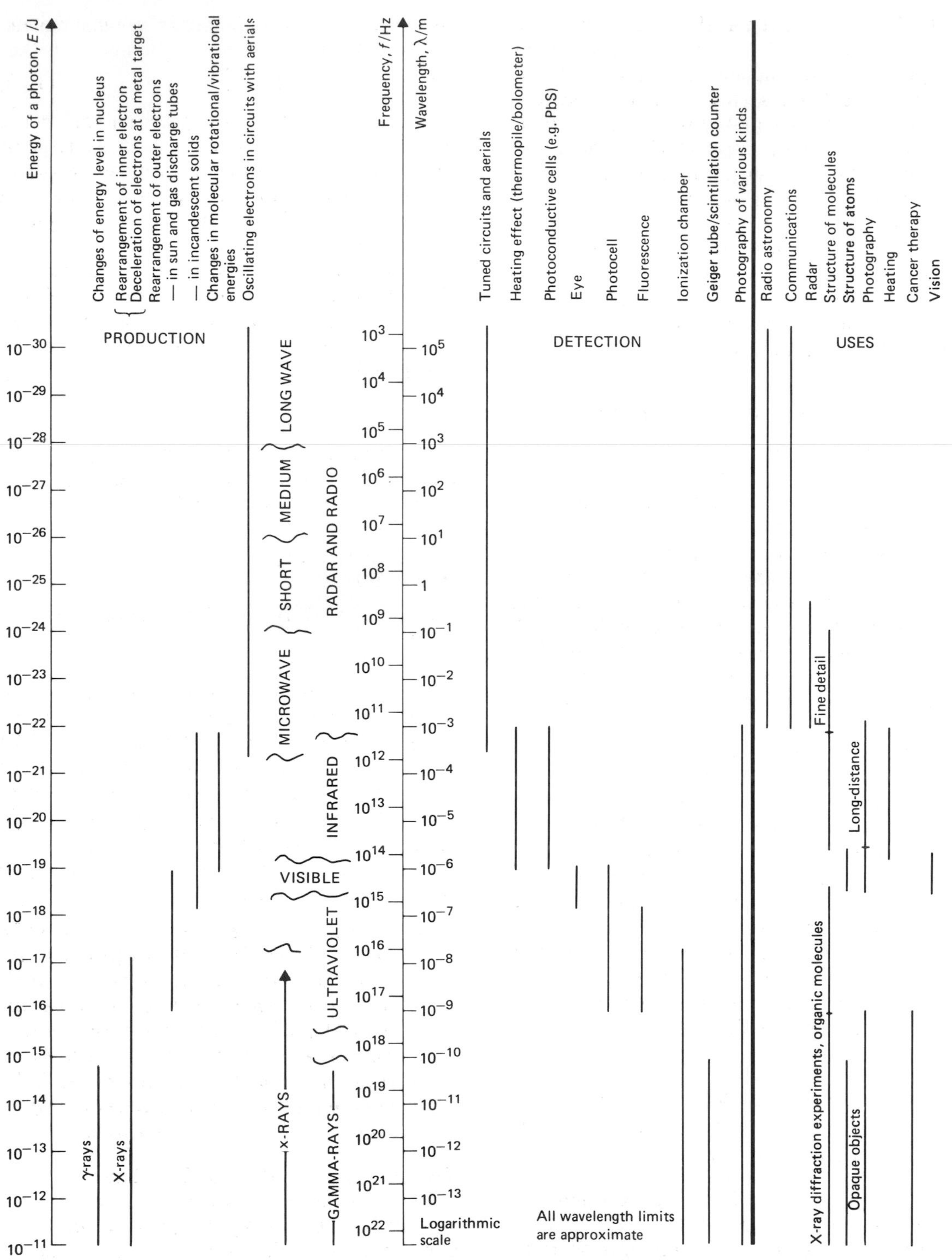

Fig. 10.5 The production, detection and uses of the electromagnetic spectrum.

emitted electrons acted as charge carriers in the gap. For more details of the establishment of the properties of the electron, the reader is referred to a book such as Bennet *Electricity and Modern Physics* or Millikan *The Electron.*

Other metals, such as the alkali metals, show the same effects when illuminated by visible light. The emission of electrons from a metal, stimulated by ultraviolet or visible light, is known as the *photoelectric effect*, and it was investigated in some detail by Lenard at the turn of the century; his results were to challenge the whole of the understanding of the wave nature of light at that time.

Experiments on the photoelectric effect

Although much of Lenard's work can be done with metal surfaces in the open air, the results are more consistent if the surface from which the electrons are to be emitted—the *emitter*—and the surface at which they are to be collected—the *collector*—are contained in a vacuum enclosure. This prevents the surfaces becoming dirty or oxidized during the experiment. A simple *vacuum photo-cell* is shown in Fig. 10.6(a). In the circuit shown in Fig. 10.6(b) the emitter is electrically isolated, being connected only to the input of a very high resistance voltmeter, such as the d.c. amplifier illustrated; the current through the voltmeter is so small that it does not alter the potential difference between collector and emitter at all. When the emitter gives out electrons it becomes positively charged and thus has a positive potential with respect to the collector; this potential difference is then measured by the voltmeter.

The photo-electrons will reach the collector if the kinetic energy with which they are emitted is greater than the work they must do moving through the potential difference between emitter and collector.

Now work done = charge × voltage

$$= qV \qquad (10.1)$$

and kinetic energy $= \frac{1}{2}mv^2$

where m is the mass of the electron and v is the velocity of emission.

When the source of electromagnetic radiation is first switched on, the potential difference between emitter and collector will rise until it is just sufficient to prevent further electron flow:

i.e.
$$qV = \tfrac{1}{2}mv^2 \qquad (10.2)$$

Thus, by measuring the potential difference with the voltmeter, and knowing the charge on an electron, it is possible to calculate the kinetic energies of the emitted photo-electrons; any calculation based on Equation (10.2) ignores relativistic effects.

It is also possible to use a microammeter with the photo-cell to measure the size of the photo-current, from which the number of photo-electrons emitted per second may be calculated.

Q 10.7

An electron is brought to rest by moving through a potential difference of 3 V. Its charge is 1.6×10^{-19} C and its mass 9.1×10^{-31} kg. Calculate:

(a) its kinetic energy,
(b) its velocity,

before it entered the field. Ignore relativistic effects.

If a suitable metal, such as potassium, is used for the emitter the photo-cell can be made sensitive to visible light. It is then placed so that a spectrum falls onto the emitter as shown in Fig. 10.6(c); the small slit can be moved to select the colour entering the photo-cell, and striking the metal surface.

Experiments using this arrangement give three important results:

(1) *the energy of the photo-electrons bears a linear relationship to the frequency of the light. For a given metal, there is a threshold frequency below which no electrons are emitted at all, however great the intensity*;
(2) *the energy of the photo-electrons is independent of the intensity of the light*;
(3) *the number of electrons emitted is proportional to the intensity of the light.*

The intensity of the incident light must be reduced by stopping down the beam and not by reducing the voltage supplied to the lamp, because if the brightness of the lamp is reduced the relative intensities of colours in the spectrum are also changed.

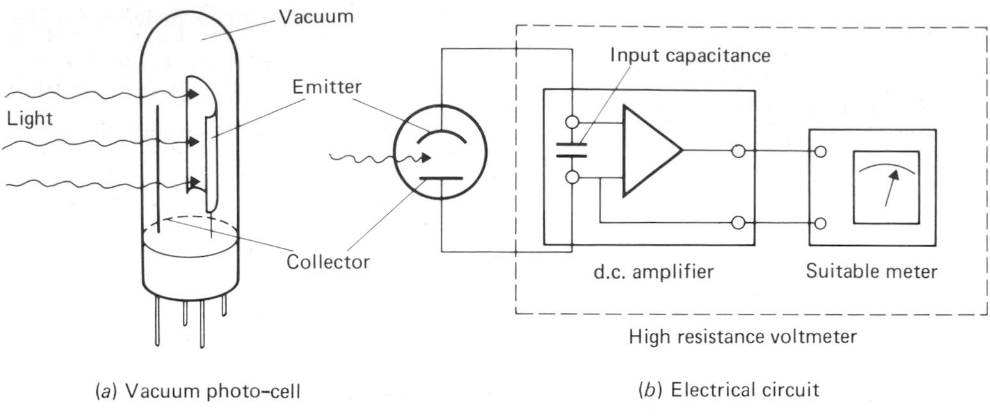

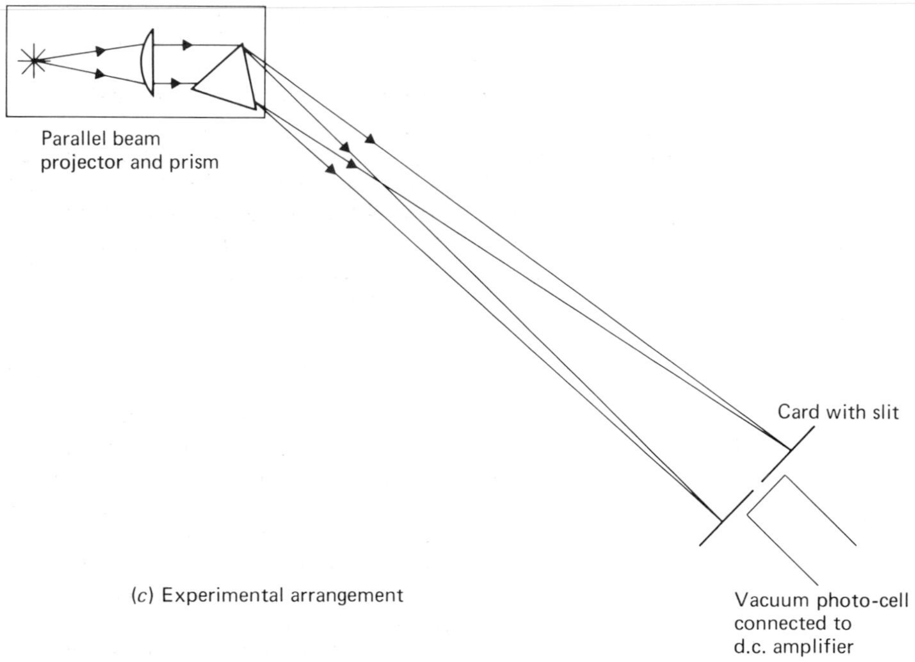

Fig. 10.6 Investigation of the photo-electric effect.

An alternative method of obtaining the above results is to allow the photo-current to flow through a microammeter and then apply a measured voltage in the opposite direction until the photo-current falls to zero. The size of the measured voltage is equal to that measured by the previous method; the circuit diagram is shown in Fig. 10.7.

Einstein's theory of photo-electricity

These results are at variance with the wave theory of light that has been developed so far. According to wave theory, high-intensity light would stimulate a lot of energetic photo-electrons whatever its frequency, and there should be no such thing as a threshold frequency. Clearly a new approach was needed.

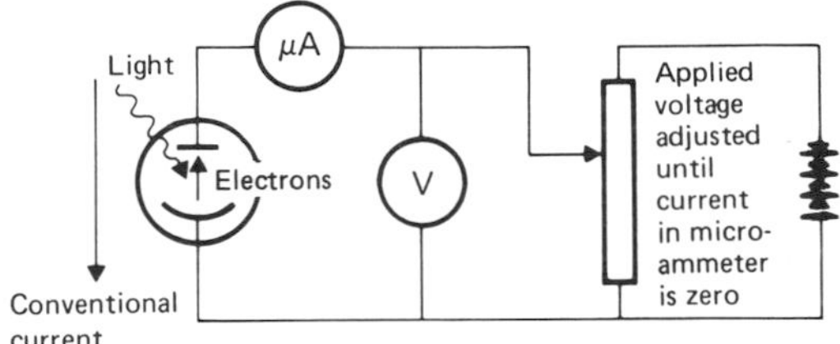

Fig. 10.7

In 1905, Einstein published three papers, each of fundamental importance to the development of physics. The best known is his paper on relativity but his Nobel prize was, in fact, later awarded for the paper on the photo-electric effect. In the extract printed below, Einstein explains the observations we have discussed above:

On a Heuristic Point of View about the Creation and Conversion of Light

'... According to the assumption considered here, when a light ray starting from a point is propagated, the energy is not continuously distributed over an ever increasing volume, but it consists of a finite number of energy quanta, localized in space, which move without being divided and which can be absorbed or emitted as a whole. ... The usual idea that the energy of light is continuously distributed over the space through which it travels meets with especially great difficulties when one tries to explain photoelectric phenomena, as was shown in the pioneering paper by Mr Lenard.
According to the idea that the incident light consists of energy quanta with an energy (hf) ..., one can picture the production of ... (electrons) by light as follows. Energy quanta penetrate into a surface layer of the body, and their energy is at least partly transformed into electron kinetic energy. The simplest picture is that a light quantum transfers all of its energy to a single electron; we shall assume that that happens. We must, however, not exclude the possibility that electrons only receive part of the energy from light quanta. An electron obtaining kinetic energy inside the body will have lost part of its kinetic energy when it has reached the surface. Moreover, we must assume that each electron on leaving the body must produce work W, which is characteristic for the body. Electrons which are excited at the surface and at right angles to it will leave the body with the greatest normal velocity. The kinetic energy of such electrons is

$$hf - W$$

If the body is charged to a positive potential V and surrounded by zero potential conductors, and if V is just able to prevent the loss of electricity by the body, we must have

$$Ve = hf - W$$

where e is the ... (charge) of the electron. ...

... If the formula derived here is correct, V must be, if drawn in Cartesian coordinates as a function of the frequency of the incident light, a straight line, the slope of which is independent of the nature of the substance studied. ...

... If every energy quantum of the incident light transfers its energy to electrons independently of all other quanta, the velocity distribution of the electrons ... will be independent of the intensity of the incident light; on the other hand ... the number of electrons leaving the body should be proportional to the intensity of the incident light ...'.

Einstein, A. (1905) *Ann. Phys.*, 17, 132. Translation in ter Haar, D. (1967) *The Old Quantum Theory.* Pergamon.

Einstein proposed that light energy came in quanta. This was not an original idea, for Planck had used the idea of 'lumps' of energy—or quanta—to develop a theory of the radiation of energy from hot bodies. In 1900 Planck had tried to derive a mathematical expression to describe the energy distribution in the heat radiation spectrum, and had eventually produced a formula to fit the experimental facts. He then realized that this empirical formula could be derived theoretically assuming that energy was emitted in 'lumps' rather than in a continuous stream; the amount of energy E, in a *quantum* was equal to hf, where f was the frequency of the radiation and h a constant, now known as *Planck's constant*:

$$E = hf \qquad (10.3)$$

Not unexpectedly, the new theory was regarded as too radical and unconvincing by most physicists of the time, and Planck himself was most disturbed by it. He preferred to view natural processes as continuous and initially he held on firmly to the 'continuous' Maxwellian wave theory of electromagnetism; the quantization process he applied only to the emission of heat radiation. Maxwell's theory had proved most successful in accounting for the properties of light and extending physicists' understanding to other parts of the electromagnetic spectrum. Planck had produced a completely new approach to the concept of the radiation of energy, which explained the facts it had been designed for, but which did not seem to be necessary elsewhere in physics.

Einstein, however, showed that the idea of the quantization of energy could be used to explain Lenard's experiments on the photo-electric effect. As a result, the theory became widely accepted as a general principle throughout physics.

Quanta of light are frequently called *photons*, and the energy of a photon of ultraviolet light will be about twice the energy of a photon of red light. It is most improbable that more than one photon can give energy to any one electron since the probability of a 'double event' is the product of the individual probabilities; thus if a metal is illuminated by light of frequency f, no one electron is likely to gain more energy than hf from the light, however great the intensity. If the intensity is increased there will be more photons of energy hf and thus *more electrons can each gain the same energy as before*.

The electrons in a metal have a continuous range of energies up to a maximum. An electron with the maximum energy still requires an energy W to escape from the metal surface; this can be supplied by any photon of energy greater than W. Thus, the incident light must have a frequency greater than, or equal to, the *threshold frequency*, f_0, where:

$$hf_0 = W \tag{10.4}$$

Less energetic electrons within the metal cannot be released by light of this frequency, and no electrons at all are emitted if the frequency is reduced. These situations are shown diagrammatically in parts (a) and (b) of Fig. 10.8; these diagrams are charts with energy as the vertical axis.

If the light is of frequency, f, which is higher than the threshold, the photons will have more energy than W. The remainder of the energy will be given to the electron as kinetic energy, $\frac{1}{2}mv^2$, and for an electron which has the maximum energy within the metal, we have:

$$(\tfrac{1}{2}mv^2)_{max} = hf - W \tag{10.5}$$

$$= hf - hf_0$$

i.e.

$$(\tfrac{1}{2}mv^2)_{max} = h(f - f_0) \tag{10.6}$$

Electrons which have less energy than the maximum within the metal will have less kinetic energy than this; these situations are shown diagrammatically in parts (c) and (d) of Fig. 10.8.

Thus, as the frequency of the light is increased the energy of the emitted electrons will also increase. The relationship is linear but *not*

(*a*) Frequency of incident light equal to threshold

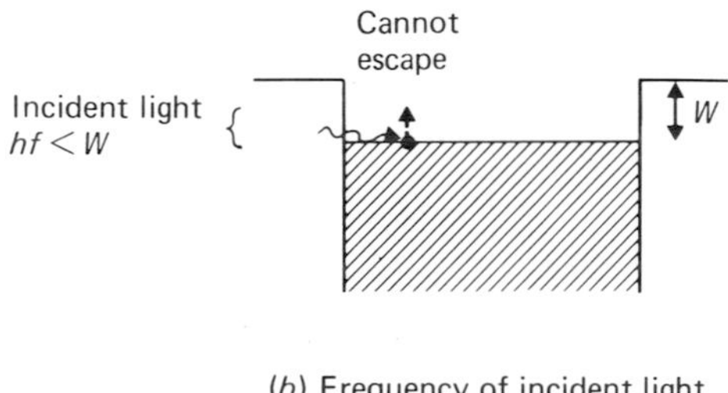

(*b*) Frequency of incident light below threshold

(*c*) Frequency of incident light above threshold

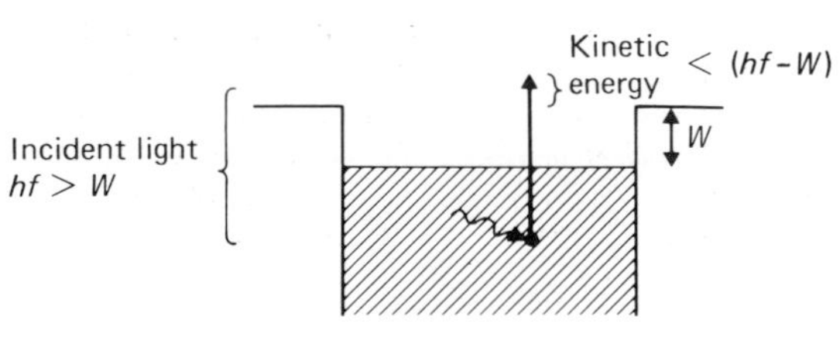

(*d*) Electron with less energy than maximum

Fig. 10.8 The emission of photo-electrons.

proportional, since the constant factor W is subtracted each time. If the intensity of the light is increased, the number of photons is increased but their energy remains constant; hence a larger number of photo-electrons will be emitted, each with the same energy as before.

The variation of the maximum energies of photo-electrons with the frequency of the incident light is shown in Fig. 10.9 for a series of metals. The intercept on the y-axis enables us to determine the energy W required to remove the most energetic electrons from each metal. The slopes of the graphs are all identical, being equal in size to Planck's constant; it is the slope of the graph described by the equation:

$$(\tfrac{1}{2}mv^2)_{max} = hf - W \qquad (10.5)$$

The graphs give a value for Planck's constant of about 6×10^{-34} J s, close to the currently accepted value of about 6.63×10^{-34} J s.

An alternative method of calculating a value of h is to take readings of the maximum kinetic energy of the emitted electrons at two frequencies for a particular metal. We then have:

$$(\tfrac{1}{2}mv^2)_1 = hf_1 - W$$

at one frequency

$$(\tfrac{1}{2}mv^2)_2 = hf_2 - W$$

at the other frequency

$$\therefore (\tfrac{1}{2}mv^2)_1 - (\tfrac{1}{2}mv^2)_2 = h(f_1 - f_2)$$

i.e.

$$h = \frac{(\tfrac{1}{2}mv^2)_1 - (\tfrac{1}{2}mv^2)_2}{(f_1 - f_2)} \qquad (10.7)$$

In each case $\tfrac{1}{2}mv^2$ is the maximum kinetic energy.

Q 10.8

In an experiment, illustrated in Fig. 10.6(b), with a potassium photocell, a maximum voltage of 0.37 V is measured when the cell is illuminated by blue light of frequency 6.5×10^{14} Hz; ultraviolet light of frequency 7.8×10^{14} Hz produces a voltage of 0.91 V. Calculate Planck's constant and the minimum energy, W, required to remove an electron from potassium. The charge on an electron is 1.6×10^{-19} C.

In the experiments described above the kinetic energy of the electrons was measured because the collector was negative with respect to the emitter. If the maximum potential

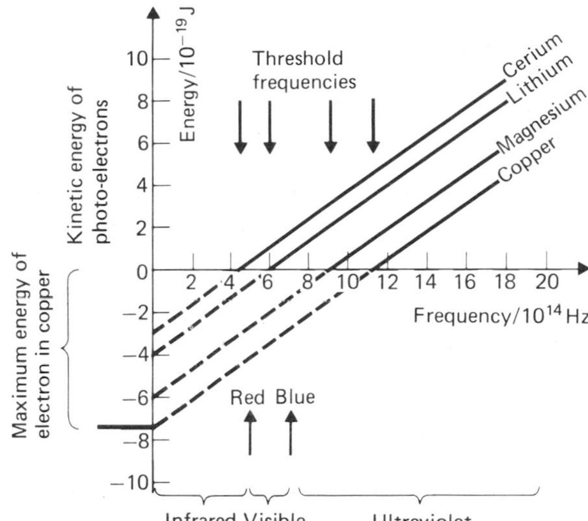

Fig. 10.9

difference through which the electrons can move is V, and the charge on an electron is q, we have:

$$(\tfrac{1}{2}mv^2)_{max} = qV \qquad (10.2)$$

Instead of measuring the energy in joules, we sometimes use the *electron-volt* (eV); an electron-volt is the energy an electron gains in being accelerated from rest through a potential difference of 1 volt. The amount of energy in an electron-volt is therefore given by:

$$\begin{aligned} \text{energy} &= qV \\ &= (1.6 \times 10^{-19}) \times 1 \\ &= 1.6 \times 10^{-19}\ \text{J} \end{aligned}$$

The kinetic energy of an electron that can just pass through an opposing potential difference of V volts, is V electron-volts.

The minimum energy required to remove an electron from a metal, W, is given by:

$$W = hf_0 \qquad (10.4)$$

where f_0 is called the threshold frequency. An amount of work, W, is done on the electron and this is equivalent to moving the charge, q, through a potential difference of Φ volts. Φ is called the *work function* of the metal, and is measured here in volts. If V is the retarding potential difference the equations we used earlier may be re-written in terms of potential differences.

The kinetic energy of a photo-electron was shown to be:

$$(\tfrac{1}{2}mv^2)_{max} = hf - W \qquad (10.5)$$

where the kinetic energy was related to the stopping potential V by the equation:

$$(\tfrac{1}{2}mv^2)_{max} = qV \qquad (10.2)$$

Hence $\qquad qV = hf - W \qquad (10.8)$

where $\qquad W = q\Phi$ from above (10.9)

Therefore $\qquad qV = hf - q\Phi$

$$V = \frac{hf}{q} - \Phi \qquad (10.10)$$

Q 10.9
Calculate the energy of an electron, in electron-volts and joules:

(*a*) if it is accelerated from rest through a potential difference of 6 V:
(*b*) when emitted from a metal if it is then brought to rest by a retarding potential difference of 2.33 V.

10.4 The properties of photons

Energy of photons

Einstein and Planck showed that the energy of a photon, E, is given by the equation:

$$E = hf \qquad (10.3)$$

where f is the frequency of the radiation and h is Planck's constant, 6.6×10^{-34} Js. We can use this to work out the energy of photons of different typical types of radiation; values given in Table 10.2 are approximate.

It is possible, given the energy of a photon, to calculate how many are emitted in one second by a source of a given power, as shown in Fig. 10.10. For example, for a 100 W light bulb:

$$\text{number of photons} = \frac{\text{power}}{\text{energy of a photon}} = \frac{100}{10^{-18}} = 10^{20} \text{ photons per second}$$

A weak γ-ray source, on the other hand, may emit 10^4 γ-photons in one second, each of energy 2×10^{-13} J. The power emitted is given by:

$$\text{power} = (\text{number of photons/second}) \times (\text{energy/photon}) = 10^4(2 \times 10^{-13}) = 2 \times 10^{-9}\text{ W}$$

Laboratory sources are weak; if there were 10^{20} photons per second as in the case of the light bulb, there would not be much container left:

$$\text{power} = 10^{20}(2 \times 10^{-13}) = 2 \times 10^7\text{ W} \equiv 10\,000 \text{ two-bar electric fires!}$$

Some other comparisons of this type are to be found in the chart and in the questions.

Q 10.10
A powerful transmitter of Radio 2 VHF emits 2×10^{30} photons per second at a frequency of 90 MHz. What is its power output; to how many two-bar fires is it equivalent?

Table 10.2

	Wavelength/m	Frequency/Hz	Energy of photon/J
Radio 2	1.5×10^3	2.0×10^5	1.3×10^{-28}
BBC London	3.2×10^{-1}	9.5×10^8	6.3×10^{-25}
Infrared	4.2×10^{-3}	7.1×10^{10}	4.7×10^{-23}
Green light	5.6×10^{-7}	5.4×10^{14}	3.5×10^{-19}
Ultraviolet	2.5×10^{-7}	1.2×10^{15}	8.0×10^{-19}
X-rays	1.5×10^{-10}	2.0×10^{18}	1.3×10^{-15}
γ-rays	1.0×10^{-12}	3.0×10^{20}	2.0×10^{-13}

Similar information is conveyed by the left-hand scales in Fig. 10.5 and Fig. 10.10.

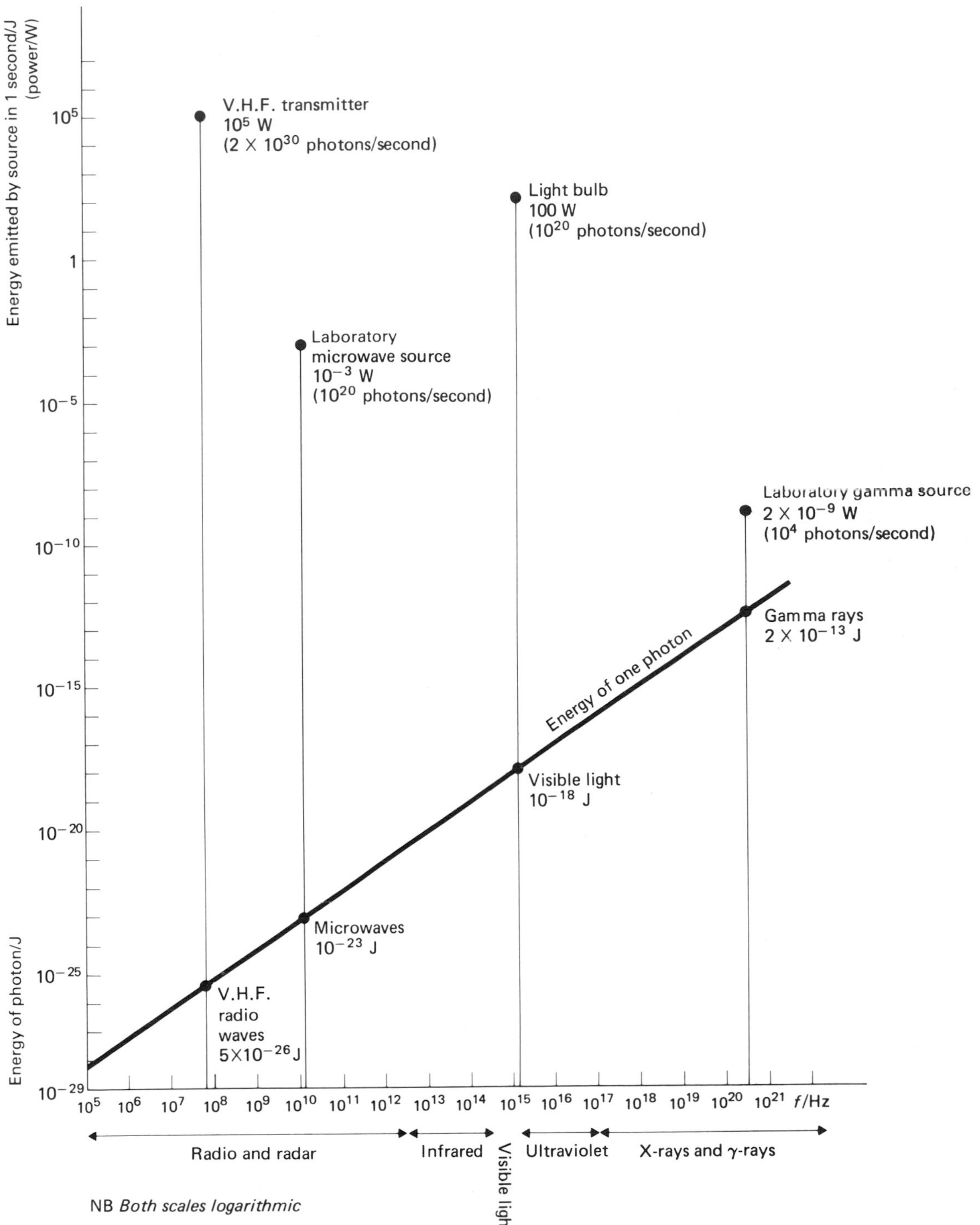

Fig. 10.10 This graph shows the variation in photon energy with frequency, together with the power of some typical sources of electromagnetic waves.

The number of photons given out by a light bulb may sound very large; how far apart are they? Suppose your eye is 1 m from a 100 W light bulb, and has a pupil of area 4 mm^2:

fraction of emitted photons entering eye

$$= \frac{\text{area of pupil}}{\text{area of 1 m sphere } (4\pi r^2)}$$

$$= \frac{4\times 10^{-6}}{4\pi}$$

$$\approx 3\times 10^{-7}$$

$$\approx \frac{1}{3\,000\,000}$$

number of photons entering eye per second

$$\approx \frac{10^{20}}{3\times 10^6}$$

$$\approx 3\times 10^{13}$$

Thus it is impossible for us to distinguish between individual photons as they enter the eye. Photons travel at 3×10^8 m s^{-1}; if all the photons entering the eye did so in a single stream we would have:

'distance between photons'

$$= \frac{\text{distance travelled by photon in 1 second}}{\text{no. of photons entering eye in 1 second}}$$

$$\approx \frac{3\times 10^8}{3\times 10^{13}}$$

$$\approx 10^{-5}\text{ m}$$

$$\approx 10^{-2}\text{ mm}$$

Q 10.11
How 'far apart' are γ-photons if they enter a counter at one thousand per second; assume they are in a single stream.

The momentum of photons

All particles that we meet in mechanics have momentum, defined as the product of the mass and velocity of the particle. Can a photon, which has no rest mass, have momentum? Maxwell predicted that electromagnetic waves would exert a pressure on an obstacle and in 1899, Lebedev succeeded in measuring it, in an elegant experiment using a mirror and a torsion balance. It is now known that the pressure at the earth's surface due to the sun's radiation is 10^{-6} N m^{-2}, compared with a pressure due to the atmosphere of 10^5 N m^{-2}. The pressure exerted by the sunlight, like that exerted by a jet of water from a hose pipe, is proportional to the rate of change of momentum of the beam of photons.

The theory of relativity can be used to relate the momentum of photons, p, to their energy, E; it can be shown that:

$$p = \frac{E}{c} \qquad (10.11)$$

$$= \frac{hf}{c}$$

i.e.

$$p = \frac{h}{\lambda} \qquad (10.12)$$

where f is the frequency and λ the wavelength of the photons.

Thus, although a photon has no rest mass, when it is moving with its velocity, c, it can be regarded as having a *mass-equivalent* given by:

$$m = \frac{p}{c}$$

i.e.

$$m = \frac{E}{c^2} \qquad \text{from (10.11)}$$

or

$$E = mc^2$$

This is the same equation as is used elsewhere in the Theory of Relativity to relate mass and energy.

Q 10.12
If the pressure at the earth's surface due to sunlight is 10^{-6} N m^{-2} calculate the number of photons arriving per second and hence the power per square metre incident at the earth's surface. This will, of course, be less than the power per square metre incident on the atmosphere because of absorption. Take $\lambda = 5.5\times 10^{-7}$ m.

10.5 The nature of light?

The next obvious step in our argument is to state which theory is correct, but unfortunately it's not quite so simple as that! Physical optics and Maxwell's electromagnetic theory provide evidence for a wave theory; the photo-electric effect and related phenomena provide evidence for a quantum picture. It is true that the quanta have a frequency but they behave in many ways like particles; the essence of a wave, on the other hand, is that the energy distribution is continuous.

Cut two fine slits, about 5 mm long, close together in a piece of thin card. Take it outside on a dark night and look through the slits at a street light, preferably of the gas-discharge type. If the card is held close to your eye, you will be able to see Young's slits fringes. Now look at distant house lights to see if the fringes can still be observed. You will probably find that they can just be seen at distances up to 1.5 km away.

Let us suppose that each slit measures $5\ \text{mm} \times 0.2\ \text{mm}$. The fraction of the emitted photons that pass through either slit will be given by:

$$\text{fraction} = \frac{\text{area of slits}}{\text{area of sphere radius 1.5 km}}$$
$$= \frac{2 \times 5 \times 0.2 \times 10^{-6}}{4\pi(1500)^2}$$
$$= 7 \times 10^{-14}$$

If the light bulb that is visible is rated at 100 W, the maximum possible number of photons emitted per second is 10^{20}; in fact, much of the electrical energy is converted into heat and the number of photons of visible radiation will be less than this.

$$\therefore \text{ maximum number of photons entering the eye} \approx 10^{-13} \times 10^{20}$$
$$\approx 10^7 \text{ per second}$$

Hence, assuming that the photons arrive in a single beam:

$$\text{separation of photons} \approx \frac{3 \times 10^8}{10^7}$$
$$\approx 30\ \text{m}$$

The significant point about this result is that only one photon is likely to be between the slits and the retina at once; with what does it interfere? If a photon is a particle it can go through either one slit or the other, but not through both. How is it that these 'single' photons hit the retina in exactly the right places to build up an interference pattern?

Let us consider the energy arriving at a point in an interference pattern from the points of view of both theories. According to the 'particle' theory, the energy arriving is determined by the equation:

$$\text{energy arriving at a point/s} = \text{number of photon/s} \times \text{energy of a photon}$$

The wave theory predicts that:

$$\text{energy arriving at a point/s} \propto (\text{amplitude of wave at that point})^2$$

The experiment described above shows us that all photons do not follow identical paths. Each photon must go to a point in the interference pattern, and more must arrive where the bright fringes are to be and less where there is to be almost total darkness. It would be impossible to predict where a particular photon will go but the large numbers that arrive over a period of time will be correctly distributed to give the expected pattern.

We can say that it is *more probable* that a photon will go to a bright fringe than to a dark one; the concept of probability is very important in quantum physics. Hence, instead of combining the two relationships above to give:

$$\textit{number} \text{ of photons arriving/s} \propto (\text{amplitude of wave at that point})^2$$

we say:

$$\textit{probability} \text{ of a photon arriving} \propto (\text{amplitude of wave at that point})^2$$

This process of building up a predictable pattern from the chance arrival of photons is illustrated in Fig. 10.11. Six photographs of a woman were taken, with a wide range of light intensities. When a relatively small number of photons had arrived, the random nature of their arrival was apparently important, and the picture confused. When a very large number had arrived, the expected pattern was clearly formed. It is like throwing dice; throw one and you cannot tell whether it will be a 'six' or not, but throw six hundred, and you will get very nearly one hundred 'sixes'.

Thus, both theories appear to be necessary to explain all the properties of electromagnetic waves, although as we have seen they may appear to be contradictory. In a photo-electric experiment, the quantum theory is needed to

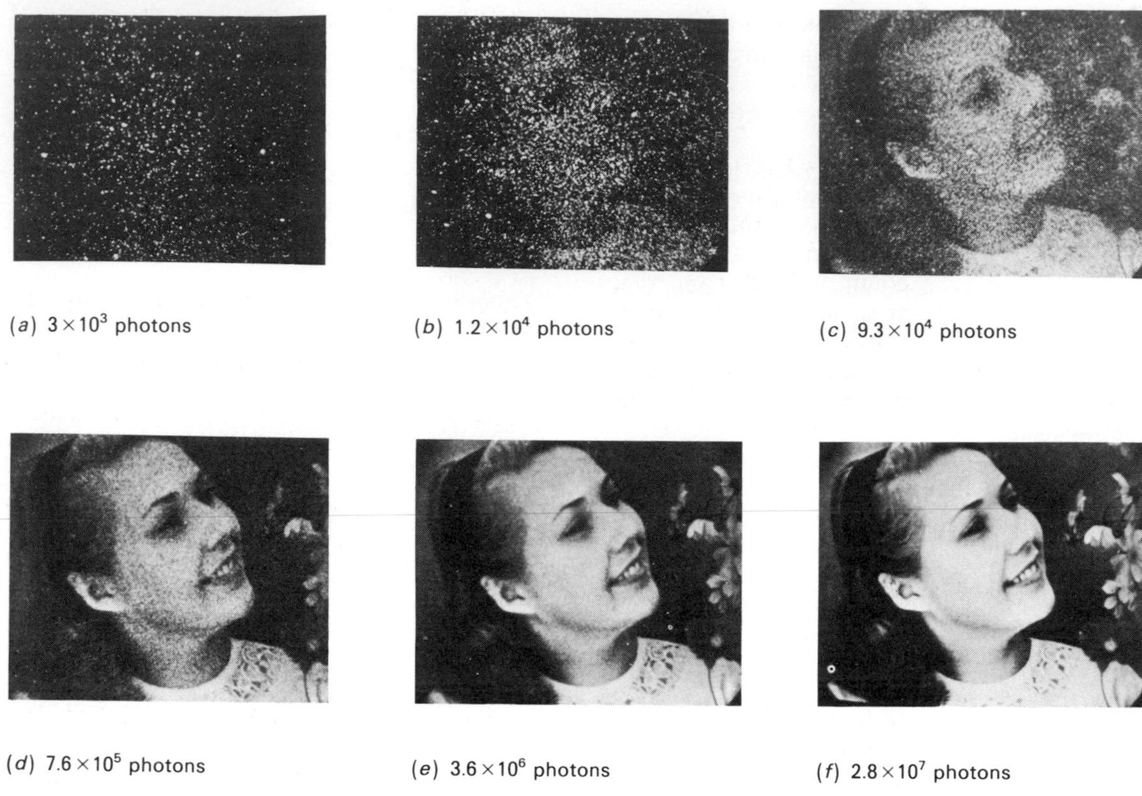

(*a*) 3×10^3 photons (*b*) 1.2×10^4 photons (*c*) 9.3×10^4 photons

(*d*) 7.6×10^5 photons (*e*) 3.6×10^6 photons (*f*) 2.8×10^7 photons

Fig. 10.11 Six photographs of a woman's face taken with increasing numbers of photons.

explain the absorption of photons and emission of electrons; nevertheless, the value of the frequency of light used in calculations is determined from measurements of the wavelength using wave superposition methods. The diffraction of X-rays may be analysed experimentally by considering their wave nature; however, the X-rays are detected by a Geiger tube which produces pulses as each X-ray photon arrives.

One possible over-simplification is to say that the particle theory is used to discuss interaction with matter, and the wave theory used to explain propagation. The truth is perhaps that there is no conceptual picture adequate to the reality and we must resort to the abstractions of the mathematics of quantum mechanics to try to provide one theory that will explain all situations.

'Things on a very small scale behave like nothing that you have any direct experience about. They do not behave like waves, they do not behave like particles, they do not behave like clouds, or billiard balls, or weights on springs, or like anything that you have ever seen.

Newton thought that light was made up of particles, but then it was discovered . . . that it behaves like a wave. Later, however . . . it was found that light did indeed sometimes behave like a particle. . . . Now we have given up. We say, "It is like *neither*".

Because atomic behaviour is so unlike ordinary experience, it is very difficult to get used to and it appears peculiar and mysterious to everyone, both to the novice and the experienced physicist. Even the experts do not understand it the way they would like to, and it is perfectly reasonable that they should not, because all of direct human experience and of human intuition applies to large objects. We know how large objects will act, but things on a small scale just do not act that way.'

Feynman, R.P. (1963) *Lectures on Physics.* McGraw-Hill.

Bibliography for Chapters 9 and 10

General accounts

Boulind, H.F. (1972) *Waves or Particles?* Longman; a short reader covering much of this ground at a straightforward level.

Mason, P. (1981) *The Light Fantastic.* Pelican; a most readable account of the development of optics and of our understanding of light.

PSSC (1981) *Physics* 5th Edition. Heath; Chapter 25 provides a well-illustrated account of the experimental evidence for the different theories of light.

Revised Nuffield Advanced Physics (1986) *Teachers' and Students' Guides.* Longman; Unit J in Chapter 2 includes details of suitable experiments on the topics covered in these chapters, as well as much of the theory.

Sobel, M.I. (1987) *Light.* University of Chicago Press; contains an excellent account of the development of our ideas about light together with much else.

Van Heel, A.C.S. and Velzel, C.H.F. (1968) *What is light?* Weidenfield and Nickolson; a more popular, but thorough, discussion of the nature of light.

Specific topics

Three first-rate college texts which consider the development of physics from an historical point of view:

Arons, A.B. (1965) *Development of Concepts of Physics.* Addison-Wesley, Chapters 26, 34.

Holton, G. (1973) *Introduction to Concepts and Theories in Physical Science.* Addison-Wesley, Chapters 23, 25, 26.

Sources of original writings on the nature of light:

Hertz, H. (1893) *Electric Waves.* Reprinted by Dover.

Huygens, C. (1678) *Treatise on Light.* Reprinted by Dover.

Magie, W.F. (1935) *Source Book in Physics.* McGraw-Hill; selected papers.

Millikan, G.H. (1917) *The Electron.* Reprinted by Dover.

Newton, Sir I. (1704) *Opticks.* Reprinted by Dover.

Segre, E. (1984) *From Falling Bodies to Radio Waves.* Freeman.

Two surveys of both early and modern experiments including, in the second case, reprinted papers:

Froome, K.D. and Essen, L. (1969) *Velocity of Light and Radio Waves.* Academic Press.

Sanders, J.H. (1965) *Velocity of Light.* Pergamon.

Books giving more detail of the discovery of the properties of the electron:

Bennet, G.A.G. (1974) *Electricity and Modern Physics* (2nd edn.). Edward Arnold.

Caro, D.E., McDonell, J.A., Spicer, B.M. (1978) *Modern Physics* (3rd Edn.). Edward Arnold.

Software

G.S.N. Educational Software *Physics Suite* Velocity of Light; enables the operation of Foucault's experiment to be simulated and the taking of readings discussed.

Longman Micro Software *Photoelectric Effect*; simulates the photoelectric emission experiments and enables them to be related to the photoelectric equation.

11
Waves and the nature of matter

11.1 Types of spectra

The last two chapters have been concerned with the nature of light, but in this final chapter we will apply our knowledge of light and understanding of waves to a discussion of the structure of matter. We saw earlier that a wide range of electromagnetic waves can be used to investigate atomic structure, and we shall begin by seeing what can be learnt from the study of spectra.

Continuous and line spectra

Newton passed rays of sunlight through a prism and observed a *continuous spectrum* with colours ranging from violet to red. All very hot objects give out light; that from a piece of common salt vaporized in a bunsen flame is always yellow, whilst that from a hot poker is red. A fluorescent tube, on the other hand, which is comparatively cold, also gives out a lot of light; the energy in this case comes from an electrical discharge. In all these cases there is an energy transformation, for instance from nuclear, heat or electrical energy, into light energy.

Thomas Melville, in 1752, was one of the first physicists to realize that luminous gases gave a different type of spectrum from that produced by hot objects, such as the sun. Luminous gases emit spectra that consist only of a number of discrete lines, or in some cases bands; each line corresponds to a particular wavelength of light. William Herschel, in 1823, proposed that it might be possible to identify the presence of elements by identifying their *characteristic line spectrum*, for he realized that different gases emit different lines. This technique is an important part of modern chemical analysis; it is, for example, possible to detect the presence of as little as 10^{-12} kg of sodium in any compound vaporized in a bunsen flame.

Hot objects, such as the sun or a poker, emit continuous spectra in which the wavelength varies continuously. Light from the sun ranges over the whole of the visible spectrum and more, whereas most of the energy from a red hot poker falls near the red end of the visible spectrum.

Examples of different types of spectra are shown in Fig. 11.1.

The two examples of line spectra shown are of hydrogen and mercury. The hydrogen is contained at a low pressure (2 mm Hg) in a sealed tube and the molecules are split into separate atoms by the discharge between electrodes in the tube; a potential difference of about 5000 V is applied and a characteristic pale pink colour produced. The spectrum is formed with either a prism or a diffraction grating; in spectroscopic work the diffraction grating is normally used and the wavelength of the lines may be calculated as explained in Section 7.6. The spectrum of mercury vapour is also shown and can be seen to contain different lines from those in the hydrogen spectrum.

Band spectra

The molecular nitrogen spectrum consists not just of single lines but of *bands*; each of these consists of a series of closely spaced lines which become progressively closer together towards the *head of the band*. This is called a *band emission spectrum* and it is produced only when the discharge is too weak to dissociate the molecules into atoms. Band spectra tend to be characteristic, not of the elements present, but of a particular type of bond, such as an O—H bond.

Fraunhofer lines and absorption spectra

Woollaston, in 1802, was the first person to see that the continuous spectrum produced by the sun was in fact crossed by a series of dark lines, although he failed to appreciate their significance. However, in 1814, Joseph Fraunhofer was examining the sun's spectrum and decided to use these dark lines as reference points in his investigation of other spectra. He discovered and labelled about 700 dark lines,

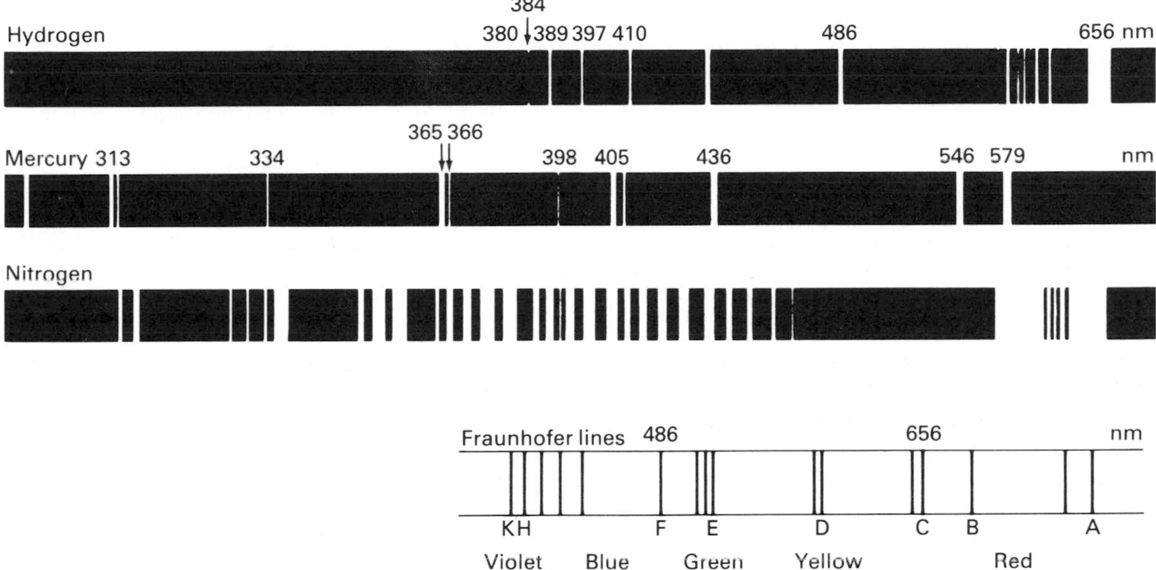

Fig. 11.1 Fraunhofer dark lines in the visible spectrum of sunlight. Only a few of the prominent lines are shown (adapted from A. B. Arons *Development of the Concepts of Physics* Addison-Wesley, 1965).

although improved techniques have since revealed that there are about 15 000; they are now referred to as *Fraunhofer lines*. It was 1859 before Kirchhoff observed that the wavelength of many of these dark lines corresponded exactly with those of emission lines for various elements. He proposed that the dark lines arose because *atoms which could emit light energy of a certain wavelength could also absorb energy at that wavelength*. Therefore if, say, hydrogen were present in the outer layers of the sun, it would be able to absorb energy from the continuous spectrum at the same wavelengths at which it emits light when excited in a gas discharge.

This effect can be seen in the simple laboratory demonstration in Fig. 11.2. A car headlamp bulb emits a continuous white-light spectrum; this may be seen by viewing the bulb through a direct-vision spectroscope as shown in Fig. 11.2(*a*). A sodium pencil placed in the bunsen flame in Fig. 11.2(*b*) emits yellow light in all directions; the characteristic yellow lines of the sodium spectrum are seen. If the headlamp bulb is placed behind the bunsen, as shown in Fig. 11.2(*c*), the spectrum observed is the continuous spectrum of the bulb but with dark lines where the sodium yellow lines were; this can be seen most effectively by switching the bulb on and off alternately. The sodium vapour has absorbed light at those wavelengths at which it also emits light; when the light is reradiated, however, it is emitted *in all directions* and thus there is a net loss of intensity *in the forward direction*. The spectrum seen in this case is known as the *absorption spectrum* of sodium. Because of Fraunhofer's original labelling of the dark lines in the solar spectrum, the very close prominent yellow lines are known as the 'sodium D lines'.

Fraunhofer lines are used to analyse the elements present in the outer layers of the sun, and in 1870 a set of lines was found that did not correspond to any known element. The existence of a new element called helium (Gk; *helios = the sun*) was proposed and twenty years later it was identified in a laboratory as one of the inert gases. Absorption spectra are also used in spectrographic analysis on earth. In order to explain these various types of spectra we must first look at the ways in which atoms can absorb and emit energy.

Q 11.1
Which, if any, of the elements hydrogen and mercury are shown to be present in the outer layers of the sun by the evidence of the photographs in Fig. 11.1?

11.2 Spectra and energy levels

The section on the photo-electric effect showed that light can only cause the emission of electrons from a particular metal if the frequency of the incident light is greater than a minimum threshold frequency; this is because the energy of the light photon has to be greater than the energy required by the electron to escape from the surface.

When we analyse line spectra we find that atoms of a given element can only absorb and emit light of certain frequencies; that is to say, these atoms can only absorb or emit photons of certain energies. Thus the atoms, instead of being able to have *any* energy, can exist only in certain energy states, or *energy levels*; the energies of the photons must be equal to the difference between two energy levels. For light of frequency, f, the energy of a photon, E, is given by:

$$E = hf \tag{10.3}$$

where h is Planck's constant. When a photon is absorbed the atom moves to a higher energy level, as shown in Fig. 11.3, and when energy is emitted it moves to a lower energy level. The energy level structure of an atom is analogous to a flight of stairs, where your potential energy can only change by an amount equal to the difference in energy between two steps.

According to this theory, if an atom had two energy levels 5.6×10^{-19} J apart, light of frequency f could be absorbed and emitted where:

$$\text{photon energy} = \frac{\text{difference between}}{\text{energy levels}}$$

i.e.

$$hf = 5.6 \times 10^{-19}\ \text{J}$$

$$f = \frac{5.6 \times 10^{-19}}{6.6 \times 10^{-34}}$$

$$= 8.5 \times 10^{14}\ \text{Hz}$$

It is usual, in fact, to express atomic energies in electron-volts:

$$\begin{array}{c}\text{energy difference}\\ \text{in electron-volts}\end{array} = \frac{\text{energy in joules}}{\text{charge on an electron}}$$

$$= \frac{5.6 \times 10^{-19}}{1.6 \times 10^{-19}}$$

$$= 3.5\ \text{eV}$$

Q 11.2

The wavelength of the light emitted as a result of the transition between two energy levels in the krypton-86 is 1/1 650 764 metre. Calculate, as accurately as you can, what the difference in energies must be between these two levels, giving the answer in joules and in electron-volts. The velocity of light is $2.997\,925 \times 10^{8}$ m s^{-1}, the charge on an electron is $1.602\,10 \times 10^{-19}$ C, and Planck's constant is $6.625\,59 \times 10^{-34}$ J s.

To explain the types of spectra that are observed we have introduced the idea of energy levels. Before developing this idea further we need to see whether it can be applied to other phenomena; as we pointed out in Chapter 10, a theory that only explains one phenomenon is unlikely to attain much credibility. Thus we look for *independent* evidence for energy levels, in other words evidence which is not dependent upon spectral phenomena.

The Franck–Hertz experiment

Energy is given to a cricket ball by hitting it with a bat, and an alternative way of supplying energy to atoms might be to hit them with other, fast-moving particles. If we found that energy was taken in from such collisions only in quanta of certain energy, then this would provide independent evidence for the existence of energy levels.

Most collisions between atomic, or subatomic, particles are *elastic*; the total kinetic energy of the particles is the same after the collision as it was before. In contrast, the collision between a lump of plasticine and a wall is *inelastic*; all the kinetic energy has been converted into other forms of energy such as heat. In 1914 J. Franck and G. Hertz, the nephew of H. Hertz, investigated to see whether

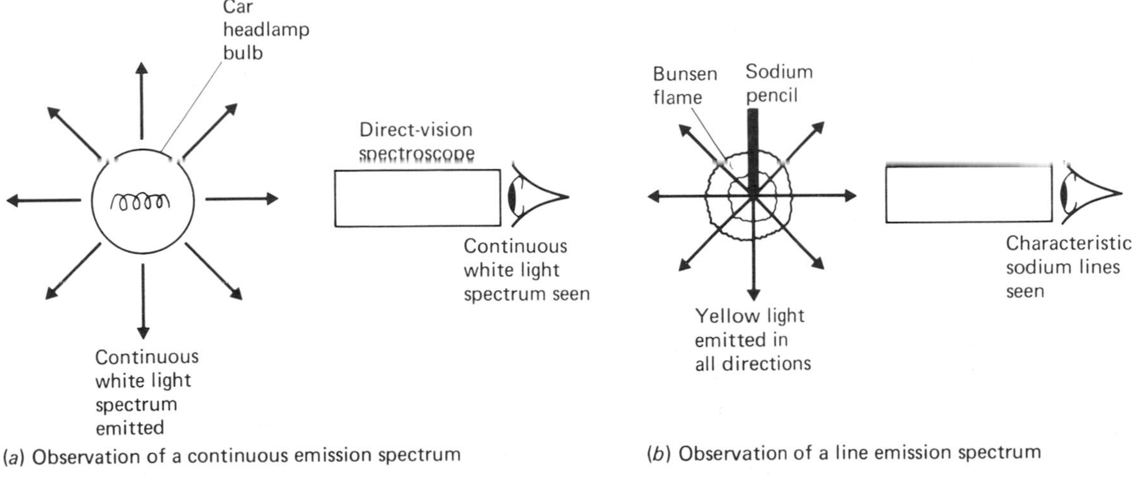

(*a*) Observation of a continuous emission spectrum

(*b*) Observation of a line emission spectrum

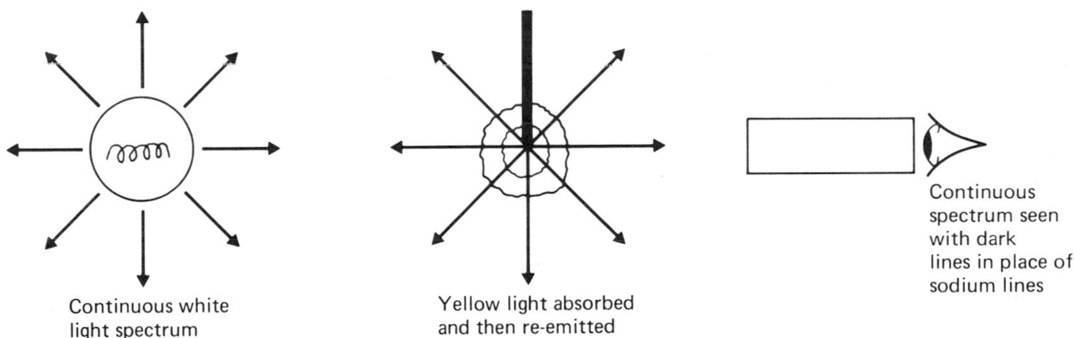

(*c*) Observation of a line absorption spectrum

Fig. 11.2

inelastic collisions would occur between moving electrons and atoms and, if so, whether energy was absorbed by the atoms in quanta of specific energy.

The energy gained by the atoms in such collisions is equal to the kinetic energy lost by the electrons, and Franck and Hertz measured the loss in kinetic energy of the electrons during the collisions. The experiment described below

Fig. 11.3

is a modern version of the original apparatus and is illustrated in Fig. 11.4.

The electrons are produced by indirect heating of a cathode K, where they are released by the process known as *thermionic emission*. They are then accelerated by two grids, g_1 and g_2, kept at voltages positive with respect to the cathode; the accelerating voltage can be varied up to about 30 V. The whole glass envelope contains mercury vapour at very low pressure and the accelerated electrons are likely to collide with the mercury atoms in the enclosure.

When we discussed the photo-electric effect, we saw how a retarding voltage could be used to measure the kinetic energies of electrons, and the same technique is used in this experiment. The collector A is made 1 V *negative with respect to* g_2 and electrons must therefore have a minimum kinetic energy at g_2 of 1 eV if they are to reach the collector.

In the discussion following, we assume that the collector current is determined partly by the accelerating potential difference between the cathode and g_2. If this potential difference is made too large it ceases to have any further effect on the current, which then depends only upon the rate of emission at the cathode; all the emitted electrons are reaching the collector and the tube is said to be *saturated.* The potential differences in this experiment are much smaller than those required for saturation.

Thus, as the potential difference between the cathode and g_2 is increased the collector current increases, as shown at the beginning of the graph in Fig. 11.5(*a*). However, when the potential difference reaches 4.9 V the current drops considerably before rising again; further drops occur at 9.8 V, and at succeeding 4.9 V intervals.

At the start of the experiment the vast majority of the mercury atoms are in their lowest energy level—this energy level is called the *ground state.* The results suggest that the *first excited energy level* is 4.9 eV above the ground state; when the electrons have been accelerated through 4.9 V they have just enough energy to excite the atoms. They use all their kinetic energy to do this and have none left to overcome the potential difference between g_2 and the collector; as a result the collector current falls suddenly at an accelerating potential difference of 4.9 V. Once this potential difference is greater than 5.9 V the electrons have enough energy to give up 4.9 eV to a mercury atom and still have 1 eV left to reach the collector; the current rises again. However, electrons with an energy of 9.8 eV have just sufficient energy to excite two mercury atoms in succession and, therefore, at this voltage the current falls for a second time.

The early experiments were not very sensitive and produced simple results like those in Fig. 11.5(*a*). The existence of a single energy level is shown and its energy in joules is given by:

$$\begin{aligned}\text{energy} &= 4.9\ \text{eV}\\ &= (1.6\times 10^{-19})\times 4.9\ \text{J}\\ &= 7.8\times 10^{-19}\ \text{J}\end{aligned}$$

Thus, a mercury atom can accept energy in a quantum of 4.9 eV (7.8×10^{-19} J).

We shall see in Section 11.3 that exciting an atom into higher energy states can be associated with moving one or more electrons into orbits that are further from the nucleus. Removing an electron from an atom completely is called *ionization*; the energy required to do this, called the *ionization energy*, is shown in Fig. 11.5(*b*) together with a number of other energy levels for mercury atoms.

When discussing energy levels it is usual, in fact, to regard the energy as zero when the atom is ionized; all other states are then expressed as negative energies and the ground state of mercury, for instance, is then at (−10.3 eV). The advantage of this approach is that the zero from which energies are measured is the same for all atoms. The energy level structure of mercury from this point of view is shown in Fig. 11.5(*c*).

This experiment provides independent evidence for the existence of energy levels in a mercury atom, and we will now examine a mercury spectrum to see if it predicts the same energy level structure.

Q 11.3

The ionization energy for a mercury atom is seen to be 10.3 eV higher than the ground state. What is this energy in joules and how fast would an electron have to travel to ionize a mercury atom? Ignore relativistic effects and state at the end whether this assumption is justified. The mass of an electron is 9.1×10^{-31} kg.

The spectrum of mercury vapour

A suitable experiment to investigate the spectrum of mercury vapour is shown in Fig. 11.6. A mercury discharge lamp shines through a small slit in a screen onto a concave reflection grating; the reflected spectrum falls onto fluorescent paper attached to the screen. The lamp has a hole drilled in it so that the ultraviolet light will not be absorbed by the glass, and the slit in the screen is about 1 mm wide. The concave grating is placed at a distance from the screen equal to its radius of curvature (typically about 0.5 m) and orientated so that the zero-order line is above the slit, with a spectrum to either side. It is important to distinguish which

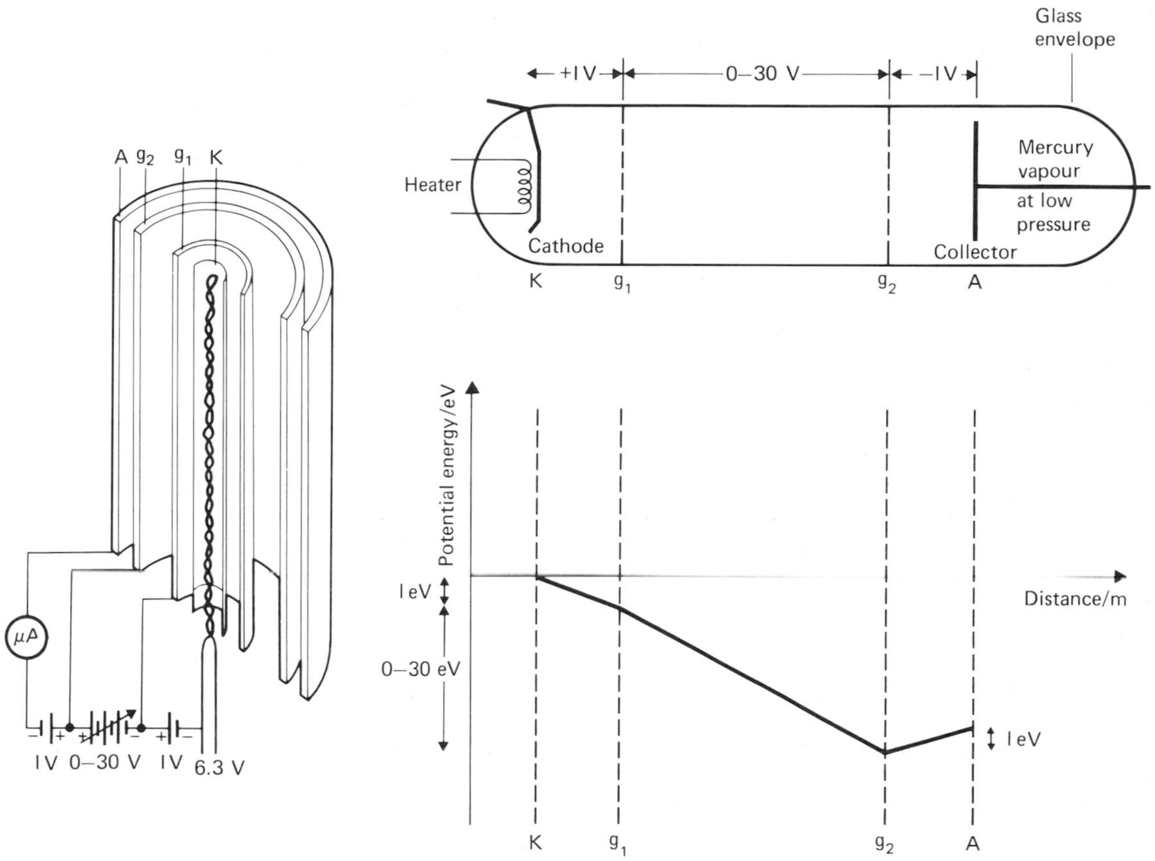

Fig. 11.4 The Franck–Hertz experiment.

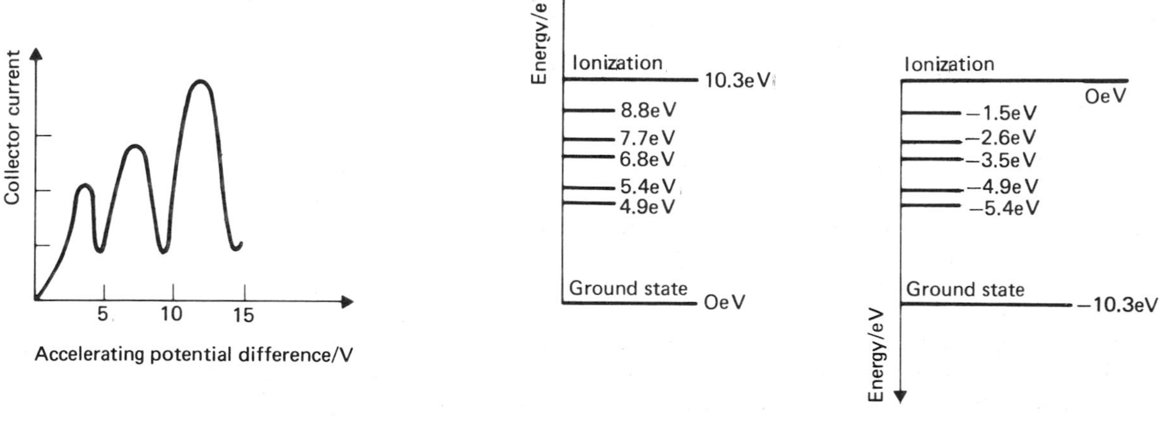

(*a*) Results of Franck-Hertz experiment

(*b*) Simplified energy level structure for mercury

(*c*) Simplified energy level structure with zero energy at ionization

Fig. 11.5

lines lie in the ultraviolet and which do not; a piece of glass may be inserted in the beam to absorb the ultraviolet light, leaving only the visible spectrum on the screen.

If the lamp is run at low power, the spectrum contains six lines. Line A is the zero-order line and lines C, E and F are also at visible wavelengths; they form the visible part of the first-order spectrum and are, in fact, the violet, green and yellow lines respectively. Lines B and D are absorbed by a piece of glass and must, therefore, lie in the ultraviolet part of the spectrum. As ultraviolet light has a shorter wavelength than violet light, line B is clearly the first-order ultraviolet line and line D is the second-order line for the same wavelength.

Our thesis is that each line corresponds to a jump from one energy level to another, and if that is the case, the lines of shortest wavelength and highest frequency, will correspond to the largest jumps, since we are assuming the relationship $E = hf$. We wish to see whether the first energy level is 4.9 eV above the ground state; if so, photons of energy 4.9 eV will be emitted when the atom changes from this state to the ground state. From the evidence above we cannot, however, tell which lines correspond to transitions to the ground state; to do this we look for an absorption effect using cold mercury atoms, which have a very high probability of being in the ground state.

A small quantity of mercury vapour is puffed through the light from the lamp by squeezing a small polythene bottle containing mercury; *the bottle must not be warmed to increase the amount of vapour, as it is very poisonous.* The ultraviolet lines vanish; light of this frequency must have been absorbed by the cold mercury atoms. Hence, it is the ultraviolet line that corresponds to transitions between the ground state and a higher energy level.

The wavelength of the ultraviolet line is measured by using the usual formula for a diffraction grating and is found to be 2.53×10^{-7} m. Alternatively, if the wavelength of the green line immediately adjacent to the second-order ultraviolet line has been determined in this way, the wavelength of the ultraviolet line may be obtained by comparison. Therefore

$$\text{frequency} = \frac{\text{velocity of light}}{\text{wavelength}} = \frac{3 \times 10^{8}}{2.53 \times 10^{-7}} = 1.19 \times 10^{15}\ \text{Hz}$$

Hence

$$\begin{aligned}\text{energy of photon, } E &= (\text{Planck's constant}) \times (\text{frequency}) \\ &= (6.6 \times 10^{-34}) \times (1.19 \times 10^{15})\ \text{J} \\ &= 7.83 \times 10^{-19}\ \text{J} \\ &= \frac{7.83 \times 10^{-19}}{1.6 \times 10^{-19}}\ \text{eV}\end{aligned}$$

i.e. $E = 4.89$ eV

Hence, the energy level involved is 4.9 eV above the ground state; this agrees with the information deduced from the Franck and Hertz experiment. These two independent experiments present strong evidence for the existence of discrete energy levels in atoms, and hence for the truth of the quantum hypothesis. In general, the agreement between energy level measurements by collision experiments and by

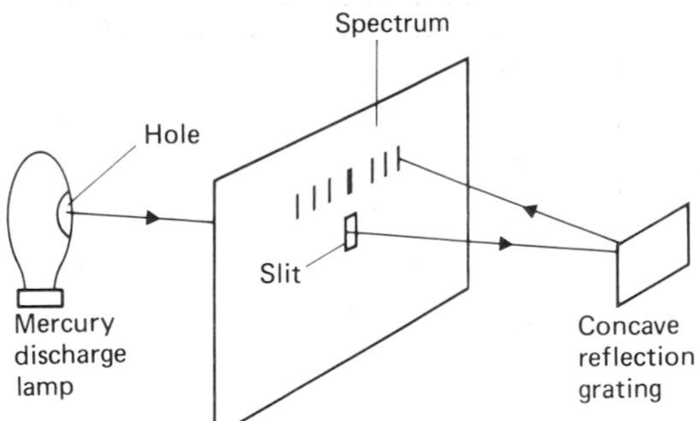

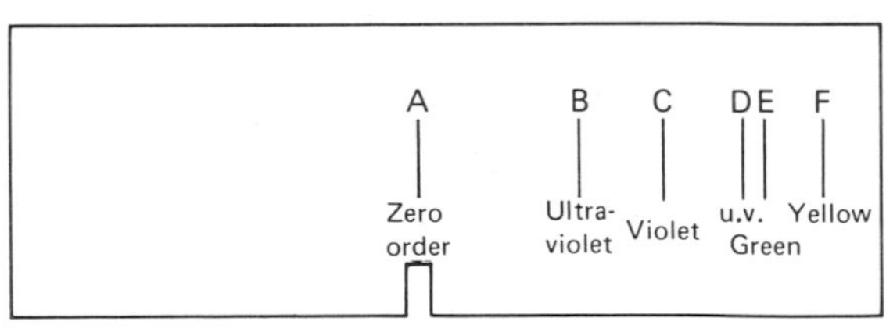

Fig. 11.6 The spectrum of mercury vapour.

spectral analysis is rarely as precise as is suggested above; the spectral information is the more accurate and is thus used to investigate the energy level structure of atoms and molecules in detail.

Q 11.4
An ant of mass 1 mg lives on a flight of irregular stairs. The first step is 1 mm above the ground. Supposing that all the potential energy an ant loses could be converted into a photon, calculate the frequency of the resulting electromagnetic wave when the ant steps down from the first step to the ground. What type of electromagnetic photon would it be?

We see from this question that the energy level differences in an atom are very small indeed. How many mercury atoms would each have to give out an ultraviolet photon to emit the same amount of energy as the ant? What mass of mercury would that be? The relative atomic mass of mercury is 200 and Avogadro's number is 6×10^{23} mol^{-1}.

Fig. 11.7 shows how the mercury spectrum described in the experiment is related to the simplified energy level structure mentioned earlier; there are, of course, a great many other lines not visible with the elementary apparatus used, and the complete energy level structure is much more complex than that shown. Remember that the energy of the photons, and hence the frequency of the emitted light, is determined by the difference in energy levels; a transition between two high energy levels will cause a photon of low energy to be emitted if, as is often the case, the energy levels lie close together.

Q 11.5
Copy and complete the chart for the green line in Fig. 11.7 and identify the energy levels between which the corresponding transition must occur.

The interpretation of other spectra

Spectra of many types can be interpreted in terms of an energy level structure; they can be used to identify substances or interatomic bond arrangements. Some uses of spectroscopy at different wavelengths are shown in Fig. 11.8.

11.3 Atomic energy levels and the Bohr theory

In the last section we showed that the energy level structure of a mercury atom predicted from the spectrum of mercury vapour coincided with that predicted by the Franck and Hertz collision experiments. We shall now try to explain briefly the energy level structure of atoms in terms of the arrangement of the elec-

Colour	Wavelength λ/nm	Frequency $f/10^{14}$ Hz	Photon energy	
			$hf/10^{-19}$ J	hf/eV
Ultraviolet	253	11.9	7.8	4.9
Violet	435	6.9	4.5	2.8
Green	546	5.5		
Yellow	579	5.2	3.2	2.0

(1nm = 10^{-9} m)

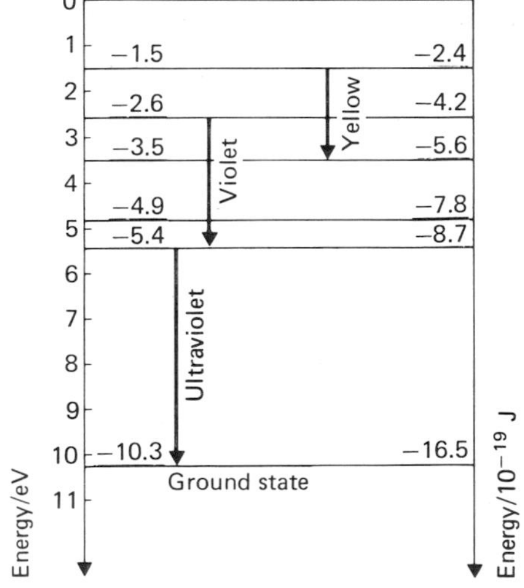

Fig. 11.7 The spectrum and energy level structure of mercury.

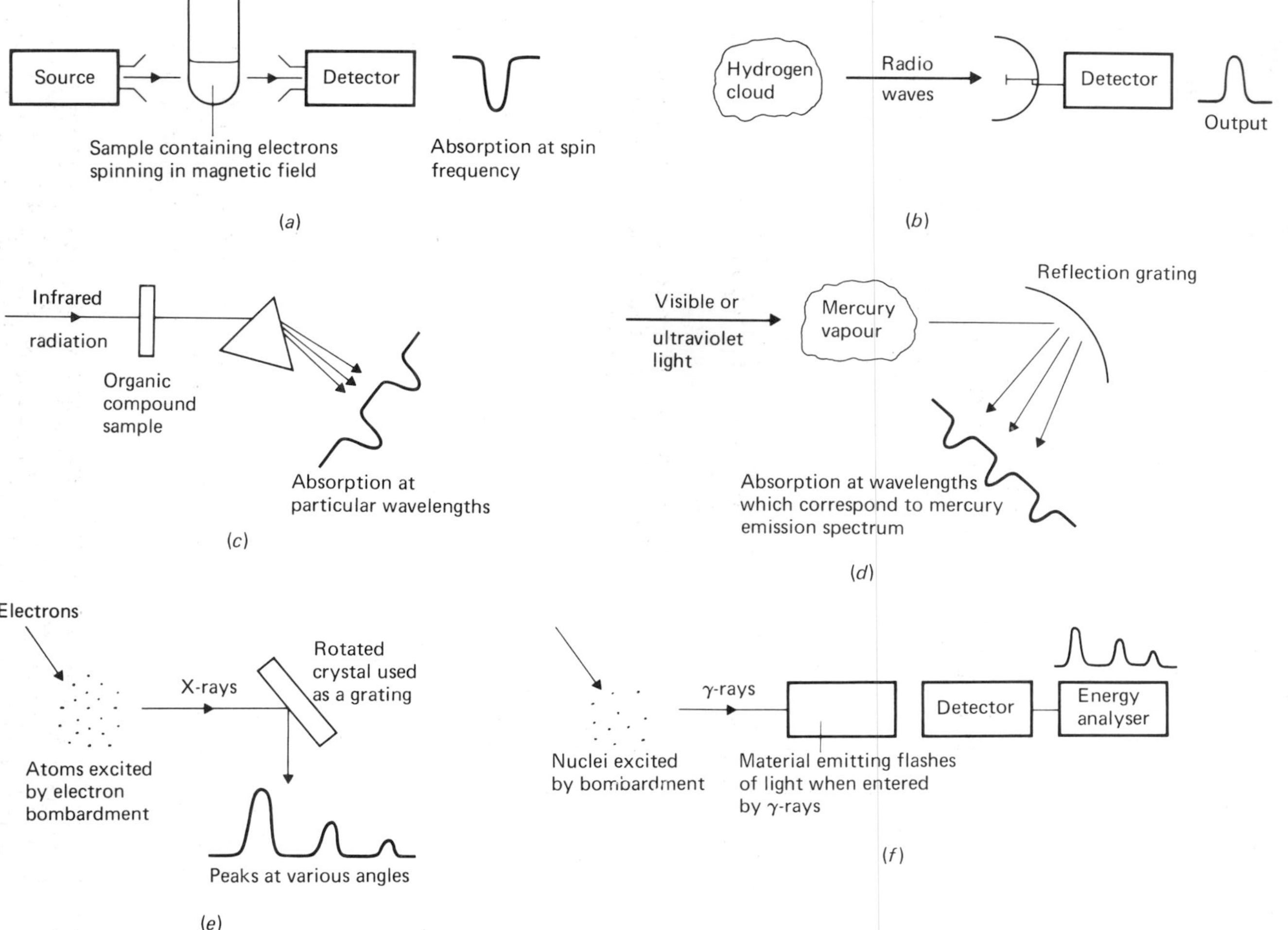

Fig. 11.8 Some uses of spectroscopy. (Adapted with permission, from Nuffield Advanced Physics (1971) Teachers' Guide, Unit 8, *Electromagnetic Waves.* Longman). (a) *Electron spin resonances* ($\lambda \approx 1$ m): radiation absorbed by electrons in the presence of a magnetic field. Information obtained about the electron-environment in the sample. (b) *Radio astronomy* ($\lambda \approx 0.2$ m): this wavelength is emitted from a hydrogen cloud in our galaxy, because of small rearrangements of electrons in the hydrogen atoms. (c) *Infra-red spectroscopy* ($\lambda \approx 10^{-6}$ m): radiation absorbed in the oscillations of atoms bonded into molecules. Information gained about interatomic force constants, molecular structure or substances present. (d) *Visible and ultra-violet spectroscopy* ($\lambda \approx 10^{-7}$ m): the absorption or emission of radiation by electrons changing energy levels. Information gained about energy level structures, especially of outermost electrons. (e) *X-ray spectroscopy* ($\lambda \approx 10^{-10}$ m): radiation from electron transitions deep within atoms, giving evidence of deep energy levels. (f) *γ-ray spectroscopy* ($\lambda \approx 10^{-12}$ m): analysis of the radiation from bombarded nuclei gives information about nuclear energy levels.

trons that are outside the nucleus. This is a complex subject and we shall consider only the simplest case, that of hydrogen, which has just one electron outside a nucleus consisting of one proton. For a more complete discussion, the reader is referred to books on atomic physics such as Bennet *Electricity and Modern Physics* or Caro, McDonnell and Spicer *Modern Physics*.

The spectrum of hydrogen

The lines in a hydrogen spectrum form a number of series, in each of which the lines get closer and closer together towards the short-wavelength end. The first progress towards understanding these series was made in 1885 by Balmer who was able to produce an empirical mathematical formula which predicted the wavelength of the lines in the series that now bears his name; this series is shown in Fig. 11.9. The formula derived by Balmer is:

$$\frac{1}{\lambda}=R\left(\frac{1}{2^2}-\frac{1}{n^2}\right) \quad \text{Balmer series } n>2 \qquad (11.1)$$

where n is an integer greater than 2, R is a constant and λ is the wavelength. The formulae for the wavelengths of lines in two of the other series, the Lyman and the Paschen, are:

$$\frac{1}{\lambda}=R\left(\frac{1}{1^2}-\frac{1}{n^2}\right) \quad \text{Lyman series } n>1 \qquad (11.2)$$

and

$$\frac{1}{\lambda}=R\left(\frac{1}{3^2}-\frac{1}{n^2}\right) \quad \text{Paschen series } n>3 \qquad (11.3)$$

The same constant, R, is used in each formula; it is now known as the *Rydberg constant* and is found to have the value $1.096\,78\times10^7\ \text{m}^{-1}$. The frequency of the lines can be calculated using the equation:

$$\text{frequency}=\frac{\text{velocity}}{\text{wavelength}}$$

or, directly, by using formulae of the form:

$$f=cR\left(\frac{1}{n_1^2}-\frac{1}{n_2^2}\right) \qquad (11.4)$$

$$\begin{cases}\text{Lyman series: } n_1=1, n_2>1\\ \text{Balmer series: } n_1=2, n_2>2\\ \text{Paschen series: } n_1=3, n_2>3\end{cases}$$

The wavelengths predicted by Balmer using his formula were extremely close to those measured by Ångström as shown in Table 11.1.

Balmer was impressed by the close correspondence:

> '(this) is really a striking evidence for the great scientific skill and care with which Ångström must have gone to work!'

So far we have taken the various lines in the hydrogen spectrum and arranged them in series, each accurately described by Equation (11.4).

Table 11.1

Wavelength/10^{-7} m	
Prediction	Observation
6.564 67	6.562 10
4.862 72	4.860 74
4.341 71	4.340 1
4.102 92	4.101 2

The next stage in the discussion is to relate the frequencies of the lines to the energy level structure of hydrogen. For this we must recall

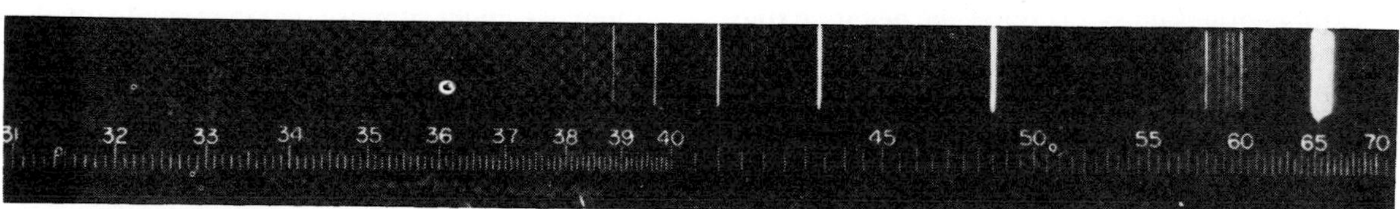

Fig. 11.9 Hydrogen spectrum: the Balmer series.

that:

energy of a photon = (Planck's constant) × (frequency)

where the energy of the emitted photon is, of course, equal to the change in energy of the atom. Thus, if an atom changes its energy state from E_n to E_1 we have:

$$E_n - E_1 = hf$$

where f is the frequency of the emitted light.

Let us consider the Lyman series, which is described by Equation (11.2), rewritten in the form:

$$f = cR\left(\frac{1}{1^2} - \frac{1}{n_2^2}\right) \qquad (11.2)$$

where n_2 can be any whole number greater than 1. This equation can be rewritten:

$$f = \frac{cR}{1^2} - \frac{cR}{n_2^2}$$

and if we are considering an energy transition from E_n to E_1 we then have:

$$E_n - E_1 = \frac{hcR}{1^2} - \frac{hcR}{n_2^2}$$

It is possible to explain the Lyman series of lines by postulating that E_1 represents the ground state, or lowest possible energy level, of hydrogen where:

$$E_1 = -\frac{hcR}{1^2}$$

The transitions to the ground state may be from a series of other states whose energy is determined by the value of n_2:

$$E_n = -\frac{hcR}{n_2^2}$$

Both energies must, of course, be negative since the ionization energy is taken to be zero, as explained in the last section.

The energy level structure giving rise to the Lyman lines is shown in Fig. 11.10. The highest frequency line in the series will be caused by a transition from the ionization level to the ground state.

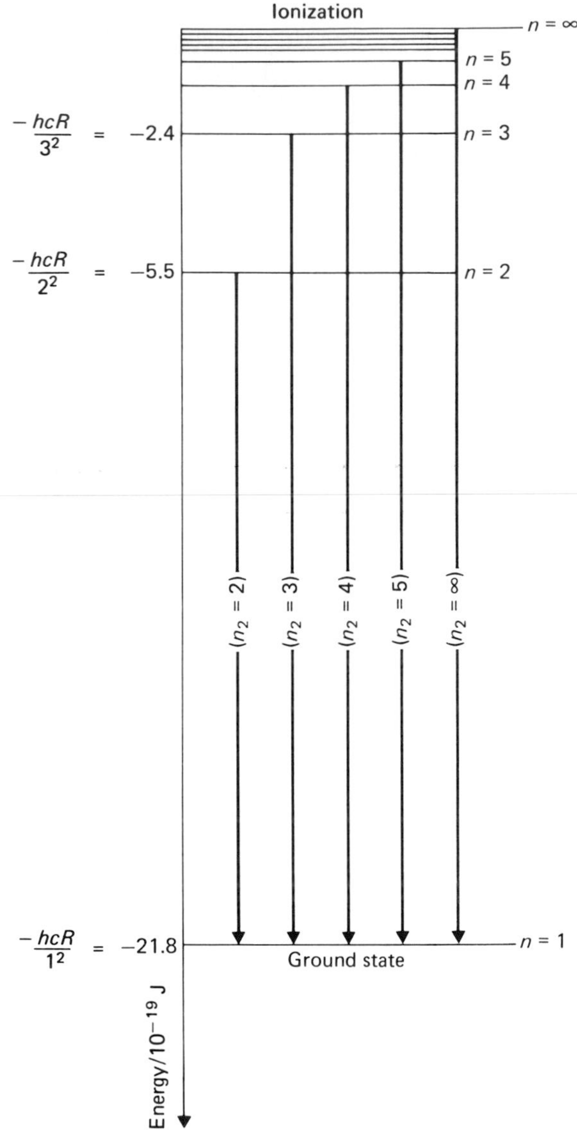

Fig. 11.10 Hydrogen spectrum: the Lyman series.

The next important thing to realize about these three series is the relationship between them; we might suppose that they would involve three different sets of energy levels but this is not, in fact, the case. Let us use Equation (11.1), (11.2) and (11.3) to write down expressions for some of the energy differences in the series (Table 11.2). We see immediately that the constant lower energy for transitions in the Balmer series ($-hcR/2^2$) is just the first excited energy

Table 11.2

Lyman	Balmer	Paschen
$(n_1 = 1)$	$(n_1 = 2)$	$(n_1 = 3)$
$hcR\left(\frac{1}{1^2}-\frac{1}{2^2}\right)$		
$hcR\left(\frac{1}{1^2}-\frac{1}{3^2}\right)$	$hcR\left(\frac{1}{2^2}-\frac{1}{3^2}\right)$	
$hcR\left(\frac{1}{1^2}-\frac{1}{4^2}\right)$	$hcR\left(\frac{1}{2^2}-\frac{1}{4^2}\right)$	$hcR\left(\frac{1}{3^2}-\frac{1}{4^2}\right)$
$hcR\left(\frac{1}{1^2}-\frac{1}{5^2}\right)$	$hcR\left(\frac{1}{2^2}-\frac{1}{5^2}\right)$	$hcR\left(\frac{1}{3^2}-\frac{1}{5^2}\right)$
⋮	⋮	⋮
$hcR\left(\frac{1}{1^2}-\frac{1}{\infty}\right)$	$hcR\left(\frac{1}{2^2}-\frac{1}{\infty}\right)$	$hcR\left(\frac{1}{3^2}-\frac{1}{\infty}\right)$

level for the Lyman series; the Balmer lines involve transitions not to the $n = 1$ level, but to the $n = 2$ level. Similarly the Paschen series lines involve transitions to the $n = 3$ level, as shown in Fig. 11.11. Thus, the same sets of energy levels are used but with a different lower level for each series; because the energy differences are much smaller for all Balmer lines they are of much lower frequency than Lyman lines.

Having related the frequencies of lines in the hydrogen spectrum to energy levels, we now need to discuss how these energy levels can arise within the atom. We shall look at two theories of how this can be explained; the first was proposed by Bohr, and the second by Schrödinger.

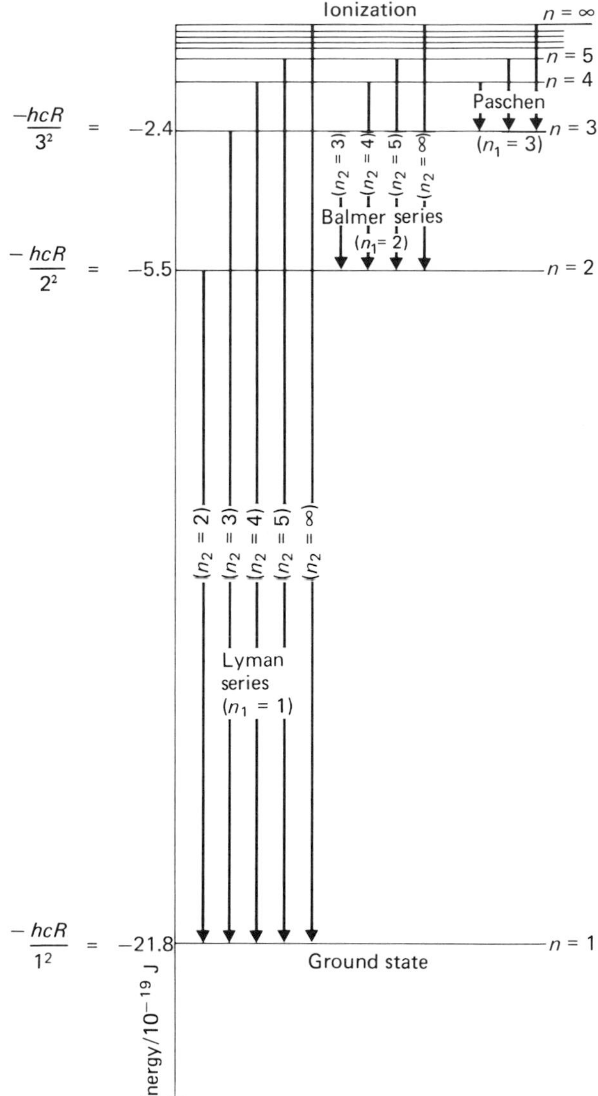

Fig. 11.11 Hydrogen spectrum: Lyman, Balmer and Paschen series.

Bohr's theory of the atom

In 1911 Geiger and Marsden bombarded thin sheets of gold with α-particles. Analysis of the angles at which these particles were scattered enabled Rutherford to show that an atom consists of a relatively small, but massive, positively charged nucleus with negatively charged electrons in the surrounding space. Newton's second law predicted that the electrons, which were attracted by electrostatic forces to the nucleus, must move in circular or elliptical orbits. Thus, an atom was seen to be like a miniature solar system, with small planet-like electrons orbiting a large sun-like nucleus. It was a convenient, simple picture which explained a considerable proportion of the atomic physics known at the time.

However, it had one serious flaw. In the last chapter, we saw that radio waves are emitted by electrons accelerated backwards and forwards in an aerial; classical physics predicts that whenever electrons accelerate they will radiate energy. An electron moving in a circle is accelerating towards the centre of the circle and hence should lose energy; detailed analysis predicts that any orbital electron will have given up all its energy and collapsed into the nucleus within 10^{-10} s.

In 1913, the Danish physicist Niels Bohr put forward his theory of the atom which, whilst leaving many questions unanswered, at least codified the problems in a number of assumptions. Once these assumptions had been made, however, it was possible to derive equations for the frequencies of the spectral lines in the hydrogen spectrum. He made the following proposals:

(1) *An orbital electron can exist only in certain orbits.* If the mass of the electron, m, is multiplied by its velocity, v, and by the radius of its orbit, r, we obtain its angular momentum. Bohr assumed that *the angular momentum of the electron was limited by the equation*:

$$mvr = n\frac{h}{2\pi} \qquad (11.5)$$

where h is Planck's constant and n an integer usually referred to as a quantum number.

(2) *An electron does not radiate energy whilst moving within the same orbit.*

(3) *When an electron changes to an orbit of lower energy, energy is emitted as a quantum of radiation of energy equal to the difference in orbit energies.*

It is important to appreciate that these assumptions were an arbitrary introduction of Planck's quantum ideas into classical mechanics but, nevertheless, could be used to derive some quantitative results. We shall restrict our discussion of these results to electrons moving in circular orbits.

In the hydrogen atom the force between the electron and the proton is given by Coulomb's law:

$$\text{electrostatic force, } F = \frac{e^2}{4\pi\varepsilon_0 r^2} \qquad (11.6)$$

where e is the charge on both electron and proton, r their separation and ε_0 the permittivity of free space. Now by Newton's second law, the force F acting on a mass m moving in a circle of radius r with velocity v is given by:

$$F = \frac{mv^2}{r}$$

Hence

$$\frac{e^2}{4\pi\varepsilon_0 r^2} = \frac{mv^2}{r} \qquad \text{from (11.6)}$$

i.e.

$$\frac{e^2}{4\pi\varepsilon_0 r} = mv^2 \qquad (11.7)$$

From Bohr's theory we have:

$$mvr = n\frac{h}{2\pi} \qquad (11.5)$$

which may be squared and rearranged to give:

$$mv^2 = \frac{n^2h^2}{4\pi^2mr^2} \qquad (11.8)$$

Equations (11.7) and (11.8) give:

$$\frac{e^2}{4\pi\varepsilon_0 r} = \frac{n^2h^2}{4\pi^2mr^2}$$

i.e.

$$r = n^2\frac{h^2\varepsilon_0}{\pi me^2} \qquad (11.9)$$

Substituting in Equation (11.9) we can calculate a value for r corresponding to $n = 1$:

$$r = 1^2 \times \frac{(6.6\times10^{-34})^2\times(8.9\times10^{-12})}{\pi\times(9.1\times10^{-31})\times(1.6\times10^{-19})^2}$$

$$\approx 0.5\times10^{-10}\text{ m}$$

This represents the radius of the atom in its ground state and is the same order of magnitude as the experimentally measured value.

Q 11.6
Calculate the radius of a hydrogen atom in its first excited state.

In our discussion of energy levels we considered an electron to have zero energy when it was completely removed from the atom, and negative energy when it was in one of the bound energy levels. Thus, the potential energy of the electron, E_p, will be given by the equation:

$$E_p = -\frac{e^2}{4\pi\varepsilon_0 r} \qquad (11.10)$$

The potential energy is zero at $r = \infty$, as shown in Fig. 11.12, and becomes more and more negative as the electron approaches the nucleus.

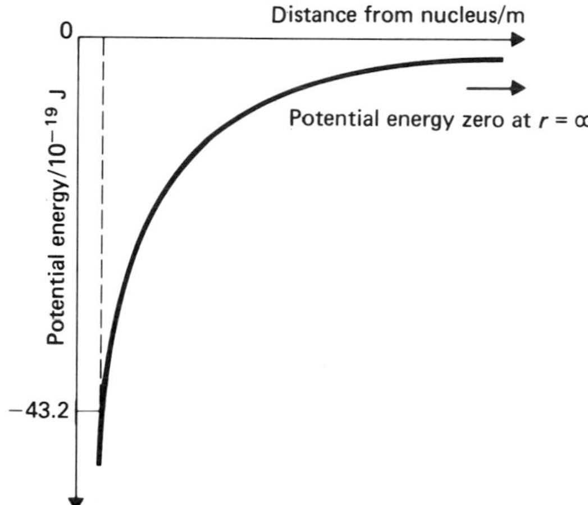

Fig. 11.12 The potential energy of the electron in a hydrogen atom.

As the electron is moving round the nucleus it will also have kinetic energy, E_k, given by:

$$E_k = \tfrac{1}{2}mv^2 \qquad (11.11)$$

$$= \frac{e^2}{8\pi\varepsilon_0 r} \qquad \text{from (11.7)}$$

Hence the total energy of an electron, E, is given by:

$$E = E_p + E_k$$

$$= -\frac{e^2}{4\pi\varepsilon_0 r} + \frac{e^2}{8\pi\varepsilon_0 r}$$

i.e.

$$E = -\frac{e^2}{8\pi\varepsilon_0 r} \qquad (11.12)$$

But we have already shown that:

$$r = n^2 \frac{h^2\varepsilon_0}{\pi m e^2} \qquad (11.9)$$

and hence:

$$E = -\frac{e^2}{8\pi\varepsilon_0} \cdot \frac{\pi m e^2}{n^2 h^2 \varepsilon_0}$$

i.e.

$$E = -\frac{1}{n^2} \cdot \frac{me^4}{8\varepsilon_0^2 h^2} \qquad (11.13)$$

Thus, we may obtain a value for the total energy of an electron:

$$E = -\frac{1}{n^2} \cdot \frac{(9.1\times10^{-31})(1.6\times10^{-19})^4}{8(8.9\times10^{-12})^2(6.6\times10^{-34})^2}$$

$$= -\frac{1}{n^2} \cdot 2.16\times10^{-18}\ \text{J}$$

In the ground state $n = 1$, and hence:

$$E_1 = -2.16\times10^{-18}\ \text{J}$$

This agrees closely with the value -2.18×10^{-18} J obtained from spectroscopic data. The discrepancy arises, in fact, from errors in quoting the constants above to two figures only.

The potential energy is related to the total energy by Equations (11.10) and (11.12) and must, therefore, equal -4.32×10^{-18} J in the ground state, as shown in Fig. 11.12.

Thus, in an energy level transition from ($n = n_2$) to ($n = 1$) the energy released by the atom will be given by:

$$\text{energy emitted} = E_2 - E_1$$

$$= -2.16\times10^{-18}\left(\frac{1}{n_2^2} - \frac{1}{1^2}\right)$$

$$= 2.16\times10^{-18}\left(\frac{1}{1^2} - \frac{1}{n_2^2}\right)$$

The frequency of the resulting spectral line is given by:

$$\text{frequency} = \frac{\text{energy emitted}}{\text{Planck's constant}}$$

Hence

$$f = \frac{2.16\times10^{-18}}{h}\left(\frac{1}{1^2} - \frac{1}{n_2^2}\right)$$

$$= 3.27\times10^{15}\left(\frac{1}{1^2} - \frac{1}{n_2^2}\right)$$

But from the formula for the Lyman series we have:

$$f = cR\left(\frac{1}{1^2} - \frac{1}{n_2^2}\right) \qquad (11.2)$$

Hence

$$R = \frac{3.27\times10^{15}}{3\times10^8}$$

$$= 1.09\times10^7\ \text{m}^{-1}$$

This compares well with the experimental value of $1.09678\times10^7\ \text{m}^{-1}$.

Thus, Bohr could predict atomic radii, the equations for the various series of hydrogen lines and a value for the Rydberg constant. In fact, his accurate predictions of the wavelengths of lines in the hydrogen spectrum were correct to one part in forty thousand. These results were sufficient to lead many to accept the assumptions, but they still formed an unhappy marriage between classical mechanics and quantum mechanics. It must also be realized that the theory worked much less well for atoms other than hydrogen, because of the interactions between the larger number of particles. With hindsight, it is perhaps surprising that so much came out right, but an even more radical understanding of the nature of matter was needed before really satisfactory theories could be developed.

11.4 Matter waves

In the last chapter we found that to account for all types of optical phenomena it was necessary to assume that light had a particle nature as well as a wave nature. In 1924, Louis de Broglie proposed in his Ph.D. thesis that particles must have a wave nature associated with them. Thus, he introduced a symmetry into modern physics which was to question all the accepted classical assumptions about the nature of matter. As we shall see in the final section, his theory enabled him to explain the quantization of electron orbits which, in Bohr's theory, had been merely an arbitrary assumption.

de Broglie proposed that a particle of momentum p had an associated wavelength, λ, given by:

$$\lambda = \frac{h}{p} \qquad (11.14)$$

which can be written, for non-relativistic speeds, as:

$$\lambda = \frac{h}{mv} \qquad (11.15)$$

for a particle of mass m travelling with velocity v. He regarded a particle as being 'localized but with a guiding wave'. The intensity of the wave at a point was proportional to the probability of finding the particle there; the intensity was, of course, also proportional to (wave amplitude)^2.

The wavelength of a 100 eV electron can be compared with that of a 20 g marble moving at a typical speed of 1 cm s^{-1}.

(i) Energy of electron in joules $= \begin{pmatrix}\text{charge on}\\\text{electron}\end{pmatrix} \times \begin{pmatrix}\text{energy}\\\text{in eV}\end{pmatrix}$

$$= (1.6\times10^{-19})\times(100)$$
$$= 1.6\times10^{-17}$$
$$= \tfrac{1}{2}mv^2$$

neglecting relativistic effects.

Hence $$v^2 = \frac{2\times1.6\times10^{-17}}{9.1\times10^{-31}}$$
$$v = 5.93\times10^6 \text{ m s}^{-1}$$

Therefore $$\lambda = \frac{h}{mv} \qquad (11.15)$$
$$= \frac{6.6\times10^{-34}}{(9.1\times10^{-31})\times(5.9\times10^6)}$$

i.e. electron wavelength $= 1.2\times10^{-10}$ m

(ii) For the marble:

$$\lambda = \frac{h}{mv} \qquad (11.15)$$
$$= \frac{6.6\times10^{-34}}{(2\times10^{-2})\times(10^{-2})}$$

i.e. marble wavelength $= 3.3\times10^{-30}$ m

The marble wavelength is very short compared with that of the electron and its wave nature is relatively unimportant.

Q 11.7

(*a*) A neutron has a mass of 1.7×10^{-27} kg. Calculate with what speed, and thus with what kinetic energy, it must be moving to have a wavelength identical to that of the 100 eV electron. Convert the energy into eV.

(*b*) Estimate the wavelength of a cricket ball bowled by a fast bowler. State clearly your assumptions.

We may use relativistic formulae for particles of zero rest mass to extend the results above to include photons. Suppose that the photons travel at velocity c and have a relativistic mass M. Their momentum p is still given by:

$$p = Mc$$

where relativity relates their energy, E, to their relativistic mass by the following equation:

$$E = Mc^2$$

Hence $$E = pc$$
$$= \frac{h}{\lambda}c \qquad \text{from (11.14)}$$

i.e. $$E = hf \qquad (10.3)$$

This is Planck's formula for the energy of a light photon or 'particle', showing that de Broglie's proposals were consistent with Planck's.

We saw, in Section 10.4, that photons have momentum; Equation (10.12), derived there from the relativistic relation between energy and momentum, is now seen to be de Broglie's equation.

Q 11.8
Calculate the 'mass' of a photon of green light ($\lambda = 5.5 \times 10^{-7}$ m) travelling with the velocity of light. Note that the 'mass' of the photon only has this value at this velocity; its rest mass is zero.

When a photon collides with a particle at rest it will, in general, lose some of its momentum and kinetic energy to the body. Hence de Broglie's equation predicts that its wavelength will increase and its frequency decrease. This effect had first been observed by Compton in 1922, and is shown in Fig. 11.13.

Evidence for the wave nature of particles

When de Broglie put forward his hypothesis it was greeted with scepticism; 'la Comédie Française' was how one physicist described it. Nevertheless, within three years, two American scientists, Davisson and Germer, had produced evidence for the wave nature of electrons.

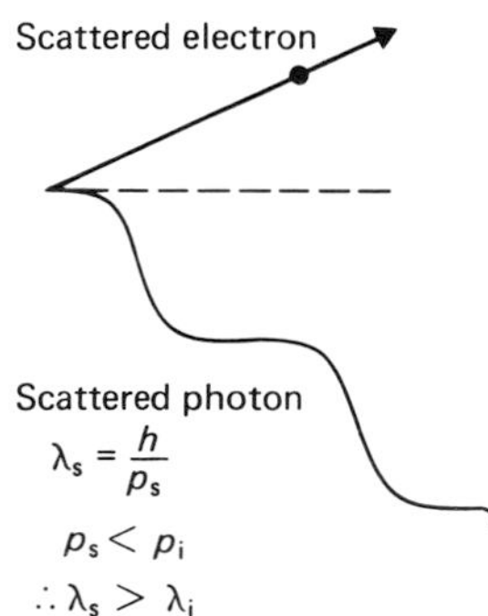

Fig. 11.13 The Compton effect.

They scattered low-energy electrons off the surface of a crystal of nickel and observed that there was an angular variation in reflected intensity similar to that observed in the light reflected by a reflection grating. They assumed that the lines of atoms acted like the ruled lines on a grating and were able to calculate the wavelength of the electrons; the value of Planck's constant obtained from this agreed with that calculated from the photo-electric effect to an accuracy of 1%. Their apparatus is shown in Fig. 11.14.

The interatomic spacing in nickel, s, is 2.15×10^{-10} m and the direction, θ, of the first reflected maximum will be given by:

$$s \sin\theta = \lambda \qquad (7.4)$$

where λ is the wavelength of the electrons. In fact, a full analysis involving multiple layers of atoms is more complex, but yields the same value for λ. Davisson and Germer found that the first maximum was at $\theta = 50°$.

$$\text{Hence} \quad \lambda = (2.15 \times 10^{-10})(\sin 50) = 1.65 \times 10^{-10}\ \text{m}$$

Now the electrons had been accelerated through a potential difference of 54 V. We therefore have:

$$\left.\begin{array}{l}\text{energy of}\\ \text{electrons}\\ \text{in joules}\end{array}\right\} = \left(\begin{array}{c}\text{charge on}\\ \text{electron}\end{array}\right) \times \left(\begin{array}{c}\text{energy in}\\ \text{electron-volts}\end{array}\right)$$

$$= (1.6 \times 10^{-19}) \times 54$$
$$= 8.64 \times 10^{-18}\ \text{J}$$

But the kinetic energy of the electrons must be equal to the energy stored in the electrons by this acceleration:

$$\therefore \quad \tfrac{1}{2}mv^2 = 8.64 \times 10^{-18}\ \text{J}$$

$$\text{Now } (mv)^2 = 2m(\tfrac{1}{2}mv^2) = 2 \times (9.1 \times 10^{-31}) \times (8.64 \times 10^{-18})$$

$$\text{i.e.} \quad mv = 3.97 \times 10^{-24}\ \text{kg m s}^{-1}$$

We can now calculate a value of Planck's constant from the wavelength obtained above,

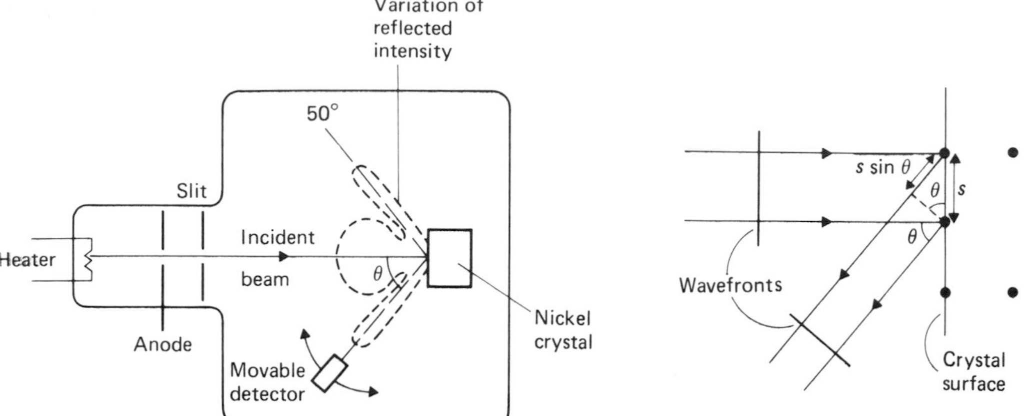

Fig. 11.14 Davisson and Germer's experiment.

using Equation (11.15):

$$\lambda = \frac{h}{mv} \text{ at non-relativistic speeds}$$

i.e. $h = \lambda(mv)$

$\therefore \quad h = (1.65 \times 10^{-10}) \times (3.97 \times 10^{-24})$

$= 6.53 \times 10^{-34}$ J s

This value of Planck's constant agrees well with the accepted value of 6.6×10^{-34} J s.

If electrons have a wave nature we would expect them to show properties such as diffraction and interference, and the experiment described above shows that this is indeed the case. A similar demonstration utilizes an electron diffraction tube, as shown in Fig. 11.15(*a*).

A beam of electrons is produced by thermionic emission and accelerated onto a fluorescent screen at the far end of the tube. If a target of graphite, for instance, is placed in the beam, a series of concentric rings is formed. The pattern is similar to that produced in the diffraction of X-rays by polycrystalline materials, shown in Fig. 11.15(*c*). Each layer of graphite atoms acts like a two-dimensional diffraction grating and produces a pattern of spots; the succession of randomly orientated 'gratings' produces circles instead of spots.

Thus, electrons show the same diffraction phenomena that electromagnetic waves do, provided that the dimensions of the aperture or grating are suitable. We expect more energetic electrons to have a larger momentum and thus a smaller wavelength; experience with other forms of diffraction leads us to suppose that the angular distribution will be reduced and this is seen to be the case in the composite photograph in Fig. 11.15(*b*).

Finally let us examine a remarkable experiment published by Möllenstedt in 1956, shown in Fig. 11.16(*a*). A fine wire is placed in a beam of electrons from an electron gun and, as expected, a shadow is produced on a photographic film. The wire is then given a positive charge so that the beams of electrons are 'bent inwards'. We would expect the shadow to disappear, and a bright region to be formed where overlap occurs. Instead, the bright region is crossed by dark and bright interference fringes, such as those obtained in a Young's slits experiment; the electron experiment is, in fact, more nearly analogous to the arrangement known as Fresnel's biprism, shown in Fig. 11.16(*b*). It would appear that in the dark fringes the electrons annihilate, but it would be more correct to say that their wave amplitudes are such that they never arrive, being deviated into the bright fringes. A simplified version of the apparatus is shown; complex electron microscope 'lenses' are needed to produce a pattern such as that in the photograph.

As before, there are many paradoxes in this 'dual' approach to the nature of electrons. An electron must go one side of the charged wire

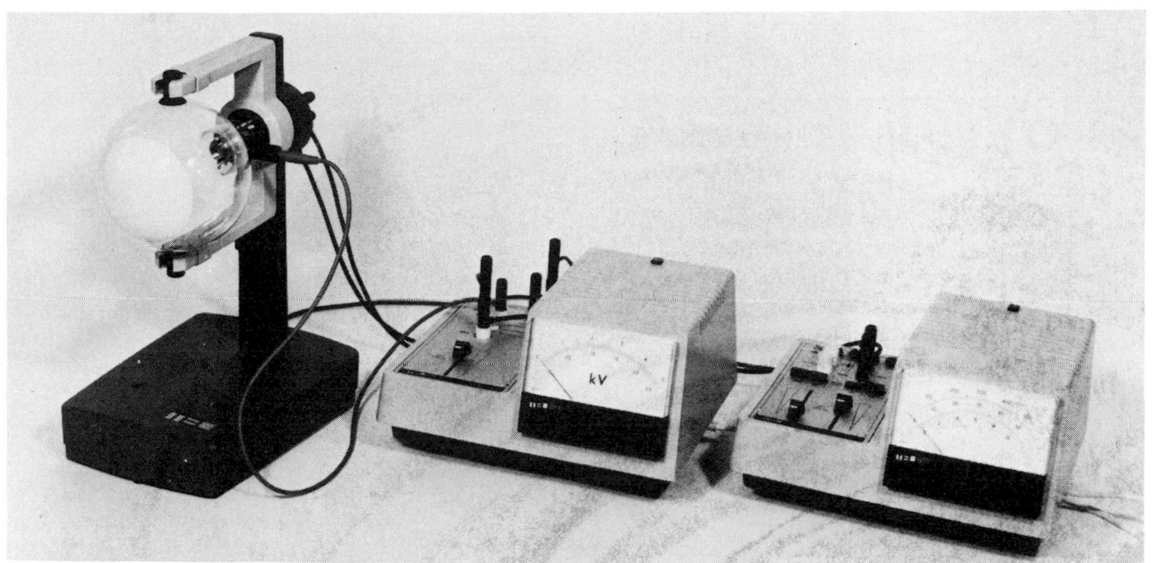

(*a*) An electron diffraction tube and power supplies

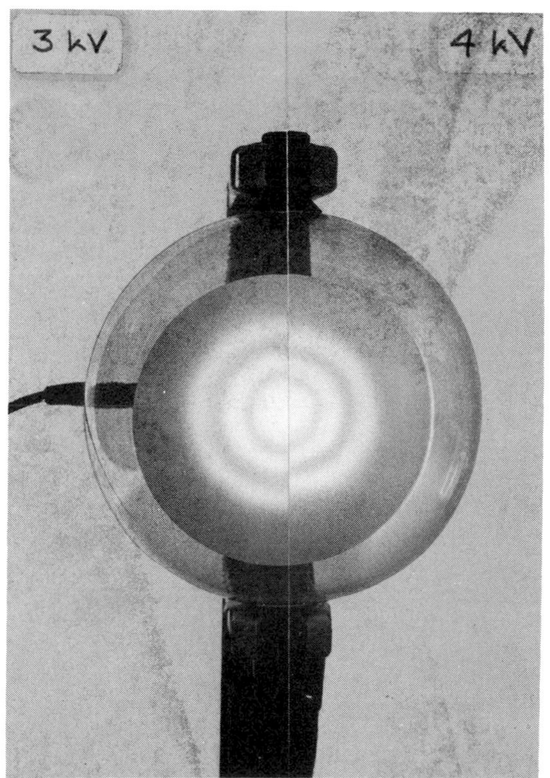

(*b*) A composite photograph showing the electron diffraction patterns obtained at different voltages.

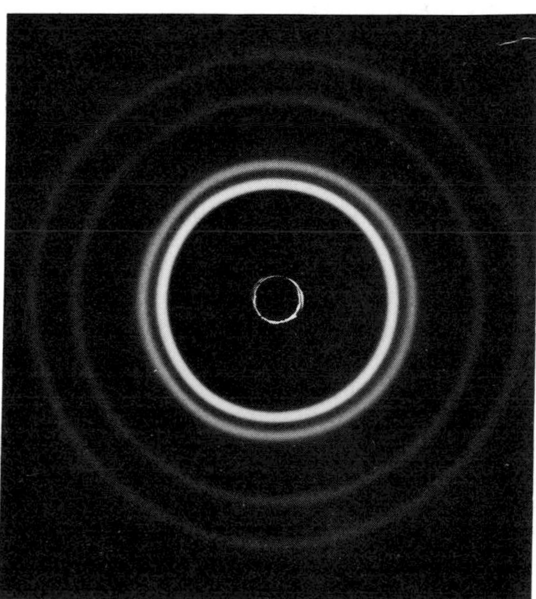

(*c*) X-ray diffraction pattern formed by a polycrystalline material

Fig. 11.15

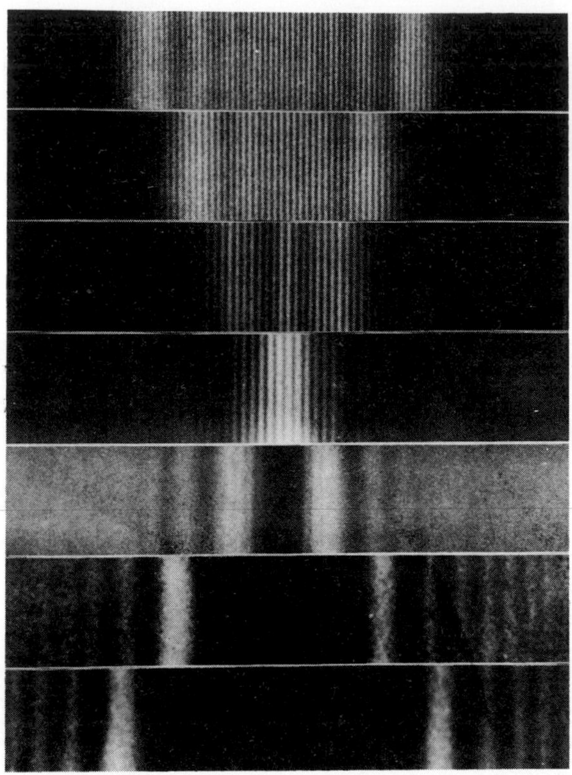

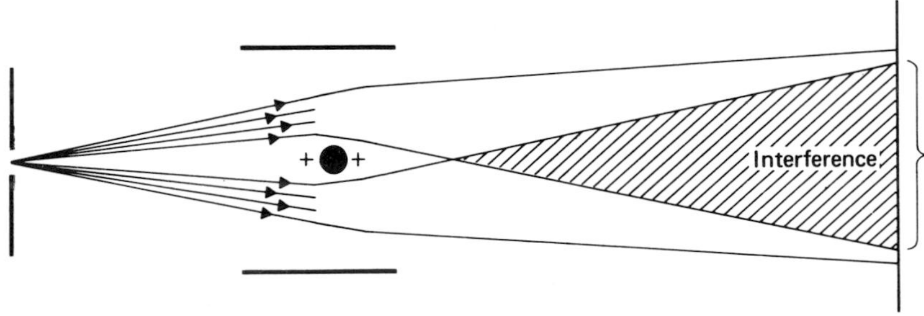

Fig. 11.16 (*a*) Möllenstedt's experiment with electrons. (Photograph first published in *Zeitschrift für Physik*, vol 145, 1956.)

or the other. If, however, those going one side are stopped by a screen, the intensity at certain points in the pattern will increase rather than decrease because an interference pattern is no longer formed. It is also possible to reduce the intensity in the experiment so that there is, on average, only one electron in the apparatus at a time; it is now most unlikely that two electrons will arrive at the screen simultaneously, but an interference pattern is still produced. These are some of the problems which are as real for electrons as they are for photons.

Waves of chance

Earlier we stated that:

$$\begin{pmatrix}\text{probability of finding}\\ \text{an electron at a point}\end{pmatrix} \propto \begin{pmatrix}\text{intensity of}\\ \text{electron wave}\end{pmatrix} \propto (\text{amplitude})^2$$

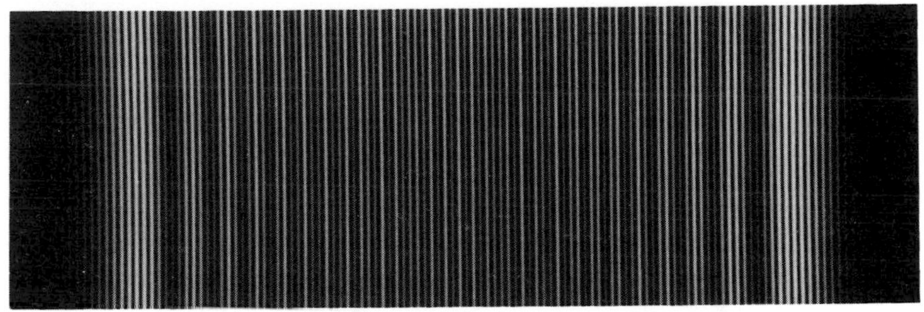

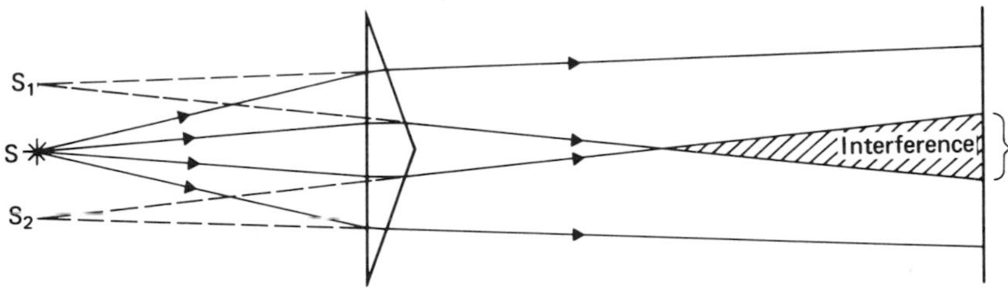

Fig. 11.16 (*b*) Fresnel's biprism with light

and hence we have:

$$\begin{pmatrix}\text{chance of an electron}\\ \text{arriving in a short time}\\ \text{interval}\end{pmatrix} \propto \begin{pmatrix}\text{intensity of}\\ \text{electron wave}\end{pmatrix}$$

$$\propto (\text{amplitude})^2$$

In these ways, we imply that the electrons behave like waves; we do not, however, say that the electrons *are* waves, but rather that phenomena such as the diffraction of electrons require the application of wave ideas. Other wave concepts, such as that of oscillation, are not used at all for they seem to have no relevance to electron waves.

It has since been possible to show that other small particles also exhibit wave properties; there is now a great deal of evidence for what are known collectively as *matter waves*. However, when a massive particle such as a cricket ball is treated in this way, the wavetrain is so specifically defined that there is no appreciable chance of the ball deviating from the path predicted by Newtonian mechanics.

11.5 Waves in atoms

Wave mechanics

de Broglie applied his concept of matter waves to the study of atomic structure by combining with it the idea of stationary waves. We saw in Chapter 4 that progressive waves are reflected at the boundaries of a medium; thus in a closed system the waves travelling in opposite directions may superpose constructively to form stationary waves. This will only occur at certain frequencies; each frequency is such that the length of the medium is a whole number of half-wavelengths, as shown in Fig. 11.17(*a*). It is impossible to set up stationary waves of other frequencies or wavelengths, because constructive superposition does not occur.

A similar argument may be put forward in the case of waves travelling round a ring. If waves travelling in opposite directions round the ring superpose constructively, stationary waves are set up as shown in Fig. 11.17(*b*). This

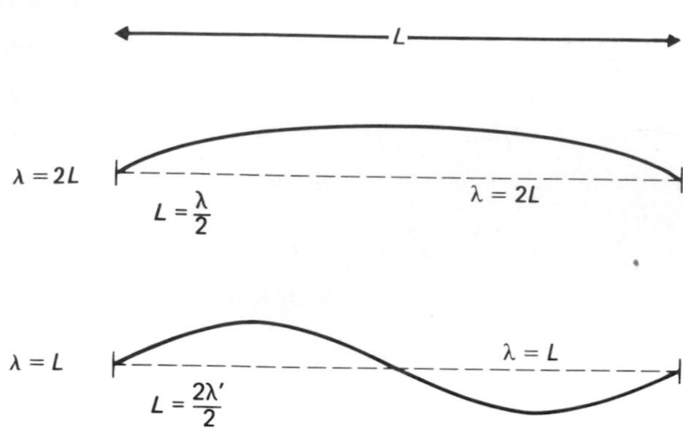

(*a*) Stationary wave on a straight rope

(*b*) A stationary wave set up on a loop of wire; in this case $2\pi r = 4\lambda$

Fig. 11.17

will occur only at frequencies for which the circumference of the ring, $2\pi r$, is a whole number of wavelengths:

$$2\pi r = n\lambda \qquad (11.16)$$

This idea, that waves confined in a region can only have certain wavelengths, can be applied to electrons within atoms. For instance, we may consider an electron wavelength fitting into the circumference of its orbit, in which case the wavelength will again be determined by Equation (11.16). If we use de Broglie's equation to connect the electron wavelength with its momentum, we can show that this condition is mathematically equivalent to that proposed by Bohr:

$$\lambda = \frac{h}{p}$$

$$= \frac{h}{mv} \quad \text{at non-relativistic speeds}$$

But $$2\pi r = n\lambda \qquad \text{from (11.16)}$$

$$= n\frac{h}{mv}$$

i.e. $$mvr = n\frac{h}{2\pi} \qquad (11.5)$$

This was the equation used in stating Bohr's quantum conditions in Section 11.3, and thus de Broglie's stationary wave conditions can be made to give the same radius for an atom as Bohr's theory. The conceptual advantage of wave mechanics is that it seems less arbitrary than did Bohr's quantum conditions; the restrictions placed upon the radius of the electron orbit follow from our classical understanding of stationary waves, whereas Bohr's restrictions, upon the angular momentum of electrons, were arbitrary assumptions chosen to fit the facts.

Bohr's theory based on the quantisation of angular momentum worked fairly well for the energy levels of the hydrogen atom; the first requirement of a new theory must therefore be that it gives the same results as the Bohr theory in this particular case. Then, if it is to be a constructive advance, it must give improved results to problems where the Bohr theory failed; de Broglie's theory of *wave mechanics* does provide much more accurate predictions for the spectra of more complex atoms, and for the fine detail in the hydrogen spectrum.

Q 11.9
This question shows how wave mechanics can be used to derive energy levels for the hydrogen atom, and shows that on this model results are obtained which are the same as those given by the Bohr theory.

(*a*) The ground state of the hydrogen atom is assumed to occur when

$$2\pi r_1 = \lambda_1$$

Write down a similar equation for the wavelength of the first excited state.

(*b*) Use de Broglie's equation to re-write this in terms of the velocity of the electron. This result can also be obtained from the Bohr theory.

(*c*) Use Newton's second law to write down an equation relating the circular motion of the electron to the electrostatic force exerted on it.

(*d*) Eliminate v between the equations in (*b*) and (*c*) to obtain an expression and value for the size of the electron orbit.

(*e*) How many times bigger is this orbit than the ground state orbit?

(*f*) Calculate the potential energy of the electron at this radius.

(*g*) Calculate the kinetic energy of the electron at this wavelength.

(*h*) Calculate the total energy of the electron in the first excited state.

(*i*) How is this energy related to the ground state energy (-2.18×10^{-18} J)?

(*j*) What is the predicted wavelength of the first line of the Lyman series? The actual wavelength is 1.2×10^{-7} m.

In fact, the stationary waves may better be assumed to fit into the 'width' of the atom and thus a half-wavelength loop must fit into the diameter, $2r$:

$$\lambda = \frac{2d}{n} = \frac{4r}{n}$$

instead of:

$$\lambda = \frac{2\pi r}{n}$$

as shown in Fig. 11.17(*a*).

This is more consistent with the idea that the significance of the wave amplitude is related to the probability of finding the electron at a particular distance from the nucleus.

Though much simplified, this picture of electron-waves within an atom enables us to calculate the kinetic and potential energies of the electron, and to predict values for the energy levels of the hydrogen atom.

We can also see why atomic radii cannot be decreased, using this theory. If the atomic diameter is reduced for a given energy state, the electron wavelength must be reduced for its stationary wave to 'fit'. Equation (11.15) implies that there would be an increase in velocity proportional to 1/radius, and hence an increase in kinetic energy proportional to $1/(\text{radius})^2$. The potential energy of the electron in an electrostatic force field depends on $(-1/\text{radius})$, and thus the kinetic energy increases more rapidly than the potential energy decreases. The electron would eventually have a total energy that was positive and would thus escape from the atom.

The next stage in developing the model is to consider the variations in electrical potential within the atom, due to the presence of the positive nucleus. This means that the electron would have more kinetic energy if closer to the nucleus, and a correspondingly smaller wavelength. The mathematical implications of this idea are beyond the scope of this text, but lead to the Schrödinger wave equation and its solution in terms of extremely precise energy levels.

The significance of electron waves

The picture of wave mechanics given above is much simplified. de Broglie's ideas were applied in detail by Schrödinger who derived his celebrated equation for the distribution of electron waves around a nucleus. He made his solutions fit into the inverse-square Coulomb force field and derived results, not only for complex atoms, but also for molecules. When modified by Dirac to allow for relativistic effects, they provided the largely correct basis for our modern understanding of atomic and molecular physics; wave mechanics explains results in fields as diverse as surface tension, chemical energy and radioactivity.

Schrödinger's wave mechanics is essentially a mathematical model. The mathematics is 'wavy' but it is difficult to attach physical significance to all aspects of the electron waves. The amplitude of the stationary waves is used to predict where an electron is likely to be; it is most likely to be in the region predicted by de Broglie's quantum equation, but it is possible, though very much less likely, for it to be at different distances from the nucleus. The deterministic approach of classical physics is reduced to a probabilistic approach where the most probable events are what 'happen' but other events are always 'possible'.

Because the electron 'orbits' are very different from the classical orbits of the planets, it is usual to avoid the term 'orbit'; the term *orbital* is sometimes used, but it is much better to use *modes of oscillation* to describe the stationary waves that can occur. Ultimately we have to accept that what happens in an atom cannot be reduced to the sort of conceptual 'picture' that we might hold in everyday life.

Heisenberg's uncertainty principle

At the same time that Schrödinger was developing wave mechanics to solve the problem of the hydrogen atom, another German physicist, Heisenberg, was working on a different approach. Just as Einstein had questioned concepts like 'simultaneous', and de Broglie had questioned concepts like 'particle', so Heisenberg questioned the idea of a completely determined line trajectory. His work enabled him to analyse the hydrogen atom using a completely different theory which gave identical results to those of Schrödinger.

Heisenberg questioned whether there was any meaning in the phrase 'a *point*-body moves along a *line*'. He imagined an ideal 'thought experiment' in which light was used to observe the trajectory of a moving object, as shown in Fig. 11.18. The cannon fires the projectile and its motion is measured by using the theodolite. The problem is that the light bounces off the object and alters its motion. According to classical theory one could reduce the intensity without limit to minimize this effect, and then use a very much more sensitive detector. However, according to quantum physics, the light arrives in quanta and the less energy these have the longer the wavelength of the light. As the wavelength of the light is increased, the particle will cause more diffraction and it will be more difficult to determine its position

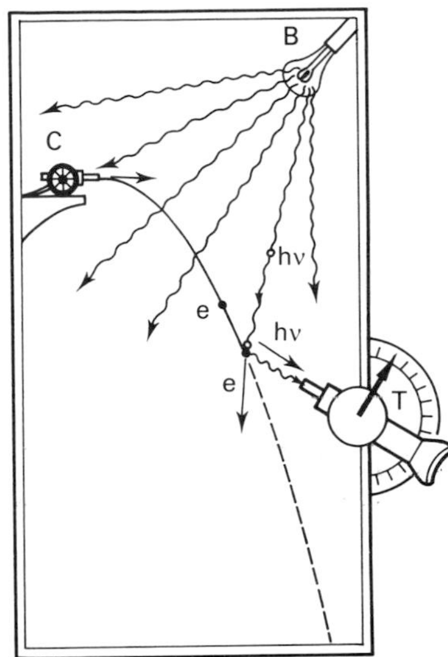

Fig. 11.18 Heisenberg's thought experiment. (Adapted, with permission, from Gamov (1966), *Thirty Years that Shook Physics*, Doubleday.)

accurately. The increased accuracy in the measurement of velocity has caused increased uncertainty in the measurement of position. Conversely, if a very small wavelength is used to minimize diffraction effects, the photon will considerably change the velocity of the particle, and uncertainty will be introduced into that measurement.

Heisenberg expressed this uncertainty in his famous equation:

$$\Delta p \Delta x \approx \frac{h}{2\pi} \tag{11.17}$$

where h is Planck's constant (again!), and Δp and Δx are the uncertainties in the momentum and position respectively. If we rewrite the equation in terms of velocity and mass, we have:

$$\Delta v \Delta x \approx \frac{h}{2\pi m} \tag{11.18}$$

where Δv is the uncertainty in the velocity, and m the mass, of the particle. We see that the effect is only important for particles of very small mass.

Q 11.10
Suppose that the uncertainty in the position of: (a) a piece of shot of mass 1 mg; (b) an atom of mass 10^{-26} kg; is in each case one nuclear radius (10^{-14} m). Calculate the uncertainty in their velocities.

For an electron bound in an atom of diameter 10^{-10} m we have:

$$\Delta v \Delta x \approx \frac{h}{2\pi m} \tag{11.18}$$

$$\approx \frac{10^{-34}}{10^{-30}}$$

$$\approx 10^{-4}$$

Now since the uncertainty in the electron's position is less than 10^{-10} m, the uncertainty in its velocity is given by:

$$\Delta v \gtrsim \frac{10^{-4}}{10^{-10}}$$

$$\gtrsim 10^{6}\ \mathrm{m\,s^{-1}}$$

But we can show from our analysis of the hydrogen atom that the velocity of the electron is only 10^{6} m s^{-1}, and thus the uncertainty in velocity is the same order of magnitude as the velocity. Clearly, if the electron were to be confined in a smaller space, the uncertainty in its velocity would give it a high possibility of having a kinetic energy so large that it might escape. With such uncertainty in both position and energy, a picture of electron orbits based on classical mechanics is clearly insufficient.

It should also be noted that we can never be certain that a particle has no energy; Heisenberg's equation tells us that if we know that the energy of a particle is exactly zero, the uncertainty in its position will be infinite. It is for this reason that we ascribe to molecules at absolute zero a *zero-point energy*.

An alternative form of Heisenberg's equation relates the uncertainties in energy and time measurement, ΔE and Δt:

$$\Delta E \Delta t \approx \frac{h}{2\pi} \tag{11.19}$$

Some changes in atomic energy levels take place very quickly indeed, in a time of the order of 10^{-10} s or less. This introduces uncertainty into our knowledge of the energies involved and the spectral lines are broader as a result; this is one application of Heisenberg's uncertainty principle that is particularly easy to observe.

11.6 Conclusion

We have traced the history of optics down the years, looking at both wave and particle viewpoints. At first it seemed that the wave theory displaced the particle viewpoint; certainly wave theory can explain not only the physical properties of light, but also electromagnetism, providing the key link between electricity and light. The need for photons, and the existence of matter waves, however, demonstrates that we cannot restrict what exists to concepts that we find easy to grasp. Scientists have often fought against the uncertainty inherent in modern physics, and some have suggested that there is a deterministic reality that exists but cannot be measured. Even the objectivity of physics is questioned when we see, in Heisenberg's thought experiment, that the observer may affect the observed.

> 'The quantum mechanics is very imposing. But an inner voice tells me that it is still not the final truth. The theory yields much, but it hardly brings us nearer to the secret of the Old One. In any case, I am convinced that He does not play dice.'

A. Einstein (1926)

Bibliography

Books

Bennet, G.A.G. (1974) *Electricity and Modern Physics* (2nd edn.). Edward Arnold; Chapter 13 sets the development of our understanding of energy levels in the context of increased knowledge about the electrons.

Bolton, W. (1986) *Patterns in Physics* (2nd Edition). McGraw-Hill; Chapter 18 takes some of the ideas about electron-waves in atoms a little further whilst keeping the treatment non-mathematical.

Boulind, H.F. (1972) *Waves or Particles?* Longman; a short reader covering some of this ground at a straightforward level.

Breithaupt, J. (1987) *Understanding Physics for A-level.* Hutchinson; contains a discussion of the wave nature of particles and the uncertainty principle in Chapter 31.

Caro, D.E., McDonell, J.A., Spicer, B.M. (1978) *Modern Physics* (3rd edn.). Edward Arnold; Chapter 9 traces wave mechanics further and gives a more mathematical treatment of the basic ideas. The book as a whole is an excellent and most readable account of the development of twentieth century physics.

PSSC (1981) *Physics* 5th edition, Heath; Chapter 27 gives a well-illustrated, non-mathematical account well backed up by experimental evidence.

Revised Nuffield Advanced Physics (1986) *Teachers' and Students' Guides.* Longman; Unit L in Volume 2 includes details of suitable experiments on the topics covered in this chapter, as well as some of the theory. It includes an introduction to more advanced ideas such as the solution of the Schrödinger Wave Equation, and a series of applications in both physics and chemistry.

Rogers E.M. (1960) *Physics for the Inquiring Mind.* Oxford University Press, Chapter 44; a non-mathematical treatment of much of this subject.

Books which consider the development of the ideas of the wave nature of matter from an historical point of view.

Gamov, G. (1966) *Thirty Year that Shook Physics.* Heineman.

Holton, G. (1973) *Introduction to Concepts and Theories in Physical Science.* Addison-Wesley, Chapters 28, 29.

Software

Garland Computing *Models in Physics* 'Line Spectra' enables the relationship between energy level transitions and the emitted spectra to be discussed.

Heinemann/Five Ways *Hydrogen Spectrum*; demonstration and exercise programmes illustrating the line emission spectrum of hydrogen.

Longman Micro Software *Dynamic Modelling System* 'HATOM'; relates the precisely defined energy levels of the hydrogen atom to the mathematical model of the Schrödinger Equation.

Longman Micro Software *Nuffield A-Level Software Pack* 'Solving the Schrödinger equation'; enables the student to identify the various energy levels for the hydrogen atom and shows how well-defined those energy levels are.

Appendices

1 Radian measure

There are two units in which angles are measured—degrees and radians. The reader will be most familiar with the *degree* (°) for that is the unit used in most situations. There are 360° in one complete revolution, and hence 90° in one right angle, as shown in Figure A.1. Thus, the basis of degree measure is to subdivide one revolution into 360 equal parts.

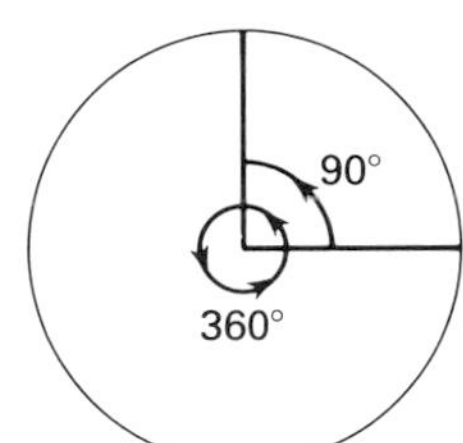

(*a*) Degree measure

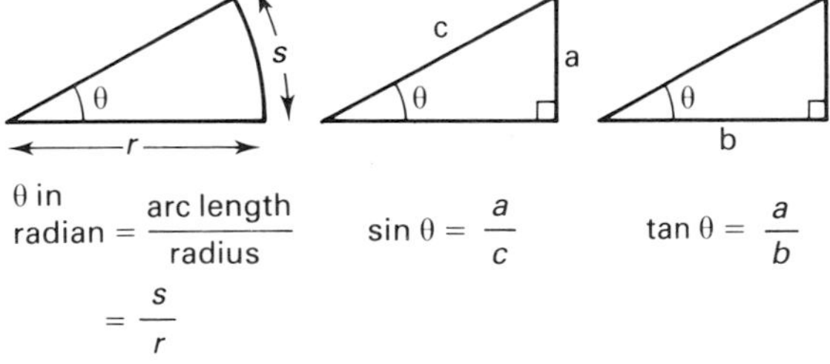

θ in radian $= \frac{\text{arc length}}{\text{radius}} = \frac{s}{r}$ $\quad \sin\theta = \frac{a}{c}$ $\quad \tan\theta = \frac{a}{b}$

(*b*) Radian measure

Fig. A.1

The alternative unit is the *radian*, which is defined in terms of the ratio of two distances—the arc length and the radius, as indicated. When an arc length is equal to its radius, the angle subtended by the arc is one radian. To measure an angle in radians, an arc is drawn across the angle and the arc length and radius measured:

$$\text{angle } \theta \text{ in radians} = \frac{\text{arc length}}{\text{radius}} = \frac{s}{r}$$

The significance of this method of measuring angles is that it is related to the definitions of trigonometrical relationships, which also involve the ratios of distances. For small angles, both the sine and tangent functions are nearly equal to the size of the angle measured in radians, as shown in the table on page 126.

The circumference of a circle of radius r is, of course, $2\pi r$ and hence one complete revolution, 360°, is equivalent to an angle, θ, measured in radians that is calculated by

$$\theta = \frac{\text{arc length}}{\text{radius}} = \frac{2\pi r}{r} = 2\pi$$

Thus, there are 2π radians in 360°, π radians in 180°, $\pi/2$ radians in 90° and so on; it is usual to express angles in radians as a multiple or submultiple of π.

If the angle θ is equal to one radian, the arc length must equal the radius of the circle. Hence, if θ is measured in degrees:

$$\frac{\theta}{360^\circ} = \frac{r}{2\pi r} = \frac{1}{2\pi}$$

i.e.

$$\theta = \frac{360^\circ}{2\pi} = 57^\circ\ 18'$$

One radian is equal to 57° 18′.

Because of the small angle approximations that can be made between θ, $\sin\theta$ and $\tan\theta$, radian measure is widely used in scientific work. Conversions from degrees to radians and vice versa are made using proportion.

Q A-1
Convert the following angles in degrees into angles in radians:
(*a*) 270°;
(*b*) 45°;
(*c*) 30°.

Convert the following angles in radians into angles in degrees:

(*d*) $\pi/3$ radian;
(*e*) $\pi/4$ radian;
(*f*) $\pi/2$ radian.

2 Trigonometrical formulae

There are a number of trigonometrical formulae to which reference is made in the text and these are listed below, together with other useful relationships. Proofs of these may be found in a good trigonometry text. P and Q represent any two angles where P is greater than Q.

$$\sin(P+Q) = \sin P \cos Q + \cos P \sin Q \quad *$$

$$\sin(P-Q) = \sin P \cos Q - \cos P \sin Q$$

$$\cos(P+Q) = \cos P \cos Q - \sin P \sin Q$$

$$\cos(P-Q) = \cos P \cos Q + \sin P \sin Q$$

$$\sin P + \sin Q = 2\sin\frac{P+Q}{2}\cos\frac{P-Q}{2} \quad *$$

$$\sin P - \sin Q = 2\cos\frac{P+Q}{2}\sin\frac{P-Q}{2}$$

$$\cos P + \cos Q = 2\cos\frac{P+Q}{2}\cos\frac{P-Q}{2}$$

$$\cos P - \cos Q = -2\sin\frac{P+Q}{2}\sin\frac{P-Q}{2}$$

$$\sin P \cos Q = \tfrac{1}{2}[\sin(P+Q) + \sin(P-Q)] \quad *$$

$$\cos P \sin Q = \tfrac{1}{2}[\sin(P+Q) - \sin(P-Q)]$$

$$\cos P \cos Q = \tfrac{1}{2}[\cos(P+Q) + \cos(P-Q)]$$

$$\sin P \sin Q = -\tfrac{1}{2}[\cos(P+Q) - \cos(P-Q)]$$

$$\sin\theta = \cos(\theta - 90°) \quad *$$

$$= \cos(\theta - \pi/2)$$

$$\cos\theta = \sin(\theta - 90°)$$

$$= \sin(\theta - \pi/2)$$

$$\sin^2 P + \cos^2 P = 1 \quad *$$

$$\sin^2 P = \frac{1-\cos 2P}{2}$$

$$\cos^2 P = \frac{1+\cos 2P}{2} \quad *$$

Q A-2
Take $P = 60°$ and $Q = 30°$ and check each of the starred equations quoted above.

3 The analysis of diffraction using phasors

In Chapter 1 we saw how the rotation of a phasor could be used to represent a simple harmonic oscillation, and how the angle between two phasors could represent their phase difference. The study of interference and diffraction frequently requires us to consider the resultant effect of different light waves arriving at one point on a screen; in Chapter 7 we used the path differences between rays from, for instance, two coherent point sources to predict the position of maxima and minima but the methods given there do not make it easy to calculate the light intensity at intermediate points. In this Appendix, phasors are used to represent the amplitude and phase of the light from different slits, or parts of a slit, and methods identical to those for vector-addition are used to find the resultant of a number of phasors; the reader is reminded that the resultant intensity is proportional to the square of the resultant phasor amplitude.

Single-slit diffraction

When we add two or more phasors, the resultant depends upon the phase difference between them, as well as upon the amplitudes, as shown in Fig. A.2; the resultant is found by constructing a phasor diagram similar in principle to a vector triangle.

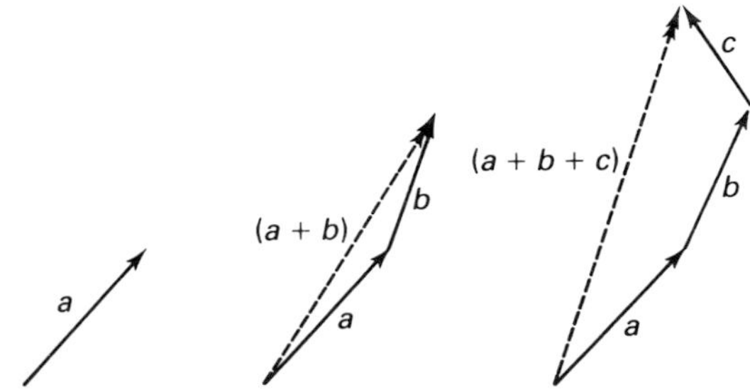

Fig. A.2

Now when we analysed the diffraction pattern caused by a single slit, as explained in Section 7.5, we divided the slit into a large number of imaginary, narrow strips as shown in Fig. A.3. Suppose that in a given direction, the path difference introduced between light from adja-

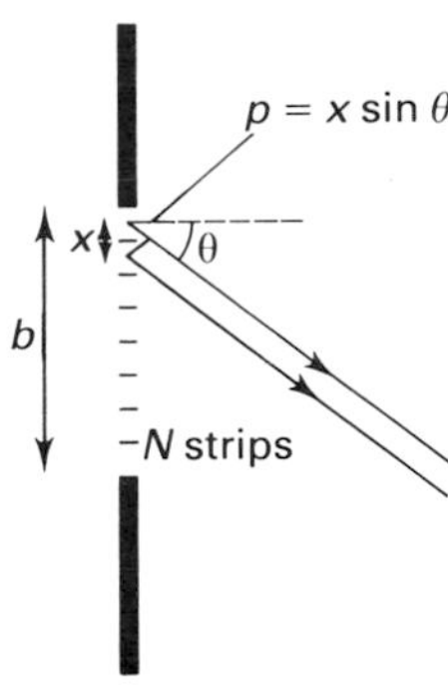

Path differences due to a single slit

Fig. A.3

cent strips is p; the phase difference is given by:

$$\phi = \frac{p}{\lambda} \times 2\pi$$

$$= \frac{2\pi p}{\lambda}$$

But if the width of each strip is x we have:

$$p = x \sin\theta$$

and $$x = b/N$$

where N is the number of imaginary strips.

Hence $$\phi = \frac{2\pi x \sin\theta}{\lambda}$$

$$= \frac{2\pi b \sin\theta}{N\lambda}$$

We see that the phase difference between successive phasors depends upon the direction θ, for which we are calculating the resultant intensity, and also upon the number of imaginary strips into which we divided the slit; however, we shall see below that the number of strips does *not* affect the resultant amplitude and intensity.

Let us suppose that the slit is divided into 8 strips. When $\theta = 0$, the phase difference between light from adjacent strips arriving at the screen will be zero, and hence a central maximum is formed in the diffraction pattern; if the amplitude of the light from each strip is given by a, the resultant amplitude is $8a$, as shown in Fig. A.4(a). The intensity is proportional to the square of the amplitude, and hence to $64a^2$.

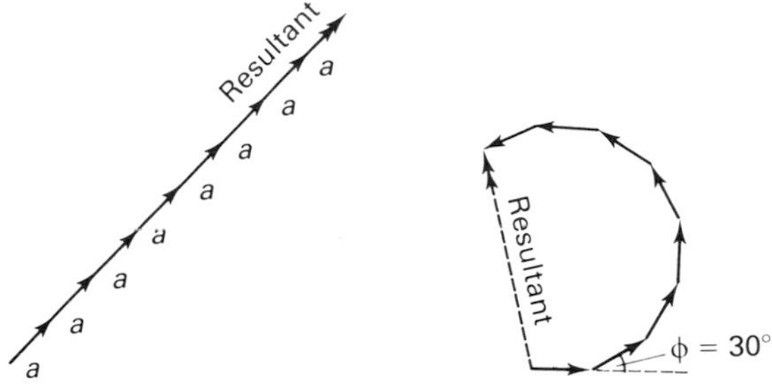

(a) Central maximum (b) Resultant: ϕ = 30°

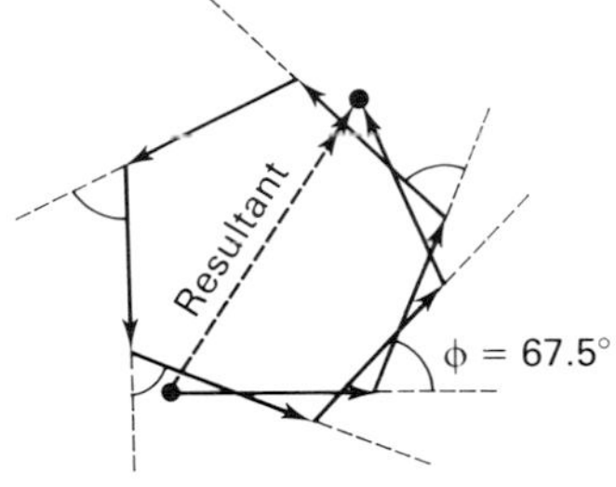

(c) Resultant: ϕ = 67.5°

The resultant amplitude in single-slit diffraction

Fig. A.4

As θ increases, the phase difference ϕ increases and the resultant changes, as shown in diagram (b). When $\phi = 2\pi/8$ (45°), the phasors form a closed octagon because the sum of the exterior angles of any polygon is equal to 2π radians, or 360°. In general, when the slit is divided into N strips, the first closed polygon will be formed when the phase difference between phasors is $2\pi/N$.

i.e. $$\frac{2\pi}{N} = \phi$$

$$= \frac{2\pi b \sin\theta}{N\lambda} \qquad \text{(A.1)}$$

$$\therefore \quad \frac{b \sin\theta}{\lambda} = 1$$

$$\sin\theta = \lambda/b$$

This is, of course, the condition we found before for the first minimum; the closed polygon of phasors has no resultant and represents zero resultant amplitude.

The second minimum occurs when the phasors form two complete revolutions and again have no resultant. The external angles must now add up to 4π radians, and for N phasors we have:

$$\phi = \frac{4\pi}{N}$$

Hence
$$\frac{4\pi}{N} = \frac{2\pi b \sin\theta}{N\lambda}$$

i.e.
$$\sin\theta = \frac{2\lambda}{b}$$

which is the condition derived previously for the second minimum.

However we can, in addition, also use the phasor method to derive the intensity half-way between these two minima, that is when $\sin\theta = 3\lambda/2b$ and the phasors form one and a half revolutions. In this case we have:

$$\phi = \frac{2\pi b \sin\theta}{N\lambda} = \frac{2\pi b}{N\lambda} \cdot \frac{3\lambda}{2b} = \frac{3\pi}{N}$$

Thus, for a slit divided into 8 strips:

$$\phi = \frac{3\pi}{8} = 67.5°$$

This is plotted out in diagram (c) and we see that there is a finite resultant which is, in fact, almost exactly a maximum. If we imagine the slit divided into a much larger number of strips, the resultant becomes the diameter of a circle whose circumference is equal to $2A/3$ where A is the total amplitude of the central maximum, equal to the length of all the phasors laid end to end.

Now
$$\text{diameter} = \frac{\text{circumference}}{\pi}$$

Therefore
$$\text{resultant} = \frac{2A/3}{\pi} = \frac{2A}{3\pi}$$

Thus, the amplitude of the first secondary maximum is $2/3\pi$ times that of the central maximum, and the intensity is thus reduced by a factor of $4/9\pi^2$ or approximately $1/25$.

Therefore, we can use the phasor method not only to predict the positions of the minima, which our previous theory would also do, but also to calculate the intensities of the subsidiary maxima; in fact, by using constructional methods, or suitable trigonometry, we can predict the intensity at any angle.

Double-slit interference

A phasor analysis of double-slit interference, like that of single-slit diffraction, enables us to calculate the resultant amplitude at all angles as well as the positions of maxima and minima. Initially, we shall ignore the variation in amplitude due to single-slit diffraction and consider merely the superposition of light from two coherent sources.

At an angle θ the light from the two slits has a path difference approximately equal to $s\theta$ as shown in Fig. A.5; hence the phase difference

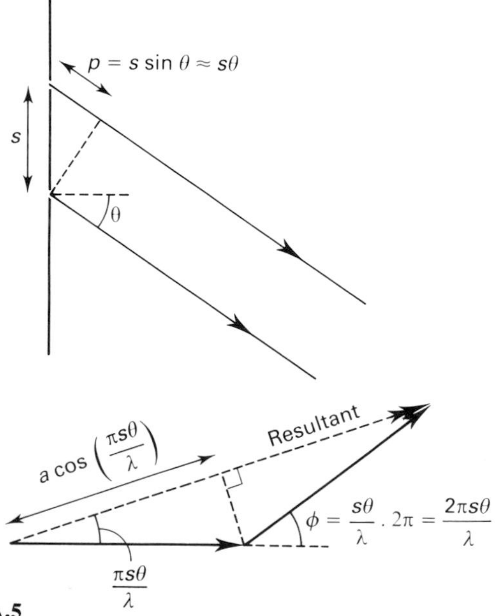

Fig. A.5

between these two rays is approximately equal to $2\pi s\theta/\lambda$ and the angle between the first phasor and the resultant must therefore be $\pi s\theta/\lambda$. Consideration of the phasor diagram shows that the resultant amplitude must be given by:

$$A = 2a \cos\left(\frac{\pi s\theta}{\lambda}\right)$$

where a is the amplitude of the light wave from one slit. Thus, if the intensity due to one slit is I_1, the variation of intensity with angle will be given by:

$$I = 4I_1 \cos^2\left(\frac{\pi s\theta}{\lambda}\right)$$

However, we have already seen how phasor methods may be applied to determine the distribution of light from individual slits, as well as to the interference of light from two slits; the analysis of single-slit diffraction given earlier may be extended to prove that:

$$I_1 - I_0 \frac{\sin^2(\pi b\theta/\lambda)}{(\pi b\theta/\lambda)^2}$$

Hence, the overall distribution of intensity with angle for a double slit, shown in Fig. 7.20, is described by the equation:

$$I = \frac{4I_0 \sin^2(\pi b\theta/\lambda)}{(\pi b\theta/\lambda)} \cos^2(\pi s\theta/\lambda)$$

The diffraction grating

Finally, phasor analysis can help us to understand the origin of the secondary maxima that occur between the principal maxima in the interference pattern due to a diffraction grating; we shall not, in this discussion, consider the effect of diffraction at the individual slits. We assume, as before, that the interference pattern is formed on a distant screen.

The phasors contributing to the central maximum are all in phase, with zero phase differences, as shown in Fig. A.6. The first order principal maximum occurs when the path difference between light from adjacent slits is λ, and the phase difference is hence equal to 2π. The phasor diagram will look identical to that for the central maxima and, ignoring single slit diffraction, the resultant amplitude and intensity will be identical.

However, at intermediate path differences, there will be a series of secondary maxima and minima, as shown in the intensity graph in Fig. A.7. If the amplitude due to one slit is a,

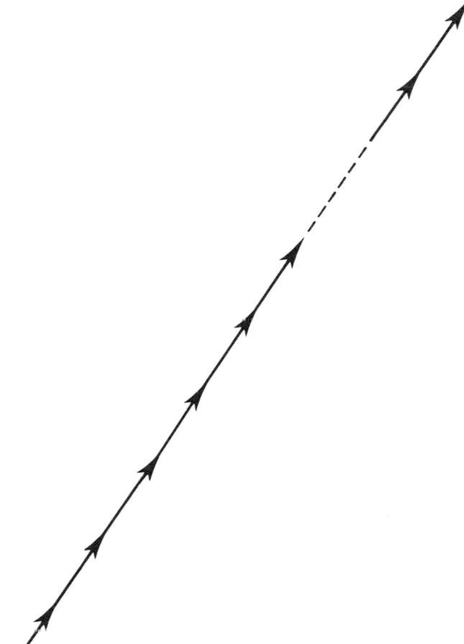

Formation of principal maxima

Fig. A.6

and there are N slits, the total amplitude for a principal maximum is Na, and the intensity, I_0, is proportional to $(Na)^2$. The amplitude of the first secondary maximum is given by the *diameter* of a circle whose *circumference* is $\frac{2}{3}Na$ since one and a half circumferences is equal to Na.

$$\text{Hence} \quad \text{resultant amplitude} = \frac{2Na}{3\pi}$$

$$\text{i.e.} \quad \text{resultant intensity} \propto \frac{4}{9\pi^2}(Na)^2$$

$$= \frac{4}{9\pi^2} I_0$$

Thus, the first secondary maxima has an intensity about 1/25 that of the principal maxima; successive secondary maxima have smaller intensities and they do not feature significantly in the interference patterns produced by gratings.

We also see that the intensity first falls to zero when the phasors form one closed circle. Hence, for a grating with N slits, the phase difference between successive phasors must then be given by:

$$\phi = \frac{2\pi}{N}$$

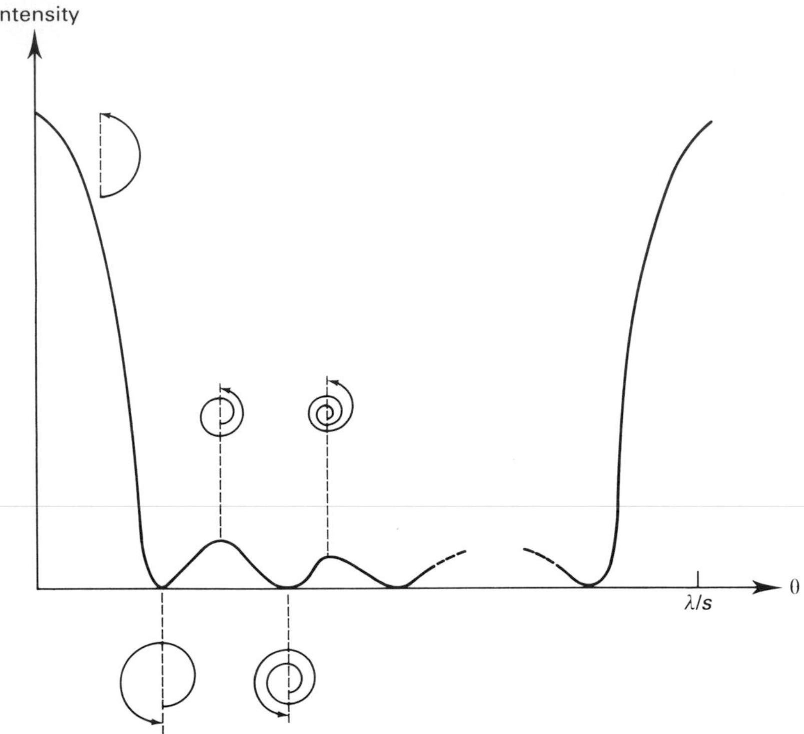

Fig. A.7

Thus, from the analysis above for a double slit, the first minimum will occur at an angle θ given by:

$$\frac{2\pi s\theta}{\lambda}=\frac{2\pi}{N}$$

i.e.
$$\theta=\frac{2\pi}{N}\cdot\frac{\lambda}{2\pi s}$$
$$=\frac{\lambda}{Ns}$$
$$=\frac{\lambda}{\text{total width of grating}}$$

Hence, for a typical grating 30 mm wide, illuminated by light of wavelength 500 nm, the angle at which the intensity first falls to zero is given by:

$$\theta=\frac{5\times10^{-7}}{3\times10^{-2}}$$
$$=1.7\times10^{-5}\text{ rad}$$
$$=0.06'$$

We see from this calculation that the principal maxima formed by diffraction gratings are very narrow indeed; on the scale of the secondary maxima in Fig. A.7, the first order principal maximum, for a grating with 10 000 lines, should be displaced about 100 m to the right!

Q A.3
Use phasor arguments to find the number of secondary maxima between each pair of principal maxima, in a grating with N slits.

Exercises

Additional questions on the subject matter of each chapter are gathered together in this section of the book. The questions have been sorted into three grades of difficulty—**A**, **B** and **C**. Questions in the **A** grade are straightforward introductory applications of new ideas; **C** grade questions are intended as a stimulus to stretch the more able student. Please note that only answers to odd-numbered numerical questions are given in the Answers section.

Unless otherwise stated take:

Strength of the Earth's gravitational field, $g = 10\ \text{N kg}^{-1}$
Speed of light in vacuo, $c = 3.0 \times 10^{8}\ \text{m s}^{-1}$
Planck constant, $h = 6.6 \times 10^{-34}\ \text{J s}$
Charge on an electron, $e = -1.6 \times 10^{-19}\ \text{C}$

Chapter 1

A

Q 1.16
On the same time axis, sketch graphs to show how the displacement, velocity, and acceleration of a simple harmonic vibrator vary with time. (W)

Q 1.17
A long pendulum performs small oscillations and the position of the bob at 0.10 s intervals is shown below. The diagram shows exactly one half cycle of its motion.

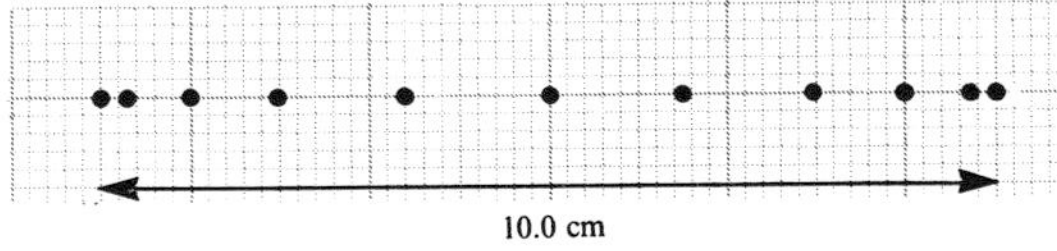

(*a*) (i) On graph paper plot the displacement of the bob from its equilibrium position against time.
(ii) Explain how you would show that the variation is sinusoidal.
(*b*) Give values for the amplitude and the period of the oscillation.
(*c*) Use your answers to (*b*) to calculate the maximum speed of the bob. (AEB, 1986)

Q 1.18
A body of mass 0.30 kg executes simple harmonic motion with a period of 2.5 s and an amplitude of 4.0×10^{-2} m.

Determine:

(i) the maximum velocity of the body;
(ii) the maximum acceleration of the body;
(iii) the energy associated with the motion. (S)

Q 1.19
A helical spring requires a force, *k*, to produce unit extension. When it is suspended vertically a mass, *M*, attached to its lower end performs *simple harmonic motion* with a time period, *T*, when given a small vertical displacement and released. The time period is independent of the *amplitude.*

(*a*) Define the terms in italics and write down an expression for *T* in terms of *M* and *k*.
(*b*) The diagrams below show helical springs connected in series and in parallel. All four springs are identical and each arrangement has a mass, *M*, attached to its lower end. In each case determine, in terms of *T*, the new time period and explain your reasoning.

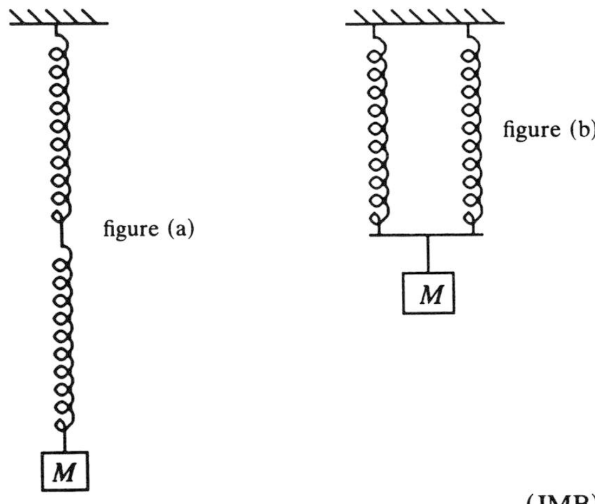

(JMB)

Q 1.20
What is the length of a simple pendulum which has a period of 2.24 ($=5\pi/7$) s?

If the pendulum bob has a mass of 1.00 kg, and the amplitude of the oscillation is 20.0 cm, what is (*a*) the speed of the bob as it passes through its equilibrium position, (*b*) the tension in the string at the same instant? (S)

Q 1.21
A horizontal plate is vibrating vertically with simple harmonic motion at a frequency of 20 Hz. What is the maximum amplitude of vibration so that fine sand on the plate always remains in contact with it? (C)

Q 1.22
The pick-up of some types of record player may be regarded as a mass *m*, attached to a cantilever of stiffness *k*, stiffness being defined as the force per unit deflection of the cantilever. The natural

frequency of vibration, *f*, of the system is given by

$$f = \frac{1}{2\pi}\sqrt{\frac{k}{m}}$$

(*a*) State the dimensions of stiffness.
(*b*) Calculate the maximum value of stiffness if *m* is 0.4 mg and the natural frequency is not to be greater than 10 Hz.
(*c*) Why is it desirable that *f* should be less than about 10 Hz? (AEB, 1985)

Q 1.23
A light spring is suspended from a rigid support and its free end carries a mass of 0.40 kg which produces an extension of 0.060 m in the spring. The mass is then pulled down a further 0.060 m and released causing the mass to oscillate with simple harmonic motion.

(*a*) Potential energy is stored in two ways in this arrangement: explain briefly what they are.
(*b*) Calculate the kinetic energy of the mass as it passes through the mid point of its motion. (L)

Q 1.24
Describe experiments you could do in a school laboratory to illustrate the meanings of the terms *free oscillation, forced oscillation,* and *resonance.* Give further examples of each of these from non-laboratory situations.

B

Q 1.25
(*a*) Explain what is meant by simple harmonic motion (s.h.m.).
(*b*) A particle executes s.h.m. of amplitude *a* and period *T*.
  (i) Explain the meanings of the terms *amplitude* and *period.*
  (ii) Write down expressions in terms of *a* and *T* (as appropriate) for the speed and the acceleration of the particle when at a displacement *x* from the midpoint of its oscillations.

  Hence derive an expression involving *a* and *T* for the total energy associated with the simple harmonic motion of a particle of mass *m*.
(*c*) A loudspeaker cone, sounding a pure note of frequency 2.5 kHz, executes s.h.m. of amplitude 2.0 mm.
  Calculate
  (i) the maximum speed of the cone;
  (ii) the maximum acceleration of the cone;
  (iii) the mean power required to maintain the cone's oscillations in the presence of a mean damping force of 0.30 N.
(*d*) Describe an experiment (using normal laboratory equipment) to confirm that the cone's oscillations are simple harmonic. (O)

Q 1.26
In this question you are asked to discuss the use of mathematical models in physics. The following statements should form the basis of your discussion in which you should give brief details of relevant theory and experiments, and practical examples.

For (*a*) discuss only *one* specific example, but for (*b*) make your discussion more general.

(*a*) Some oscillating systems can be described by the equation $d^2x/dt^2 = -\omega^2 x$.

This equation represents the behaviour of a model of the oscillating system, but in formulating the model simplifying assumptions about the oscillating system have to be made.

(*b*) A model of this kind enables analogies to be drawn between different areas of physics. Thus mathematical models have an important role in physics. (O & C, Nuffield)

Q 1.27
The period T of vertical oscillations of a mass M suspended by a spiral spring is given by

$$T^2 = \frac{A}{g}M + \frac{A}{3g}m$$

where A is a constant depending on the stiffness of the spring and m is the mass of the spring itself.

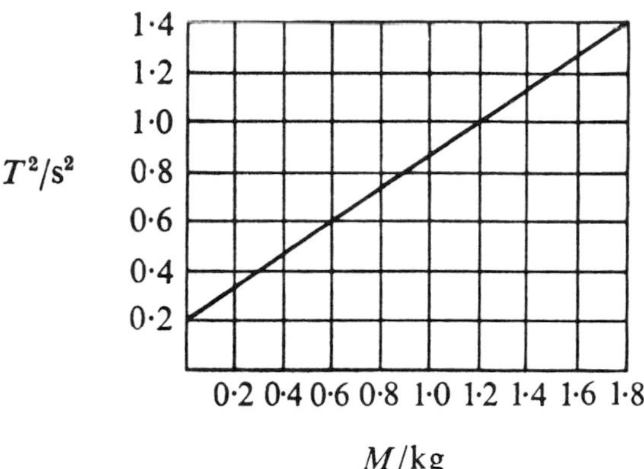

The graph shows the results of measurements of the period for various values of M. Use the graph to determine the constant A and the mass m of the spring. What are the dimensions of A? (S)

Q 1.28
The simple hydrogen iodide (HI) molecule can be imagined as two ions at the end of a bond, oscillating rather like two masses, one small and the other large, linked by a spring, as shown in the figure below.

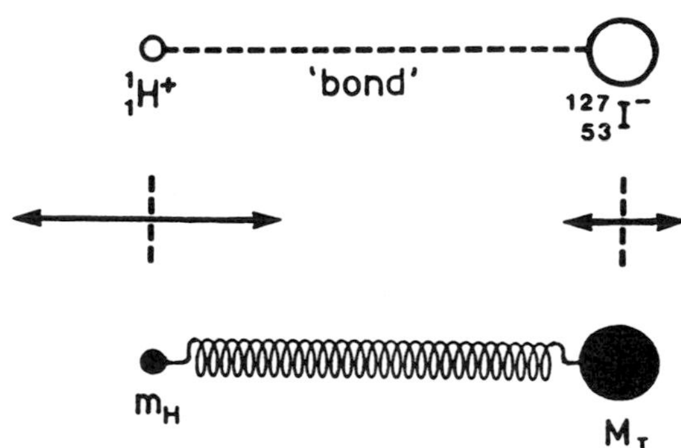

(*a*) Assuming that the movement of the iodine ion can be ignored in this mechanical model write down an equation relating f, the natural frequency of oscillation of the hydrogen ion, to m_H, the mass of the hydrogen ion, and to k, the force constant of the bond.

(*b*) When electro-magnetic radiation is passed through hydrogen iodide gas it is found that there is an absorption band with a mean wavelength of 4.5×10^{-6} m.
  (i) Calculate the mean frequency of the radiation absorbed.
  (ii) Hence calculate the force constant of the ionic bond.

(*c*) (i) Explain why it is reasonable to ignore any movement of the iodine ion.
  (ii) State one other assumption that is made in applying this model.

(*d*) Using the same model and assumptions as above, calculate the mean absorption frequency to be expected if the hydrogen ion were replaced by a deuterium ion ($^2_1H^+$).
(O & C, Nuffield)

Q 1.29
(*a*) Show that the time period, T, of small amplitude oscillations of a simple pendulum of length, l, is given by $T = 2\pi\sqrt{l/g}$, where g is the acceleration of free fall.

(*b*) (i) Describe how you would set up a simple pendulum of length 1.00 m in a laboratory and determine accurately the time period of small amplitude oscillations.
  (ii) Discuss whether or not it would be possible to detect the effect on the time period of this pendulum if its length were increased by 0.02 m.

(*c*) A simple pendulum of length 1.00 m has a bob with a mass of 0.20 kg. It is set into motion by a single sharp sideways blow to the bob which causes the pendulum to oscillate with an angular amplitude of 5.0°.
Calculate
  (i) the maximum vertical displacement of the bob from its rest position.
  (ii) the energy acquired by the bob from the initial blow.
  (iii) the maximum speed of the bob.
Neglect the effect of damping.

(*d*) Explain briefly the mechanism by which damping would cause the amplitude of the oscillations to decrease. (JMB)

Q 1.30
(*a*) What is meant by *simple harmonic motion*?
The equation $x = a \sin 2\pi ft$ can represent the motion of a body executing simple harmonic motion where x represents the displacement of the body from a fixed point at time t. Sketch two cycles of the motion beginning at $t = 0$,

clearly labelling the axes of the graph. Use the graph to explain the physical meanings of a and f.

Explain how you could obtain from the graph the speed of the body at any instant.

(*b*) In order to check the timing of a camera shutter a student set up a simple pendulum of length 99.3 cm so that the bob swung in front of a horizontal metre scale. The bob was observed to swing between the 40.0 cm and 60.0 cm marks at its extreme positions. The camera was mounted directly in front of the scale set for an exposure time (time for which the shutter is open) of 1/50 s and a photograph taken. The resulting photograph showed the bob to have moved from the 51.0 cm mark to the 51.6 cm mark while the shutter was open.

What is the percentage error in the exposure time indicated on the camera?

(The period of oscillation of a simple pendulum of length l may be taken as

$$T = 2\pi\sqrt{\frac{l}{g}})$$

(L)

Q 1.31

(*a*) Explain what is meant by linear simple harmonic motion (s.h.m.).

(*b*) A particle moves in a straight line so that its displacement x from a fixed point at time t is given by the equation

$$x = a\sin(\omega t + \varepsilon)$$

(i) Show that its motion is simple harmonic.

(ii) Explain the significance of the terms a, ω, and ε.

(iii) Write down corresponding expressions in terms of a, ω, t, and ε for the velocity and the acceleration of the particle.

(*c*) A harbour in a dockyard is in the shape of a square of side 300 m, with a narrow entrance from the sea; the bottom is horizontal and the walls are vertical. The tide causes the water-level in the harbour to perform simple harmonic motion with a period of 12.5 hours; the maximum depth is 15 m and the minimum is 10 m.

Write down expressions for:

(i) the depth of water at time t (in hours) after high water;

(ii) the rate at which the water-level is changing at time t.

Calculate:

(iii) the times when the water-level in the harbour is changing most rapidly:

(iv) the greatest rate of flow of water in cubic metres per second in the entrance.

(*d*) It is proposed that the movements of sea-water in the entrance to the harbour might be used to power turbogenerators for electrical energy. Estimate the mean power that might be generated over the whole tidal cycle, assuming that the generators were 50% efficient. State the physical principles used in your calculation.

[Take the density of sea-water to be 1050 kg m^{-3}.] (O)

Q 1.32

(*a*) Define *simple harmonic motion.*

(*b*) A light helical spring, for which the force necessary to produce unit extension is k, hangs vertically from a fixed support and carries a mass M at its lower end. Assuming that Hooke's law is obeyed and that there is no damping, show that if the mass is displaced in a vertical direction from its equilibrium position and released, the subsequent motion is simple harmonic. Derive an expression for the time period in terms of M and k.

(*c*) If $M = 0.30$ kg, $k = 30$ N m^{-1} and the initial displacement of the mass is 0.015 m, calculate:

(i) the maximum kinetic energy of the mass,

(ii) the maximum and minimum values of the tension in the spring during the motion.

(*d*) Sketch graphs showing how (i) the kinetic energy of the mass, (ii) the tension in the spring vary with displacement from the equilibrium position.

(*e*) If the spring with the same mass attached were taken to the moon, what would be the effect, if any, on the time period of the oscillations? Explain your answer. (JMB)

Q 1.33

A student who is investigating the effect of damping on an oscillatory system obtains the following values of P, the amplitude of the oscillations when n complete oscillations have been made.

n	P cm
10	12.4
20	8.3
30	5.5
40	3.7
50	2.5
60	1.6
70	1.1

The student knows that he can analyse the vibration in terms of the expression

$$P = P_0 K^{-n} \quad \text{(i)}$$

where P_0 is the initial amplitude and K a constant which determines the degree of damping in the system.

(*a*) Plot a graph of P(y-axis) against n(x-axis).
(*b*) Measure the slope S of your graph where $P = 5.0$ cm.
(*c*) Given that $-S = 2.30 \times P \times \log_{10} K$ obtain a value for K.
(*d*) Draw up a table of values of n and $\log_{10} P$ and plot a graph of $\log_{10} P$(y-axis) against n(x-axis).
(*e*) Measure the slope of your graph and using (i) obtain another value for K. Explain your working clearly. (S)

Q 1.34
Define *simple harmonic motion.* How would you investigate experimentally whether the motion of a pendulum remains simple harmonic as the amplitude of vibration is increased?

The suspension of a car may be considered to be a spring under compression combined with a shock absorber which damps the vertical oscillations of the car. Draw sketch graphs, one in each case, to illustrate how the vertical height of the car above the road will vary with time after the car has just passed over a bump if the shock absorber is

(*a*) not functioning, i.e. slides without resistance,
(*b*) operating normally.

When the driver, of mass 80 kg, steps into the car, of mass 920 kg, the vertical height of the car above the road decreases by 2.0 cm. If the car is driven over a series of equally spaced bumps, the amplitude of vibration becomes much larger at one particular speed. Explain why this occurs and calculate the separation of the bumps if it occurs at a speed of $15\ \text{m s}^{-1}$.

(The frequency of vibration of a loaded spring is $1/2\pi\sqrt{(k/m)}$, where m is the mass on the spring and k is the force required to produce unit extension of the spring.) (C)

Q 1.35
(*a*) Give one example of a system oscillating with simple harmonic motion. Explain what is meant by
 (i) natural frequency
 (ii) damping
 (iii) forced vibrations
 (iv) resonance.
(*b*) A spring for which the force constant $k = 4\ \text{N m}^{-1}$ hangs vertically from a vertically oscillating support. A mass of 0.1 kg is attached to the lower end of the spring. The frequency of the support increases steadily from 0.1 Hz to 10 Hz. Draw sketches of amplitude as a function of frequency, and phase as a function of frequency, to describe the subsequent motion of the mass on the spring.
(*c*) If the suspended mass was a lump of soft-iron suspended in a magnetic field, how would increasing the magnetic flux density from zero to a large value affect the results you have shown in the sketches? (W)

Q 1.36
'*Resonance is a Mixed Blessing.*'

Discuss this statement. Include in your answer an explanation of what is meant by resonance and describe *four* examples of resonance. *At least one* example should show how resonance can be helpful and *at least one* should show how it can be a nuisance.

In *at least one* example explain quantitatively how the resonant frequency is related to other properties of the system. (O & C, Nuffield)

C

Q 1.37
(*a*) A mass m is suspended by a vertical spring of stiffness k. The mass executes small vertical oscillations. Show that the motion is simple harmonic with period

$$T = 2\pi(m/k)^{1/2}$$

(*b*) (i) The spring has natural length a and the stiffness k has magnitude such that a force of mg doubles the length to $2a$. A second identical spring is attached to the mass m and the free ends of the two springs are fixed to points A and B distance $4a$ apart in a vertical line as shown.

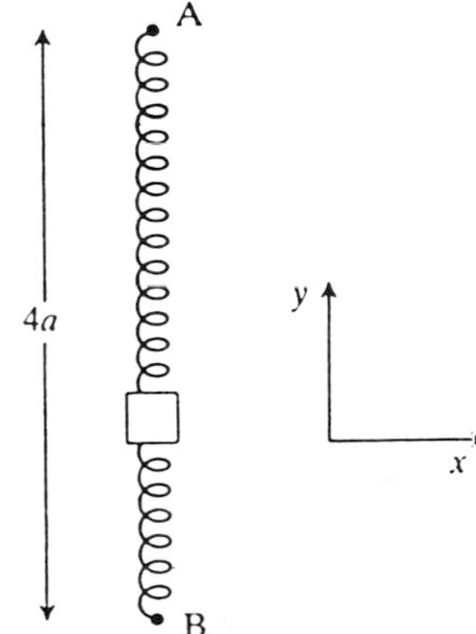

Show that the equilibrium position for the mass is $5a/2$ below A.

(ii) The mass m is given a small vertical displacement y and released. Calculate the period T_1 for the vertical oscillations.

(iii) The mass is given a horizontal displacement x where $x \ll a$. Calculate the total horizontal force exerted on the mass by the springs.

Hence calculate the period T_2 for the horizontal oscillations when the mass is released and show that

$$T_2/T_1 = (\tfrac{15}{7})^{1/2}$$

(*c*) Assuming damping is negligible, give a sketch graph of the variations with time of the kinetic energy of the mass and the total strain energy in the springs for the horizontal oscillatory motion. (O & C)

Q 1.38

This question is about the frequency of atomic oscillations in a sodium chloride crystal. Read the four statements given below, and then answer the questions about them.

(i) The average mass m of sodium and chlorine atoms is about 5×10^{-26} kg.

(ii) The spring constant k of the ionic bond is about $100\ \mathrm{N\,m^{-1}}$.

(iii) Thus, k/m is about $20 \times 10^{26}\ \mathrm{N\,m^{-1}\,kg^{-1}}$, and hence the frequency of oscillation f is approximately 10^{13} Hz, corresponding to a wavelength of 3×10^{-5} m.

(iv) This compares well with the experimental value of 6×10^{-5} m.

(*a*) What kind of experiment gives the value quoted in statement (i)? (A brief outline only is needed; experimental detail need not be given.)

(*b*) Show briefly how k may be derived from experimentally determined constants of sodium chloride.

(*c*) Show how the values of 10^{13} Hz and 3×10^{-5} m were deduced in statement (iii).

(*d*) The experimental value quoted in statement (iv) is much larger than 3×10^{-5} m. Suggest and explain *two* reasons why this is so. (O & C, Nuffield)

Q 1.39

State the conditions under which a system will perform simple harmonic motion.

A test tube of uniform cross section area A is partly filled with lead shot so that it floats upright in a fluid of density ρ contained in a large vessel. Show that, when displaced vertically, the tube will execute simple harmonic motion with angular frequency $\omega = \sqrt{g/l}$, where l is the length of the tube immersed at equilibrium.

Sketch the variation of the frequency with (*a*) the mass m and (*b*) the cross section area A of the loaded tube.

In order to isolate a record player from external vertical vibrations it is suggested to mount it in a shallow tray floating on a large area of water. Discuss whether this would eliminate the pick up of unwanted audible frequencies. Take the mass of the player as 1 kg and the area of the tray as $0.3\ \mathrm{m^2}$. (Ox. Schol.)

Q 1.40

A simple pendulum, consisting of a mass m on the end of a string of length l, performs small oscillations of angular amplitude θ. Derive from first principles an expression for the period.

What would be the period

(*a*) in a lift descending with uniform acceleration a,

(*b*) in a lorry travelling at constant speed v round the arc of a horizontal circle of radius r?

Would the periods in (*a*) and (*b*) be greater or less if the amplitude were large? Explain your reasoning.

If $m = 0.20$ kg, $l = 1.5$ m and the pendulum is set swinging in the laboratory with $\theta = 5.0°$, calculate the maximum and minimum tensions in the string.

(C)

Q 1.41

(*a*) Explain the terms *simple harmonic motion*, *damped harmonic motion*.

(*b*) A light helical spring is suspended vertically from a clamp at its upper end. A 50 g mass is attached to the lower end and the static extension of the spring is 5.0 cm. The 50 g mass is pulled down a further 3.0 cm and released. Show that the motion of the 50 g mass is simple harmonic and calculate

(i) its period,

(ii) the maximum value of the kinetic energy of the 50 g mass, and

(iii) the maximum value for the energy stored in the spring. ($g = 9.8\ \mathrm{m\,s^{-2}}$.)

(*c*) The same 50 g mass and spring system has attached to it a damping mechanism which dissipates 36% of the vibrational energy each complete oscillation. If the initial amplitude of the motion is 3.00 cm, calculate the amplitudes at the end of 1, 2, 3, 4 and 5 oscillations. Sketch a graph which shows how the *energy* stored in the system changes with the number of oscillations completed. Explain, with reference to this graph, the meaning of the term *exponential decay.*

(*d*) A ball falls from a height of 3.00 m and loses 20% of its energy each time it bounces.

Sketch a graph of height above the ground against time for the first five bounces. Discuss whether the term exponential decay is relevant here. (L)

Q 1.42

(*a*) A small mass m is constrained to move along the x-axis such that its potential energy when it is displaced a distance x from the origin is

$$U = \tfrac{1}{2}kx^2$$

where k is a constant. Show that

(i) its equilibrium position is at the origin, and

(ii) when it is displaced from the origin and released, it oscillates with simple harmonic motion.

The mass is displaced a distance A from the origin and released. Find

(iii) the angular frequency of oscillation ω, and

(iv) the kinetic energy T of the mass at displacement x.

Draw accurate graphs with suitably labelled axes of the variation of

(v) the force F acting on the mass, and

(vi) the kinetic energy T against the displacement x from equilibrium for the situation where $k = 20\ \text{N m}^{-1}$ and $A = 50$ mm.

(*b*)

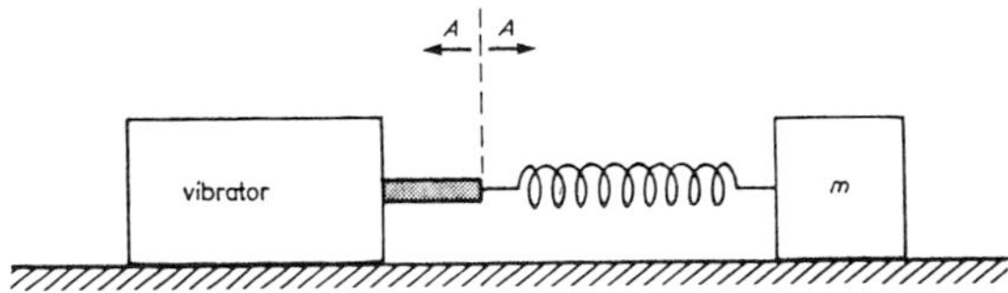

The mass m rests on a nearly smooth horizontal surface. It is attached by a light open wound spring of force constant k to a vibrator which makes *small* simple harmonic oscillations of constant amplitude A. Describe, in as much detail as you can, how the motion of the mass will alter, relative to the motion of the vibrator as the vibrator's frequency varies from zero to $(k/m)^{1/2}$. Include in your answer sketch graphs of how you would expect the amplitude and phase of the motion of the mass to vary relative to the motion of the vibrator. (O & C)

Q 1.43

(*a*) Explain the phenomenon of resonance. Give one example in each case of the occurrence of resonance effects with (i) sound waves, (ii) radio waves, (iii) light waves.

(*b*) With the aid of suitable sketch-graphs, describe the behaviour of an oscillatory system subject to various degrees of damping. Explain what is meant by *critical damping*, and give one example of a system that is normally critically damped.

(*c*) A spin-dryer drum has a mass of 12 kg. It is flexibly mounted on rubber bearings which exert a restoring force of 125 N when the axis of the drum is pushed sideways in a radial direction by 1 cm.

(i) Calculate the period of radial oscillations of the drum.

(ii) The drum is packed non-uniformly with a load of mass 4 kg and accelerated up to its normal running speed of 800 rev/min. Describe and explain the vibrations that will be observed. (O)

Q 1.44

(*a*) Define *simple harmonic motion.* The energy associated with the simple harmonic motion of a mechanical system oscillates between two potential energy states and through a kinetic energy state. Trace the details of these energy transformations in the case of a mass oscillating vertically at the lower end of a helical spring.

(*b*) Explain the term *damped harmonic motion.* Identify the damping forces which drain the energy of oscillation from the mass attached to the helical spring and state, with reasons, whether these forces are steady or periodic.

(*c*) Discuss whether or not an exponential decay law applies in each of the following situations:

(i) the periodic motion of a simple pendulum,

(ii) the discharge of a capacitor through a resistor, and

(iii) the decay of a radioactive specimen.

(*d*) A mechanical system has a natural frequency of oscillation f_0. It is acted on by a periodic driving force of constant amplitude F and frequency f. Sketch three curves on one graph to show how the amplitude of the oscillations (when the system reaches equilibrium) varies with the frequency f. Draw one curve each for light, medium and heavy damping. Explain

(i) why the three curves coincide at low and high values of f,

(ii) why resonance is sharpest when damping is light (Hint: think about the energy content), and
(iii) why the rate of energy transfer to the system is a maximum at resonance. (L)

Q 1.45
This question is about the isolation of sensitive apparatus from floor vibrations.

Isolation is achieved by mounting the apparatus and table (total mass M) on a flexible suspension, which may be modelled by a single spring as shown. The floor vibrates at a frequency f (angular frequency $\omega = 2\pi f$) and with an amplitude x_f. The natural frequency of the mass-spring system is f_0 (angular frequency $\omega_0 = 2\pi f_0$), and the damping constant for the oscillation is k.

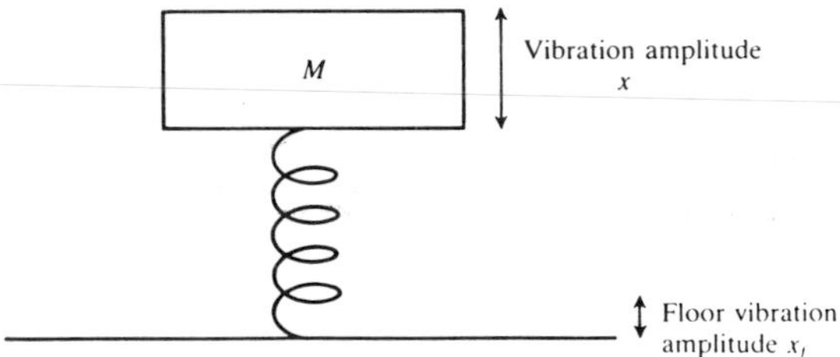

The amplitude of vibration of M is given by

$$x = \frac{\omega_0^2 x_f}{\{(\omega k)^2 + (\omega^2 - \omega_0^2)^2\}^{1/2}}.$$

(*a*) In analysing the effectiveness of isolation mountings, engineers use two dimensionless quantities R and D, defined as:

Frequency ratio $R = \omega/\omega_0$
Damping ratio $D = k/\omega_0$

Write down an expression for x in terms of R and D. Show that, as the floor frequency f and hence the frequency ratio is varied, the maximum amplitude of vibration of M occurs when $R^2 = 1 - (D^2/2)$.

(*b*) Use your expression for x from part (*a*) to show that for very small values of R the amplitude of oscillation of M becomes equal to the amplitude of vibration of the floor.

Explain in physical terms how it is that under these conditions the spring is behaving like a rigid rod.

(*c*) The ratio x/x_f is called the transmissibility T. Write down the approximate expressions for T in terms of R and D for the cases when:
(i) $R \to 0$;
(ii) $R \simeq 1$;
(iii) $R \gg 1$.

Use these results to sketch graphs on the same axes showing how T varies with frequency for
(iv) $D = 0.1$;
(v) $D = 0.9$.

(*d*) For many situations, damping is small, and can be neglected except near resonance. Write down an expression for T which would be valid everywhere except near resonance. Show that isolation would be effective ($T < 1$) only for $R > \sqrt{2}$.

(*e*) An analytic balance mounted on a slate slab is found to be impossible to use to its required accuracy because the floor is vibrating with an amplitude of 1.6 μm. From experience, it is known that this balance would be satisfactory if the amplitude of vibration is reduced to 0.2 μm. The main source of vibration is a nearby underground railway, and the predominant frequency is 22 Hz. Calculate the natural frequency of the isolation system which would just enable the satisfactory working of the balance. If the mass of the balance plus slab is 50 kg, what is the required spring constant of the suspension? (O & C, Nuffield)

Q 1.46
Write an essay on the phenomenon of resonance. Your essay should include consideration of the following three points:

(*a*) the meaning of and the relationship between the terms *resonance, natural frequency, forcing or driving frequency* and *damping*;
(*b*) the wide variety of technological situations and physical phenomena in which resonance plays a part;
(*c*) the similarities and differences between examples of resonance from different areas of science and technology. (O & C, Nuffield)

Q 1.47
(*a*) Give an account of some phenomena which can arise when two simple harmonic motions are superposed.

A rectangular drawing-board is suspended at its corners from a horizontal ceiling by four parallel strings each of length 3.60 m. As the board swings to and fro in the x-direction, the strings move in planes of constant y with amplitude 0.050 rad. To a point P above the board is attached a simple pendulum of length 400 mm. A small pen is fastened to the bottom of the pendulum bob and at all times remains in contact with paper pinned to the board. At the moment when the drawing-board is at its extreme position in the positive

x-direction, the bob is released from rest having been drawn aside through 60 mm,
(i) in the positive x-direction,
(ii) in the negative x-direction,
(iii) in the positive y-direction.
Sketch in each case, indicating dimensions where possible, the traces left by the pen on the paper.

(*b*) Sketch the trace that would result if case (i) were repeated but with the point P being made to move steadily at 100 mm s^{-1} in the positive y-direction. Computation of the trace at six suitably chosen points will give sufficient accuracy. Use this result to give an improved answer for case (i). (Cam. Schol.)

Chapter 2

A

Q 2.13
A small piece of cork in a ripple tank oscillates sinusoidally up and down as ripples pass it. If the ripples travel at 0.20 m s^{-1}, have a wavelength of 15 mm and an amplitude of 5.0 mm, what is the maximum velocity of the cork? (Assume that the cork moves up and down in a straight line.) (L)

Q 2.14
A sound wave in air, having a frequency 680 Hz and a wavelength 0.5 m, is travelling northwards parallel to the earth's surface. The amplitude of the vibration of the air particles is 1 μm.

(*a*) What is the direction of motion of the air particles?
(*b*) What is the speed of the sound wave?
(*c*) What is the maximum speed of the air particles? (S)

Q 2.15
Explain briefly why it is impossible for a pure transverse wave to be propagated by a fluid. (C)

Q 2.16
The speed c of longitudinal waves in a wire is given by the expression $c = \sqrt{(E/\rho)}$ where E is the Young modulus for the material of the wire and ρ is its density. Show that this equation is dimensionally correct.

The extension e, of a wire of cross-sectional area A and of initial length L, is measured for various extending forces F and a graph of F against e is plotted. How would you find a value of c from this graph? What other quantity would you need to measure? (L)

Q 2.17
The diagram shows a loudspeaker, L, which emits a continuous sound of frequency 400 Hz, and a line to represent the positive x-direction. The graph shows the displacements of the air masses along x from their undisturbed positions at one instant.

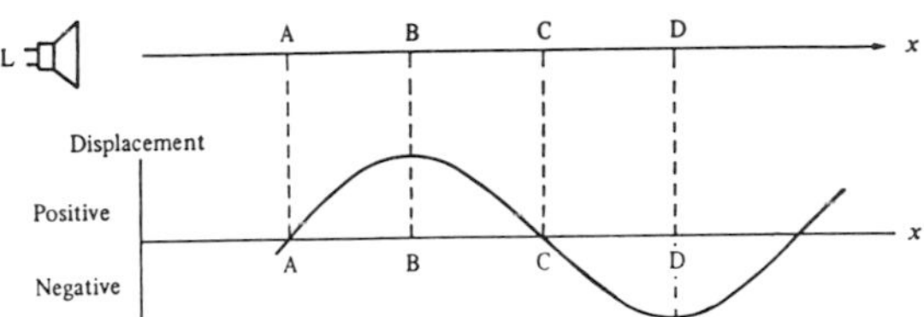

(*a*) At which of the four points, A, B, C, D is the instantaneous *pressure* at its peak value? Explain your answer.
(*b*) Calculate the time interval required for the sound to travel the distance AD. (L)

Q 2.18
The speeds of longitudinal and transverse waves on a stretched wire are

$$\sqrt{\frac{E}{\rho}} \quad \text{and} \quad \sqrt{\frac{T}{M}}$$

respectively, where E is the Young modulus, ρ is the density, T is the tension, and M is the mass per unit length, of the wire. Find the value of the stress that would be required if the transverse and longitudinal waves were to have the same speed, and comment on the result. (C)

Q 2.19
The velocity v of a transverse wave on a stretched string is given by

$$v = \sqrt{\frac{T}{m}}$$

where T is the tension in the string and m is its mass per unit length.

(*a*) Justify the equation by the method of dimensions.
(*b*) Use the equation to derive an expression for the fundamental frequency of the string in terms of its length l. (W)

Q 2.20
The equation $y = a \sin(\omega t - kx)$ represents a plane wave travelling in a medium along the x-direction, y being the displacement at the point x at time t. Deduce whether the wave is travelling in the positive x-direction or in the negative x-direction.

If $a = 1.0 \times 10^{-7}$ m, $\omega = 6.6 \times 10^3\ \text{s}^{-1}$ and $k = 20\ \text{m}^{-1}$, calculate (a) the speed of the wave, (b) the maximum speed of a particle of the medium due to the wave. (J)

B

Q 2.21

Using the apparatus shown below, a student attempts to measure the speed of sound in a metal rod of length 0.6 m. He drops the rod a short distance onto a massive metal plate, allows it to bounce once, and catches it. For the short time that the rod is in contact with the plate, the output of the signal generator is connected to the C.R.O.

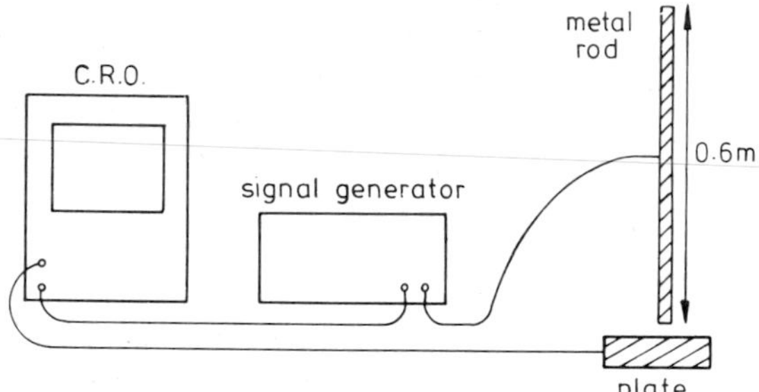

With the signal generator set at 20 kHz the trace shown below was obtained.

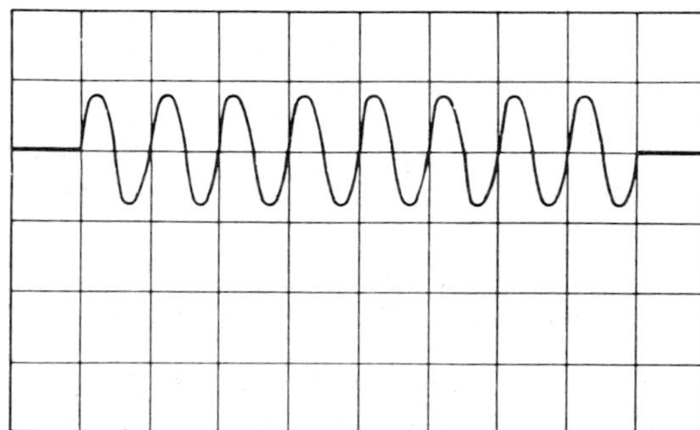

(*a*) (i) What was the time of contact? Show the steps in your calculation.
(ii) Use this result to calculate the speed in the metal rod. Show the steps in your calculation.

(*b*) The student repeated the experiment, using a rod of the same length but made of a metal in which he knew that the speed of sound had twice the value it had in the first rod. He decided to change the signal generator setting to 10 kHz but left the C.R.O. controls unchanged.

State what the effect would be on
(i) the time of contact of the rod on the plate,
(ii) the time period of a single oscillation of the signal generator output.

(*c*) Sketch carefully, on a copy of the axes provided above, the trace he should now obtain. (O & C, Nuffield)

Q 2.22

For each of the following statements:
state the physical principles which are relevant, and
show how they apply to the situation described. Also
make calculations and/or give equations where possible to show the relationship between the quantities involved.

(*a*) By analogy with a simple model consisting of masses joined together by springs, and employing suitable numerical data, it is possible to predict the speed of sound in a copper rod. The measured value of $3.8 \times 10^3\ \text{m s}^{-1}$ does not agree exactly with the prediction.

Data: mass of copper atom $= 1.1 \times 10^{-25}$ kg
interatomic spacing in copper $= 2.6 \times 10^{-10}$ m
bond stiffness in copper $= 33\ \text{N m}^{-1}$

(*b*) The string which produces the note of lowest frequency on a guitar has about four times the diameter of the string which gives the highest frequency, and is used at approximately the same tension.

Q 2.23

Explain the principle of the method of checking an equation by dimensions.

One of the following equations shows how the speed v of ripples on a deep liquid depends on the wavelength λ, the surface tension γ and the density ρ:

(i) $v = A\dfrac{\gamma}{\rho}\sqrt{\dfrac{1}{g\lambda^3}}$,

(ii) $v = B\dfrac{\gamma}{\rho}\sqrt{\dfrac{g}{\lambda}}$,

(iii) $v = C\sqrt{\dfrac{\gamma}{\lambda\rho}}$,

where g is the acceleration of free fall, and A, B and C are dimensionless constants. Identify any equation that is wrong dimensionally.

The following results were obtained in an experiment with water in a ripple tank:

v/mm s^{-1}	230	270	340	450	590
λ/mm	8.5	6.2	3.9	2.2	1.3

(*a*) Explain how the wavelength and speed of the ripples might be determined.
(*b*) Draw a graph of log v against log λ and find its slope. Which equation is correct? Find the value of the appropriate constant *A*, *B* or *C*.

[$g = 9.8$ m s^{-2}; density of water $= 1.0 \times 10^3$ kg m^{-3}; surface tension of water $= 7.2 \times 10^{-2}$ N m^{-1}.] (C)

Q 2.24
(*a*) $y = a \sin(kx - \omega t)$ is an equation of a progressive wave. What are the names given to k and ω? If $k = 4.5$ m^{-1} and $\omega = 1.5 \times 10^3$ s^{-1} what are the values of (i) the wavelength and of (ii) the period, of the wave?
(*b*) If the equation represented a sound wave in air what physical quantity would y represent? How would this physical quantity change with time at some fixed value of x? (W)

Q 2.25
In this question you are asked to write an essay about waves.

Waves and wave motion occur in many situations in physics, technology and in nature. Choose **four** examples, which are as different as possible, which illustrate the great variety of waves which exists. Your answer should include descriptions of these four examples and should show how they differ in origin, physical nature, speed, frequency and wavelength. Mention also any features which any of the various waves have in common. (O & C, Nuffield)

Q 2.26
This question is about the behaviour of water waves.

The diagram shows a wave in water deep enough for the wavelength to be much less than the depth. As the wave moves to the right, the water at P acquires in succession the velocities v_1, v_2, v_3, v_4 and v_5 ($v_5 = v_1$) and it can be shown that it moves in a circle with constant speed, where the radius of the circle is equal to the wave amplitude A.

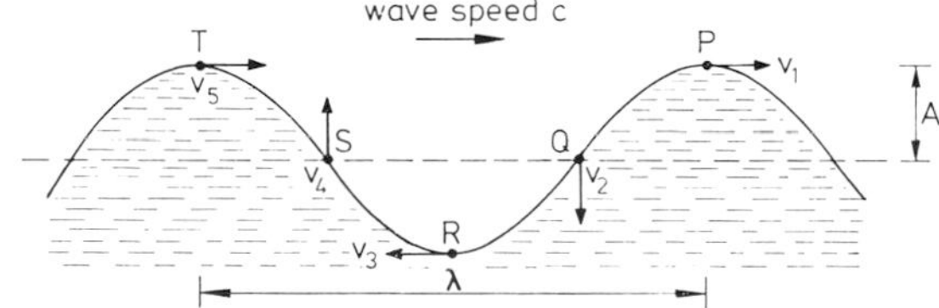

(*a*) (i) How would you find the time it takes for a water particle to go once round in a circle?
(ii) If the water particles are moving in circles, how would you find the magnitudes, and what are the directions, of their accelerations at P and at Q?
(*b*) The speed c of deep water waves is given by $c^2 = \lambda g/2\pi$.

Sketch graphs of (i) speed c, and (ii) frequency, against wavelength λ, for deep water waves in the range $\lambda = 1$ m to $\lambda = 100$ m. Indicate on your sketch the orders of magnitude of speed and frequency for the wavelengths 1 m and 100 m. What would you plot in order to obtain straight line graphs relating wavelengths, λ, to (i) speed, (ii) frequency?
(*c*) (i) Suppose a storm at sea generates waves of wavelengths in the range 1 m to 100 m. What wavelengths of waves from the storm will be felt by a ship 100 km from the storm during the 24 hours following the onset of the storm?
(ii) A small boat at sea has to ride such waves. What will be the speed of the water around circles in waves of wavelengths 100 m and amplitude 10 m? Describe the motion of such a boat (short compared to the wavelength) which rides such waves and also travels forward in the direction of travel of the waves at a mean speed of about 2 m s^{-1}.
(iii) What will be the maximum vertical acceleration of the water in such waves? How will the maximum force on the yacht causing this acceleration compare with the weight of the yacht?
(*d*) If the wavelength λ is larger than the depth d, the water no longer moves in circles and the wave speed for such shallow water waves is given by $c^2 = gd$.

Calculate the wave speed for waves of wavelengths 1 m and 100 m in a sea of depth 100 m.

Combine these results and those from (*b*) above to draw a rough sketch-graph to show how wave speed will vary with wavelength in the range $\lambda = 1$ m to $\lambda = 1000$ m for a sea of depth 100 m so that the waves are 'deep' water waves for short λ and 'shallow' waves for long λ. Label and explain the main features of your sketch. (O & C, Nuffield)

C

Q 2.27
(*a*) The velocities of transverse waves on a string and longitudinal waves in a solid are given

respectively by $v_T=(T/m)^{1/2}$ and $v_L=(E/\rho)^{1/2}$, where T is the tension in the string and E the Young modulus of the solid. Explain physically why these latter two quantities affect the velocities. What are the dimensions of m and ρ?

A 'slinky' is a helical spring with many turns of large radius, is made of thin wire, and may be extended to many times its unstretched length, L_0. Such a spring can support both transverse amd longitudinal waves. If the force constant of the spring is k and its total mass M, find the effective values of T, m, E and ρ when the total length of the spring is L, and hence determine v_T and v_L in terms of k, M, L_0 and L. Assume a constant cross-sectional area A for a turn of the slinky.

(i) A transverse pulse travels the length L of a uniformly extended slinky spring, for which L_0 is negligible, in a time t_0. If the spring is now uniformly extended to length λL, what is the new travel time?

(ii) Under what condition would a small disturbance of a general kind reach the other end of an extended slinky spring undistorted?

(*b*) A slinky spring is freely hung from one of its ends. What length does it assume? Describe qualitatively how the velocity of a transverse pulse would vary as it travels down the suspended slinky. (Cam. Schol.)

Q 2.28

A heavy rope of length l is suspended to hang vertically under gravity. At the same time as a transverse disturbance is propagated from the top end, a mass is released from the same level. At what distance down the rope does the mass catch up with the disturbance? (W)

Q 2.29

The velocity of propagation v of ripples on the surface of a liquid in a shallow tank is given by

$$v^2=\frac{\lambda g}{2\pi}+\frac{2\pi\gamma}{\lambda\rho}$$

where λ is the wavelength, g the acceleration due to gravity, γ the surface tension of the liquid, and ρ its density.

(*a*) In a certain experiment using water the wavelength was 5.0 mm. Assuming the surface tension to be $0.075\ \mathrm{N\,m^{-1}}$, calculate the velocity of the ripples and their frequency.

(Density of water $=1.0\times10^3\ \mathrm{kg\,m^{-3}}$)

(*b*) Describe the experimental arrangement you would set up in order to find a value for the surface tension of water using this equation. Indicate clearly how the necessary measurements are made.

(*c*) In the light of your calculations in (*a*) discuss quantitatively the experimental conditions under which the term $\lambda g/2\pi$ might be neglected.

(*d*) Show that v^2 is a minimum when

$$\lambda^2=\frac{4\pi^2\gamma}{\rho g}$$

(JMB)

Q 2.30

(*a*) Sound waves in air are either *progressive waves* or *stationary waves*. Explain the differences between these waves.

(*b*) Show that the displacement y from their mean positions of particles in air transmitting a sound wave of amplitude a, frequency f at a speed c can be represented by an equation of the form

$$y=a\sin 2\pi f(t-x/c)$$

or by a similar equation.

Hence deduce expressions in terms of a, f, t, x and c for (i) the velocity, (ii) the acceleration of the particles as they oscillate about their mean positions. (O)

Q 2.31

Suppose a (not very knowledgeable) critic of physics were to say:

'A wave is just a lot of water rushing up the beach and running down again. There doesn't seem to be much worth studying in that.'

The above is inadequate as a description of what a physicist means by a wave. Explain why and say what a physicist does mean by a wave, giving examples. Use these examples and others to indicate the wide variety of wave problems which might concern either a physicist or an engineer or both. (You can use examples of standing waves as well as of travelling waves.) (JMB)

Chapter 3

A

Q 3.10

Explain what is meant by Huygens' principle. Use the principle to show that a plane wave incident obliquely on a plane mirror is reflected

(*a*) as a plane wave

(*b*) so that the angle of incidence is equal to the angle of reflection. (JMB)

Q 3.11
Describe Huygens' construction. Use this construction to deduce Snell's law of refraction. (JMB)

Q 3.12
Light is refracted on passing across the boundary between two media in which its velocities arc v_1 and v_2. Show that the observed facts are consistent with the statement that '*all points on a wavefront take the same time to travel to a new position on that wavefront*'. (L)

Q 3.13
An ambulance sounds a warbling signal comprised of alternate notes of frequency 400 Hz and 500 Hz. What will be the apparent frequencies of these notes when heard by:
(i) a stationary observer when the ambulance approaches at $20\ \mathrm{m\ s^{-1}}$;
(ii) a moving observer receding at $20\ \mathrm{m\ s^{-1}}$ from the ambulance when stationary?

[Speed of sound in air = $330\ \mathrm{m\ s^{-1}}$] (O)

Q 3.14
(*a*) A sound of frequency 500 Hz is emitted from an alarm system in a vehicle. The frequency of the noise as heard by a stationary observer changes as the vehicle accelerates away. What will be the value of this frequency when the vehicle reaches a speed of $10\ \mathrm{m\ s^{-1}}$? [Speed of sound in air = $340\ \mathrm{m\ s^{-1}}$.]
(*b*) Assume the vehicle is capable of accelerating up to the speed of sound. Sketch a graph showing how the frequency of the sound heard by the stationary observer changes as the speed of the vehicle increases from 0 to $340\ \mathrm{m\ s^{-1}}$. (L)

Q 3.15
An equation for the Doppler effect for sound is $f' = f(1 \pm v/c)$. Identify the symbols in this equation. A line in the ultraviolet spectrum of a distant nebula is found to occur at a wavelength of 377 nm, instead of at 373 nm when the same spectrum is examined in the laboratory. What can be declared about the motion of the nebula relative to Earth? (C)

B

Q 3.16
(*a*) Explain briefly Huygens' method for constructing wavefronts.

A parallel beam of light is projected onto the surface of a plane mirror at an angle of incidence of about 70°. Draw a diagram showing clearly how Huygens' method can be used to determine the direction of the reflected beam.
(*b*) What is meant by *critical angle*? Under what conditions will a wave be totally reflected on meeting a boundary between two media, both of which will allow passage of the wave?

A beam of light travelling through a transparent medium A is incident on a plane interface into air at an angle of 20°. If the speed of light in the medium is 60% of that in air, calculate the angle of refraction in air.

When the beam is shone through another transparent medium B and the incident angle is again 20°, it is found that the beam is just totally reflected at a plane interface with air. Calculate the speed of light in B as a percentage of the speed of light, *c*, in air. (L)

Q 3.17
(*a*) The diagram illustrates part of an oil exploration test in which a beam of longitudinal (compressional) waves, generated at the Earth's surface, is directed through various strata in the Earth's crust. The frequency of the waves is 75 Hz.

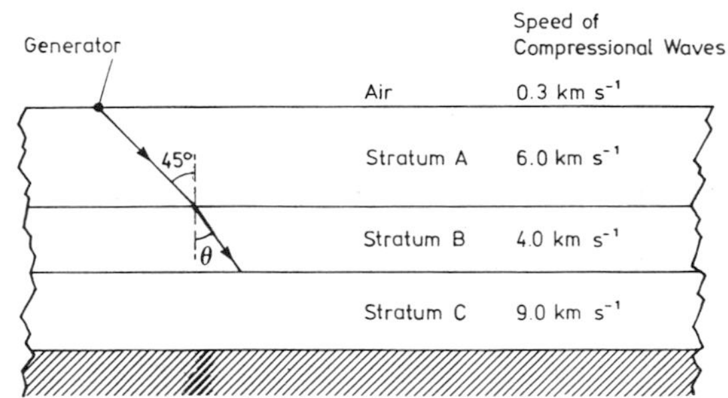

Using the information provided in the figure, calculate:
(i) the wavelength of the waves in stratum A;
(ii) the angle θ to the vertical at which the waves enter stratum B.

Explain, with the aid of a suitable calculation, why the waves do not enter stratum C.

(*b*) (i) Reproduce the diagram to show the subsequent paths of the waves within strata A and B, taking account of both refraction and reflection.
(ii) Discuss whether the waves will eventually emerge from stratum A into the atmosphere. (O)

Q 3.18

(*a*) Describe the Doppler effect and explain how it arises. Derive the formula for the effect in the case of a source of sound waves moving towards a stationary observer, explaining each step in your argument.

Describe an experiment to test this formula.

(*b*) A man stands at the edge of a straight road, whilst an ambulance operating a siren at 1000 Hz speeds past at 30 m s^{-1} in the centre of the road. The speed of sound in air is 330 m s^{-1}. Draw a sketch graph of the frequency of sound received by the man against time, indicating on the graph

(i) the moment at which the ambulance passes,

(ii) the value of the maximum observed frequency (f_1),

(iii) the value of the minimum frequency (f_2).

(*c*) Without detailed calculation, suggest what the observer would hear when passed by a 'superambulance' which travels at a speed (i) just less than, (ii) greater than 330 m s^{-1}.

(O & C)

Q 3.19

A binary star system consists of two components of equal mass rotating about their centre of mass in a plane which includes a terrestrial observer. The period of rotation is 3.0×10^6 s. It is observed that the wavelength of a spectral line from one of the components varies by ± 0.05 nm about its mean value of 620 nm during one revolution.

(i) Explain this observation.

(ii) Calculate the speed of the stars.

(iii) What is the separation of the stars?

(iv) Find the distance of the system from the observer on Earth, if the maximum angular separation of the components is measured to be 1.4×10^{-7} radian.

(v) Draw a graph to show how the wavelength of the observed spectral line varies with time, showing clearly the scales of the axes. (O)

C

Q 3.20

A and B are two points on the surface of the earth which may be regarded as flat and parallel to the rock layer below. An explosive charge is fired at A and the sound reaches B either directly via AB through the soil or indirectly via ACDB. In the latter case the sound suffers refractions at C and D and travels the distance CD in the rock. If the speed of sound in the soil layer is v_1 and that in the rock v_2, show that when the sound reaches B

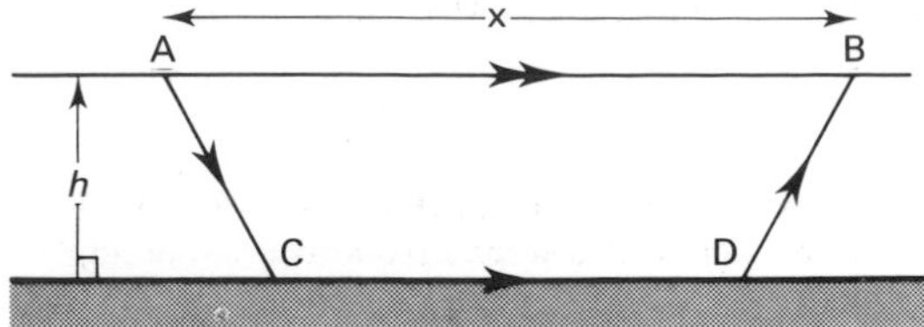

by the two paths simultaneously

$$\frac{h}{x} = \frac{(1 - \sin\theta)}{2\cos\theta}$$

where $\sin\theta = v_1/v_2$.

Sketch a graph showing, as ordinates, the time between the explosion at A and the arrival of the initial disturbance at B and, as abscissae, varying values of the distance AB. Ignore direct reflections from the surface of the rock layer. (JMB)

Q 3.21

(i) State the *principle of superposition* as it applies to wave motions. Use the principle to explain what is heard when two sound waves of equal amplitude but slightly different frequencies interfere.

(ii) One line in a textbook, under the section on the Doppler effect in sound, reads as follows:

$$f' = \frac{f_0}{1 \pm v/c} \quad \text{(moving source, stationary observer)}$$

Identify the symbols in the equation and explain the significance of the $\pm$ sign.

(iii) A twin-engined light aircraft flies directly over a stationary observer on a still day. The engines are running at constant but slightly different speeds, so that the observer detects a low-frequency fluctuation in sound intensity. When the aircraft is approaching but is still very distant, the frequency of the fluctuation is 5.0 Hz. When it has receded to a great distance, the frequency is 3.0 Hz. Calculate the speed of the aircraft. What is the difference between the angular speeds of the engines, as measured by the pilot of the aircraft?

[Take the speed of sound in air to be 340 m s^{-1}.] (Cam. Step)

Q 3.22

A body of mass M moves with a speed v (which is very much less than the speed of light c) and emits a photon of frequency f. By writing down the equations of conservation of energy and momentum, show that, if M is sufficiently large,

$$f = \frac{f_0}{1-(v/c)\cos\theta}$$

where f_0 is the frequency of the photon which would be emitted if v were zero and θ is the angle between the direction of v and the direction of the emitted photon.

(The momentum of a photon of frequency f is hf/c. The body will change speed and direction due to recoil.)

Using this formula, and the laws of mechanics, show that it is possible to decide whether the rings of Saturn are solid or accumulations of a large number of small particles.

One of the lines in the hydrogen spectrum (from a stationary lamp) occurs at a wavelength of 434 nm. This line is observed in the spectrum of a star, but shifted 2 nm towards the red end of the spectrum. Assuming the star is moving radially relative to the earth, calculate its velocity. Explain whether it is moving away from, or towards the earth. (W)

Q 3.23

Show that in acoustics the simple Doppler treatment for cases (a) of the source moving towards a stationary observer, and (b) of the observer moving towards a stationary source, lead to different expressions for the apparent change in frequency.

A car carrying a siren on its roof emitting a note of frequency 420 Hz is driven towards a wall at 15 m s^{-1} so that an echo is produced. If the speed of sound in air is 330 m s^{-1} what is heard by the car driver?

The special theory of relativity states that the speed of electromagnetic radiation is always equal to a constant, c, regardless of any relative motion between the source and the observer. It also states that if the relative speed between the source and the observer is u then a time interval Δt in the source is observed as an interval $\Delta t'$ by the observer where

$$\Delta t' = \frac{\Delta t}{\sqrt{(1-u^2/c^2)}}$$

By considering Δt as the interval between two 'peaks' in a light signal emitted by a star receding from the earth at a speed u and allowing for the extra distance travelled by the second 'peak' show that the frequency f' observed on the earth is given by

$$f' = f\sqrt{\frac{(1-u/c)}{(1+u/c)}}$$

where f is the frequency emitted by the source.

By considering a star receding at a speed $c/2$ illustrate the difference obtained by using the simple Doppler treatment to that obtained by using the above result for a receding source. (L)

Q 3.24

The Doppler broadening of a spectral line increases with the rms speed of the atoms in the source of light. Deduce which lamp gives the narrower spectral line: a 198-mercury lamp at a temperature of 300 K or an 86-krypton lamp at a temperature of 77 K. (W)

Chapter 4

A

Q 4.9

Explain what is meant by the 'principle of superposition' in wave motion. Illustrate your answer by drawing a sketch on graph paper showing the superposition of two waves of equal amplitudes which have wavelengths in the ratio 2:1. (W)

Q 4.10

The diagram shows two point sources of sound, S_1 and S_2 which are close together. They emit sound waves which are in phase, and of the same

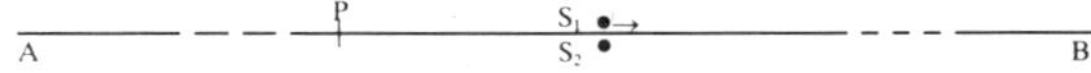

wavelength and intensity. When S_1 is slowly moved parallel to AB in the direction of B, the sound intensity, I, detected at P shows a series of maxima and minima. As the displacement, x, of S_1 from its initial position increases, the intensities at the maxima decrease and the intensities at the minima increase.

(i) Explain this variation in I.
(ii) Explain why the greatest observed value of I occurs when x is zero, and why I becomes constant when x is very large.
(iii) *Indicate* how the variation in I with x could be used to determine the wavelength of the sound waves.
(iv) Describe and explain how I would have varied with x if the sound waves emitted by S_1 and S_2 had differed in phase by π (180°). (JMB)

Q 4.11

A point source A emits spherical sound waves. State how the intensity of sound varies with position around the source, assuming that there is no absorption in the medium.

A second identical source B is placed near A, the two sources being in phase. Explain why there will be positions of maximum and minimum intensity near the sources.

If the wavelength for each source is 0.40 m, and the rate at which sound energy is emitted by A is 1.44 times the rate of that emitted by B, explain why the sound intensity is found to be zero at a point 13.2 m from A and 11.0 m from B. (JMB)

Q 4.12

(*a*) State two differences between *stationary* and *progressive* waves, referring specifically to the terms *phase* and *amplitude.*

(*b*) A wire 1.0 m in length has a fundamental frequency of 50 Hz when the tension in the wire is 5.0 N.

(i) Calculate the velocity of transverse waves in the wire.

(ii) Determine the fundamental frequency if the tension is doubled and the length halved. (S)

Q 4.13

A stretched wire is fixed at both ends and plucked at its centre. Draw a diagram to represent the mode of vibration which gives rise to the lowest possible frequency. Draw a second diagram to represent another mode of vibration of the wire. Write down the relationship between the frequencies of the two modes. (C)

Q 4.14

Figure (*a*) shows a loudspeaker L connected to an oscillator and placed a few metres from an extensive reflecting surface R. The oscillator output is fixed at 500 Hz and the speed of sound at the time is 320 m s^{-1}.

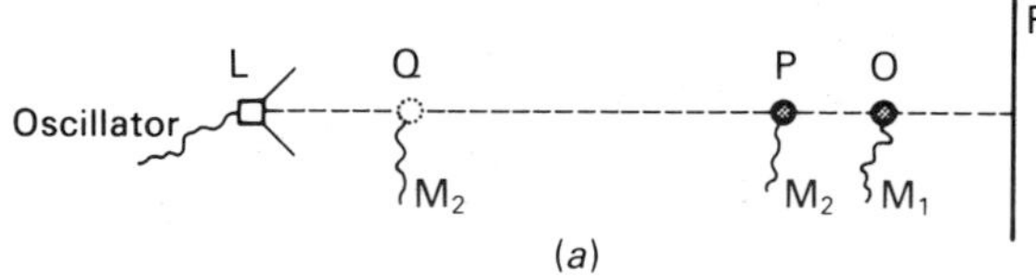

(*a*)

A small microphone M_1 is moved away from the wall along the normal towards L and fixed at a point O where the amplitude of the detected signal is a maximum. A second small microphone M_2 is moved from a point close to O towards L. It reaches a *first* maximum at P and a *fourth* maximum at Q. Calculate

(*a*) the wavelength of sound from the loudspeaker, and

(*b*) the distance PQ.

The microphones M_1 and M_2 are connected to the *X*-plates and *Y*-plates respectively of an oscilloscope whose time base is switched off. Figures (*b*) and (*c*) show the traces on the oscilloscope screen with M_2 at P and Q

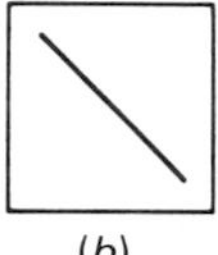
(*b*)

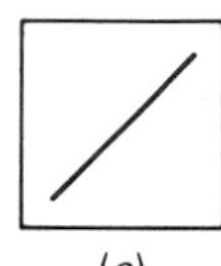
(*c*)

respectively. Explain the change in direction of the oscilloscope trace. (L)

Q 4.15

An organ pipe is approximately 1 m long, and is closed at one end. One of the harmonics in the sound it produces has frequency 425 Hz. Show on a diagram the positions in the pipe where the vibration of the air in this harmonic is a maximum. On your diagram, mark by arrows the direction of movement of the air particles at these positions at an instant when the particle speed is a maximum. (Speed of sound in air = 340 m s^{-1}.) (S)

Q 4.16

Two pure notes at frequencies f_1 and f_2 are sounded together. In what circumstances will beats be audible?

If $f_1 = 300$ Hz and the beat frequency is 3 Hz, what are the possible frequencies of f_2? Draw a sketch showing how the amplitude of the resultant sound varies with time and mark a suitable scale on the time axis. (L)

Q 4.17

(i) Explain the formation of *beats.*

(ii) A microphone connected to a cathode-ray oscilloscope receives simultaneously two sound waves of frequencies 1000 Hz and 1100 Hz and equal amplitudes. Sketch and explain what is seen on the screen of the oscilloscope if the time-base repetition frequency is set at 50 Hz. (O)

B

Q 4.18
What is meant by the *period* T and the *wavelength* λ of a wave?

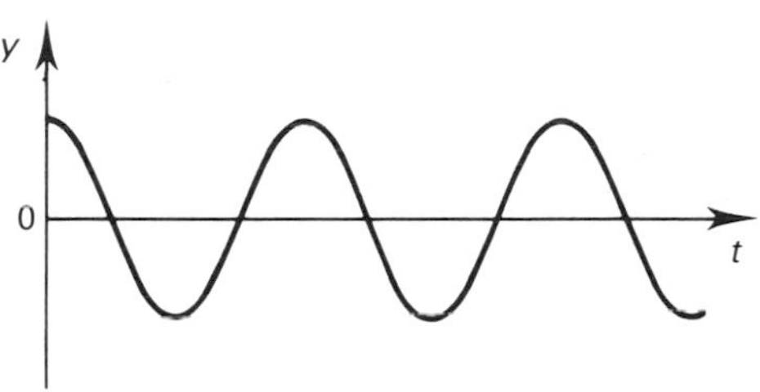

The graph shows the displacement y of a particle in a sinusoidal wave as a function of time t. Write an equation that represents the displacement of the particle in terms of t and T, explaining any other symbols used.

A second particle is situated a distance $\lambda/4$ from the first, measured in the direction in which the wave is travelling. What is the phase angle between the vibrations of the two particles? Draw a sketch graph to illustrate the variation with time of the displacement of the second particle. Give an equation to represent this displacement.

Two continuous sound waves, each of amplitude A and wavelength λ, meet at a point such that their phase difference is $\pi/3$. Show, by means of a phasor diagram or otherwise, that the amplitude of the resultant wave is approximately $1.7A$. Hence find the ratio of the intensity of the resultant wave to the sum of the intensities of the component waves.

Discuss your result with reference to the principle of conservation of energy. (C)

Q 4.19
(*a*) (i) How are standing (stationary) waves formed?
(ii) State *four* features that distinguish a travelling (progressive) wave from a standing (stationary) wave.
(*b*) (i) With the aid of a diagram, describe a laboratory method for producing standing waves in air and for measuring their wavelength. You may assume that a source of sound of known frequency is available.
(ii) How could your results be used to determine the speed of sound in air?
(*c*) The diagram shows a sinusoidal standing wave produced on a stretched cord by a vibrator

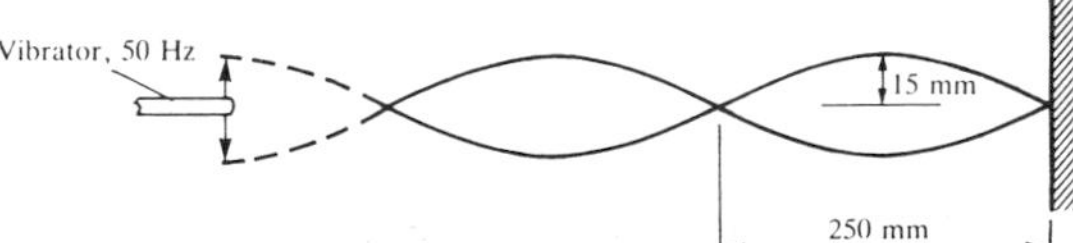

oscillating at a frequency of 50 Hz. The mass per unit length of the cord is 0.080 kg m^{-1}. Using the data from the diagram, calculate:
(i) the tension in the cord,
(ii) the horizontal separation of a node and the nearest particle of the cord whose amplitude of oscillation is 7.5 mm.
(*d*) Describe and explain what will be seen if the cord is viewed in light from a stroboscope set at a frequency of:
(i) 50 Hz;
(ii) 101 Hz. (O)

Q 4.20
What adjectives can be used to describe the sound waves in an organ pipe sounding its fundamental note? A closed pipe of variable length has 18.0 cm and 56.0 cm as the two shortest lengths for which it will resonate to a tuning fork of frequency 440 Hz. Sketch a graph showing how the amplitude A of displacement of air molecules in the pipe varies with distance x from the closed end when it is resonating to the fork and its length is 56.0 cm, labelling significant points and features on the graph.

Use the data to calculate the speed of sound in air. (S)

Q 4.21
(*a*) (i) Explain how stationary transverse waves form on a stretched string when it is plucked.
(ii) State the factors that determine the frequency of the fundamental vibration of such a string and give the formula for the frequency in terms of these factors.
(*b*) When two notes of equal amplitude but with slightly different frequencies f_1 and f_2 are sounded together, the combined sound rises and falls regularly.
(i) Explain this, and draw a diagram of the resulting wave form.
(ii) Show that the frequency of these variations of the combined sound is $|f_1 - f_2|$.
(*c*) A car engine has four cylinders, each producing one firing stroke in two revolutions of the engine. The exhaust gases are led to the atmosphere by a pipe of length 3.0 m.
(i) Assuming that vibration anti-nodes occur near each end of the pipe, calculate the

lowest engine speed (in revolutions per minute) at which resonance of the gas column will occur.

(ii) What may happen at higher speeds?

(iii) Where, in the gases in the pipe, will the greatest fluctuations of pressure take place at resonance?

[Take the speed of sound in the hot exhaust gases to be 400 m s^{-1}.] (O)

Q 4.22

A copper wire is subjected to increasing tension until it breaks. Sketch the graph of tension against extension for the wire, and identify any important features. Give a brief explanation of these features in terms of the microscopic structure of the metal. Show how the Young modulus may be deduced from the graph.

An equation for the speed v of longitudinal waves in a thin rod of an elastic solid of Young modulus E and density ρ is $v=(E/\rho)^{1/2}$. Using the base units or dimensions of the quantities involved, show that this equation is dimensionally homogeneous.

A thin copper rod, 800 mm long, is clamped at one end. It is made to vibrate by an oscillator of variable frequency. This produces *longitudinal* waves in the rod. As the frequency is varied it is found that the rod resonates: the two lowest resonant frequencies are 1190 Hz and 3570 Hz. In all resonant modes the clamped end is a displacement node and the free end is an antinode. The diagram below illustrates the position of the node (N) and the antinode (A) for the resonant frequency of 1190 Hz.

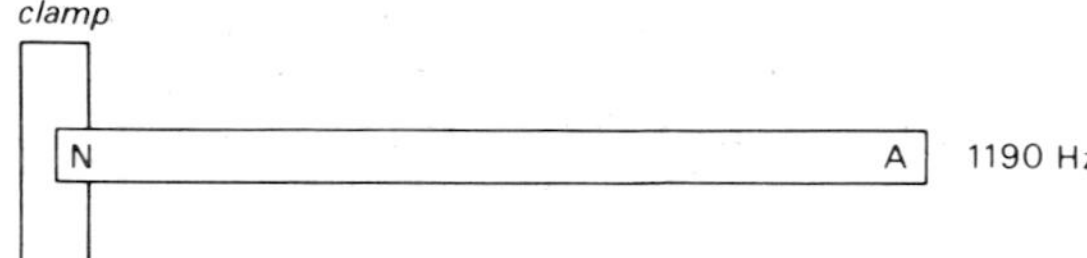

(*a*) Draw a labelled sketch showing the positions of the displacement nodes and antinodes for the resonant frequency 3570 Hz. (Indicate clearly which end of the rod is clamped.)

(*b*) Find two frequencies *higher* than 3570 Hz at which the rod might also be expected to resonate.

(*c*) The density of copper is 8.94×10^3 kg m^{-3}. Calculate the Young modulus of copper. (C)

Q 4.23

A car travels at a speed of 25 m s^{-1} along a straight road towards a distant wall. Mounted on the roof of the car, a loudspeaker emits a steady sound in the forward direction at a frequency of 1000 Hz. The speed of sound is 340 m s^{-1}.

(*a*) How far does a sound wave travel in 1 ms?

(*b*) Calculate the wavelength of the sound arriving at the wall.

(*c*) How many waves reach the wall in 1 s? Show the steps in your calculation.

(*d*) The sound reflected from the wall is received at the car and mixed with the source frequency. Showing the steps in your calculation, find the beat frequency. (O & C)

Q 4.24

(*a*) Give *two* examples of observations that utilize the Doppler effect of light waves. In each case state clearly the information that is derived from the observation.

(*b*) In a police speed-check, radar waves of frequency 1.0 GHz (10^9 Hz) are directed towards oncoming traffic. The reflected waves return to the transmitter where they combine with a signal of the original frequency to produce a beat-note in the audio range. The frequency of the beat-note is recorded with a frequency meter.
[Take the speed of radar waves in air to be 3.0×10^8 m s^{-1}.]

(i) Calculate the frequency of the beat-note obtained when checking the speed of a vehicle approaching at a speed of 16 m s^{-1}.

(ii) Calculate the percentage error in a speed measurement where the police observations were taken from an overhead bridge at a height of 10 m above the oncoming vehicle when it was at a distance of 50 m from the transmitter. Would the recorded speed be higher or lower than the true speed? (O)

C

Q 4.25

(*a*) What conditions must be fulfilled before interference effects are observed between the radiations from two sources producing waves of equal frequency?

(*b*) (i) Describe how you would explore in the laboratory the spatial distribution of interference effects produced by two small loudspeakers energized by the same signal generator.

(ii) Draw a diagram showing quantitative details of the interference pattern observed when using sound of frequency 9.0 kHz.

[Take the speed of sound in air to be 330 m s^{-1}.] (O)

Q 4.26

(*a*) Explain what is meant by the *principle of superposition.*

(*b*) Show how a progressive wave travelling in the x-direction is represented by an equation of the form

$$y = a \sin 2\pi\left(ft - \frac{x}{\lambda}\right)$$

(*c*) Describe the reflection of such a wave incident normally on a boundary:

(i) with no change of phase;

(ii) with a change of phase π.

Give expressions for the displacements resulting from the superposition of the incident and reflected waves and illustrate your answers with wave diagrams. State an example of a physical situation in which each type of reflection occurs. (O)

Q 4.27

Explain the difference between *stationary* waves and *progressive* (travelling) waves. An organ pipe of length L can support a stationary wave with a node at its closed end and an antinode at its open end. Show that its possible frequencies of oscillation are given by

$$f = (c/2L)(n - \tfrac{1}{2}),$$

where c is the velocity of sound and n is a positive integer.

Two sinusoidal progressive waves, of the same frequency and of equal amplitudes, but moving in opposite directions, coincide at time $t = 0$. Make separate sketches of these waves at times $t = 0$, $T/6$, $T/4$ and $T/3$, where T is the wave period, and hence show that, when superposed, they constitute a stationary wave. How are its amplitude, frequency, and the positions of its nodes and antinodes, related to properties of the two original progressive waves? How are the phases of the two progressive waves related (i) at a stationary-wave antinode, and (ii) at a stationary-wave node?

Any stationary wave in the organ pipe described above may be treated as resulting from a progressive wave which travels continuously forwards and backwards along the pipe and suffers reflection at the ends. What must be the phase change on reflection at (i) the open end and (ii) the closed end? Taking these phase changes into account, find the total phase change of the progressive wave in one complete transit, forwards and backwards, as a function of n. Comment on your answer. (Cam. Schol.)

Q 4.28

State under what conditions you would be able to observe interference effects with wave motions and explain why the conditions are necessary.

A small sound transmitter T radiates uniformly in all directions and at four times the power of two other similar small transmitters placed 0.25 m on either side of T along a north-south line through T. The central transmitter is wired to be out of phase with the other two, and all three emit a 200 kHz signal. A small receiver is placed 10 m due east of T and slowly moved eastward. Where will the maximum and minimum signal responses occur? At the position of a maximum the two outer transmitters are switched off for a while; by what factor does the power received fall?

What can you say qualitatively about the response if the receiver were placed 10 m north of $T°$ and then moved slowly northwards?

[Velocity of sound in air = 330 m s^{-1}.] (C)

Q 4.29

What do you understand by the *period* and *frequency* of a wave motion? State the relationship between them.

What are the conditions necessary for audible beats to be formed from two separate sound sources?

Deduce the expression for the frequency of such beats in terms of the frequencies of the two sources. (Either a graphical or an algebraic argument is acceptable.)

A microphone and a cathode-ray oscilloscope with a calibrated timebase are used to display the wave-form produced by two audiofrequency signal generators. With the timebase set so that one complete sweep takes 50 ms, a steady trace is obtained, as shown in the graph below.

Estimate

(*a*) the period of the wave motion,

(*b*) the beat frequency,

(*c*) the frequencies of the two signal generators.

Hence discuss why it is inadvisable for two nearby radio transmitters to employ frequencies separated by less than about 15 kHz, a typical upper frequency limit for audibility. (C)

Q 4.30
Show that as a result of the Doppler effect the frequency f heard by an observer moving with speed u towards a source of sound of frequency f_0 is given by $f = f_0(c+u)/c$, where c is the speed of sound in air.

A hooter emitting sound at 500 Hz is situated 30 m from the foot of a sheer vertical cliff face which can be regarded as a perfect reflector of sound. A cyclist, about 100 m from the foot of the cliff, is approaching the cliff face normally towards a point 1.0 km from the hooter at a speed of 6.0 m s. Describe what he hears, and show that the phenomenon may be regarded either as an interference pattern through which he is moving, or as a beat between two sources with slightly different Doppler effects.

(You may assume without proof that you can use the formula deduced in the first part of the question, where u is the component of the cyclist's velocity towards the source of sound.)

[The speed of sound in air is 343 m s^{-1}.]

(Ox. Schol)

Chapter 5

A

Q 5.13
(*a*) What facts concerning the nature of sound may be deduced from the fact that an electric bell inside an evacuated bell jar is inaudible?
(*b*) Give one piece of experimental evidence that sound of all wavelengths travels at the same speed in air under constant conditions. Compare this behaviour of sound with that of light of different wavelengths travelling through glass. (JMB)

Q 5.14
When a violin string is plucked a musical note is produced. Explain how (*a*) the note is produced by the violin and (*b*) the sound reaches the ear.

(JMB)

Q 5.15
The lowest resonant frequency of a guitar string of length 0.75 m is 400 Hz. Calculate the speed of transverse waves on the string. (C)

Q 5.16
A resonance tube containing air at 16°C was closed at one end. It resonated to a tuning fork of frequency 512 Hz when its length was 15.9 cm and again when its length was increased to 49.7 cm.

(*a*) What is the velocity of sound (i) at 16°C, (ii) at 0°C?
(*b*) Estimate the diameter of the tube. (W)

B

Q 5.17
Explain what is meant by the statement that '*sound is propagated in air as longitudinal progressive waves*,' and outline the experimental evidence in favour of this statement. Compare the mode of propagation of sound in air with that of (*a*) waves travelling along a long metal rod, produced by tapping one end, (*b*) water waves.

Two loudspeakers face each other at a separation of about 100 m and are connected to the same oscillator, which gives a signal of frequency 110 Hz. Describe and explain the variation of sound intensity along the line joining the speakers. A man walks along the line with a uniform speed of 2.0 m s^{-1}. What does he hear?

[Speed of sound = 330 m s^{-1}.] (C)

Q 5.18
In this question you are required to estimate the tension in a violin string which vibrates at a natural frequency of 650 Hz. The string is made of steel.

(*a*) Starting from the formulae $c = f\lambda$ and $c = \sqrt{(F/\mu)}$ as given on your formula sheet, obtain an expression for the frequency (f) in terms of the tension (F), the length (L), the cross-sectional area (A) and the density (ρ) of the string.
(*b*) Estimate values for the quantities you will need to know in order to calculate the tension using the above expression.
(*c*) Combine your estimates in order to obtain a value for the tension in the string.
(*d*) Say in a few words how you decide on the appropriate number of significant figures to give in the answer. (O & C, Nuffield)

Q 5.19
The diagram shows a standing wave pattern of a steel guitar string stretched between two supports, called the nut and the bridge, on a guitar.

The fundamental standing wave pattern shown produces a note of frequency 280 Hz.

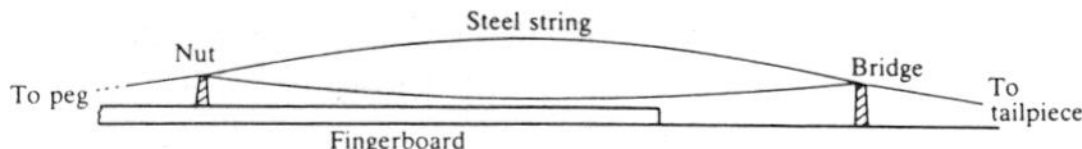

(*a*) By placing a finger lightly at certain places on the string it is possible to produce further standing wave patterns with other specific frequencies.

(i) Sketch on a copy of the diagram one of these standing wave patterns, and
(ii) state its frequency.

(*b*) The speed, c, of a transverse wave along a stretched string is given by $c=\sqrt{(T/\mu)}$, where T is the tension and μ the mass per unit length of the string.

Show that the fundamental frequency f is given by $f=1/2l\sqrt{(T/\mu)}$, where l is the vibrating length of the string between nut and bridge.

(*c*) Assuming that both l and μ remained constant, calculate the frequency of the new fundamental mode of vibration if the tension were halved.

(*d*) In practice μ, the mass per unit length, changes because the string contracts when the tension is reduced.

Consider a situation in which the tension is halved.

(i) If the strain reduction produced were 0.4% what would be the percentage change in μ? State both the size and sign of the change.
(ii) Write down the percentage error this would cause in your answer to (*c*). State, giving your reasoning, whether the actual frequency would be higher or lower than that you calculated. (O & C, Nuffield)

Q 5.20

(*a*) Blowing across the top of a test tube can produce a note. Explain why this is so. State how the frequency of the note would be affected by (i) using a longer test tube, (ii) raising the temperature of the air in the tube.

How would the sound produced be affected by blowing harder?

(*b*) A small loudspeaker is placed opposite and close to the open end of a tube 0.20 m long, closed at the other end. The loudspeaker is connected to an audio-frequency signal generator. When the frequency of the generator is varied, the air in the tube is found to resonate at several different frequencies. Assuming the speed of sound in air is 330 m s^{-1}, calculate approximate values of the two lowest of these frequencies. Show clearly how these values were obtained.

(*c*) Assuming that the signal generator is accurately calibrated, state what main causes of error would exist if the experiment were used to determine the speed of sound in air.

Describe what you would expect to observe if the frequency were varied with the loudspeaker close to
(i) the same tube with both ends open,
(ii) a small hole in a tube of the same length which is bent into a circle to form a continuous tube.

Calculate the value of the fundamental resonant frequency in (i). (L)

Q 5.21

Summarize the stages involved in the complete process of recording sound to its eventual reproduction.

Write an account of the various stages, including two methods of storing the information. Emphasis must be placed on the physical principles involved, rather than on technical details of construction or electronic circuitry. (W)

Q 5.22

(*a*) Describe briefly the factors which control the frequency of the note emitted by each of the following instruments: violin, drum, triangle, trumpet.

If the same basic note were played in turn by each of these instruments, explain why an observer would have no difficulty in identifying the different instruments by ear alone. How would you demonstrate the validity of your explanation?

(*b*) Explain what is meant by ultra-sonic waves and describe their production by a method based on either magnetostriction or piezo-electricity.

State one application of ultra-sonic waves. (W)

Q 5.23

In testing railway lines for faults an ultrasonic probe is placed on a rail, as shown in the diagram.

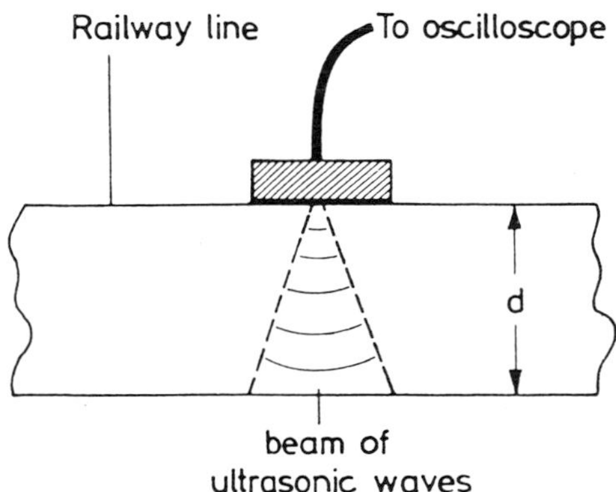

The probe, which acts as both emitter and receiver, is connected to an oscilloscope. The probe emits short bursts of ultrasonic radiation of frequency 3.0 MHz each millisecond. The speed of this radiation in steel is 6000 m s^{-1}. The trace shown below is obtained, with the emitted pulse appearing on the left of the screen.

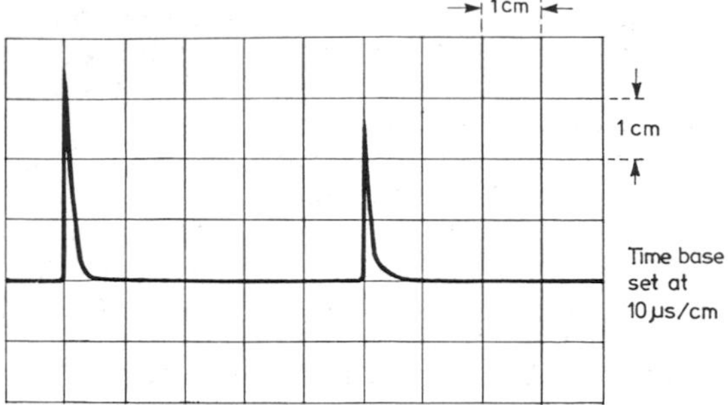

(*a*) Making a measurement from the diagram of the screen, calculate the depth, *d*, of the rail.
(*b*) When the probe is placed at another point on the rail the trace shown below is obtained.

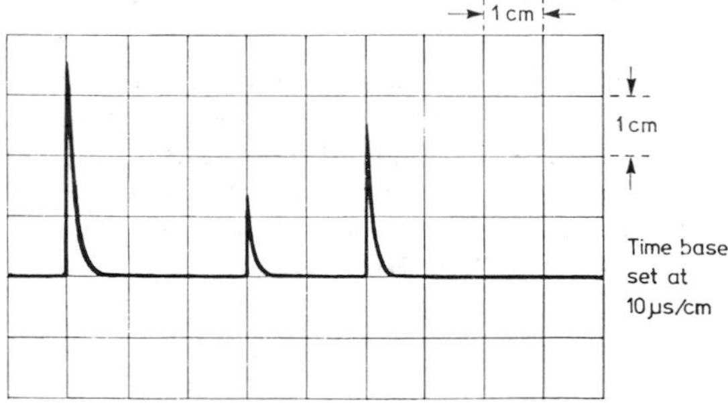

Suggest a possible cause of the extra pulse now seen. Justify your answer quantitatively.

(*c*) Calculate the wavelength of the waves in steel.
(*d*) Give a reason why such a high frequency is used.

(O & C, Nuffield)

Q 5.24
Explain what is meant by the *pitch*, *quality* and *loudness* of a musical note.

Write down an expression, in terms of pressure and density, for the speed of sound in a gas. Define any other quantity in this expression and explain its significance. Consider how the speed depends (i) on temperature and (ii) on pressure. Estimate the change in room temperature which causes the speed of sound to increase by 1%.

(*a*) Describe a non-laboratory piece of evidence which shows that the speed of sound cannot vary greatly with frequency.
(*b*) A point source of sound, in mid-air, radiates uniformly in all directions. Sketch, and explain, diagrams showing typical surfaces of equal loudness when the surrounding air is (i) stationary and (ii) moving with uniform velocity. It is a common observation that sound waves are more easily heard downwind than upwind. Show, by taking a typical wind speed, that the considerations of (ii) will not explain this.

[The speed of sound in air is approximately 330 m s^{-1}.]

(W)

Q 5.25
Describe, with the aid of diagrams where appropriate, what is heard when

(*a*) a vibrating tuning fork is placed above a long vertical tube containing water which slowly runs out at the lower end,
(*b*) two sound waves of slightly different frequencies reach the ear simultaneously,
(*c*) a car, sounding its horn continuously, passes close by a stationary observer. (Include a physical explanation of this phenomenon.)

The speed of sound v in an ideal gas is given by the expression

$$v = k\sqrt{T}$$

where T is the thermodynamic temperature and k is a constant. A parallel beam of sound passing through an ideal gas at 17°C makes an angle of incidence of 50° on a plane thin membrane separating the gas from another sample of the same ideal gas at 100°C. What will be the angle of refraction of the beam? (You may assume that the membrane itself produces no deviation.)

If the incident beam is to be totally internally reflected at the boundary, calculate the minimum temperature to which the gas on the other side of the membrane must be raised.

(C)

Q 5.26
(*a*) Sound produced by a loudspeaker diaphragm vibrating at a frequency of 2 kHz is propagated in the air as a longitudinal progressive wave.
  (i) Explain what is meant by '*a frequency of 2 kHz*'.
  (ii) Describe carefully how the longitudinal progressive waves are propagated through the air.
  (iii) State the effect, if any, on the speed of the sound of first increasing the pressure of the air and then increasing its temperature.

(*b*) The diagram shows an experimental arrangement which can be used for the determination of the velocity of sound in air. The signal generator produces a vibration of known frequency in the loudspeaker diaphragm, and the sound produced is directed towards the metal plate. A small microphone connected to the Y-plates of an oscilloscope is moved from the metal plate towards the loudspeaker. The time-base of the oscilloscope is switched off.

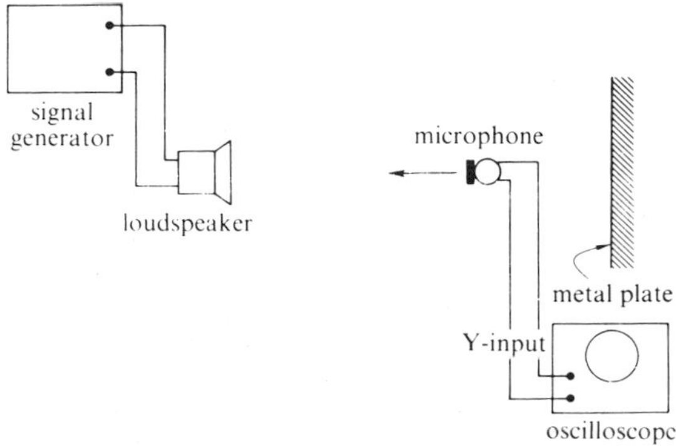

(i) Describe and explain the oscilloscope responses you would expect as the microphone is moved.
(ii) Describe how the experimental observations could be used to determine the speed of sound in air. State the observations which would be made, and show how the final value would be calculated.

(*c*) The diagram below shows a similar experimental arrangement to that above except that the metal plate has been removed

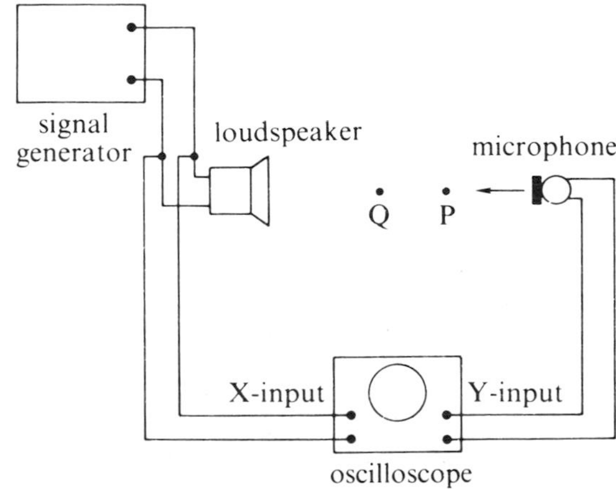

and the output from the signal generator has been connected to the X-plates of the oscilloscope as well as to the loudspeaker.

As the microphone is moved towards the loudspeaker there are points such as P where the trace on the oscilloscope is a sloping straight line with a positive gradient, indicating that the signal to the X-plates is in phase with the signal to the Y-plates. As the microphone is moved to point Q, the trace changes to a sloping line with a negative gradient.
(i) Explain these observations.
(ii) When the frequency of the signal generator is 2.0 kHz, the distance PQ is 82 mm. Calculate a value for the speed of sound in air.
(iii) Describe and explain the trace you would expect when the microphone is positioned about half-way between P and Q.

(AEB, 1986)

Q 5.27

Define *wavelength* and *frequency*. Deduce a relation between these quantities and the speed of propagation of a wave.

Two microphones, M_1 and M_2, are positioned at a distance d apart in still air. A source S of sound of fixed frequency is placed on the line M_1M_2 but beyond M_2 as shown in the diagram. The outputs of the two microphones are monitored using a cathode ray oscilloscope.

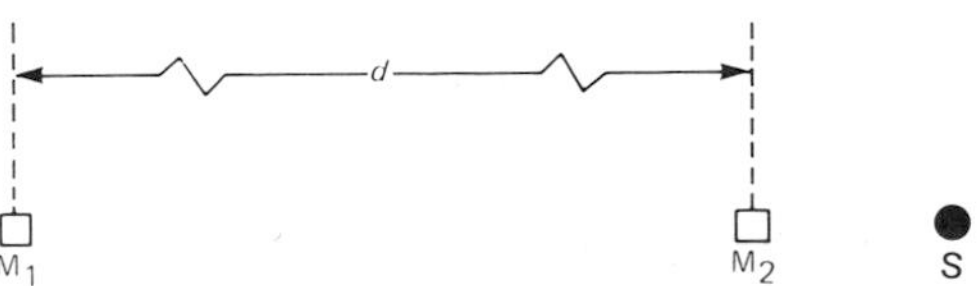

(*a*) What connections would you make and how would you adjust the cathode ray oscilloscope in order to measure the phase difference between the signals from the microphones?
(*b*) Draw diagrams to illustrate the traces observed when the phase angle between the two signals is
(i) 0,
(ii) π rad.
(*c*) Describe how the apparatus could be used to measure the speed of sound in air by moving M_1 along the line M_1M_2.

The speed v_s of sound waves in air varies with thermodynamic temperature T according to the relation

$$v_s = bT^{1/2},$$

where b is a constant. If the temperature of the air changes from 18°C to 20°C, what is the resulting fractional change in v_s? Discuss whether this change could be measured by this method if the wavelength at 18°C were 500 mm. (C)

Q 5.28

(*a*) A note of frequency 600 Hz is sounded continuously over the open upper end of a vertical tube filled with water. As the water is slowly run out of the bottom the air in the tube resonates, first when the water level is 130 mm below the top of the tube and next when the level is 413 mm below the top of the tube. Calculate
  (i) the speed of sound in the air in the tube,
  (ii) the position of the water level when the third resonance occurs.

(*b*) Describe the motion of the air particles at various points along the axis of the tube when the air first resonates. (JMB)

Q 5.29

(*a*) (i) What is meant by *resonance* in vibrating systems?
  (ii) Describe a laboratory demonstration of resonance. State the apparatus used, the procedure followed, and describe the results obtained.

(*b*) What is meant by the *quality* of a musical note?

(*c*) An oscilloscope and a microphone are used to display the waveform of a pure (sinusoidal) note of frequency 200 Hz and trace amplitude 30 mm.
  (i) On a sheet of squared paper provided, draw accurate scale diagrams showing the waveforms of:
    this pure note;
    the note modified by the addition of a pure harmonic of frequency 600 Hz and amplitude 10 mm, initially in phase with the original note.
  (ii) Determine the form and length of an air column suitable for producing the two notes of frequencies 200 Hz and 600 Hz as its lowest two resonant frequencies.
  [Take the speed of sound in air to be 320 m s^{-1}.] (O)

Q 5.30

(*a*) Explain with the aid of suitable diagrams how you would produce experimentally in the laboratory
  (i) a longitudinal pulse in a long horizontal spring in tension and fixed at one end,
  (ii) a transverse stationary wave in a stretched rubber cord fixed at one end.

(*b*) (i) Describe with the aid of suitable diagrams how sound waves are propagated in air. Describe how you would measure the speed of sound in free air.
  (ii) A small loudspeaker emits sound energy uniformly into a hemispherical region in front of itself. If the total power of the sound emitted is 80 mW, what is the sound intensity (energy flux) at a distance of 3 m in front of the loudspeaker? What would be the distance from the loudspeaker at which the sound intensity was half of this value? You may assume that the loudspeaker behaves as a point source of sound energy. (L)

Q 5.31

Give a concise account, with particular reference to the way in which energy is transformed at each stage, of the method of *recording* and subsequent *reproduction* of sound by *either* (*a*) a gramophone disc *or* (*b*) a tape recorder. Illustrate your answer with suitable diagrams.

Define the *decibel.* Two sounds of the same frequency are respectively 45 and 35 decibels above the intensity level for minimum audible loudness. Find the ratio of the power outputs required to produce the sounds. (L)

C

Q 5.32

The key on a piano corresponding to the note of frequency 880 Hz is depressed very gently. The hammer does not strike the string but the damper is lifted and the string is free to vibrate. When the key corresponding to the note of frequency 440 Hz is struck firmly, it is found that the 880 Hz string also vibrates. Give a brief explanation of this phenomenon.

Suppose the 220 Hz string had also been free to vibrate. What frequencies, if any, would it have emitted when the 440 Hz key was struck? (Cam. Step)

Q 5.33

(*a*) (i) In what ways do the motions of air molecules propagating a 1 kHz travelling sound wave through air differ from those of air molecules sustaining a 1 kHz stationary sound wave in a closed air column?
  (ii) Describe an accurate experiment to determine the wavelength in air of a constant frequency note produced by a small loudspeaker.

(*b*) (i) Two sound waves of equal amplitude but slightly different frequencies arrive at a point simultaneously. Give a mathematical calculation illustrating what a listener at this point will hear, identifying all frequencies heard.

(ii) Carefully sketch the waveform that would be observed on the screen of an oscilloscope (with a suitable X time-base) whose Y-input is connected to a microphone placed at this point.

(*c*) In a warship's sonar system ultrasonic transmitters in a linear array are placed with equal spacing *d* along the keel, as illustrated in the diagram.

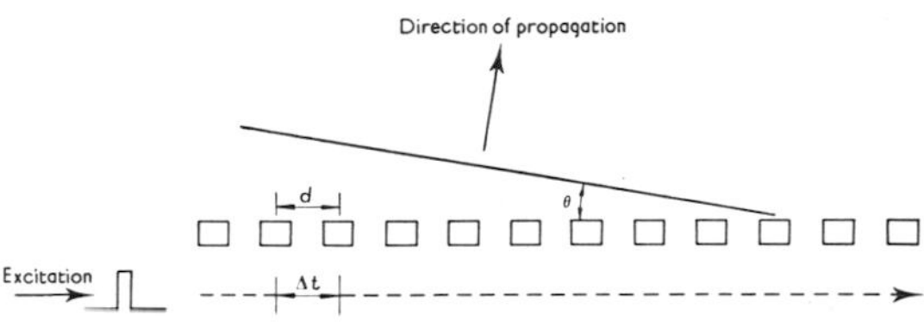

(i) Show that if these transmitters each produce a single pulse of sound in rapid succession at equal time-intervals a plane ultrasonic wavefront will result.

(ii) Given that the speed of ultrasonic waves in seawater is *c* and that the transmitters are energized at equal time-intervals Δt, derive an expression for the angle θ at which the plane wavefront leaves the array. (O)

Q 5.34

The intensity I of a sound wave is defined as the energy per second crossing a surface of one square metre normal to the direction of the sound wave. Given that I is related to the air density ρ, the frequency f, the amplitude a and the speed c of the sound wave by an expression

$$I = \tfrac{1}{2}a^2 c^p f^q \rho^r,$$

determine the unknown powers p, q and r.

Relative intensity R is measured in decibels (dB) and is defined by the equation

$$R = 10 \log (I_1/I_2)$$

where I_1 and I_2 are the two sound intensities being compared.

Given that the amplitude of vibration in a sound wave having the lowest intensity detectable by the human ear is 5.0×10^{-11} m, and that the safe dynamic range of the ear is 120 dB, calculate the amplitude of vibration in the loudest sound wave of the same frequency to which the ear should be subjected. (O)

Q 5.35

Distinguish between the *amplitude* and the *intensity* of a wave.

Sound waves of frequency f and amplitude A are transmittted through a gas of density ρ. By considering each molecule of the gas to be undergoing simple harmonic motion, find D, the energy per unit volume due to the sound wave, in terms of f, A and ρ.

Hence show that the intensity I at a point in the path of the sound wave is given by $I = 2\pi^2 f^2 A^2 \rho v$ where v is the speed of the wave.

A point source of frequency 3.0 kHz radiates sound energy uniformly in air at a rate of 1.0 mW. At this frequency an observer can hear the sound clearly when standing 150 m from the source. Assuming no absorption or reflection of the sound energy and using approximate values of any quantities involved, estimate the amplitude of the wave at the observer.

Comment on your result. (C)

Chapter 6

A

Q 6.25

State the simple relation between the refractive index of a transparent medium and its real and apparent depths. In what circumstances may this relation be used?

A concave mirror of radius of curvature 30 cm and small aperture is immersed in a liquid with its reflecting face upwards, its axis vertical and its pole 6 cm below the horizontal surface of the liquid. A small luminous object on the axis of the mirror at a height of 15 cm above the surface of the liquid coincides with its image. Indicate how this occurs and calculate the refractive index of the liquid. (JMB)

Q 6.26

Calculate the critical angle for sound waves travelling from air into water, across a plane boundary. The speed of sound in air is 340 m s^{-1} and the speed in water is 1500 m s^{-1}.

Someone standing on the side of a swimming pool shouts to a friend swimming under water. Give one reason why the swimmer is unlikely to hear the shout. (L)

Q 6.27

The critical angle for light incident upon a glass-air boundary is approximately 40°. Use this fact to determine a value for the speed of light in glass. (C)

Q 6.28
The figure shows a section of a glass prism having angles A = C = 32°, B = 116°. A ray of (*a*) red light, (*b*) blue light is incident normally on the side

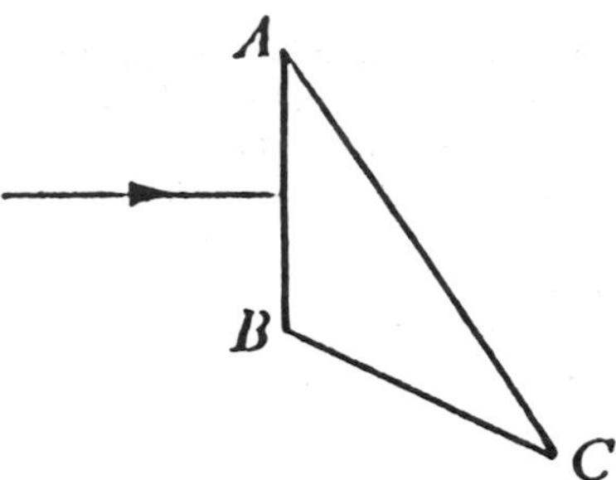

AB of the prism. Determine the deviation of the emerging ray in each case. (Refractive index of glass is 1.86 for red light and 1.92 for blue light.) (S)

Q 6.29
Draw a diagram showing the path of a ray of monochromatic light through a glass prism of refracting angle 60° at minimum deviation.

If the angle of minimum deviation is 36° calculate the refractive index of the glass. (L)

Q 6.30
A prism of glass of refractive index 1.63 has an angle *A* between two of its faces. If a ray of light is incident normally on one face of the prism, for what range of values of *A* will the ray emerge from the second face? (C)

Q 6.31
A coin of diameter 20 mm lies at the bottom of a beaker, and a converging lens of focal length 15 cm is mounted 25 cm vertically above it. What is the position and size of the image of the coin formed by the lens?

If the beaker is now filled with water, what is the (qualitative) effect on the position and size of the image? (S)

Q 6.32
Explain what is meant by (*a*) a virtual image, (*b*) a virtual object, in geometrical optics. Illustrate your answer by describing the formation of (i) a virtual image of a real object by a thin converging lens, (ii) a real image of a virtual object by a thin diverging lens. In each instance draw a ray diagram showing the passage of TWO rays through the lens for a non-axial object point. (JMB)

B

Q 6.33
Define the *focal length* of a spherical mirror.

Using the laws of reflection, derive a relation between the object and image distances and the radius of curvature of a concave spherical mirror. Define *linear magnification* and obtain an expression for it. Make a concise statement of the sign convention you are using.

Explain what is meant by a virtual image and a virtual object. How is a virtual object produced? Draw suitable ray diagrams to show how a concave mirror can produce (*a*) a virtual image of a real object and (*b*) a real image of a virtual object.

A concave mirror forms on a screen a real image which is one third the linear dimensions of a virtual object. By adjusting the positions of the virtual object and the screen, the linear magnification of the image is doubled. Draw ray diagrams (not to scale) illustrating the two cases and calculate the focal length of the mirror if the distance through which the screen has been moved is 0.05 m. (W)

Q 6.34
What do you understand by *linear magnification*? Prove that linear magnification produced by a concave mirror is equal to the ratio of the image distance to the object distance.

A coin 2.54 cm in diameter held 254 cm from the eye just covers the full moon. What is the diameter of the image of the moon formed by a concave mirror of radius of curvature 1.27 m?

Describe carefully what happens to this image if the aperture of the mirror is reduced. (C)

Q 6.35
The diagram shows a narrow parallel horizontal beam of monochromatic light from a laser directed towards the point A on a vertical wall. A semi-circular glass block G is placed symmetrically across the path of the light and with its straight edge vertical. The path of the light is unchanged.

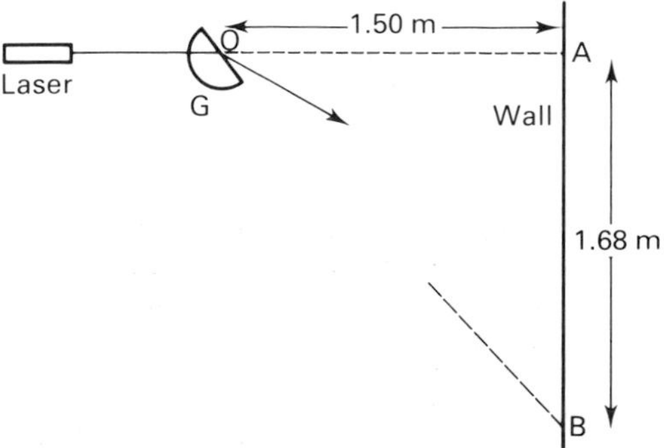

The glass block is rotated about the centre, O, of its straight edge and the bright spot where the beam strikes the wall moves down from A to B and then disappears.

OA = 1.50 m   AB = 1.68 m

(*a*) Account for the disappearance of the spot of light when it reaches B.
(*b*) Find the refractive index of the material of the glass block G for light from the laser.
(*c*) Explain whether AB would be longer or shorter if a block of glass of higher refractive index was used. (L)

Q 6.36
Explain the terms *critical angle* and *total internal reflection* applied to the refraction of light.

Explain why a ray of light entering one face of a rectangular block of glass of refractive index 1.5 will not leave by an adjacent face.

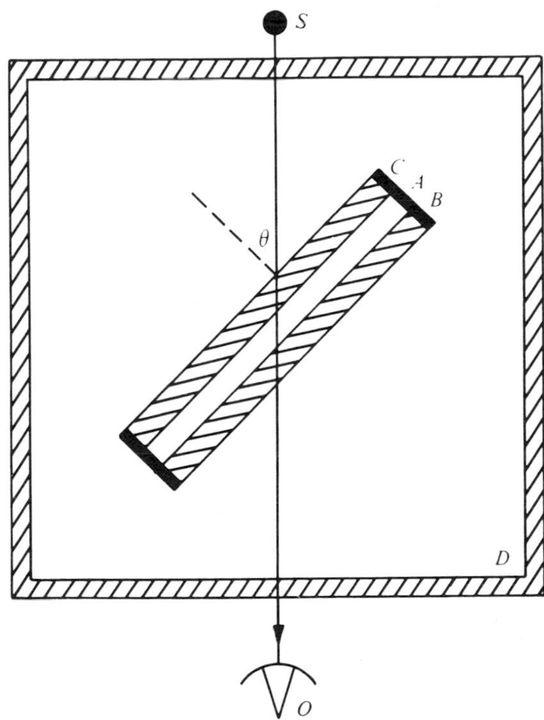

A method of measuring the refractive index of a liquid is shown in the diagram. The liquid is contained in a square glass tank *D*. A cell, consisting of a thin layer of air *A* between parallel glass plates *B*, *C*, is mounted in the liquid in such a way that it can be rotated about an axis perpendicular to the plane of the diagram, and the angle it is turned through can be measured. The refractive index of the glass plates is *n*. An observer *O* views a monochromatic light source *S* through the cell. It is found that the light from *S* is cut off if the angle of incidence θ on *C* is greater than 48.6°. What is the refractive index of the liquid?

When the air cell is set at the critical angle, the observer sees one half of the cell dark and the other half illuminated. Explain why this is so, and state which half of the cell appears illuminated when the cell is oriented as shown in the diagram. Explain the appearance of colours at the boundary between the dark and light fields when a white light source is used. (S)

Q 6.37
A ray of light strikes face AB of a triangular glass prism of section ABC, enters the prism and next strikes face AC. Angle BAC is 62.00°. What is the minimum angle of incidence on face AB which will allow light to emerge from face AC if the refractive index of the glass is 1.520?

What is the minimum deviation which this prism can cause for light entering face AB and leaving via face AC? (S)

Q 6.38
The diagram shows the path of a light ray through a prism. The angle of deviation, *D*, is measured for a range of values of the incident angle i_1. A graph of *D* against i_1 is plotted. Sketch the shape of this graph and explain how you would use it to determine the refractive index of the material of the prism.

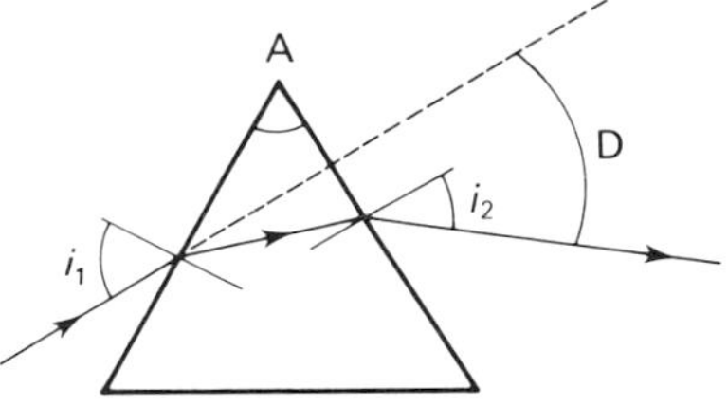

What would the fact that the direction of light is reversible tell you about a graph of *D* against i_2? (L)

Q 6.39
(*a*) A student performs an experiment to measure the focal length of a converging lens. In the experiment a series of object and image distances (u and v) is obtained and then a graph drawn of uv against $(u+v)$. This graph is shown.

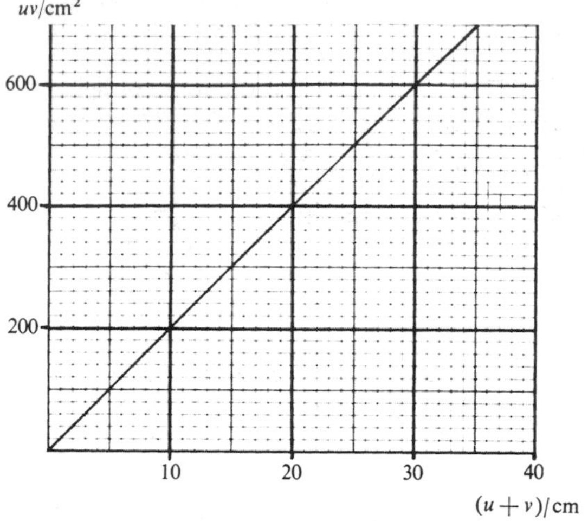

(i) Show that the slope of such a graph is equal to the focal length of the lens.
(ii) From the graph, obtain a value for this focal length.

(*b*) Explain why the images produced by this lens may be slightly coloured. (S)

Q 6.40

An illuminated disc 1 cm diameter is placed at a distance 1 m from a screen. An image of the disc 4 cm diameter is formed on the screen by a converging lens placed between the disc and the screen. What is the distance of the lens from the disc, and the focal length of the lens?

At what other position of the lens would a sharp image be formed on the screen, and what would be the size of the image? (S)

Q 6.41

An opaque card is pierced by two small holes 4.0 mm apart and strongly illuminated from one side. A lens on the other side of the card focuses images of the holes, 16.0 mm apart, on a screen 125 cm from the card. Find

(*a*) the position of the lens, and
(*b*) the focal length.

Why are measurements of the object and image sizes in this experimental arrangement unlikely to yield a reliable value for the focal length of the lens? (L)

Q 6.42

A lamp and a screen are 80 cm apart and a converging lens placed midway between them produces a focused image on the screen.

A thin diverging lens is placed 10 cm from the lamp, between the lamp and the converging lens. When the lamp is moved back so that it is 30 cm from the diverging lens, the focused image reappears on the screen. What is the focal length of the diverging lens? (L)

Q 6.43

Describe with the aid of labelled diagrams how:

(*a*) A point source of light forms sharp shadows.
(*b*) A spherical source of light forms the shadow of a spherical object on a wall. (The source and the object are the same size, and the object is midway between the source and the wall, which is at right angles to the line joining source and object.)
(*c*) A converging lens can form an image the same height as the object.
(*d*) A diverging lens can form an image half the height of the object.
(*e*) A concave parabolic mirror of large aperture can produce a smaller spot of reflected sunlight than can a concave spherical mirror of similar aperture and focal length. (C)

Q 6.44

Explain what is meant by the focal length of a lens.

Derive an expression for the focal length of a thin spherical lens in terms of (i) the radius of curvature of its faces and its refractive index, (ii) object and image distances. State clearly the assumptions made in your derivation, and explain the sign convention used.

When an equi-concave lens is placed at a certain distance from an object illuminated with monochromatic light, a virtual image lies midway between lens and object and a real image coincides with the object. Deduce the refractive index of the lens material.

If, without moving lens or object, the latter is now illuminated with light of a shorter wavelength, will the real and virtual image positions change, and if so, in what way? (W)

Q 6.45

The formula for the focal length f of a thin lens in terms of the refractive index n of its material and the radii of curvature r_1 and r_2 of its surfaces is

$$1/f = (n-1)(1/r_1 \pm 1/r_2)$$

where the sign in the second bracket depends on the convention chosen.

(*a*) Describe in outline how you would measure the focal length, and the radii of curvature of the faces, of a thin biconvex lens.

(*b*) For an equiconvex thin lens, of which the radii of curvature of the faces are each 50 cm, the refractive indices for red light and for blue light are: $n_{red} = 1.57$ and $n_{blue} = 1.62$. A small white light source is placed on the axis, 50 cm from the lens. Calculate the positions of the images formed in red light and in blue light.

(*c*) Given that the aperture of the lens is 10 cm in diameter, describe and explain as fully as you can what you would expect to see on a white screen if it were placed in succession at each of these image distances from the lens. (O)

Q 6.46
The focal length of a thin equiconvex glass lens is given by the formula $f = R/2(n-1)$ where n is the refractive index of the glass and R is the radius of curvature of each face. Describe the experiments you would perform in order to verify this formula for a lens of known refractive index.

For a thin equiconvex glass lens, $R = 100$ cm and the refractive indices for red light and for blue light are 1.630 and 1.640 respectively. Find the positions of the red and blue images of a small white-light source placed on the axis at a distance of 100 cm from the lens.

Explain why this dispersion effect is not observed when a converging lens used as a magnifying glass forms a virtual image. (O)

C

Q 6.47
Write down a formula connecting the distances of the object and image from a spherical concave mirror, explaining clearly the quantities involved and the sign convention used. Deduce a formula relating the distances p and q of the object and image respectively from the centre of curvature.

The imaging properties of a system consisting of a thin convex lens and a plane mirror are similar to those of a concave mirror. One particular system consists of a lens of focal length 0.3 m placed a distance 0.2 m, from a plane mirror.

By considering the situation where object and image are coincident show by means of a carefully drawn and annotated ray diagram where a concave mirror having the same imaging properties as the above lens-mirror system should be placed. (Numerical calculations are not required.) (Cam. Schol.)

Q 6.48
Describe a direct terrestrial method by which the speed of light may be determined.

The speed of light in air varies slightly with its frequency, blue travelling more slowly than red. Discuss what effect this will have on a pulse of white light travelling a long distance through air of uniform density.

A 200 m long glass fibre has a beam of light passing through it. The refractive indices of the glass for red and blue light are 1.50 and 1.53 respectively. Calculate

(*a*) the speed of red light in glass,

(*b*) the difference in time taken for a pulse of red and a pulse of blue light to pass down the fibre.

Very short pulses of white light are sent at regular intervals into one end of this fibre—as shown in the diagram. Sketch a corresponding graph showing the form of the pulses as they emerge from the far end. Hence estimate the maximum frequency at which very short pulses of white light could be passed through the fibre and still emerge as just separate flashes. (C)

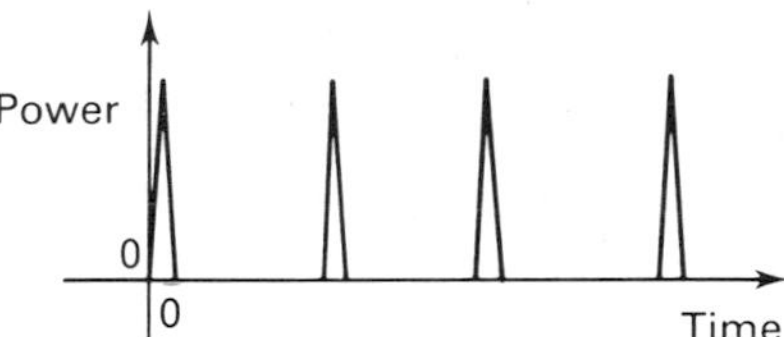

Q 6.49
Give the laws of refraction of light at the boundary between two optical media and show how these lead to the phenomenon known as total internal reflection.

An optical fibre consists of a thin (<1 mm) core of glass of refractive index n_1, clad in a uniform layer of another glass of refractive index n_2, such that $n_1 > n_2$. Show that there is a maximum angle to the axis which the light moving down the axis can have if all this light is to be confined to the inner core. Describe a practical application for such a fibre.

Such a fibre begins in a flat face perpendicular to its axis, and light is focused by a lens on to the centre of this face. Assuming the quantity $\Delta = (n_1 - n_2)/n_2$ is small, obtain an expression for the maximum angle of incidence if all light is to be confined to the core. Calculate a numerical value if $n_2 = 1.5$ and $\Delta = 0.005$. Show that whatever the angle of incidence light may leave the core but cannot escape from the cladding to the outside. (Ox. Schol.)

Q 6.50
A small object travels at a uniform velocity of 10 cm s^{-1} along the principal axis towards the optic

centre of a convex lens of focal length f. Find the velocity of the image when the object distance from the optic centre is (a) $4f$, (b) $2f$, (c) f, (d) $f/2$.

Use your results to sketch a graph indicating how the velocity of the image varies with the position of the object. (JMB)

Q 6.51

(*a*) A triangular glass prism ABC has equal angles at B and C. A wide beam of parallel monochromatic light incident upon face AB passes through the prism at minimum deviation. Draw a diagram showing the positions of wavefronts at equal time intervals as they pass through the prism. By considering the times taken by the portions of a wavefront passing through B and C and through A derive the relationship between the refractive index of the glass, the angle of the prism at A and the angle through which the beam is deviated.

(*b*) A ray of light travels through a series of parallel-sided transparent blocks of varying refractive indices, n. If n_i is the refractive index of the ith block and the angle of incidence in this block is θ_i show that $n_i \sin \theta_i$ is a constant independent of n.

If the refractive index of air varies with height, h, according to the relationship $n_h = n_0 - \alpha h$, where n_h is the refractive index at a height h above ground level, n_0 is the refractive index at ground level and α is a constant, show that, neglecting the curvature of the earth, the maximum height reached by a strong beam of light shone upwards at a small angle θ to the horizontal is given by $n_0\theta^2/2\alpha$.

$$\left(\cos\theta = 1 - \frac{\theta^2}{2!} + \frac{\theta^4}{4!} - \cdots\right) \qquad \text{(L)}$$

Q 6.52

(*a*) The lens shown may be used to magnify a thin flexible transparent biological specimen placed on the surface *A*. The spherical surfaces *A* and *B* have a common centre *C*.

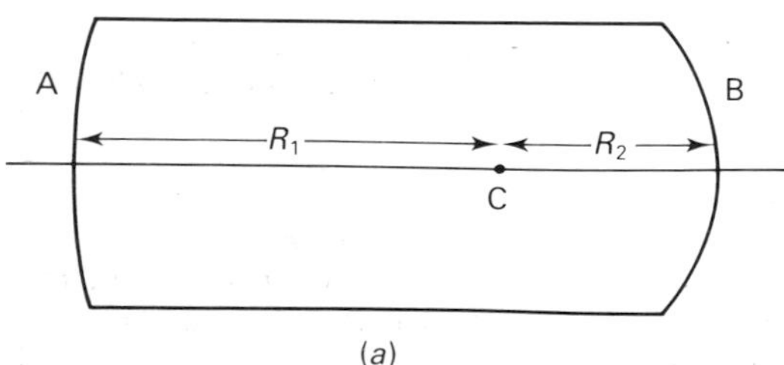

(*a*)

Show that for a certain value of R_1/R_2, a point object on *A*, illuminated by monochromatic light, will be seen at infinity by an observer looking into *B*. Find the angular magnification for a small object on *A*, when $R_1 = 0.80$ cm and the observer's least distance of distinct vision is 24 cm.

(*b*) The direct vision prism shown is made of two prisms *A* and *B* of glasses which for sodium light have refractive indices respectively 1.710 and 1.515. Find the value of the angle α for which sodium light incident normally on the vertical face of *A* is undeviated.

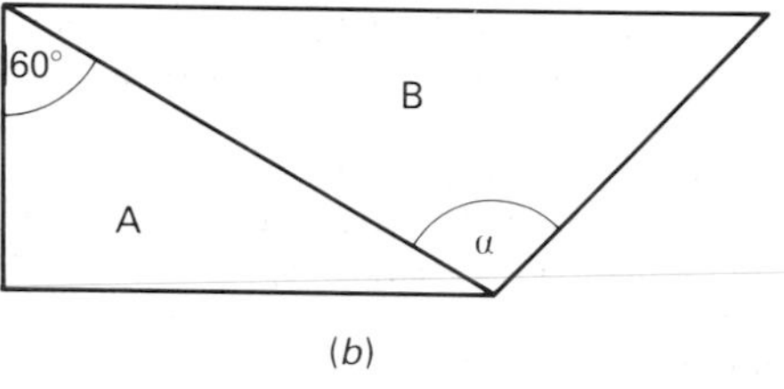

(*b*)

(L)

Q 6.53

A student used the distant object method to determine the focal length of a converging lens known to be about 30 cm. Estimate how far away the distant object must be if his result is to be reliable to 1%.

If you used the window in your school laboratory as the distant object to determine the focal length of the same lens, what percentage accuracy would you expect to obtain?

Describe briefly the method you would use to determine accurately the focal length of a converging lens in the first five minutes of a practical session. Justify your choice of method. (JMB)

Q 6.54

Why is chromatic aberration not usually a noticeable defect of a simple magnifying glass?

An achromatic telescope objective is to be constructed of a flint glass and a crown glass lens cemented together. One external surface of the doublet is to be plane, and the other convex. The focal length of the doublet is to be 1 m.

The refractive indices of flint and crown glass are given below.

	Blue light	Yellow/green light	Red light
Flint glass	1.6350	1.6250	1.6150
Crown glass	1.5040	1.5000	1.4960

Calculate the radius of curvature of (*a*) the convex outer surface of the doublet, (*b*) the surfaces of the two components which are cemented together.

Which surface (plane or convex) of the objective should face the distant object, and why? (S)

Chapter 7

A

Q 7.20

(*a*) What conditions must be satisfied if two wave-trains are to interfere with each other to produce a stationary interference pattern?
(*b*) Explain why it is possible to produce a stationary interference pattern using as sources two separate radio transmitters, but impossible to produce such a pattern using two separate sodium-vapour lamps. (O)

Q 7.21

Two narrow parallel slits, distance d apart, act as coherent sources of monochromatic light of wavelength λ. Interference fringes are observed in a plane parallel to the plane of the slits and a distance D from it.

(*a*) Derive an expression for the distance apart of the centres of adjacent dark fringes near the centre of the pattern, if D is very much greater than d.
(*b*) Describe an arrangement for producing fringes in this way, indicating appropriate dimensions for the apparatus. Explain how you would use the arrangement to determine the wavelength of the light.
(*c*) Explain how the fringe pattern would change if it were possible to immerse the whole arrangement in water. (JMB)

Q 7.22

S, S_1 and S_2 are slits. The filter transmits light of wavelength 0.60 μm, $d = 0.25$ mm and $D = 1.5$ m.

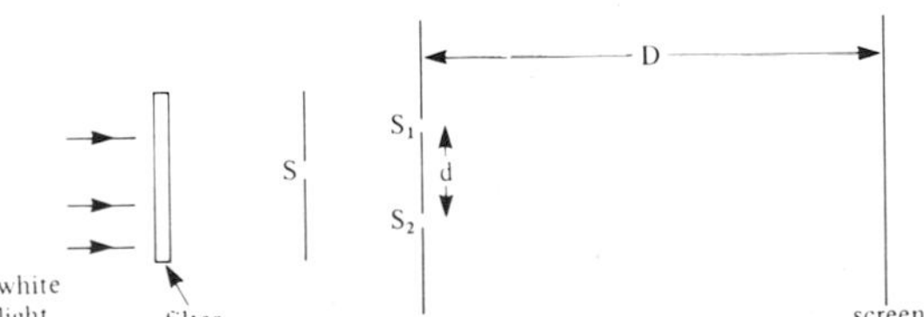

(*a*) Calculate the distance between consecutive bright fringes seen on the screen.
(*b*) What is the purpose of the slit at S?
(*c*) Explain briefly what is seen when the filter is removed. (W)

Q 7.23

With the aid of a clearly labelled diagram, explain how thin film interference effects can be used to test the flatness of a surface. (L)

Q 7.24

Two plane glass plates are placed so that a wedge air film of small angle is formed between them. When the film is illuminated near-normally with monochromatic light of wavelength λ, a system of interference fringes is observed. The spacing between adjacent dark fringes is 4.0 mm.

(*a*) Sketch a side view of the plates, indicating how interference fringes are formed. (Ignore any refraction in the plates.)
(*b*) Write down the condition for a dark fringe to occur at a position where the thickness of the film is t.
(*c*) Sketch a plan view of the plates, showing the pattern of dark fringes that is observed.
(*d*) What happens to the fringe pattern when liquid of refractive index 1.33 is introduced into the wedge so that it completely fills it. (C)

Q 7.25

Monochromatic light of wavelength λ is incident normally on a narrow slit of width D and is focused on a screen by a lens of focal length f, situated just beyond the slit. Sketch a graph to illustrate in detail the intensity distribution of the resulting diffraction pattern. State the width of the central fringe. (C)

Q 7.26

Sketch a diagram of apparatus arranged for the observation of the interference pattern produced in Young's two slits experiment. State suitable values for the relevant dimensions. Explain the parts played by diffraction and by the principle of superposition in accounting for

(*a*) the overall width of the pattern, and
(*b*) the alternation of bright and dark bands within the pattern. (L)

Q 7.27

Red light is incident normally on a plane diffraction grating having 5×10^5 lines per metre. Estimate the angular deviation of the first order spectrum. (C)

Q 7.28
A distant bright point-source of sodium light is viewed normally through a uniform fine wire-mesh interwoven at right angles.

(i) Describe what is observed.
(ii) Given that the first-order diffracted images lie in a direction making an angle of 0.20° with the normal, calculate the distance between the axes of adjacent wires in the mesh.

[Take the wavelength of sodium light to be 589 nm.] (O)

Q 7.29
A beam of electromagnetic waves of wavelength 3.0 cm is directed normally at a grid of metal rods, parallel to each other and arranged vertically about 2.0 cm apart. Behind the grid is a receiver to detect the waves. It is found that when the grid is in this position, the receiver detects a strong signal but that when the grid is rotated in a vertical plane through 90°, the detected signal strength falls to zero. What property of the wave gives rise to this effect? Account briefly in general terms for the effect described above. (L)

Q 7.30
A beam of plane-polarized microwaves is incident upon an aerial which is initially positioned to give maximum response. In a storm, the aerial is rotated about the direction of the incident waves until it makes an angle of 30° to the plane of polarization, as shown in the diagram.

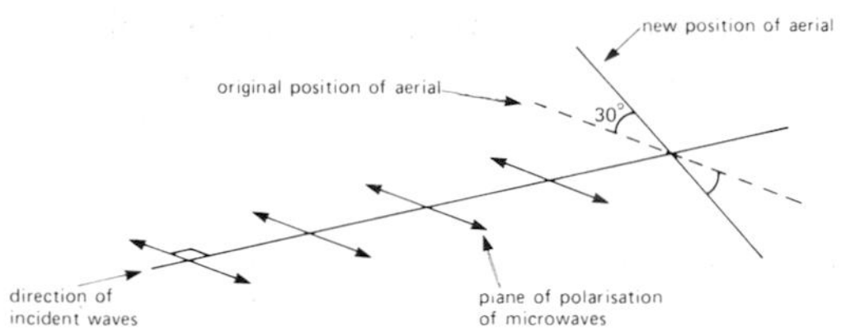

Calculate the percentage reduction in the amplitude of the signal now received from the aerial. (C)

B

Q 7.31
State the conditions which must be satisfied for interference to be observed between waves from two sources.

An arrangement of a narrow line source, a double slit of separation 5.0×10^{-4} m and a screen on which interference fringes can be observed is shown on the diagram. The slits and line source are parallel and the distance between the double slit and the screen is 1.0 m.

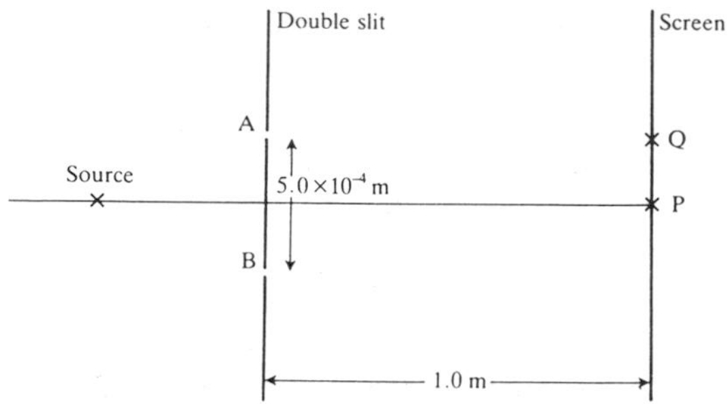

(i) Explain why a regular pattern of fringes is produced on the screen when the source of light is monochromatic.
(ii) Calculate the fringe separation when the wavelength of the light is 600 nm.
(iii) Describe and explain, qualitatively, how the fringes for monochromatic light will change as the width of the source is increased.
(iv) When a source of white light is used a central bright white fringe flanked by coloured fringes is observed. Explain this observation.
(v) One of the slits forming the double slit is now covered with a thin sheet of transparent material of refractive index 1.4. The position of the bright white fringe, previously at P, is now at Q. Explain why the position has changed. Which slit, A or B, is covered by the transparent material?
(vi) The distance PQ is equal to the distance between 41 fringes when the system is illuminated with the source of wavelength 600 nm. Calculate the thickness of the sheet of transparent material. (O & C)

Q 7.32
(*a*) Describe briefly how you would measure the wavelength of a monochromatic light source using a double slit. Your account should include
  (i) a labelled diagram showing the apparatus used, how it is arranged and giving rough values of the dimensions involved,
  (ii) a list of measurements that need to be taken and the means by which each is measured,
  (iii) a diagram showing the appearance of the interference pattern.
The theory of the experiment is not required.

(*b*) A beam of microwaves of wavelength 3.1 cm is directed normally through a double slit in a metal screen and interference effects are detected in a plane parallel to the slits and at a distance of 40 cm from them. It is found that the distance between the centres of the first maxima on either side of the central maximum in the interference pattern is 70 cm. Calculate an approximate value for the slit separation. Why does the formula you have used only give an approximate value?

(*c*) Light from a point in a monochromatic source is emitted in pulses lasting about 10^{-9} s.

(i) Two light beams derived from the same source can be recombined and produce stable interference patterns provided the path difference introduced between the beams is not too great. Explain why this is so and calculate an approximate value for this path difference. (Speed of light is 3.0×10^8 m s^{-1}.)

(ii) Splitting the wavefront by means of a double slit is one method of using a single source to produce two beams. Describe with the aid of a diagram an alternative method.

(iii) Explain why light from two separate monochromatic sources of the same frequency will not, when combined, result in a stable interference pattern. (L)

Q 7.33

(*a*) List four ways in which the light emitted by a laboratory laser differs from the light produced by a laboratory vapour-lamp.

What is meant when it is said that two light beams are coherent?

(*b*) Describe how you would set up a double-slit experiment to observe the formation of interference fringes, using as the light-source either a vapour-lamp or a laser.

How could the wavelength of the light be measured?

(*c*) Two radio transmitters 750 m apart on a north–south line broadcast identical signals of frequency 100 MHz. How does the strength of the radio signals received by a low-flying aircraft flying due north at 200 m s^{-1} in a region 50 km due east of the transmitters vary with time?

When one of the transmitters is switched off it is found that a radio echo is received at the other transmitter from an aircraft flying straight towards it at 200 m s^{-1}. Explain why the frequency of this echo differs from that of the transmitted signal and calculate the frequency-difference.

[Take the speed of radio waves in air to be 3.0×10^8 m s^{-1}.] (O)

Q 7.34

(*a*) When monochromatic light is incident normally on a thin film, an interference pattern of bright and dark fringes is observed in light reflected from the film. Explain with the aid of a diagram why such an interference pattern is produced.

State what information about the film can be deduced from

(i) the shape of the fringes,

(ii) the separation of the fringes.

(*b*) An air wedge is made by separating two plane sheets of glass (microscope slides) by a fine wire at one end. When the wedge is illuminated normally by light of wavelength 5.9×10^{-7} m a fringe pattern is observed in the reflected light. The distance measured between the centre of the 1st bright fringe and the centre of the 11th bright fringe is 8.1 mm. Calculate the angle of the air wedge.

(*c*) As a soap film supported on a vertical frame slowly drains, patterns of horizontal alternate bright and dark fringes are observed both when viewed in reflected monochromatic light and when viewed in transmitted monochromatic light.

Explain why

(i) the contrast between fringes seen in transmitted light is less than that seen in reflected light,

(ii) a particular thickness of film which produces a dark fringe in reflected light produces a bright fringe in transmitted light. (JMB)

Q 7.35

State two conditions necessary for the superposition of two waves to give rise to a well-defined interference pattern.

Two identical progressive waves are travelling in opposite directions. Explain what happens when they meet.

A monochromatic beam of light of wavelength λ is directed normally on to a front-silvered plane mirror where it is reflected. The incident and reflected beams interfere with each other. Explain how the spacing of the resultant pattern of nodes and antinodes is related to λ.

An extremely thin photographic film is placed at a small angle θ to such a mirror as shown in the diagram overleaf.

After exposure, the photographic emulsion is found to be blackened along a series of straight, parallel lines. Give an explanation for this.

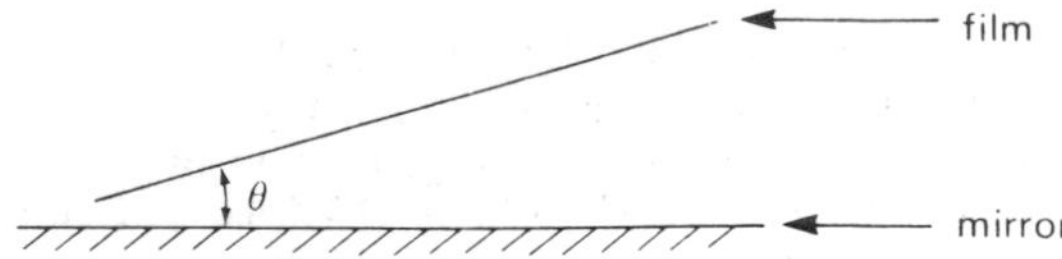

When the angle θ is approximately 10^{-3} rad, 25 of the blackened lines occupy a distance of 7 mm on the film. Estimate the wavelength of the light used.

How are your stated conditions for interference met in this particular case? (C)

Q 7.36

Describe, quoting relevant formulae, an experiment for measuring the wavelength of monochromatic light by observation of Newton's rings.

Newton's rings are formed using a plane glass surface and a lens having a surface with radius of curvature 1.75 m. The 15th bright ring from the centre (the bright ring of smallest radius being number 1) has a radius 4.0 mm. What is the wavelength of the light used?

Describe and explain the appearance of Newton's rings when white light is used instead of monochromatic light. (S)

Q 7.37

(*a*) What conditions must be fulfilled if interference between two light beams is to be observed?

(*b*) State *three* ways in which a beam of light from a laser differs from a parallel beam of light from a sodium lamp.

(*c*) A thin plano-convex lens is placed with its curved face downwards on a plane glass plate and is illuminated normally by sodium light of wavelength 589 nm. A series of circular interference fringes is observed by reflected light.

(i) Draw a diagram of a suitable experimental arrangement by which the fringes could be observed and measured.

(ii) Explain the formation of the fringes. Why is the centre of the pattern dark?

(iii) If the radius of the 20th dark ring from the centre is 4.99 mm, calculate the radius of curvature of the lens face, proving any formula you use in your calculation.

(iv) The air-space between the lens and the glass plate is now filled with water, of refractive index 1.33. Describe the changes in the fringe system, and calculate the new radius of the 20th dark ring from the centre.

(*d*) Explain why colours are observed when a thin layer of transparent liquid such as petrol spreads over a water surface and is illuminated by daylight. (O)

Q 7.38

(*a*) Draw a labelled diagram of an experimental arrangement that would enable you to examine and to make measurements on the diffraction pattern produced when a parallel beam of light from a monochromatic source is incident on a single slit. Indicate approximate dimensions of the arrangement, including the width of the slit.

(*b*) Derive an expression for the angular separation, θ, between the first minimum and the centre position of the diffraction pattern, in terms of the slit width, s, and the wavelength of the light, λ.

(*c*) Sketch a graph of relative intensity of the diffraction pattern against angular separation from the centre of the pattern. Indicate on your graph the values of the angles which correspond to the first and second minimum for a slit of width 0.20 mm and light of wavelength 6.5×10^{-7} m.

(*d*) If a white light source were used in place of the monochromatic source describe the pattern obtained and compare it with the original pattern.

(*e*) Describe how the intensity variation of part (*c*) would alter if the single slit were replaced by two slits about 1 mm apart, both of the same width as the single slit. (JMB)

Q 7.39

(*a*) The diagram shows an arrangement for observing Young's fringes.

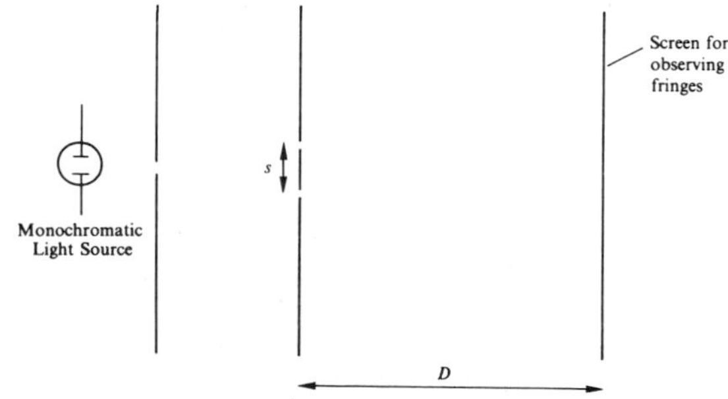

(i) Explain why both *interference* and *diffraction* of light waves are important in understanding the experiment. Describe briefly how the fringes are formed.

(ii) Derive an expression for the separation of the fringes observed on the screen in terms of s, D and the wavelength λ, of the light used. State clearly where approximations are made.

(iii) Suggest, with reasons, suitable values for s and D.

(*b*) A and B are two sources of sound waves emitted with the same wavelength (2.50 m) and the same amplitude.

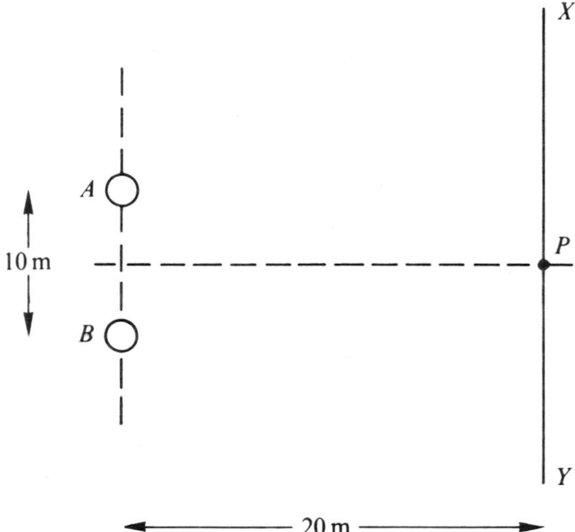

(i) An observer walks from X to Y. Describe as fully as you can what she would hear assuming that the sources are in phase. What differences would become apparent during the walk if the sources were π out of phase?

(ii) The wavelength of one of the sources is now increased to 2.52 m and the observer stands still at positive P on the perpendicular bisector of AB. Describe what would be heard.

[Velocity of sound = 330 m s^{-1}.] (S)

Q 7.40

Draw a clearly-labelled diagram, including wavelets originating from four adjacent gaps in a diffraction grating, to illustrate the formation of a second order spectrum for monochromatic light incident normally on the grating.

Calculate (*a*) the angular deviation of this spectrum for light of wavelength 589.0 nm, given that the grating has 2000 lines per centimetre; (*b*) the greatest angular deviation this grating can produce for such light at normal incidence. (S)

Q 7.41

A parallel beam of monochromatic light of wavelength 580 nm is incident normally on a diffraction grating having a large number of regular slits, each of width 0.70×10^{-6} m, as shown in the diagram below.

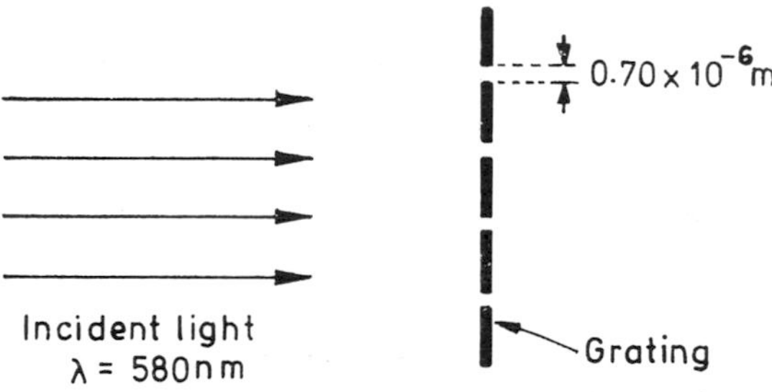

After passing through the grating the light will have, as a result of interference, intensity maxima in certain directions. Calculations predict that for the first, second and third order interference maxima, the values of the angle θ between the direction of the incident light and the directions of these maxima should be approximately 16°, 34° and 56° respectively.

(*a*) At what value of θ would the light diffracted by a *single* slit of width 0.70×10^{-6} m have its first intensity *minimum*? Show the steps in your calculation.

(*b*) Show by drawing on a copy of the axes below how the intensity of the light passing through the grating would vary over the range of θ from 0° to 60°. The angles 16°, 34° and 56° have been indicated by means of small crosses on the θ-axis.

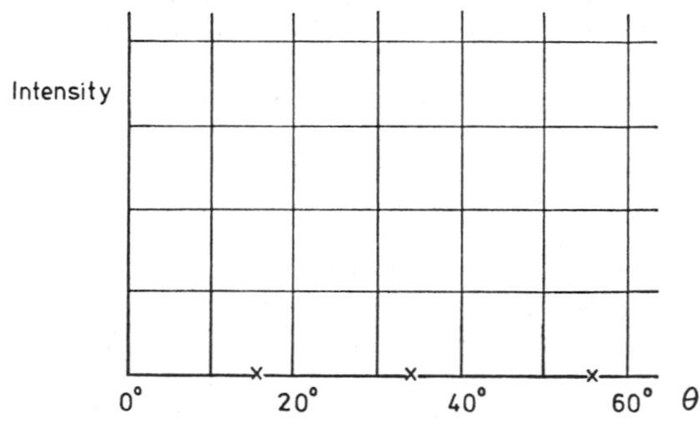

(O & C, Nuffield)

Q 7.42

A parallel beam of unpolarized light is reflected from the plane surface of a glass block. How would you ascertain whether the reflected beam is partially or completely polarized?

If complete polarization of the reflected beam occurs when the angle of incidence is 58°, what is the refractive index of the glass?

Why is polarisation not observed when sound waves are reflected? (S)

Q 7.43

Two sheets of polaroid, P and A, are placed so that their polarizing directions are parallel and vertical, as shown in the figure below; the intensity of the emergent beam is then I_0. Through what angle should A be turned for the intensity of the emergent beam to be reduced to $\frac{1}{2}I_0$? Describe the polarization of the emergent beam when this operation is carried out. (C)

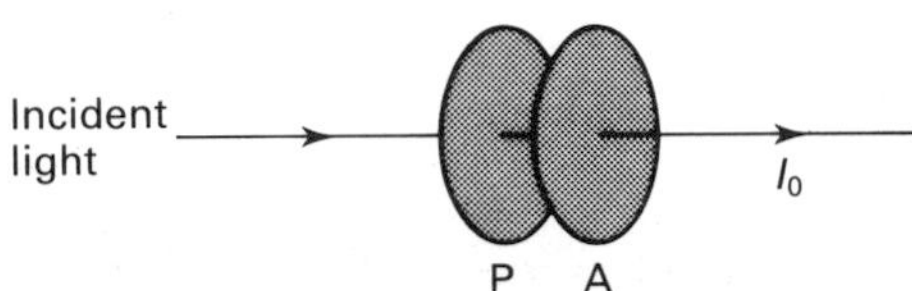

Q 7.44

Name one physical example of a *longitudinal* wave motion.

What features do longitudinal waves have in common with transverse waves?

Explain what is meant by a *plane-polarized* wave. Describe an experiment to investigate the state of polarization of a beam either of light or of microwaves.

In some crystals, light waves with different planes of polarization travel at different speeds. A thin slice of a particular quartz crystal has a refractive index of 1.553 for one plane of polarization but of 1.544 for the perpendicular plane of polarization. For light of wavelength 500 nm (in a vacuum), what is the minimum thickness of quartz which will introduce a phase difference of π radians between the two polarizations?

Explain why it is not possible to observe interference fringes between light beams which are plane-polarized in perpendicular planes. (C)

C

Q 7.45

Laser light of wavelength 630 nm is incident on a tube of length 75 mm that is partitioned along its axis into two evacuated compartments A and B. The ends of the tube are closed by thin glass plates, that at the left end being blackened except for two narrow parallel identical slits.

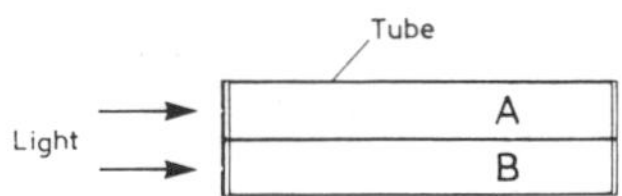

(i) Describe what is observed on a distant screen. Compartment A is now slowly filled with a gas whose refractive index n varies with pressure p (in pascal) in accordance with the relation

$$n = 1 + (2.0 \times 10^{-9})p$$

(ii) By how much must p be increased to cause the interference pattern to move laterally by one fringe-separation?

(iii) Suggest a way in which this apparatus could be modified for use as a pressure-gauge. Discuss the sensitivity that might be obtained, and the range of pressures that could be measured. (O)

Q 7.46

Explain what is meant by *phase difference* and *coherence* between two wave motions of the same frequency.

A ship 20 km from shore wishes to receive a radar signal of frequency 1.0×10^{10} Hz from a shore transmitter. If the receiver is at sea level what is the minimum height, h, that the transmitter must be above sea level? The radius of the Earth is 6.4×10^6 m.

In fact, even with the transmitter at the height h calculated, if the receiver is very close to sea level and the sea is calm the signal received is always weak. Why is this so?

As the receiver is gradually raised vertically the signal gets stronger and stronger. How far must it be raised for the signal strength to reach a maximum? The sea surface between transmitter and receiver may now be considered to be flat.

Discuss the variation in signal strength received as the vertical height of the receiver above sea level is increased beyond the value calculated above. How would the observations change if the sea were not calm? (L)

Q 7.47

Explain why soap bubbles appear brightly coloured when they are formed, but lose their colour just before they burst. (W)

Q 7.48

A microwave transmitter T is pointed at a hardboard sheet H so that the angle of incidence

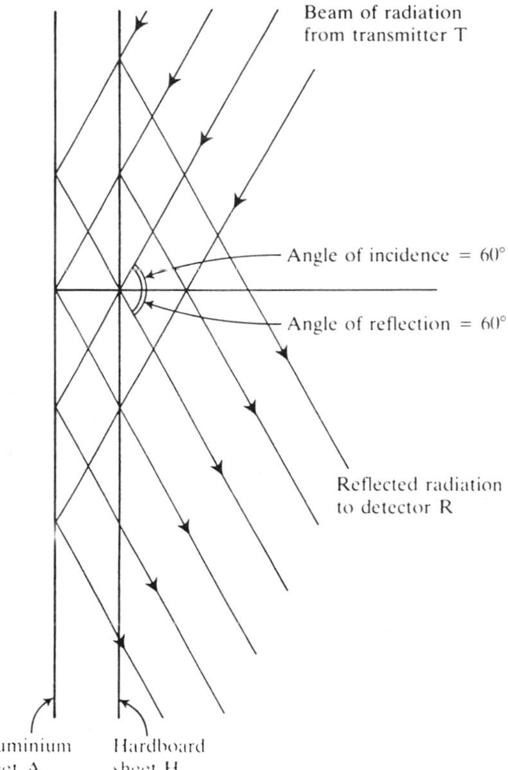

of the parallel beam of radiation is 60°, as in the diagram above. The hardboard reflects some of the incident microwave energy, and transmits the rest. An aluminium sheet A is placed a short distance behind H, and a microwave detector R receives the reflected radiation from both sheets. The strength of the reflected radiation is monitored on a meter attached to R. Both R and T are at distances from the sheets that are much greater than either the wavelength of the microwave radiation or the aluminium–hardboard spacing.

(*a*) As the distance between A and H is increased, the meter reading rises and falls periodically. Explain briefly why this occurs.

(*b*) In a second experiment, A is placed at a distance equal to about two wavelengths behind H. The wavelength of the radiation emitted from the transmitter is then increased slowly. Discuss how the meter reading varies in this case.

(*c*) The transmitter wavelength is set to some convenient value and the separation of A and H is increased from zero until the meter reads zero. The transmitter and receiver are then moved closer together to make a new angle of incidence and reflection, so that the meter now reads maximum deflection. Assuming that there are no phase changes at either reflection, calculate the new angle of incidence.

(*d*) Write a paragraph showing how the results of these experiments can be used to explain the colours formed in a thin soap film.

(O & C, Nuffield)

Q 7.49

Give the conditions necessary for the observation of interference in visible light.

Explain how interference fringes are produced when monochromatic light is incident on an air wedge between two flat glass plates, and outline how you would observe these fringes experimentally.

To test the quality of an optical 'flat' (a glass sheet with its two faces plane to much better than 500 nm) against a standard 'flat' (which may be taken to be perfect), they are set up separated by a small air gap, the standard below the test piece. The system is illuminated by diffuse light of wavelength 590 nm.

For three different flats the observed fringe pattern is as shown. The fringe separation in (*a*) corresponds to one fringe per cm travel across the flat, and the scale in (*b*) and (*c*) is the same as in (*a*).

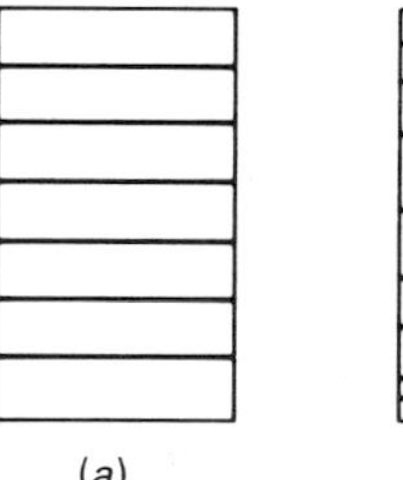
(*a*)

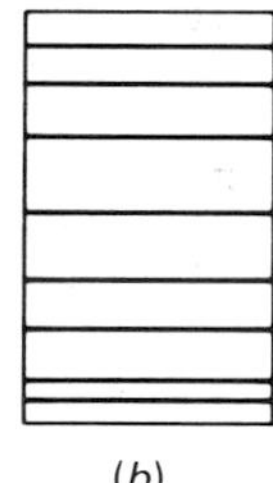
(*b*)

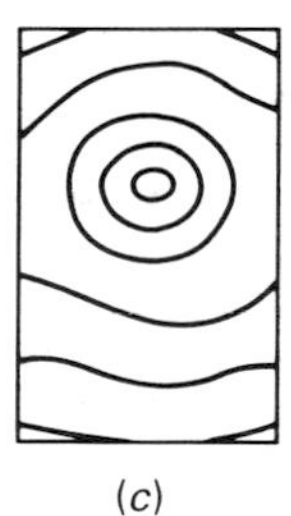
(*c*)

Suggest possible origins for these patterns in terms of any blemishes in the flats. Can the form of the blemish which produces the uneven spacing of fringes in (b), or of the circles in (*c*) be uniquely determined by these observations?

(Ox. Schol.)

Q 7.50

Describe one practical application of an interference effect at optical or radio frequencies.

In an experiment in which Newton's rings are formed in the air film trapped between a convex lens of 1 m radius of curvature and a plane block of glass, a total of 200 rings were observed out to the edge of the 20 mm diameter of the lens.

Calculate the wavelength λ of the monochromatic light source used for the experiment, and the number of rings which were visible when the space between the lens and the block was filled with water of refractive index 1.33.

Discuss the effect of replacing the monochromatic light source by one emitting two spectral lines of equal intensity and mean wavelength λ but differing in wavelength by 2%.

(Ox. Schol.)

Q 7.51

(*a*) A laser is mounted on a slowly rotating turntable and emits a parallel-sided beam of light. A distant observer sees a flash from the laser once every half-hour. A single slit of width 0.1 mm is now mounted in front of the laser. The observer now sees every half hour, not one flash, but a short series of flashes which rise to a maximum of intensity and then die away. The flashes last for 1 s each, except for the brightest flash which lasts for 2 s. Explain this phenomenon as fully as you can and find an approximate value for the wavelength of the laser light. Prove any formulae you use.

(*b*) What would the observer notice if the slit were opened up to 10 mm width and the speed of rotation of the turntable were simultaneously raised by a factor of 100?

(Cam. Schol.)

Q 7.52

(*a*) What do you understand by
  (i) diffraction
  (ii) the principle of superposition?

(*b*) The diagram shows a small opaque sphere S between a point source of white light and a

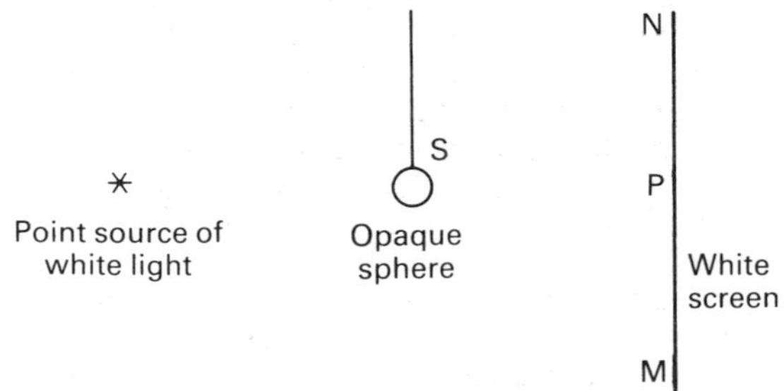

white screen MN. The screen is moved slowly away from the sphere until a bright spot becomes visible at the centre P of the shadow of the sphere on the screen.

How do you explain the formation of the bright spot at P?

Would you expect the spot to be white, or blue or red at the point where it is first visible? Explain your answer.

(*c*) Explain briefly, and without deriving any formulae, the role of the principle of superposition in accounting for
  (i) the fringes observed in Young's two-slits experiment, and
  (ii) the nodes and antinodes which form on a taut wire subjected to an oscillatory force.

Young's fringes can be seen whatever the frequency of the light which is used but the nodes and antinodes form at particular frequencies only. How do you account for this difference in behaviour?

(*d*) Discuss the evidence for the assertions that
  (i) sound waves in air are longitudinal, and
  (ii) light waves are transverse. (L)

Q 7.53

A laboratory demonstration of two-slit interference of light (Young's fringes) is to be set up, using a good quality slide projector and the arrangement shown below (not to scale).

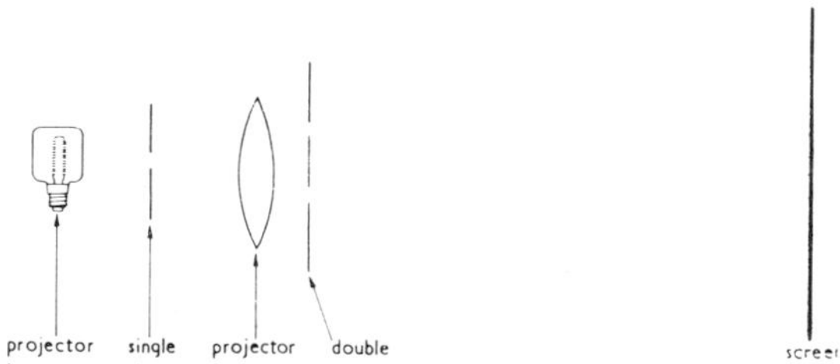

A single slit is to be placed in the projector so that a sharp image of it is cast on the screen. A double slit is to be held over the lens as shown.

[Note that part (*b*) will carry more marks than part (*a*).]

(*a*) Explain why it should be possible in principle to obtain fringes on the screen using this arrangement.
You should include an account of the function of the lens.

(*b*) You have to make the single and double slits yourself, for a screen placed about 2 m from the projector. Explain how you would decide on practicable values for:
  (i) the width of the single slit,
  (ii) the widths and spacing of the double slits so as to get fringes as large and as bright as possible on the screen. Explain the effect of making each dimension too large and too small in (i) *and* (ii).

(O & C, Nuffield)

Q 7.54

A computer, programmed to print out rows of dots, produces output similar to that shown in the diagram. A large page of such printout is to be photographed, producing a negative which is reduced by a factor of 10. This negative is to be used as a two-dimensional diffraction grating with a monochromatic visible-light source. The resulting diffraction pattern is to be focused sharply on a screen, and must be clearly visible to the naked eye.

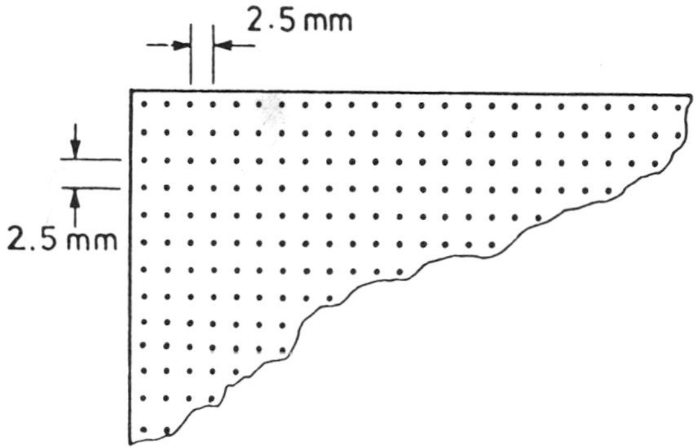

(*a*) Sketch apparatus, showing all relevant dimensions, for producing the diffraction pattern. Show how you arrive at your estimates of the dimensions.

(*b*) Sketch the form of the diffraction pattern you would expect to see. Explain its important features.

(*c*) Describe the changes you will observe when the monochromatic light source is changed for a white light source. (O & C, Nuffield)

Q 7.55

(*a*) Explain what is meant in the study of wave motion by the terms *interference* and *coherent wavetrains.*

Explain how coherent wavetrains play a part in:

(i) the production of a standing wave pattern on a taut string that has been plucked or bowed;

(ii) the production of light by a laser;

(iii) the deflection of light beams by a diffraction grating.

(*b*) In an experiment to measure the frequency of waves produced by an ultrasonic transducer, the transducer T is placed in a small glass tank of xylene on a spectrometer table as shown in the diagram. The xylene immediately forms close parallel striations, and, when illuminated

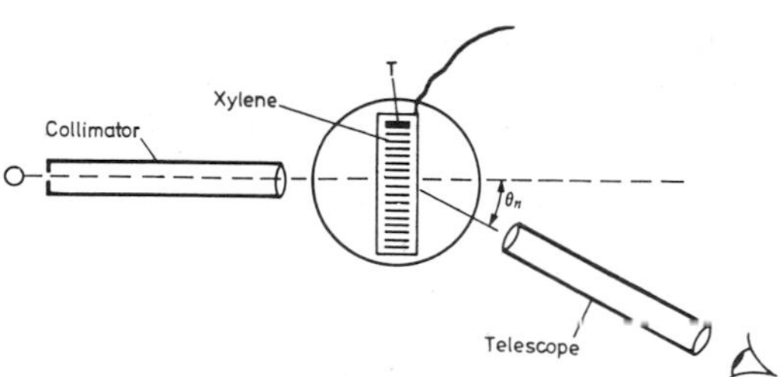

with collimated monochromatic light, deflected light beams are seen with the aid of the telescope at small angles θ with the direction of the incident light.

(i) Explain how the striations form.

(ii) Assuming that the striations act as a diffraction grating, derive an expression for the angular deflection θ_n of the nth order emergent light beam in terms of the frequency f, the speed of c of ultrasound in xylene and the wavelength λ of the monochromatic light.

(iii) The telescope can resolve (that is, see as separate) light beams with a minimum angular separation of 0.0020 rad. What is the lowest ultrasound frequency that can be measured by this method when using monochromatic light of wavelength 590 nm?

[Take the speed of ultrasound in xylene to be $1350\ \mathrm{m\ s^{-1}}$.] (O)

Q 7.56

(*a*) A parallel beam of monochromatic light is incident at an angle of incidence of about 55° onto the surface of a flat glass plate. The reflected beam strikes a second identical glass plate placed parallel to the first and is again reflected. Draw a diagram of this initial arrangement of the apparatus. When the *second* plate is rotated about the direction of the first reflected beam, the intensity of the beam reflected from the second plate varies. Explain why the intensity of the emergent light beam passes through maxima and minima.

Indicate the relative orientation of the two plates when the maxima and minima occur. How many maxima do you expect to see when the second plate is rotated through 360°?

The refractive index of the glass plates is 1.5. Is it possible to adjust the apparatus so that the minimum intensity is zero? If so, what adjustment is necessary?

(*b*) A plane transmission grating has slits 0.7 μm wide separated by opaque strips 1.4 μm wide.

The grating is placed in a spectrometer with its plane perpendicular to the collimator. The light incident on the grating contains only two wavelengths of 400 nm and 600 nm. How many lines will be observed through the telescope as it is rotated through angles of 0° to 90° with respect to the direction of the incident beam? List the angle(s) at which lines of each wavelength will be observed. (O & C)

Q 7.57
Explain the difference between unpolarized and linearly polarized light. Discuss whether it is possible to obtain linearly polarized waves of sound.

Describe and explain two methods of producing linearly polarized light.

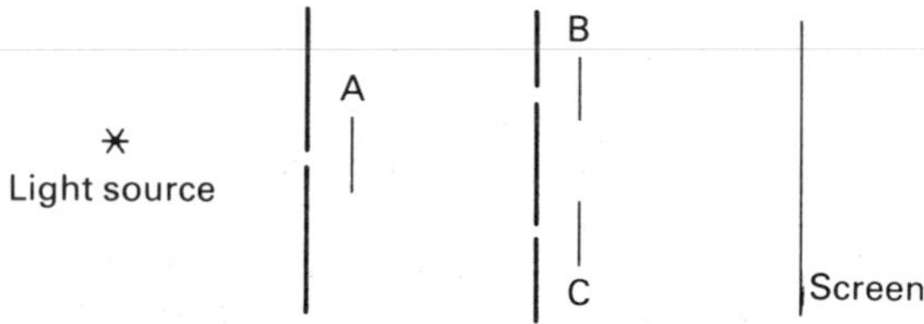

The diagram shows the schematic arrangement of an apparatus to observe Young's fringes which has been modified by the addition of three sheets of polarizing film, A, B and C, which act as linear polarizers. Describe and explain, with the help of carefully drawn sketches, the appearance of the fringes which are observed when

(i) B and C are set with their planes of polarization parallel and A is rotated.
(ii) A and B are set with their planes of polarization parallel and C is rotated,
(iii) A and B are set with their planes of polarization perpendicular and C is rotated.

(Ox. Schol.)

Chapter 8

A

Q 8.11
A distant star appears to be a point source of a wide range of electromagnetic radiation. It is observed through an optical telescope and a radio telescope. The true diameter of the star, which is at a distance of 10^{17} m, is thought to be 10^9 m, but neither telescope could confirm this.

Data:

Telescope	Aperture	Typical wavelength detected
optical	0.1 m	5×10^{-7} m
radio	20 m	1 m

Consider the statement and data given above and:

(*a*) *state* the physical principles which are relevant, and
(*b*) *show* how they can apply to the situation described. Also
(*c*) *make* calculations and/or give equations when possible to show the relationship between the quantities involved. (O & C, Nuffield)

Q 8.12
Calculate the position of the eye ring for an astronomical telescope consisting of two thin converging lenses, an objective of focal length 1.0 m and an eyepiece of focal length 20 mm, placed 1.02 m apart.

Explain the advantage of placing the eye at the eye ring position when using the telescope. (L)

Q 8.13
A thin converging lens of focal length 50 mm is to be used as a magnifying glass with the observer's eye close to the lens. If the observer can see images clearly anywhere between 250 mm from the lens and infinity, determine

(i) the range of possible object distances,
(ii) the corresponding range of magnifying powers.

(JMB)

Q 8.14
(*a*) Draw a ray diagram showing the passage of three rays from an off-axial point through a compound microscope (consisting of two simple lenses) in near-point adjustment.
(*b*) A compound microscope has an objective lens of focal length 7.5 mm separated from an eyepiece of focal length 20 mm by a distance of 250 mm. The final image is formed at a distance of 300 mm (the least distance of distinct vision) from the eyepiece. Calculate:
  (i) the distance of the object viewed from the plane of the objective lens;
  (ii) the overall magnifying power achieved.

(O)

Q 8.15
A convex camera lens is used to form an image of an object 1.00 m away from it on a film 0.050 m from the lens. What is the focal length of the lens?

If the camera is used to photograph a distant object, how far from the film would the clear image be formed? What type of lens should be placed close to the first lens in order to enable the distant object to be focused on the film if the separation of the first lens and film cannot be changed in this camera? What is the focal length of this added lens? (L)

Q 8.16
A camera has a lens, of focal length 120 mm, which can be moved along its principal axis towards and away from the film. If the camera is to be able to form perfect images of objects from infinite distance down to 1.00 m from the camera, through what distance must it be possible to move the lens? (L)

B

Q 8.17
(*a*) A parallel beam of monochromatic light incident normally on a very small circular aperture fills the aperture completely. The beam then falls on a plane surface placed approximately one metre behind and parallel to the plane of the aperture. Sketch a graph which shows the variation, due to diffraction effects, of the intensity of the pattern of light falling on the plane surface as a function of the distance from the centre. Hence explain the Rayleigh criterion for the resolution of two point sources when viewed through the aperture.
(*b*) (i) Describe the important features of the receiving dish of a radio telescope. Explain why the reflecting surface of a radio telescope does not have to be of the same high quality as that of an optical telescope.
(ii) The reflecting mirror of the Mount Palomar optical telescope has a diameter of 5 m. If a single radio telescope, working at a frequency of 300 MHz, were to have the same resolving power as the Mount Palomar telescope calculate the required order of magnitude for the diameter of its reflecting dish. Hence comment on the relative resolving powers of optical and radio telescopes. (JMB)

Q 8.18
This question is about diffraction.
The passage below presents three sets of ideas about diffraction. For each of the sections (i) to (iii) you are asked to write a more complete explanation of the ideas. Your explanation may include:
—fuller explanations of the theory
—quantitative calculations to illustrate the ideas
—discussion of possible experiments.
Take the wavelength of visible light to be about 5×10^{-7} m.

Passage
(i) Light from a point source appears to cast sharp shadows and this leads to the familiar idea that it travels in straight lines. However, this is not exactly true: the shadows are not perfectly sharp, although special experiments are needed to show the effect because it is so small. This unfamiliar property is called diffraction and is explained by a wave model of light.
(ii) The consequence is that the eye, or a camera, or even the best possible telescope, doesn't produce a perfect image. Instead it gives an image which is slightly blurred. When we try to make a telescope magnify more to show finer details of the stars this blurring effect can become an obstacle.
(iii) But diffraction can also be put to good effect. Diffraction gratings are made to enhance the effect and make use of it to give a powerful method of investigating spectra.
(O & C, Nuffield)

Q 8.19
As part of a communication system a source of radio waves is placed at the principal focal point of a dish of aperture D. The majority of the energy is emitted within an angle φ into a cone-shaped beam, as shown in the diagram.

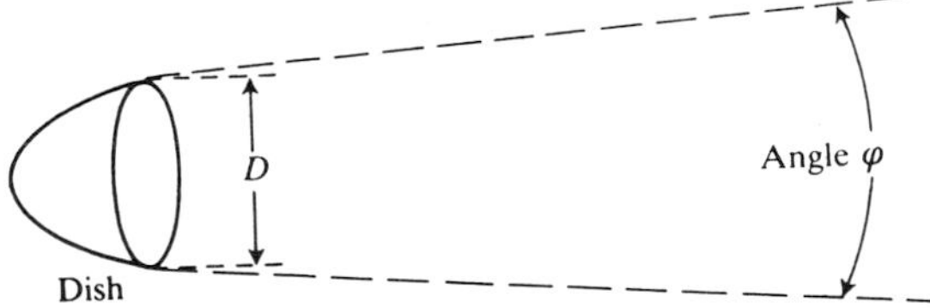

(*a*) Say why a parallel-sided beam of radiation cannot be produced whatever the shape of the dish.
(*b*) Angle φ, measured in radians, is about $2 \times 1.2\lambda/D$, where λ is the mean wavelength of the radio waves.
What is the effect on φ of doubling the carrier frequency of the signal?
(*c*) Such a source is placed on a satellite in orbit 4.0×10^4 km above the surface of the Earth. The dish is of diameter 10 m and the source

emits at a frequency of 1.5 GHz. The beam is aimed vertically downwards.

Show that the energy emitted within the angle φ is being directed towards an area of about 3×10^{12} m^2 on the surface of the Earth.

(*d*) However, due to the effect of the atmosphere, only 25% of the transmitted energy reaches this area. Calculate the minimum area required for a receiver on the Earth's surface requiring an input signal of at least 10^{-10} W, if the output power of the source is 200 W. State any assumptions you make. (O & C, Nuffield)

Q 8.20

A convex lens is used to cast an image of the full moon onto a screen.

(*a*) Assuming the moon to be in the centre of the field of view of the lens, draw a ray diagram showing clearly how the image is produced. Explain your method of construction.

(*b*) The moon is 3.5×10^3 km in diameter and is 3.8×10^5 km away from the Earth.
   (i) Calculate the angle in radian measure that it subtends at the eye of an observer on Earth.
   (ii) If the lens has a focal length of 0.30 m, what is the diameter of the image of the moon produced on the screen?
   (iii) If the observer views the image on the screen from a distance of 0.25 m, what angle will it subtend at his eye?
   (iv) Calculate the angular magnification that has been achieved by using this lens.

(*c*) Explain carefully the effect on the angular magnification of using a lens of much longer focal length.

A much greater angular magnification can be achieved by using two converging lenses, one as an objective and the other as an eyepiece. If a converging lens of focal length 0.050 m were placed 0.050 m beyond the screen (now removed) and the observer viewed the moon through both the eyepiece and the objective lenses, what would be the new angular magnification?

Apart from greater magnification, name *one* other difference between the image produced in this case and when only one lens was used. (L)

Q 8.21

An astronomical telescope, consisting of two thin lenses, is directed towards two stars which are close together, and near the axis of the instrument. Draw a ray diagram to show how the images of the stars are formed when the telescope is adjusted so that the final images are at infinity.

Define the *magnifying power* of a telescope, and show how this is related to the focal lengths of the lenses.

The lenses of this telescope have focal lengths 15 m and 75 mm respectively. If the images of the two stars seen through the telescope subtend an angle of 3° at the eye, what is the angular separation of the stars? If the eye lens is removed, and a photographic plate is placed in the focal plane of the objective, what is the separation of the images of the two stars on the plate?

What are the advantages and disadvantages of increasing the diameter of the objective? (S)

Q 8.22

Define *angular magnification* for a telescope. Explain what is meant by *normal adjustment* for a telescope.

Draw a ray diagram to show how the image of a distant object is formed by a simple two-lens astronomical refracting telescope in normal adjustment. The diagram must include rays from an off-axis point.

Derive an expression for the angular magnification of this telescope in terms of the focal lengths of its lenses.

The telescope has an overall length of 2.0 m and an angular magnification of 49. The eyepiece has a diameter of 3.0 mm. Calculate

(*a*) the focal lengths of the objective and eyepiece lenses,

(*b*) a maximum diameter for the objective so that all the light entering it parallel to the axis may pass through the eyepiece.

Explain why astronomical telescopes have objective lenses of large diameter. (O)

Q 8.23

(*a*) In terms of the wave theory of light, explain qualitatively (without using any formulae) how a thin converging lens brings a beam of light from a distant source to a focus, and why the focal length depends on the refractive index of the glass and the radius of curvature of each face.

(*b*) A thin converging lens (focal length 50 cm) and a thin diverging lens (focal length 10 cm) are mounted so that their axes coincide and their separation can be varied, and light from a distant object falls on the converging lens. Explain how the system serves as a Galilean telescope when the separation of the lenses is about 40 cm. Define *magnifying power* for the telescope and obtain an expression for its value with the telescope in normal adjustment for parallel incident light.

(*c*) Give a short account of the advantages and disadvantages of the Galilean telescope as a practical instrument. (O)

Q 8.24

Ray diagrams are required for all parts of this question.

(*a*) Describe a reflecting telescope of the Cassegrain type. Without considering the effect of the eyepiece, show, by drawing two parallel rays from a non-axial point on a distant object, how the telescope objective produces an inverted image. State the nature of the image.

Giving reasons for your answer, state *two* advantages, apart from those involving aberrations, which a reflecting telescope has over a refracting telescope of similar magnifying power.

(*b*) (i) Show how a concave spherical mirror produces spherical aberration in the image of a distant point object. Explain how spherical aberration may be eliminated in a reflecting telescope.

(ii) Explain what is meant by chromatic aberration. State, with reasons, whether chromatic aberration occurs in the objective of a reflecting telescope.

(JMB)

Q 8.25

(*a*) A compound microscope consisting of two thin converging lenses is set up by an observer to view a small object.

(i) Show, by means of a labelled ray diagram, how the instrument forms an image at *infinity*. The diagram should show the paths through the microscope of two rays from a non-axial point on the object.

(ii) Derive an expression for the *magnifying power* of the instrument as arranged in (i), in terms of the focal lengths of the two lenses, the distance of the intermediate image from the objective lens, and the least distance of distinct vision of the observer.

(iii) An object placed 4.0 cm from the objective lens, of focal length 3.0 cm, is viewed by an observer whose least distance of distinct vision is 24.0 cm, and the microscope is adjusted to give the final image at infinity. The object is then viewed with the same instrument by a second observer whose least distance of distinct vision is 20.0 cm. The object is moved slightly closer to the objective lens and the distance between the lenses adjusted until the final image is again at infinity. If the magnifying power is the same as that for the first observer, calculate by how much the distance between the lenses has been increased.

(*b*) A compound microscope normally has the final image at the near point of the observer. Explain, without calculation or a ray diagram, how the instrument described in (i) above can be adjusted to achieve this. (JMB)

Q 8.26

(*a*) A parallel beam of light from an illuminated slit consists of one red wavelength and one blue wavelength. It is incident normally on a diffraction grating placed on a spectrometer turntable.

(i) Explain why the spectrometer telescope must be adjusted to receive parallel light in order to view a pure spectrum of the source.

(ii) If the telescope is set initially in the straight-through position and then turned in one direction only, give the colour sequence of the first and second spectral 'lines' seen in the first order, and explain why this is so.

(iii) At a diffracting angle of 54.8° a blue 'line' and a red 'line' coincide. If the grating has 600 rulings per mm and the blue wavelength is 454 nm, deduce the order numbers for both colours, and also the red wavelength.

(*b*) The same diffraction grating is used to view a monochromatic light source placed 50.0 cm away from the grating. The observer holds the grating close to his eye, and observes an image on either side of the source. Explain, with the aid of a simple diagram, why this occurs and calculate the wavelength of the light if the images are 37.8 cm apart. (JMB)

Q 8.27

(*a*) Draw a diagram showing the path of three parallel light rays from an off-axial direction through a lens telescope in normal adjustment. Show on your diagram the ideal position for the observer's eye.

(*b*) The objective lens of a small telescope is a plano-convex lens of focal length 100 mm and refractive index 1.60.

(i) Calculate the radius of curvature of the curved face of this lens.

(ii) With the aid of suitable diagrams, explain the optical advantage of using a plano-convex lens.

(*c*) Such a telescope with lenses of focal lengths 100 mm and 20 mm forms parts of a diffraction grating spectrometer in which the grating, used at normal incidence, has 500 lines per millimetre. The collimator slit is illuminated with a vapour lamp producing two close spectral lines of wavelengths 510 nm and 517 nm.

Calculate:

(i) the number of pairs of spectral lines observed in the transmitted pattern on one side of the normal to the grating;

(ii) the angular separation at the observer's eye of the images of the lines in the second-order spectrum seen through the telescope.

(*d*) Describe the changes in the appearance of the images as a consequence of each of the following separate actions:

(i) observing an order higher than the second;

(ii) widening the collimator slit;

(iii) reducing the horizontal width of the illuminated portion of the grating. (O)

Q 8.28

(*a*) (i) An object is photographed several times with the same camera under conditions of constant illumination. If the f-number of the lens aperture is reduced for consecutive photographs explain why the exposure time must also be correspondingly reduced in order to allow the same amount of light to fall onto the film each time.

(ii) If, in (i), the exposure time is 1/60 s at $f/11$, estimate what exposure time is required at $f/2.8$.

(*b*) Explain what is meant, in photography, by (i) *depth of field*, (ii) *depth of focus.* By considering point objects on the axis of a converging lens, explain with the aid of two *separate* ray diagrams how each of these arises.

With the aid of another diagram and an axial point object as above, show qualitatively how *either* the depth of field *or* the depth of focus changes as the lens aperture is changed. (JMB)

Q 8.29

(*a*) A simple telephoto lens system consisting of a converging lens and a diverging lens is used in a camera to form on the film an image of a distant small object. By means of a diagram show how the telephoto lens produces an image. Hence explain why the image so produced is larger than the image which would have been produced if only a single converging lens, placed at the position of the diverging lens, had been used. You may assume that the image size of a distant object is directly proportional to the focal length of the lens system producing the image.

(*b*) The converging lens of the telephoto system described in (*a*) has a focal length of 72 mm and the diverging lens a focal length of 20 mm. The separation of the two lenses is 60 mm.

(i) Calculate the distance of the film from the diverging lens, when the image of the distant object is in focus on the film.

(ii) By considering similar triangles in your diagram in (*a*), calculate the effective focal length of the combination of lenses.

The telephoto combination is replaced by a single converging lens placed at the position of the diverging lens so that the image is still in focus. Calculate the ratio

$$\frac{\text{size of image with telephoto lens}}{\text{size of image with single lens}}$$

(*c*) Describe the optical system of prism binoculars and state *two* advantages they might have over an astronomical telescope when used to view a distant object on the surface of the earth. (JMB)

C

Q 8.30

The nearest star is 4 light years from the Earth, and has a diameter of about 1.5×10^9 m.

(*a*) Show that an optical telescope of minimum aperture about 16 m would be needed if the star were to be resolved as a disc.

(*b*) Calculate the corresponding aperture for a radio telescope operating on a frequency of 300 MHz.

(*c*) Comment on the possibilities of constructing each of these telescopes. (O & C, Nuffield)

Q 8.31

The *exit pupil* of an astronomical telescope is the image of the objective formed by the eyepiece. Explain why this is the best position in which to place the eye.

Prove for an astronomical telescope in normal adjustment that

$$\text{Magnifying power} = \frac{\text{Diameter of objective}}{\text{Diameter of exit pupil}}$$

If the focal length of the objective is 20.0 cm and that of the eyepiece 2.5 cm, and the diameter of the pupil of the observer's eye 0.6 cm, what is the maximum diameter of objective aperture that can be usefully employed visually?

Explain in what circumstances it would be desirable to increase the diameter of the objective above this value. (JMB)

Q 8.32

(*a*) Explain with the aid of diagrams the nature of *longitudinal spherical aberration* and *longitudinal chromatic aberration* of images formed by lenses.

(*b*) Sketch the layout of a telescope with a mirror objective, and show the paths of three rays from a distant off-axial point source passing through the instrument when in normal adjustment.

Describe a method of measuring the magnifying power of a lens or a mirror telescope that cannot be dismantled into its component parts.

(*c*) Accurate spectroscopic observations of the light originating from the Sun and reflected by the plant Jupiter show that when Sun, Earth and Jupiter are in alignment the solar helium spectral line of wavelength 587.56 nm is reflected from one equatorial edge of Jupiter with a wavelength of 578.61 nm and from the opposite edge with a wavelength of 587.51 nm. Given that the diameter of Jupiter is 143×10^3 km and that its axis of rotation is perpendicular to the line joining Jupiter and the Sun, estimate the rotational period of Jupiter about its axis.

(*d*) Details of planets become more apparent when viewed through large aperture telescopes than when viewed through small aperture telescopes of equal magnifying power. Explain why. (O)

Q 8.33

Distinguish between *magnification* and *magnifying* power for a microscope.

Draw a diagram to show the rays through a compound microscope from an object point not on the axis when the image is coplanar with the object.

It is desired to design such a microscope for fine construction work with the image in the same plane as the object at the distance of 25 cm from the eyepiece. If the microscope is to have a magnifying power of 12 and if the objective is to be 2.0 cm from the object, what focal length lenses should be used? (JMB)

Q 8.34

(*a*) A telescope is focused on part of the moon and set in normal adjustment. Explain the separate actions of the objective lens and eyepiece in forming the final image.

What is meant by the angular magnification of the telescope?

(*b*) The limit of resolution of the human eye is about one minute of arc. Explain this statement.

(*c*) Two strong lines in the mercury spectrum have wavelengths 579.0 nm and 577.0 nm respectively. A prism has refractive index 1.5794 for the longer wavelength and its refractive index increases almost linearly by 0.0016 for a decrease in wavelength of 20 nm. The refracting angle of the prism is 60.00°.

The diagram shows a narrow vertical slit S of length 1 cm illuminated by light from a mercury lamp a few metres from a spectrometer table. The filter F absorbs all light except the two wavelengths noted above. The prism is set in the minimum deviation position for light of wavelength 579.0 nm and then the telescope is moved out of the way. Two images of the slit are seen, with the unaided eye, close together at S′.

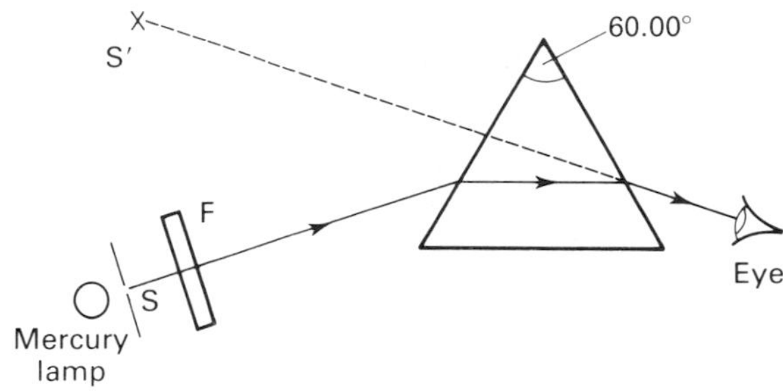

(i) Calculate the angle of deviation, D_{min}, for light of wavelength 579.0 nm.

(ii) Estimate the angular separation between the two images of the slit and state any assumption that you make.

Describe the visual appearance of the images at S′ seen by the unaided eye.

(*d*) The spectrometer telescope is swung round and focused on the images at S′. Its angular magnification in these circumstances is 15.0. Describe the visual appearance of the images at S′ seen through the telescope and account for any changes in

(i) brightness,
(ii) clarity,
(iii) separation.

The telescope is centred on one image and then moved sideways to centre on the second. Through what angle is the telescope moved?

(*e*) Calculate the angular separation of the two first order images of the slit if light from the slit, having passed through the filter F, arrives normally at a diffraction grating with 500 000 lines per metre.

Give *two* reasons for occasionally preferring a prism to a diffraction grating when examining the spectra of stars. (L)

Q 8.35

(*a*) Show that, for a plane diffraction grating of grating interval s, used at normal incidence, the angle of diffraction, θ, is related to the wavelength, λ, of the light by the equation

$$s \sin \theta = n\lambda$$

where n is an integer. What is the significance of n?

(*b*) Assuming that you were provided with a spectrometer which had been adjusted correctly and on which a plane diffraction grating was mounted vertically, describe and explain the procedure you would follow to ensure that light falls normally on the grating.

(*c*) A grating spectrometer is arranged so that light from the collimator falls normally on the grating. When the second order diffracted image of the slit due to a wavelength λ is viewed on the two sides of the normal to the grating, the readings on the telescope scale are respectively α and β. The slit is first illuminated by sodium light of wavelength 589.3 nm and the values of α and β respectively are found to be 162° 37′ and 252° 39′. The sodium source is then replaced by a discharge tube containing a mixture of gases and the values of α and β (still second order) are recorded for three lines as follows:

Line	α	β
1	171° 57′	243° 19′
2	160° 48′	254° 28′
3	155° 41′	259° 35′

Identify the gases in the tube using the data in the table below which shows the wavelengths, in nm, of the lines emitted by various gases.

Carbon dioxide	*Hydrogen*	*Helium*	*Oxygen*
451	410	319	408
483	434	389	441
520	486	447	521
561	656	588	617
608		668	

Show that, whatever the spacing of the grating used, the third order diffracted image of the shortest wavelength line in the spectrum of one of the gases identified will occur at a smaller diffracted angle than the second order image of the longest wavelength line in that gas's spectrum. (L)

Q 8.36

(*a*) It is claimed that a newly developed type of glass will be suitable for all of the following uses: (i) window glass, (ii) ovenware, (iii) electric lamps, (iv) cut glassware, and (v) achromatic lens construction.

Discuss the properties desirable for a glass to be used for each of these purposes and indicate, with reasons, whether or not it is likely that any single glass type can have all of these properties.

(*b*) A simple fixed lens box camera has an appreciable depth of focus, i.e. clear pictures of objects at a range of distances will be produced. Explain why this is so. Illustrate your explanation by considering the following example:

Such a camera has a converging lens of focal length 4.0 cm with a shutter in front of it of aperture 0.25 cm. The film is placed in the focal plane of the lens and clear images on the film are formed of objects placed at all distances greater than 1.0 m from the camera. (L)

Q 8.37

Describe briefly a pinhole camera. How would the image of a point source of light change as the pinhole is made (*a*) larger; and (*b*) smaller and smaller? What determines the optimum size of the pinhole? What are the advantages and disadvantages of replacing the pinhole by a lens?

A simple camera consists of a lens of focal length 0.1 m and of diameter 0.02 m, 0.1 m in front of a photographic film. Calculate by how much the lens-to-film distance should be adjusted if a photographer wishes to use the camera to obtain a focused picture of a subject 1 m in front of the lens.

Unfortunately the photographer finds it impossible to make this adjustment and instead has to rely on reducing the diameter of the area of the lens to be used, in order to sharpen the image and obtain a reasonable picture. As a criterion, he decides that the quality of the image should be such that he can resolve a distance of 5 mm in the object. Estimate the maximum diameter of the circular lens area that he can use.

If the required exposure would have been $\frac{1}{100}$ s with the full diameter lens, what will it be with the reduced diameter? (Ox. School)

Chapter 9

A

Q 9.6
'*Observed differences in the speed of light in different media support the wave theory of light.*' Discuss this statement qualitatively by considering the deviation of a beam of light on passing from one medium to another. (L)

Q 9.7
A girl swims across a river from A to B a distance of 100 m. The water flows in all parts of the river at a speed of 0.5 m s^{-1} in the direction shown. The girl can swim at an average speed of 1.0 m s^{-1} in still water.

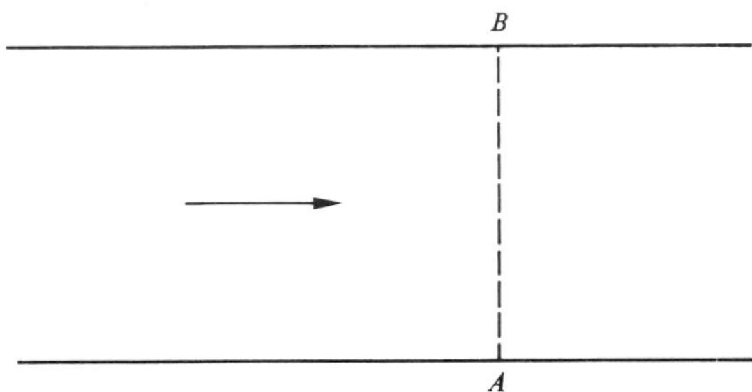

(*a*) Draw a vector diagram to show the direction in which she should aim herself.
(*b*) Determine the time it takes her to cross the river if she swims at her usual speed. (S)

Q 9.8
The diagram shows a river in which the water runs from west to east at 2.5 m s^{-1}. Two men, A and B, leave a point O on the southern bank at the same instant. A cycles due east along the bank at 8.0 m s^{-1}. B, who is in a motor boat, has a resultant motion due north across the river at 6.0 m s^{-1}.

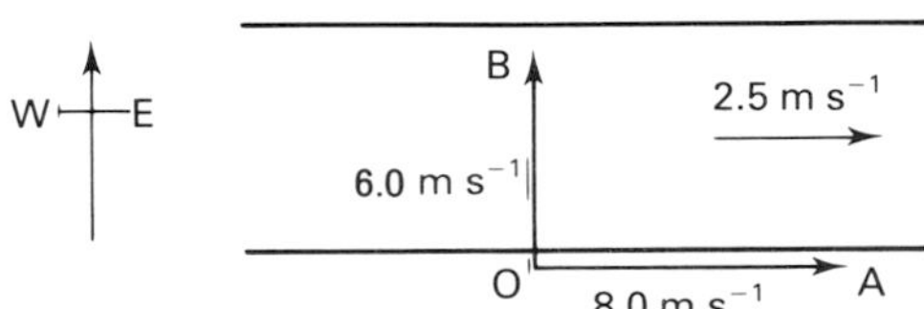

Use clearly labelled vector diagrams to find the speed and direction of B

(*a*) relative to A,
(*b*) relative to the water. (L)

B

Q 9.9
Discuss briefly the arguments by which the speed of light in glass may be expressed in terms of its speed in air and the refractive index of the glass (*a*) from the point of view of the wave theory of light, (*b*) from thc point of view of Newton's corpuscular theory of light.

Describe an experimental method of determining the speed of light in air.

The diagram represents a plane wave-front AB striking a plane glass surface in air. The refractive index of the glass is 1.5, and the speed of light in air is 3×10^8 m s^{-1}. The distance BC is 3 cm. Taking the time from the instant shown in the diagram, and considering only refraction,

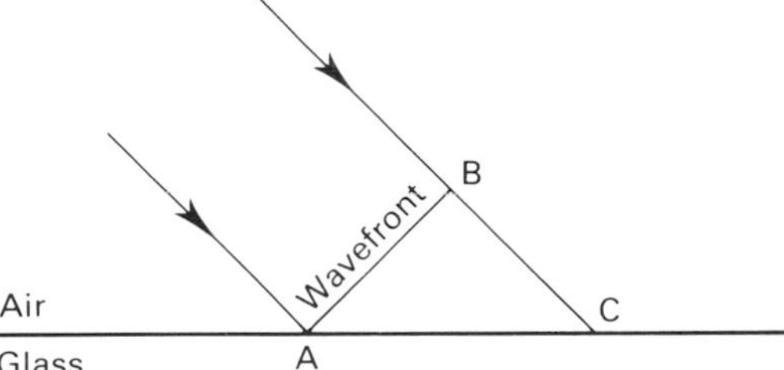

(*a*) construct accurately the wavefront at time 10^{-10} s;
(*b*) draw the position of the wavefront at time 2×10^{-10} s;
(*c*) draw the position of the wavefront at time 5×10^{-11} s. (O)

Q 9.10
(*a*) Explain in terms of a wave model how a beam of light is refracted as it crosses an interface between two transparent media. Hence derive Snell's law of refraction in terms of the speeds of light in the media.
(*b*) Describe the phenomenon of total internal reflection and explain what is meant by the critical angle. How is the critical angle related to the speeds of light in the media involved?
(*c*) A portion of a straight glass rod of diameter d and refractive index n is bent into an arc of a circle of mean radius R, and a parallel beam of light is shone down it, as shown overleaf.
  (i) Derive an expression in terms of R and d for the angle of incidence i of the central

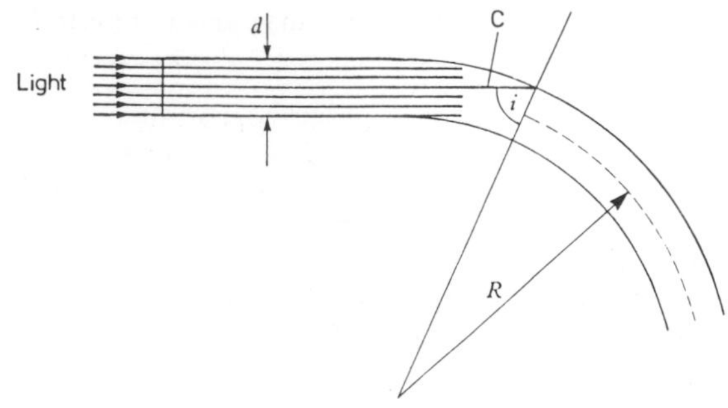

ray C on reaching the glass-to-air surface at the circular arc.

(ii) Show that the smallest value of R which will allow *all* the light to pass around the arc is given by

$$R = \frac{d(n+1)}{2(n-1)}$$

(iii) Use this result to explain why glass fibres, rather than rods, are used to carry optical signals around sharp corners.

(*d*) A glass fibre of refractive index 1.5 and diameter 0.50 mm is bent into a semi-circular arc of mean radius 4.0 mm, and a beam of light is shone along it.

(i) Show that no light escapes from the sides of the fibre.

(ii) Show by a suitable calculation that if the fibre is immersed in oil of refractive index 1.4 some light will escape.

(iii) Suggest an application for such a device. (O)

Q 9.11

Describe a terrestrial method for measuring the speed of light in air.

If it were suggested that there might be a time delay between light falling on the distant mirror and being re-emitted (reflected), how would you demonstrate that this is not the case?

Show how the refraction of light at a plane surface is explained by the wave theory of light, and derive a relation between refractive index and speed of light. (S)

Q 9.12

(*a*) Describe a terrestrial experiment to measure the speed of light in air. Your description should include an explanation of the principle of the experiment, details of the measurements required and an indication of how the speed of light is calculated.

(*b*) Suggest how the speed of light in water could be measured, either by modification of your previous method or otherwise.

In such an experiment it is found that the speed in water is 2.25×10^8 m s^{-1}, whereas in air it is 3.00×10^8 m s^{-1}. Calculate the refractive index at a water–air interface.

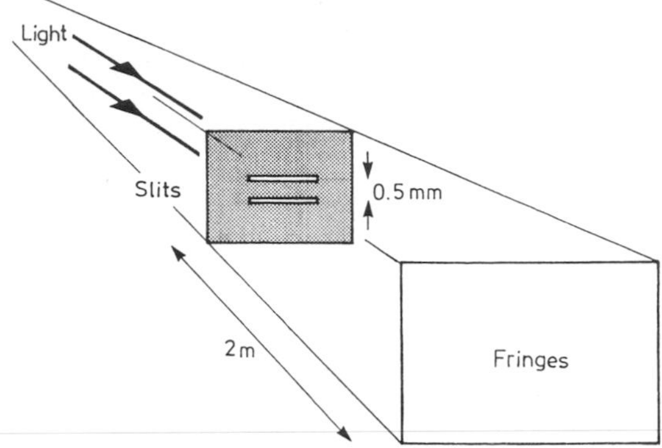

(*c*) A parallel beam of monochromatic light (wavelength 600 nm) falls normally on an opaque sheet having two narrow parallel horizontal slits 0.5 mm apart, as shown in the diagram. Describe the pattern observed on a screen placed normal to the beam a distance 2 m from the slits. Calculate the spacing of the fringes.

The upper slit is covered by a parallel sided film of water, initially 1 mm thick. It evaporates steadily so that its thickness decreases at a uniform rate of 0.5 mm per hour. In what direction do the fringes move on the screen, and at what speed? (O & C)

Q 9.13

(*a*) The basic components of the Michelson interferometer are shown in the diagram. S is

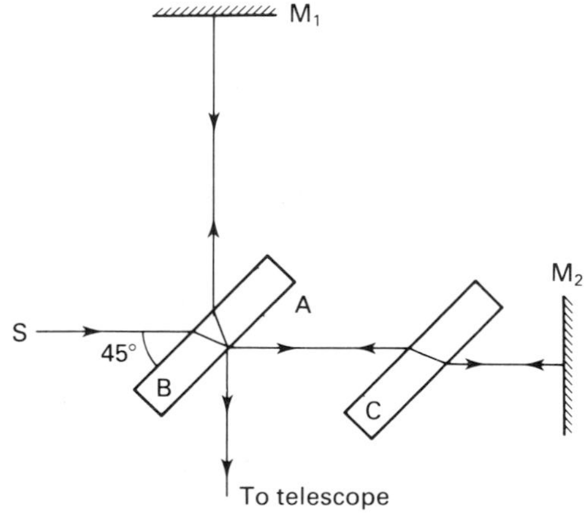

an extended source of monochromatic light; B and C are identical glass plates with surface A of plate B being partially silvered; M_1 and M_2 are plane surface silvered mirrors. The virtual image of M_2 formed by reflection at A is very close to M_1 and almost parallel to the surface of M_1.

(i) State what an observer would see in the field of view of the telescope.
(ii) Explain why plate B is at an angle of 45° to the initial direction of the light.
(iii) Explain why the plates B and C are identical in size, made of the same material and placed parallel to each other.
(iv) Describe how the interferometer can be used to measure very small distances in terms of the wavelength of the light used, and estimate the order of magnitude of the distances which can be measured.

(*b*) Michelson used the interferometer in an attempt to detect the motion of the earth through a hypothetical ether. Indicate what measurements were made and what conclusions were deduced from the experiment. Comment on the importance of the result. (JMB)

C

Q 9.14
Two alternative methods of establishing the path taken by a ray of light between two points are:

(*a*) the wave theory, in which each point on a wave front is considered as a source of secondary wavelets;
(*b*) the principle of least time, which states that the path of a light ray between two fixed points A and B across a fixed boundary is such that the time taken is a minimum.

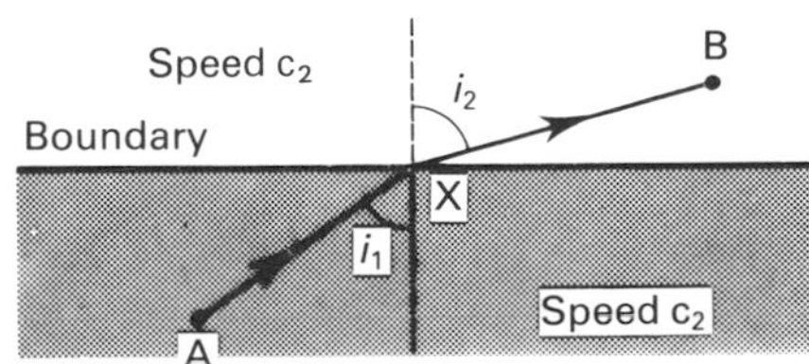

Show that both of these approaches lead to the relationship

$$\frac{\sin i_1}{\sin i_2} = \frac{c_1}{c_2}$$

for a light ray passing from one medium into another, where i_1 is the angle of incidence at the boundary between the media and i_2 is the angle of refraction, c_1 and c_2 being the speeds of light in the two media respectively.

Show also that the least time principle is consistent with the path taken being that which contains the least number of waves. (L)

Q 9.15
This question is concerned with optical effects which occur in the atmosphere.

(i) The refractive index of air increases as the air density increases. The density of the atmosphere varies continuously with height but, as a simple model, we can assume that the atmosphere is stratified and consists of a series of layers, each layer having a constant density corresponding to the actual density at its mid-height.

Ignoring the curvature of the earth, consider diagrammatically what happens to rays of light leaving a point above the earth's surface when the density of the air (*a*) increases and (*b*) decreases with height.

Explain fully the production of mirages.

(ii) Parallel monochromatic light is incident on a spherical water drop, of refractive index 1.32 for the incident light. The diagram shows a section containing a diameter and the path of a ray, internally reflected, which is typical of the type we want to study. You should confine your calculations to rays in the plane of the paper.

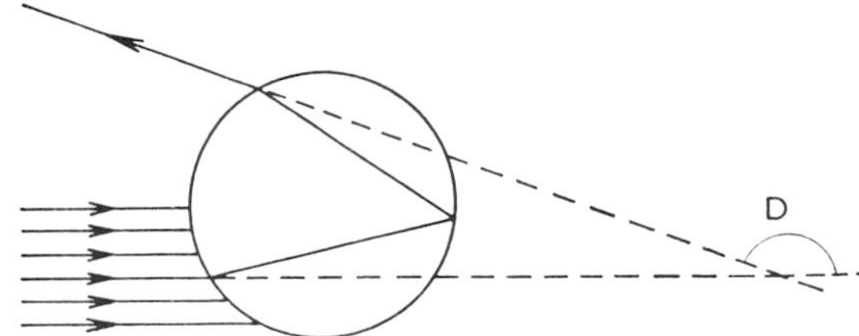

Obtain an expression for the deviation D (see diagram) in terms of the angle of incidence. Determine the minimum angle of deviation and sketch the graph of D against the angle of incidence. Consider whether the internal reflection which takes place at minimum deviation is due to total internal reflection.

Explain fully and clearly the origin and shape of rainbows. (W)

Q 9.16
Explain Huygens' Principle and use it to derive a relationship between the velocity of light in a

medium and the refractive index of the medium (defined by Snell's Law).

An electro-optical device is used to modulate the intensity of a continuous light beam. The transmission of the device is a square wave function of time with equal alternate periods of full and zero transmission of the incident light. The light, after reflection at a distant mirror, passes back through the device and its intensity is then measured. An extinction is obtained at a modulating frequency of 6 MHz, and if the frequency is increased the next one occurs at 10 MHz. With the frequency set at 6 MHz, a glass sheet ($n = 1.5$) is inserted between the device and the mirror, and it is found that the extinction can be recovered by decreasing the frequency by 1 kHz. Calculate the thickness of the glass sheet. (Cam. Schol.)

Q 9.17
The following is adapted from a letter written to a scientific journal by a physicist (R. C. Warren). '*It has been proposed that the fundamental unit of length be redefined in such a way that the velocity of light becomes a numerical constant 299 792 458 m s^{-1}. But why choose such a cumbersome number? Why not make it a round 3×10^8 or better still an even rounder 10°?*

'If the velocity of light is redefined as 10^9 new units of length per second, the new unit of length becomes 0.299 792 458 metre. What should the new unit of length be called? I propose the name "foot".'

Either, write a report for a scientific journal, supporting the use of this new unit,
or, write a report for the same journal, suggesting why you consider the change to be undesirable.

In either case give suitable numerical illustrations with your report to demonstrate the effect of the change on the layman as well as the scientist. Note that Dr. Warren considers the velocity of light should be 10^9 units s^{-1} by adjusting the unit of length. However, you are free to consider, should you wish, a new unit of time (possibly called the blink), instead. (O & C, Nuffield)

Q 9.18
The following experiment is carried out on the flat deck of an aircraft-carrier. Two vertical walls which reflect sound according to the laws of reflection, are set up at right angles to each other and an observer is situated at a point equi-distant from them. The observer emits a *very short pulse* of sound and he finds that the two reflected pulses reach him at exactly the same moment when there is no wind and the ship is stationary.

Calculate the difference in arrival times of the pulses if (a) the ship is stationary and a wind is blowing at constant speed v in a direction at right angles to one wall and (b) there is no wind and the ship travels with constant speed s in a direction at right angles to one wall. You may assume that v and s are both very much less than the velocity of sound in air.

Describe the principles and results of a similar experiment carried out with light. (The Michelson–Morley experiment.)

Write a concise account of the special theory of relativity. (W)

Q 9.19
Explain how it could be shown experimentally that the yellow light emitted by a sodium lamp has the following properties:

(a) it is some form of wave.
(b) the wave is transverse.
(c) there are two components of wavelength 589.0 nm and 589.6 nm.
(d) it propagates at approximately 3×10^8 m s^{-1} in air.
(e) it propagates more slowly in an optically dense medium such as glass or water. (Ox. Schol)

Chapter 10

A

Q 10.13
Is it possible to polarize (a) sound waves, (b) radio waves? Explain briefly why it is or is not possible in each case. (C)

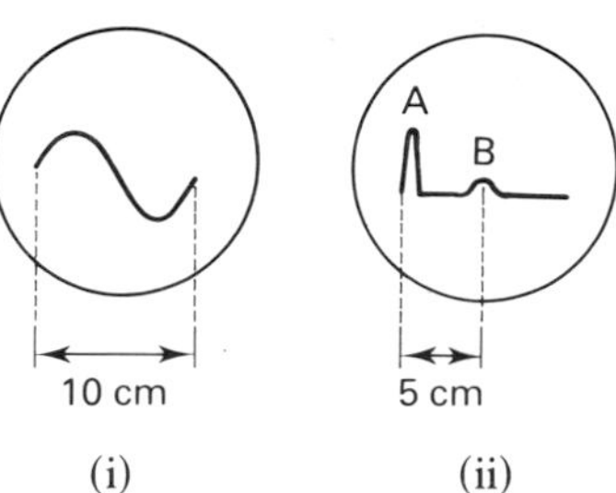

Q 10.14
When a sine-form voltage of frequency 1250 Hz is applied to the Y-plates of a cathode ray oscilloscope the trace on the tube is as shown in diagram (i).

If a radar transmitter sends out short pulses, and at the same time gives a voltage to the Y-plates of the oscilloscope, with the time-base setting unchanged, the deflection A is produced as shown in diagram (ii). An object reflects the radar pulse which, when received at the transmitter and amplified, gives the deflection B. What is the distance of the object from the transmitter? (L)

Q 10.15
Two radio transmitters emit vertically polarized electromagnetic waves of frequency 9×10^7 Hz. The speed of the waves is 3×10^8 m s^{-1}.

(*a*) Calculate the internodal distance in the standing wave set up along the line joining the transmitters.

A mobile receiver moves along the straight line joining the transmitters at a speed of 6×10^2 m s^{-1}.

(*b*) Calculate the rate at which nodes in this standing wave are passed by the moving receiver. (S)

Q 10.16
The work functions (in electronvolts) of the alkali metals are

lithium	sodium	potassium	rubidium	caesium
2.28	2.29	2.26	2.10	1.90

Which of these elements could be used in a photoelectric cell that is required to respond to light of wavelength 589 nm? (C)

Q 10.17
When light of frequency 8.22×10^{14} Hz is incident on the surface of caesium metal, it is found that the maximum kinetic energy of the emitted electrons is 2.00×10^{-19} J. Calculate the threshold frequency for photoelectric emission from caesium.

In which region of the electromagnetic spectrum does the incident frequency lie? (C)

Q 10.18
The diagram shows a clean metal disc A, which is irradiated with ultra-violet light, and a photoelectron current-collecting ring, B. C and D are terminals to which a d.c. source of e.m.f. 4 V is to be connected. E measures photoelectric current.

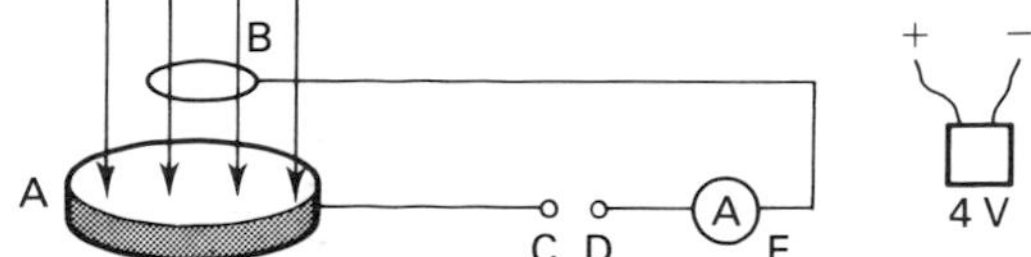

(*a*) Assuming that the stopping potential is less than 4 V, explain whether terminal C should be connected to the positive terminal of the d.c. source or to its negative terminal if photoelectric current is to flow.
(*b*) Explain why, with the d.c. source correctly connected, photoelectric current flows only if the frequency of the incident radiation is above a threshold value. (L)

Q 10.19
A clean metal surface is illuminated by a beam of monochromatic ultra-violet light and the electrons are collected by an electrode. The potential difference, V, between the electrode and the metal may be varied and the current, I, is measured. The unbroken line labelled A in the graph shows how the current I varies with the potential difference, V, across the cell.

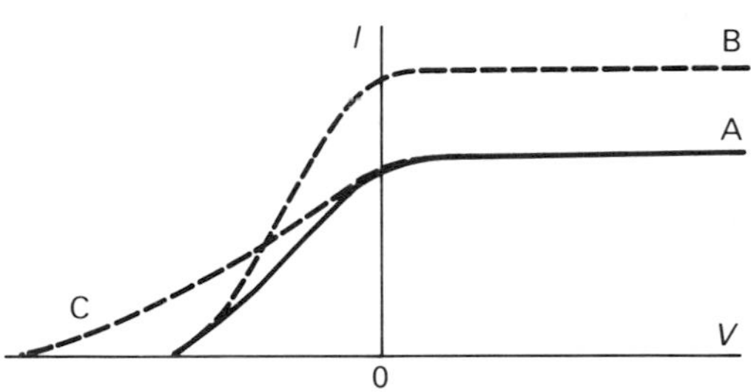

State and explain in each case, in what way the beam of ultra-violet light must be modified to achieve the photoelectric cell characteristics shown by

(*a*) the broken line B,
(*b*) the broken line C.

(L)

Q 10.20
Find the energy of a photon of electromagnetic radiation of wavelength 0.15 nm. In which region of the electromagnetic spectrum is this radiation? (C)

B

Q 10.21
With reference to the electromagnetic spectrum

(*a*) mention *two* properties common to all radiations,
(*b*) place the radiations as usually classified in order of increasing wavelength [ignore overlapping regions],

(*c*) give some idea of the magnitude of the range of wavelengths which comprise the spectrum,
(*d*) select *two* radiations, outside the visible spectrum, characterized by different physical behaviour and for each one indicate briefly the principles involved in its (i) production, (ii) detection, (iii) use in a given application. (L)

Q 10.22

Define *wavelength, wave velocity* and *frequency.* Deduce the connection between them.

List the various regions of the electromagnetic spectrum, giving the order of magnitude of wavelength for each region.

Describe how you would demonstrate an electromagnetic stationary wave, assuming you have a generator of 3 cm waves. The generator should not be described but you should mention the principle of the detector.

Assuming that electromagnetic waves are efficiently launched from an aerial when a stationary wave is set up within it, estimate the shortest possible length of aerial to radiate electromagnetic waves

(i) of 3 cm wavelength, in air;
(ii) of frequency 50 Hz, in air;
(iii) of frequency 2×10^4 Hz, from an aerial where the effective wave velocity is 2×10^8 m s^{-1}.

(C)

Q 10.23

Most people do not realize that the space around us is filled with electromagnetic vibrations of a very wide range of frequencies, just a little of which we notice as 'heat' and 'light'.

Write a short essay about electromagnetic radiation, including

—evidence to show the common properties of this wide range of frequencies
—why the radiation is called electromagnetic
—some important applications of electromagnetic radiation of different frequencies.

(O & C, Nuffield)

Q 10.24

The Einstein equation for the photoelectric emission of electrons from a metal surface can be written $\frac{1}{2}mv^2 = hf - hf_0$ where f_0 is the threshold frequency. Explain the physical process described by this equation, and the meaning of each term in the equation.

Describe an experiment to determine the value of the Planck constant *h*.

A monochromatic source emits a narrow, parallel beam of light of wavelength 546 nm, the power in the beam being 0.08 W. How many photons leave the source per second? If this beam falls on the cathode of a photocell, what is the photocell current, assuming that 15% of the photons incident on the cathode liberate electrons?

(S)

Q 10.25

(*a*) (i) State Einstein's equation for the photoelectric effect, and explain the significance of each of its terms.
(ii) Describe a method for the measurement of the Planck constant *h*, using the photoelectric effect.
(iii) Describe briefly a device that utilizes the photoelectric effect.

(*b*) The memory of a computer element can be erased by exposing it to ultraviolet radiation of wavelength 360 nm for a period of 20 minutes. The memory is contained in an insulated silicon film of exposed area 1.5×10^{-9} m^2, and the intensity of the ultraviolet light is 20 W m^{-2}.

Calculate:
(i) the number of photons incident on the film in 1 second;
(ii) the charge acquired by the film in a 20 minute exposure, if 1% of the incident photons cause photoemission of electrons.

Explain why erasure would be slower if the memory were exposed to sunlight of the same total intensity. (O)

Q 10.26

(*a*) What are the dimensions of (i) energy, (ii) momentum, (iii) the Planck constant, *h*?

It can be shown that the total energy *E* of a particle is related to its momentum *p* and its mass m_0 when at rest, by the equation

$$E^2 = p^2c^2 + m_0^2c^4$$

where *c* is the speed of electromagnetic radiation in vacuo. Show this equation to be dimensionally correct.

(*b*) Given that m_0 is zero for a photon, use the above equation to derive an expression for the momentum of a photon in terms of its wavelength.

(*c*) Estimate the potential difference across which an electron should be accelerated if it is to have the same energy as a photon of ultraviolet radiation. Support your estimate with a calculation, by choosing a suitable value for the wavelength of the electromagnetic radiation. You may assume that the value of the product *hc* is 2.0×10^{-25} J m. (JMB)

Q 10.27

In answering part (*a*) of this question you will need to estimate various quantities and then combine them in order to obtain the required answer.

Show clearly your estimates and all the steps in the calculation. Remember to include the units in estimates as well as in the answer, and to work throughout to an appropriate number of significant figures.

(*a*) A person is sitting at a table reading a book which is illuminated only by a *conventional filament lamp.*

Calculate roughly how many photons of light will fall, per square metre, per second, onto the book.

(*b*) Would your answer be *significantly* different if the filament lamp were replaced by a fluorescent tube lamp of the same rated electrical power? Give your reasoning. (O & C, Nuffield)

Q 10.28

Write *short answers* of a few lines only to *each* section of this question. Where appropriate give simple labelled diagrams. Equations may be quoted without proof.

Give *brief* accounts of experimental evidence in support of the following statements:

(*a*) light travels in straight lines;
(*b*) light does not travel in straight lines;
(*c*) light is a transverse wave;
(*d*) red light has a greater wavelength than blue light;
(*e*) light has a particle-like nature;
(*f*) the energy of a quantum of blue light is greater than the energy of a quantum of red light. (AEB, 1975)

C

Q 10.29

(*a*) In a fibre-optical communication system, data is transmitted through thin glass fibres in the form of nanosecond (10^{-9} s) light pulses, as shown in the diagram.

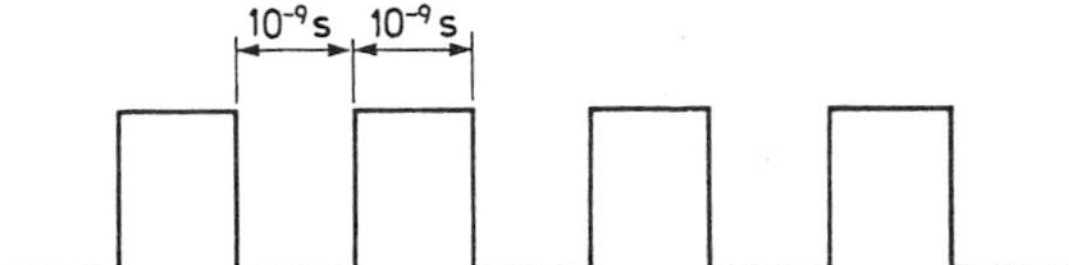

The light is nearly monochromatic and its mean wavelength in the glass is 586.6 nm. The glass has a refractive index for this mean wavelength of 1.452, and it absorbs the light slightly so that its intensity diminishes by a factor 0.020 per kilometre of distance travelled.

(i) What class of electromagnetic radiation is being used?
(ii) Calculate the mean frequency of the radiation contained in the pulses.
(iii) How far will the light travel before its intensity is reduced to one-half of the original intensity?

(*b*) In practice the radiation has a small but significant bandwidth (wavelength range).

(i) Draw sketches to show how the waveforms of the original pulses will change as they travel through the fibre. Explain your answer.
(ii) Given that the rate of change of refractive index n with wavelength λ has a value

$$\frac{dn}{d\lambda} = -1.5\times10^4\ \text{m}^{-1}$$

estimate how far data can be transmitted before it becomes unintelligible, if the bandwidth of the radiation is 0.10 nm.

(*c*) Discuss the advantages of using glass fibres for data-transmission compared with using copper cable and electrical pulses. (O)

Q 10.30

Write an article on '*The Electromagnetic Spectrum*' for a magazine which presents science seriously to the layman. Aim to convey as much information as you can as clearly as possible. Explain the meaning and significance of any equations you use. (O & C, Nuffield)

Q 10.31

Discuss how the photoelectric effect supports the photon model of light.

Light of frequency f shines on a metal surface of threshold frequency f_0. Obtain an expression for the maximum speed of emission of liberated electrons of mass m.

The metal is incorporated in a photocell. The potential required to stop emission of the electrons is called the stopping potential and has a value V_s. Obtain an equation relating V_s and f, and sketch a graph showing the relationship.

A spectrometer is set up to look at a source of light whose spectrum contains only wavelengths of 434 nm, 486 nm and 656 nm. A diffraction grating with 700 lines per mm is placed on the spectrometer turntable at right angles to the collimated light beam. The eyepiece of the telescope is replaced by a photocell of threshold frequency 5.00×10^{14} Hz. Find the angles at which the photocell detects a diffracted beam. Calculate the corresponding stopping potentials V_s.

(O & C)

Q 10.32

In an experiment, using the circuit below, to investigate the photoelectric effect,

I is the photoelectric current,
P is the light intensity and f is its frequency,
V is the potential difference between the photocathode and the anode, and
V_0 is the stopping potential.

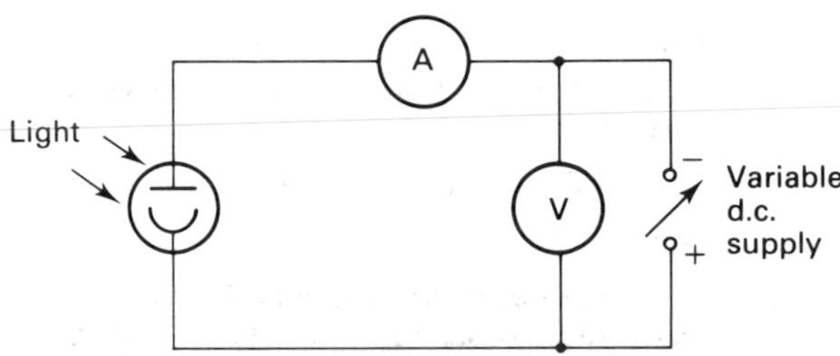

(*a*) Sketch the following graphs:
  (i) I against time, t, with P, V and f constant and $V < V_0$;
  (ii) I against P with f and V constant and $V < V_0$;
  (iii) V_0 against f;
  (iv) I against V, with f and P constant, with V taking both positive and negative values. Indicate the value of V_0 on your graph. On the same axes sketch the graph which would be obtained if light of a shorter wavelength were used.

(*b*) Which of the graphs cannot be explained by the classical theory of light? Explain why.

(*c*) Explain the hypothesis put forward by Einstein to overcome the problems in (*b*), and give the relevant equations. In particular, use these ideas to explain the shape of graph (iii).

(*d*) Experiment shows that when a particular photocathode is illuminated with the anode at a positive potential, a current flows only if the wavelength of the light is less than 560 nm. When 1.30 mW of light of wavelength 430 nm falls on this photocathode a current of 50 μA flows in the cell. What is the maximum kinetic energy of the emitted electrons and what fraction of the incident quanta cause electrons to be emitted? (L)

Q 10.33

It is possible to demonstrate the photoelectric effect in a school laboratory by directing a beam of ultra-violet radiation onto a magnesium electrode, as shown in the diagram.

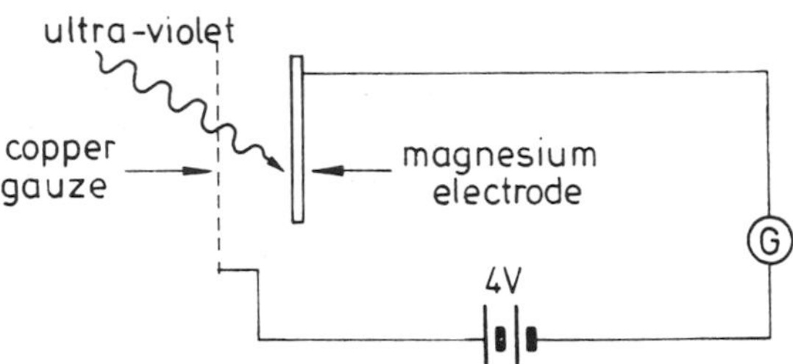

(In practice the current which flows is very small and is detected by an electrometer.)

(*a*) It is usual to explain the resulting current in terms of the ejection of photoelectrons from the magnesium, and their subsequent attraction towards the copper gauze. But alternative explanations might be considered; for example, the current might be caused by the ionization of the air by the ultra-violet radiation.
  (i) Describe how this explanation may be rejected experimentally.
  (ii) Suggest *two* more possible explanations, and show how they too may be rejected experimentally.

(*b*) Explain how this and similar experiments provide evidence for the particulate nature of light.

(*c*) When an ultra-violet lamp rated at 25 W is placed 20 cm from the magnesium electrode, a current of 10^{-12} A flows; this is much less than one would expect if all the energy of the lamp were converted into electrical energy. Discuss in detail the factors responsible for this low value of current. Wherever possible make your discussion quantitative, making realistic estimates of any quantities you need which are not printed on the front of the question paper.

(O & C, Nuffield)

Q 10.34

What evidence does the photoelectric effect provide for the existence of the photon? What other evidence is there?

Monochromatic radiation of wavelength 300 nm and intensity 2.0 W m^{-2} is incident normally on the surface of a metal whose work function is 4.0 V. The metal reflects 80% of the incident radiation and the photoelectric current density is

2.0×10^{-4} A m^{-2}. Determine the pressure on the metal's surface due to the photons, and show that the pressure due to the photoelectrons is only a few per cent of this value.

[The momentum of a photon of wavelength λ is given by h/λ.] (Cam. Schol.)

Q 10.35
The retina of the human eye can easily detect energy arriving at a rate of about 10^{-12} W. If light of wavelength 600 nm arrives at the eye at the rate of 10^{-12} W, calculate the number of photons per second received at the retina and the average distance between the photons. (W)

Chapter 11

A

Q 11.11
State how you would produce (*a*) a line spectrum and (*b*) a continuous spectrum in the laboratory. Describe the appearance of each spectrum when viewed through a grating spectrometer. (C)

Q 11.12
Explain, referring to the energy levels of atoms, the difference between a *continuous* and a *line* spectrum.

Why are a number of dark lines seen in the spectrum of light from the sun? (S)

Q 11.13
Light from a tungsten filament lamp passes through a Bunsen flame containing common salt and it is then viewed in a spectroscope. The two sodium yellow lines appear bright. The current to the lamp is increased and the lines become dark. Explain the physical mechanism of the observation. (W)

Q 11.14
The three lowest energy levels of the electron in the hydrogen atom have energies

$$E_1 = -21.8 \times 10^{-19}\ \text{J}$$

$$E_2 = -5.45 \times 10^{-19}\ \text{J}$$

$$E_3 = -2.43 \times 10^{-19}\ \text{J}.$$

The zero of energy is taken to be when the electron is at rest at a great distance from the nucleus. What is the wavelength of the H_α line in the hydrogen spectrum, which arises from transitions between the levels E_3 and E_2?

Through what potential difference must an electron be accelerated if it is to be capable of (*a*) ionizing a hydrogen atom, (*b*) causing the emission of H_α radiation from a normal hydrogen atom? (S)

Q 11.15
A hydrogen atom emits light of wavelength 121.5 nm and 102.5 nm when it returns to its ground state from its first and second excited states respectively.

Calculate

(*a*) the corresponding photon energies, and
(*b*) the wavelength of light emitted when the atom passes from the second excited state to the first.

(Speed of light, $c = 3.00 \times 10^8$ m s^{-1}.
Planck's constant, $h = 6.63 \times 10^{-34}$ J s.) (L)

Q 11.16
The energy levels of the hydrogen atom are given by the expression

$$E_n = -2.16 \times 10^{-18}/n^2\ \text{J}$$

where n is an integer.

(*a*) What is the ionization energy of the atom?
(*b*) What is the wavelength of the H_α line, which arises from transitions between $n=3$ and $n=2$ levels? (S)

Q 11.17
The difference in the energy levels $E_{n+1} - E_n$ of the hydrogen atom is given by

$$E_{n+1} - E_n = 2.18 \times 10^{-18}(1/n^2 - 1/(n+1)^2)\ \text{J}$$

where n is an integer. What is the first excitation potential, and the wavelength of the radiation emitted? (S)

Q 11.18
Electron diffraction experiments show that the wavelength associated with a certain electron beam is 0.15 nm. Find the momentum of an electron in the beam. Through what potential difference should the electrons be accelerated from rest to acquire this momentum? (C)

Q 11.19
'*The electron in the hydrogen atom has a de Broglie wavelength of about 10^{-10} m.*' Explain briefly what is meant by this statement.
(O & C, Nuffield)

B

Q 11.20
An element in the vapour or gaseous state may be caused to emit a characteristic line spectrum. It

may also yield a line absorption spectrum which is related to, but not identical with, the line emission spectrum.

(*a*) Explain the terms '*line spectrum*', '*characteristic*' and '*line absorption spectrum*'.
(*b*) Describe, in outline, the experiments you would make
  (i) to test whether the source of an unwanted line in the spectrum of a mercury arc lamp is a particular impurity;
  (ii) to demonstrate a line absorption spectrum.
(*c*) Explain, qualitatively and in outline only, how line emission and line absorption spectra are interpreted in terms of atomic structure. Formulae may be quoted without proof.
(O & C, Nuffield)

Q 11.21
Cars are parked in a street which is lit at night by sodium lamps which emit monochromatic yellow light.

(*a*) What colour would (i) a white, (ii) a green, (iii) a yellow car appear to be?
(*b*) Explain these apparent colours in terms of atomic concepts of emission and absorption of light. (W)

Q 11.22
This question is about the ionization of xenon atoms when bombarded by electrons which have been accelerated in a glass tube filled with xenon gas, as shown in the diagram. The line OABC, in the graph, shows the variation of the anode current with the potential difference between the cathode and the anode.

The line OABD shows the results obtained with a similar tube containing no gas.

(*a*) Using information taken from the graph, calculate the energy, in *joules*, required to ionize an atom of xenon.
(*b*) When the potential difference is 15 V, read off and record the current flowing in
  (i) the xenon filled tube, in mA

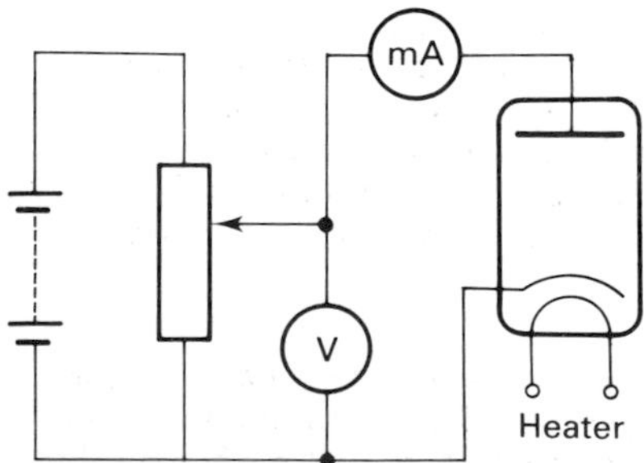

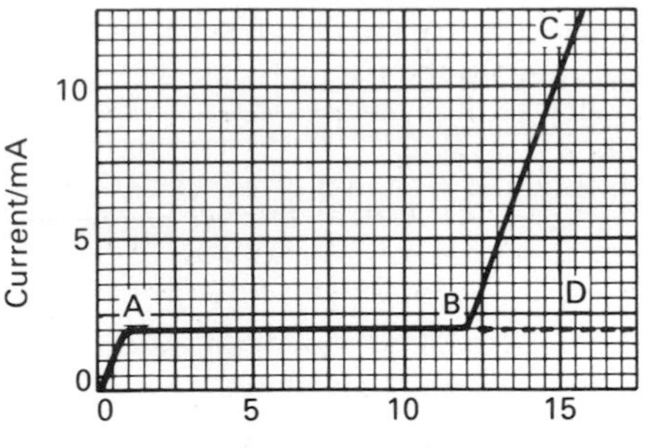

  (ii) the evacuated filled tube, in mA
  Hence show that the number of atoms being ionized per second is 5×10^{16}. You may assume that each atom becomes a singly-charged ion.
(*c*) (i) Use your answer to (*a*) and information supplied in (*b*) to calculate the energy required, per second, to ionize the xenon atoms when the potential difference is 15 volts.
  (ii) The rate of supply of energy from the battery to the tube is greater than this. Give one possible reason for the difference in energy. (O & C, Nuffield)

Q 11.23
The diagram shows the results of a Franck–Hertz experiment. Draw a circuit diagram of the apparatus which gives these results.

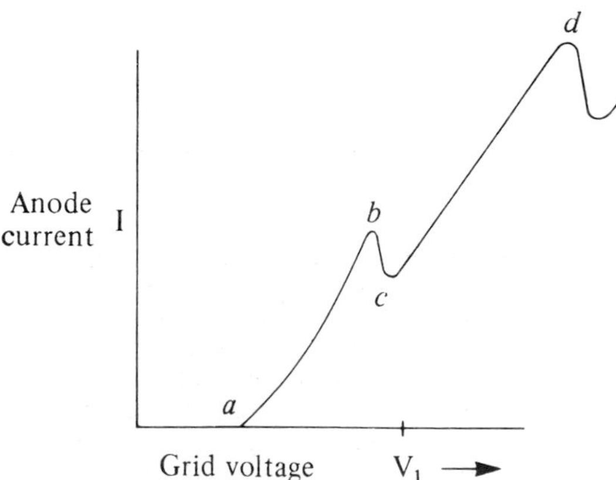

The current I varies with V_1 as shown in the diagram, where I is the current collected at the anode and V_1 is the potential difference between grid and cathode.

Account for the shape of the curve, paying particular regard to

(i) the value of V_1 at a;
(ii) the increase ab in I;
(iii) the peak at b leading to the decrease bc;
(iv) the second peak at d.

When the experiment is done on sodium vapour the difference in V_1 between b and d is 2.10 V. Calculate one wavelength which will occur in the spectrum of hot sodium vapour. (W)

Q 11.24

(*a*) Explain briefly what is meant by an emission spectrum. Describe a suitable source for use in observing the emission spectrum of hydrogen.

(*b*) The diagram, which is to scale, shows some of the possible energy levels of the hydrogen atom.

(i) Explain briefly how such a diagram can be used to account for the emission spectrum of hydrogen.

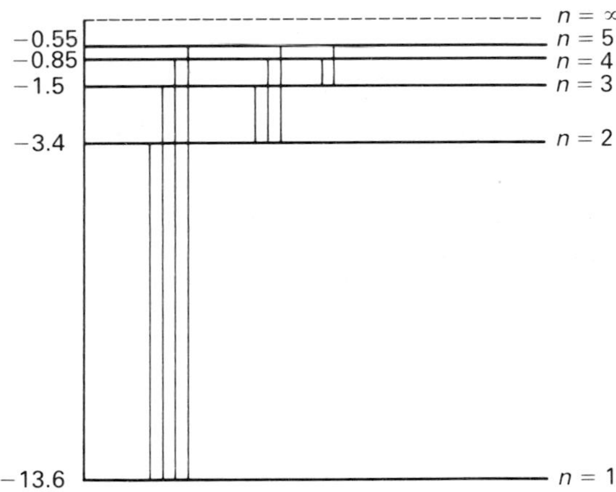

(ii) When the hydrogen atom is in its ground state, 13.6 eV of energy are needed to ionize it. Calculate the highest possible frequency in the line spectrum of hydrogen. In what region of the electromagnetic spectrum does it lie?

(The frequency range of the visible spectrum is from 4×10^{14} Hz to 7.5×10^{14} Hz.)

(iii) A number of transitions are marked on the energy level diagram. Identify which of these transitions corresponds to the lowest frequency that *would be visible.*

(*c*) The wavelengths of the lines in the emission spectrum of hydrogen can be measured using a spectrometer and a diffraction grating.

If the grating has 5000 lines per cm, through what angle would light of frequency 4.6×10^{14} Hz be diffracted in the first order spectrum? (L)

Q 11.25

What is a *photon*? Show that the energy E of a photon and its wavelength λ are related by $E\lambda = 1.99 \times 10^{-16}$ J nm.

The diagram represents part of the emission spectrum of atomic hydrogen. It contains a series of lines, the wavelengths of some of which are marked. There are no lines in the series with wavelengths less than 91.2 nm.

(*a*) In which region of the electromagnetic spectrum are these lines?

(*b*) Using the relation between E and λ given above, find the photon energies equivalent to all the wavelengths marked.

(*c*) Use this information to map a partial energy level diagram for hydrogen. Show, and label clearly, the electron transitions responsible for the emission lines labelled in the diagram.

(*d*) Another line in the hydrogen spectrum occurs at a wavelength of 434.1 nm. Identify and label on your diagram the transition responsible for this line.

Emission spectra are often produced in the laboratory using a discharge lamp containing the gas to be investigated. Explain the physical processes occurring within such a lamp which lead to the excitation of the gas and the emission of light. (C)

Q 11.26

This question is about the evidence for energy-levels in atoms.

The passage below presents three statements about the evidence for electron energy levels in atoms. For each of the statements (i) to (iii) you should write out as full an explanation as you can of the various aspects of the statements. Your explanation should include, where applicable,
—explanations of the *theory* involved
—discussion of the relevant *experimental evidence*
—*quantitative calculations* to illustrate the points made,

Passage

(i) When an electron collides with an atom, we would expect the electron to lose very little energy if the collision were like that of a little marble hitting a big marble. However, it is noteworthy that while some collisions give no energy loss, others give big energy loss and ionization, and yet others give smaller energy loss with no ionization.

(ii) A model of definite energy levels in an atom can be used to explain the set of line spectra that an atom exhibits. For example, we know that amongst the lines from hydrogen are five at the following frequencies.

31×10^{14} Hz
29×10^{14} Hz
24×10^{14} Hz
5×10^{14} Hz
2×10^{14} Hz

These data can be used both to confirm an energy level model and to make further predictions to test the model.

(iii) The fact that electron energy changes of a few electron-volts are associated with emission or absorption of light in the visible region can be used, with the photon model, to yield an approximate value of the Planck constant. It is significant that this value is found to agree with the value found by other kinds of experiment. (O & C, Nuffield)

Q 11.27

(*a*) (i) Describe briefly the Bohr model for the hydrogen atom, and
(ii) state the important assumptions that Bohr made.

(*b*) The table shows some of the energy levels for the hydrogen atom.

	Energy/eV
a	0
b	−0.54
c	−0.85
d	−1.51
e	−3.39
f	−13.6

(i) Define the electron volt.
(ii) Explain why the energy levels are given negative values.
(iii) How might the atom be changed from state e to state c?
(iv) State which level corresponds to the ground state.
(v) Calculate the ionization energy of atomic hydrogen in joules.
(vi) The result of the Bohr theory for the hydrogen atom can be expressed by

$$\frac{1}{\lambda}=R_H\left[\frac{1}{n_1^2}-\frac{1}{n_2^2}\right]$$

where n_1 and n_2 are whole numbers: for the ground state $n_1=1$.
Calculate the value of R_H. (W)

Q 11.28

Draw a labelled diagram of an apparatus which may be used to demonstrate the phenomenon of electron diffraction.

Sketch a typical diffraction pattern obtained in this experiment, and explain qualitatively why it has this form.

Explain very briefly the significance of this experiment.

Two large plane metal electrodes are arranged parallel to each other in an evacuated tube. One (the collector) is at a positive potential with respect to the other (the emitter). Starting from rest, an electron leaves the middle of the emitter and moves perpendicularly to the plates towards the collector. Sketch clearly-labelled graphs showing how the following quantities depend on the distance x from the emitter:

(*a*) the electric potential energy E_p of the electron,
(*b*) its kinetic energy E_k,
(*c*) its speed v,
(*d*) its associated wavelength λ.

If the accelerating potential is 150 V, find the wavelength associated with the electron as it reaches the collector. (C)

Q 11.29

This question is about clarifying the idea of wave-particle duality.

Physicists claim *both* that things commonly regarded as particles can behave like waves *and* that things commonly regarded as waves can behave like particles.

(*a*) Outline experimental evidence which supports the claim that particles can behave like waves.
(*b*) Outline experimental evidence which supports the claim that waves can behave like particles.
(*c*) Explain why wave properties of particles may be important for electrons but not for tennis balls, and why particle properties of waves may be important for light waves but not for radio waves.
(*d*) Answer a critic who objects that these ideas are absurd because something cannot be both a wave and a particle at the same time.

(O & C, Nuffield)

Q 11.30

(*a*) '*X-rays are used to investigate the atomic structure of solids.*'

What can you conclude from this statement about the wavelength of the X-rays used?

(*b*) '*Sometimes, as for example in the case of rubber, electrons with a De Broglie wavelength of about 0.11 nm are used instead of X-rays.*'

Showing the steps in your calculations, find
(i) the momentum of such electrons,
(ii) the potential difference through which the electrons would have to be accelerated to give them this momentum or wavelength.

(*c*) '*A stream of neutrons could be used instead of the electrons.*'

How would the speed of neutrons with a De Broglie wavelength of 0.11 nm compare with that of the electrons used in part (*b*)? Show how you get your answer.

(O & C, Nuffield)

Q 11.31

This question is about the ideas and evidence in the wave theory of atoms.

The passage below consists of numbered statements (i) to (v). For each of these you are asked to give arguments which *explain* and *support* them.

The arguments that you use may be of many kinds and you should indicate what kind they are—for example theories, models, calculations, evidence, etc.

Passage

(i) It can be shown that electrons have wave properties, having a wavelength λ related to their momentum by $mv = h/\lambda$.

(ii) If a wave-like electron is confined in a 'box' of size 10^{-10} m, its momentum can't be less than a certain size, and so its kinetic energy has a lower limit too. The smaller the box, the bigger the kinetic energy of the electrons.

(iii) If the atom is to be stable, the sum of potential and kinetic energy of the electron must be negative. We know that the electrical potential energy of an electron is -10 eV when it is about 10^{-10} m from a proton. All this leads to an explanation of why atoms cannot be much smaller than about 10^{-10} m.

(iv) Consideration of its spectrum shows that a hydrogen atom has a whole series of energy levels with energies given by c/n^2, where c is a constant and n has values 0, 1, 2, 3, . . . etc.

(v) The simplest idea is that electrons in hydrogen behave like standing waves on a string of length, *r*, as shown in the diagram.

This would explain why there are discrete values of kinetic energy associated with integers n but it would give the wrong rule for the way the energy depends on n.

(O & C, Nuffield)

C

Q 11.32

(*a*) (i) Explain the nature of an *electromagnetic wave.*
(ii) Give the main classifications of electromagnetic waves and the approximate wavelength limits of each class. State the properties that all the waves throughout the whole electromagnetic spectrum have in common.

(*b*) Outline the experimental evidence that:
(i) atoms emit electromagnetic waves in discrete packets (photons) rather than as a continuous stream of radiation,
(ii) the energy of a photon is proportional to the frequency of the wave.

(*c*) (i) Spacecraft are being used increasingly for astronomical observations. Specify those parts of the electromagnetic spectrum that can be examined far more effectively from spacecraft than from the Earth's surface, and state the reasons why.
(ii) A telescope mounted in a spacecraft will record radiation from distant objects provided that the mean rate of incidence of photons upon an area normal to the radiation is at least $100\ \mathrm{s^{-1}\ m^{-2}}$. Calculate the maximum distance at which the telescope will detect a monochromatic 250 W point source emitting radiation of wavelength 300 nm uniformly in all directions.

[Take the Planck constant h to be 6.63×10^{-34} J s, and speed of light in vacuum c to be $3.00 \times 10^8\ \mathrm{m\ s^{-1}}$.] (O)

Q 11.33

The following is a shortened version of a letter that appeared in *The Listener* in 1979. Consider carefully the arguments given, and using examples and calculations where appropriate, explain whether each point made is, in your view, correct or not.

[About half the available marks will be given to answers to (*d*) and (*e*).]

The letter

'Sir,

(*a*) Mr. Theocharis has pointed out that much modern science is simply unverified and speculative work resting on very dubious theories. The situation is far worse than he says.

(*b*) In about 1600, Galileo proved that for a pendulum, the time of swing does not depend on the size or amplitude of the swing. For vibrating bodies, the energy is proportional to the square of the amplitude. Hence the period is independent of the energy.

(*c*) Light is a kind of vibration, yet Planck put forward his equation: energy of a light wave equals h (Planck's constant) times frequency, a blatant contradiction of Galileo's irrefutable experiment which anyone can repeat.

(*d*) Furthermore, a photograph of the spectrum of, say, iron in an arc is found to consist of lines of colour, separated by gaps. Each line was allocated a particular value of energy, a procedure fully justified by electron bombardment experiments. Since the energy value is defined as the square of the amplitude, the position of the lines in the spectrum can be allocated this value of amplitude squared.

(*e*) Yet, for reasons beyond understanding, the frequency of a wave was used to specify the line's position. But a wave of particular frequency could have a dozen different amplitudes, and therefore a dozen different energies, and so have a dozen different positions in the spectrum, and all would have to be allocated the same frequency. This nonsense and confusion was added to by the allocation of a different frequency to every line position.

Yours etc.'

(O & C, Nuffield)

Q 11.34

The spectrum from a laboratory gas discharge lamp shows a number of lines. Explain the transformations of energy taking place in the lamp which lead to this emission spectrum.

The spectrum of light from a lithium discharge lamp is examined using a spectrometer fitted with a diffraction grating ruled with 7000 lines per centimetre. When used at normal incidence in the second order, one series of lines in the spectrum is found to give maxima at diffraction angles of 69.90°, 26.91°, 22.57°, 21.02°, 20.28°, ...

(*a*) Find the wavelengths corresponding to these diffraction angles.

(*b*) Explain the practical advantage of taking measurements on the second-order spectrum, rather than on the first.

(*c*) A theory suggests that the wavelengths λ of the lines in this series are given by the equation

$$\frac{1}{\lambda}=R\left(\frac{1}{(1+s)^2}-\frac{1}{(n+p)^2}\right)$$

where $n=2, 3, 4, 5, \ldots$, R is $1.097\times10^7\ \mathrm{m^{-1}}$, and s and p are constants. It is known that $p=-0.041$.

By plotting a suitable graph, or otherwise, investigate how well the experimental results fit the theory. Deduce the value of the constant s.

(*d*) Use your results to map to scale on graph paper a partial energy level diagram for lithium. Indicate on your diagram the transitions responsible for the emission of the various wavelengths. (Cam. Step)

Q 11.35

State the assumptions upon which the Bohr model of the hydrogen atom is based.

Positronium consists of a negative electron and a positive electron (positron) of the same mass in a bound state of relative motion. Using the Bohr postulates, obtain expressions for the energy levels and radii of the orbits of this system. Hence calculate the ionization energy and the radius of the ground state orbit of positronium.

Obtain an expression for the frequency of the photon emitted when positronium undergoes a transition from the state $n=n'$ to the state $n=n'-1$ and show that, when n is large, this frequency is equal to the frequency of revolution of the system in the quantum state n. Comment on this result. (W)

Q 11.36

Outline briefly the ideas involved in Bohr's explanation of the spectrum of a hydrogen atom.

A very massive star has 3 planets rotating about it in circular orbits. The masses of the planets are m, m and $10m$, while their angular momenta about the star are L, $2.5L$ and L respectively. Find the ratio of

(i) the radii of their orbits,
(ii) their speeds,
(iii) their total energies (i.e. kinetic plus potential).

Why is quantization not used in the treatment of planetary orbits?

[You may quote without proof the fact that the potential energy of a mass m at a distance d from a mass M due to gravitational effects is $-GMm/d$, where G is the gravitational constant.]

(Ox. Schol)

Q 11.37
Explain why electron microscopes can be used to study objects which are too small to be resolved by ordinary optical microscopes.

Indicate how Planck's constant h can be determined from measurements made on the behaviour of electrons.

In a bubble chamber an electron is observed to move in an approximately circular path. Would you expect the wave properties of the electron to affect the appearance of the bubble chamber track? (Cam. Schol.)

Q 11.38
The graph shows the experimental results produced when a beam of 420 MeV electrons is scattered by the nuclei of $^{12}_{6}C$.

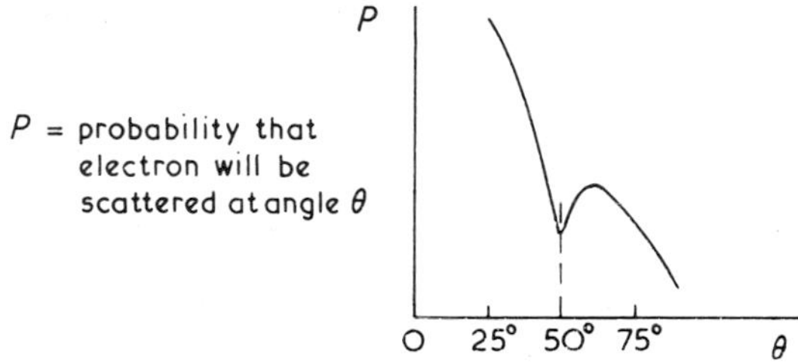

(*a*) Calculate the de Broglie wavelength of 420 MeV electrons, assuming that relativistic effects may be neglected.

(*b*) Assuming that the scattering of these electrons is analogous to the diffraction of light waves by spherical objects, we may use the equation:

$$\sin \theta = \frac{1.22\lambda}{d} = \frac{0.61\lambda}{R}$$

Here, λ is the wavelength of the electrons and R the radius of a carbon nucleus. Calculate the value of R.

(*c*) Comment on the assumption in (*a*) that relativistic effects are negligible. Explain how these effects would affect your calculated value for R.

(*d*) Nuclear radii may also be measured in alpha-scattering experiments. Calculate the closest possible approach of 5 MeV alpha particles to carbon nuclei.

(*e*) Alpha-scattering experiments with carbon are unsatisfactory in practice. Suggest reasons for this.

(*f*) Why do the results of alpha-scattering experiments not show minima in a similar way to those from electron-scattering experiments? (O & C, Nuffield)

Q 11.39
Some polymer molecules are long and thin, like a length of string. It is possible for an electron to be trapped in the molecule as a whole, as if in a long flat bottomed box. The electron waves are spread out along the whole length of the box, with a

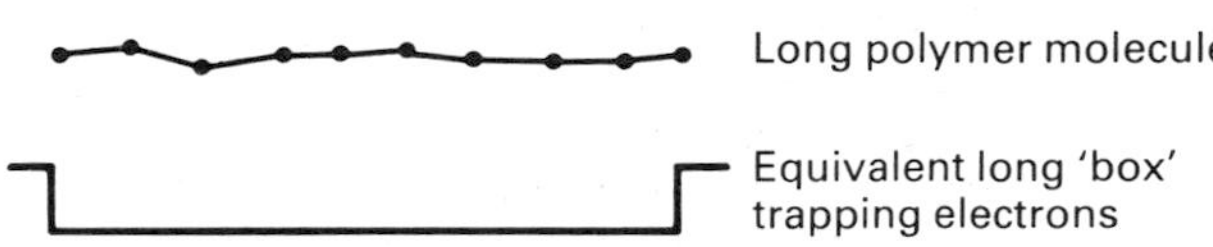

wavelength the same at all places inside the box.

Explain how standing wave ideas could be used to show that, if such a molecule is long enough, these electrons can absorb energy in the visible region of the spectrum (so that such molecules are useful in paints and dyes). (O & C, Nuffield)

Q 11.40
Write an essay entitled '*Particles, waves, energy levels and the nature of atoms*', which

(*a*) brings out the experimental evidence for the ideas you give,

(*b*) shows how theoretical arguments help explain the phenomena you mention.

(Indicate experiments in outline, with a diagram where appropriate, without giving practical details.) (O & C, Nuffield)

Q 11.41
'*Physics is essentially the process of modelling natural phenomena in terms of everyday experience and does not pretend to produce absolute explanations. Hence there is no contradiction in using different models to illustrate different aspects of the same phenomenon.*'

Discuss this statement. Illustrate you answer by reference to suitable examples such as those listed below or other examples of your own choice:

behaviour of light,
atomic structure,
nuclear structure,
structure matter. (L)

Solutions to exercises

Answers are given to all numerical questions set in the text, and to the odd-numbered questions from the Questions section.

Answers are generally given to the correct number of significant figures although occasionally a further figure is given to enable students to check their arithmetic.

The answers given are the responsibility of the author and not of the examining board concerned.

Chapter 1

1 (*a*) 1 Hz (*b*) 10 Hz (*c*) 0.1 Hz (*d*) 0.02 Hz (*e*) 1 s (*f*) 5 s (*g*) 0.04 s (*h*) 0.02 s
5 (*a*) $s=10\cos\pi t$ (*b*) $v=-10\pi\sin\pi t$ (*c*) $a=-10\pi^2\cos\pi t$ (*d*) $a=-\pi^2 s$
6 (*a*) 54 cm s^{-1} (*b*) 63 cm s^{-1} (*c*) 140 cm s^{-2} (*d*) 49 cm s^{-2}
7 (*a*) 38 s^{-1} (*b*) 3.1 s^{-1} (*c*) 3 Hz; 0.33 s
9 2 s; 0.5 Hz; 3.2 s^{-1}
10 9.0 s
11 0.5 Hz; 0.35 m s^{-1}; increase
12 Potential energy 0.0 J, 0.1 J, 0.4 J; Kinetic energy 0.4 J, 0.3 J, 0.0 J
13 0.51; 3.7×10^{-4} cm
17 (*b*) 5.0 s; 2.0 s (*c*) 16 cm s^{-1}
19 (*b*) $\sqrt{2}\,T$; $T/\sqrt{2}$
21 0.63 mm
23 (*b*) 0.12 J
25 (*b*)(ii) $2\pi^2 mA^2/T^2$
(*c*)(i) 31 m s^{-1} (ii) 4.9×10^5 m s^{-2}
(iii) 6.0 W
27 $A=6.7$ m kg^{-1}; $m=0.90$ kg; $[A]=LM^{-1}$
29 (*c*)(i) 3.8 mm (ii) 7.6×10^{-3} kg
(iii) 0.28 m s^{-1}
31 (*c*)(i) $12.5+2.5\cos\dfrac{2\pi t}{12.5}$
(ii) $-\dfrac{5\pi}{12.5}\sin\dfrac{2\pi t}{12.5}$
(iii) 3h 7m, 9h 22m after high water
(iv) 1.1×10^5 m^3 s^{-1}
(*d*) 270 kW
33 (*b*) −0.2 cm (*c*) 1
35 Resonance occurs at 1 Hz
41 (*b*)(i) 0.45 s (ii) 4.4 mJ (iii) 31 mJ
(*c*) 2.4 cm, 1.9 cm, 1.6 cm, 1.2 cm, 0.98 cm
43 (*c*)(i) 0.2 s
(ii) resonance at 533 revolutions per minute
45 (*e*) 7.3 Hz; 1.1×10^5 N m^{-1}

Chapter 2

2 (*a*) 3.15 MHz (*b*) 0.375 m (*c*) 4×10^{14} Hz (*d*) 3×10^{18} Hz (*e*) 5.1 m; 0.5
4 (*a*) 1.7 m s^{-1} (*b*) 1.8 m s^{-1}
5 7×10^{10} N m^{-2}
6 4250 m s^{-1}
7 1.06×10^{-25} kg; 23.7 N m^{-1}; 1.6 : 1
9 2.1 m s^{-1}; 3.3 kg m^{-1}; 2.1 m s^{-1}
10 (*a*) 4.8 m s^{-1}; 2.8 m s^{-1}; 2.2 m s^{-1}
(*b*) 24 m, 14 m, 11 m
11 (*a*) 0.1 m (*b*) 0.0 m (*c*) −0.1 m
(*d*) 0.06 m (*e*) 0.0 m (*f*) −0.1 m
(*g*) 0.0 m (*h*) 0.1 m (*i*) 0.0 m (*j*) −0.1 m
(*k*) 0.0 m (*l*) 0.07 m
12 (*a*) $\pi/2$, 2 m (*b*) π, 0 m
(*c*) 1.25π, −1.41 m (*d*) 1.5π, −2 m
(*e*) π, 0 m (*f*) 0.7π, 1.62 m
(*g*) 2π, 0 m (*h*) 0.75π, 1.41 m
13 0.42 m s^{-1}
17 1.88 m s
21 (*a*)(i) 400 μs (ii) 3000 m s^{-1}
23 (*b*) slope $=-\frac{1}{2}$; equation (iii) is correct; C = 2.5
27 (*a*) $v_T=\sqrt{\dfrac{kL_0(L-L_0)}{M}}$; $v_L=\sqrt{\dfrac{kL_0^2}{M}}$
(i) time is unchanged
(*b*) $Mg/2k$
29 (*a*) 0.32 m s^{-1}; 64 Hz

Chapter 3

6 45°
7 (*a*) 341 Hz (*b*) 268 Hz (*c*) 336 Hz (*d*) 264 Hz
8 3.2×10^6 m s^{-1}
9 (*a*) +200 Hz (*b*) −400 Hz
13 (i) 426 Hz, 532 Hz (ii) 376 Hz, 470 Hz
15 The nebula is receding at 3.18×10^6 m s^{-1}
17 (*a*)(i) 80 m (ii) 28°
19 (ii) 2.42×10^4 m s^{-1}
(iii) 2.3×10^{10} m
(iv) 1.7×10^{17} m
21 (iii) 85 m s^{-1}; 7.5π rad s^{-1}
23 A note with amplitude varying at 40 Hz

Chapter 4

2 (*a*) 0.0 m (*b*) −0.1 m (*c*) 0.0 m (*d*) 0.1 m (*e*) 0.0 m (*f*) 0.07 m
19 (*c*)(i) 50 N (ii) 42 mm
21 (*c*)(i) 2000 r.p.m.
23 (*a*) 0.34 m (*b*) 0.315 m (*c*) 1080 (*d*) 160 Hz
29 (*a*) 5 ms (*b*) 40 Hz (*c*) 200 Hz and (200 ± 40) Hz

Chapter 5

1 33 m
2 384 Hz; 768 Hz; 1.5
3 1.7×10^{-4} kg m^{-1}
4 (*a*) 16.5 m (*b*) 16.5 mm (*c*) 1.65 mm
5 (*a*) 330 m s^{-1} (*b*) 991 m s^{-1} (*c*) 1 octave and 1 fifth
6 333 m s^{-1}
7 328 m s^{-1}; 1 cm
8 1000 Hz
9 By a factor of 2
10 100 dB, 120 dB
11 43 phons
12 (i) 4.5×10^{-6} W m^{-2} (ii) 4.5×10^{-5} W m^{-2}, (iii) 4.5×10^{-4} W m^{-2}
15 600 m s^{-1}
17 The man hears a maximum every 0.75 s
19 (*c*) 198 Hz (*d*) +0.4%; 0.2% lower
23 (*a*) 150 mm (*c*) 2 mm
25 60°; 221°C
27 1/290
29 (*c*)(ii) closed; 0.4 m
31 10:1
33 $\sin\theta = \dfrac{c\Delta t}{d}$
35 2×10^{-10} m

Chapter 6

1 10^{-6} rad; 0.2 second
2 (*a*) 0.01%, 1%, 10%, 27% (*b*) 0.006%, 0.5%, 5%, 14%
6 (*a*) 0.060 m, virtual (*b*) 62.5 mm, real (*c*) 0.67 m, real (*d*) 105 mm, real (*e*) 16.7 mm, concave (*f*) 62.5 mm, convex
8 15 cm
10 (*a*) 19°28′ (*b*) 22°05′ (*c*) 26°19′ (*d*) 22°05′
11 15 cm below the surface; 4.5 cm above the surface
12 (i)(*a*) 41°49′ (*b*) 48°45′ (*c*) 62°27′
14 2.45; ±1.5%
15 1.49
17 (*a*) 0.2 m; real (*b*) 0.038 m; virtual (*c*) 100 mm; real (*d*) 100 mm; virtual (*e*) 46.7 mm; converging (*f*) 90 mm; diverging
18 (*a*) 2 dioptre, converging (*b*) 1 dioptre, diverging
20 (*a*) 0.12 m (*b*) 0.47 m
21 20 cm
22 30 cm
23 57 mm
24 75 mm
25 1.6
27 1.9×10^8 m s^{-1}
29 1.5
31 38 cm above the lens; 30 mm in diameter
33 0.15 m
35 (*b*) 1.50
37 32.77°; 41.05°
39 20 cm
41 (*a*) 25 cm from the card (*b*) 20 cm
45 $v_{red} = 357$ cm; $v_{blue} = 208$ cm
49 $I_{max} = n_2\sqrt{2\Delta}$; 8.6°
53 30 m

Chapter 7

1 6×10^5; 6×10^5
2 (*a*) 5×10^{-4} rad (*b*) 5×10^{-3} rad (*c*) 2.5×10^{-4} rad (*d*) 1.75×10^{-3} rad (*e*) 0.25 mm (*f*) 2.5 mm (*g*) 0.125 mm (*h*) 0.875 mm

3 $(6.06 \pm 0.16) \times 10^{-7}$ m
4 Second order red coincides with third order blue
5 1.6
6 5×10^{-7} m
7 5.06×10^{-7} m
9 5.0×10^{-4} rad
10 (*a*) 0.5 m (*b*) 1.6 mm
(*c*) 2.5 μm (*d*) 3.2×10^{-3} rad
12 1.1×10^{-7} m
13 (*a*) 10^{-3} rad (*b*) 10^{-3} rad (*c*) 10^{-2} rad
14 0.05 mm
15 5.0×10^{-7} m
16 36°52′; 5.7×10^{-3} degree
18 (*a*) 573 (*b*) 53°4′
19 45°; 1:1; 0:1
27 17°
31 (ii) 1.2 mm (vi) 60 μm
33 (*c*) It detects one maximum per second; 133 Hz
35 5.8×10^{-7} m
37 (*c*)(iii) 2.11 m (iv) 4.33 mm
39 (*b*)(i) She would hear maxima every 5.0 m
(ii) Beats of frequency 1 Hz
41 (*a*) 56°
43 45°
45 4.2×10^3 Pa
51 (*a*) 350 nm
(*b*) Brightest flash now lasts 2×10^{-4} s
55 (*b*)(iii) 2.3 MHz

Chapter 8

1 4.1×10^4 rad^{-1}; 2.4×10^{-5} rad
2 1.5×10^{-4} rad; 6.6×10^3 rad^{-1}
3 2.3×10^5; 1.4 km
4 4; 1 m
5 8
6 21 mm; 1.25 mm
8 166
9 (*a*) 3.5 (*b*) 2.5
10 (*a*) 61, 21.3 mm (*b*) 58, 21.7 mm
13 (i) 42 mm to 50 mm (ii) 5 to 6
15 0.048 m; 0.002 m; concave lens, focal length 1.0 m
17 (*b*)(ii) $\sim 10^7$ m
19 (*b*) ϕ is halved (*d*) 6 m^2
21 0.015 rad; 225 mm
25 1.8 cm
27 (*b*)(i) 60 mm
(*c*)(i) 3 pairs (ii) 0.035 rad
29 (*b*)(i) 30 mm (ii) 180 mm; 6
31 4.8 cm
33 1.7 cm, 22.2 cm
35 hydrogen, carbon dioxide
37 Increased by 0.01 m; 5 mm; about 1/6 s

Chapter 9

1 (*a*) 38.4 cm (*b*) 1.28 ns
(*c*) 39.4 cm (*d*) 1.31 ns
2 3.2×10^{-5} s
3 6×10^4 km (nearly twice the earth's circumference); 0.067 s
4 $(2.5 \pm 0.7) \times 10^8$ m s^{-1}
5 1.33 hour; 1.15 hour
7 (*b*) 115 s
13 From about $\lambda/2$ upwards
15 $D = 180 - 4r + 2i$
where $\sin r = n/\sin i$; 136°

Chapter 10

1 10^{-4}; 8×10^{-5}
2 107.5 cm
3 30 cm
4 4.8 m
5 3.3×10^8 m s^{-1}
6 10^6 km; 3.3 s
7 (*a*) 4.8×10^{-19} J (*b*) 1.03×10^6 m s^{-1}
8 6.6×10^{-34} Js; 3.7×10^{-19} J
9 (*a*) 6 eV; 9.6×10^{-19} J
(*b*) 2.33 eV; 3.73×10^{-19} J
10 119 kW; about 60
11 300 km
12 8.3×10^{20} photons m^{-2} s^{-1}; 300 W m^{-2}
15 (*a*) 1.7 m (*b*) 360 per second
17 5.19×10^{14} Hz; visible light
25 (*b*)(i) 5.5×10^{10} per second
(ii) 1.1×10^{-7} C
29 (*a*)(ii) 3.5×10^{14} Hz (iii) 34 km
(*b*)(ii) about 200 km
31 434 nm at 17.7°, 37.4°, 65.7°;
486 nm at 19.9°, 42.9°;
656 nm is not detected;
stopping potentials are 0.79 V for 434 nm, 0.48 V for 486 nm
35 3×10^6 photons per second; 99 m

Chapter 11

2 3.27892×10^{-19} J; 2.04664 eV
3 1.65×10^{-18} J; 1.9×10^6 m s^{-1}; yes

4 1.5×10^{25} Hz; γ-ray; 1.3×10^{10} atoms; 4×10^{-15} kg
5 -2.6 eV to -4.9 eV
6 about 2×10^{-10} m
7 (*a*) $3.2\times10^{3}\ \mathrm{m\,s^{-1}}$; 8.9×10^{-21} J; 5.5×10^{-2} eV (*b*) About 6.6×10^{-35} m (depending on the assumptions made)
8 4.0×10^{-36} kg
9 (*d*) 2.1×10^{-10} m (*e*) Four times
(*f*) -1.1×10^{-18} J
(*g*) 5.5×10^{-19} J (*h*) -5.5×10^{-19} J
(*i*) one quarter as negative (*j*) 1.2×10^{-7} m
10 (*a*)$10^{-14}\ \mathrm{m\,s^{-1}}$ (*b*) $10^{6}\ \mathrm{m\,s^{-1}}$
15 (*a*) 1.64×10^{-18} J, 1.94×10^{-18} J (*b*) 663 nm
17 10.2 V; 121 nm
23 589 nm
25 (*b*) 2.18×10^{-18} J, 2.09×10^{-18} J, 2.04×10^{-18} J, 1.94×10^{-18} J, 1.66×10^{-18} J
27 (*b*)(v) 2.18×10^{-18} J
(vi) $1.1\times10^{7}\ \mathrm{m^{-1}}$
35 1.1×10^{-18} J; 6.8 eV

Appendices

QA.1 (*a*) $3\pi/2$ (*b*) $\pi/4$ (*c*) $\pi/6$
(*d*) 60° (*e*) 45° (*f*) 90°
QA.3 (N − 2)

Index